Advances in the Mechanics of Plates and Shells

SOLID MECHANICS AND ITS APPLICATIONS
Volume 88

---

*Series Editor:* G.M.L. GLADWELL
*Department of Civil Engineering*
*University of Waterloo*
*Waterloo, Ontario, Canada N2L 3GI*

*Aims and Scope of the Series*

The fundamental questions arising in mechanics are: *Why?*, *How?*, and *How much?* The aim of this series is to provide lucid accounts written by authoritative researchers giving vision and insight in answering these questions on the subject of mechanics as it relates to solids.

The scope of the series covers the entire spectrum of solid mechanics. Thus it includes the foundation of mechanics; variational formulations; computational mechanics; statics, kinematics and dynamics of rigid and elastic bodies: vibrations of solids and structures; dynamical systems and chaos; the theories of elasticity, plasticity and viscoelasticity; composite materials; rods, beams, shells and membranes; structural control and stability; soils, rocks and geomechanics; fracture; tribology; experimental mechanics; biomechanics and machine design.

The median level of presentation is the first year graduate student. Some texts are monographs defining the current state of the field; others are accessible to final year undergraduates; but essentially the emphasis is on readability and clarity.

*For a list of related mechanics titles, see final pages.*

# Advances in the Mechanics of Plates and Shells

The Avinoam Libai Anniversary Volume

Edited by

D. DURBAN

D. GIVOLI

*Faculty of Aerospace Engineering,*
*Technion, Haifa, Israel*

and

J.G. SIMMONDS

*Department of Civil Engineering,*
*University of Virginia,*
*Charlottesville, Virginia, U.S.A.*

SPRINGER SCIENCE+BUSINESS MEDIA, LLC

A C.I.P. Catalogue record for this book is available from the Library of Congress.

ISBN 978-1-4020-0380-6 ISBN 978-0-306-46954-1 (eBook)
DOI 10.1007/978-0-306-46954-1

---

*Printed on acid-free paper*

For Avinoam Libai
Colleague, Teacher and Friend
on the occasion of his
seventieth birthday

## Table of Contents

## Dedication

Avinoam Libai was born on September 17, 1929 in Tel-Aviv, where he also grew up. In 1953 he graduated Summa cum Laude, at the Technion in Civil Engineering. He joined the Public Works Department of the Government of Israel for two years taking part in the important civil engineering project "Yarkon Bridge". He then went to the United States to pursue higher studies at Purdue University, and received a M.Sc. in Structures in 1956 and a Ph.D. in Structures (Engineering Science) in 1959. Upon completing his Ph.D. Avinoam took an Assistant Professor position at Johns Hopkins University until 1961. His first paper, published soon after in the Journal of Aerospace Sciences (1962), examined the nonlinear elastokinetics of shells and beams and over the years became a key reference in the field of thin walled structures.

Libai returned to Israel in 1961, joined the Israel Aircraft Industries in Lod, and stayed there for a productive period of ten years. He emerged to become Principal Engineer for Structures and Principal Staff Scientist of the I.A.I. During that period Avinoam shared major responsibilities in all engineering projects, including the leading aeronautical platforms "Arava", "Commodore Jet" and "Kfir". There is a common agreement that Libai was one of the key figures in the Israeli aeronautical community during its formative years. It was typical of this man that despite the heavy burden of work at the I.A.I. he secured the time to teach as an Adjunct Teacher at the Technion.

In 1971, after one year as a Visiting Associate Professor, Avinoam joined the Faculty of Aerospace Engineering at the Technion as a full Professor, and remained there until his retirement, becoming Professor Emeritus in 1997. He served for five years as Head of the Structures Laboratory, three years as Dean of the Faculty and participated in and chaired numerous Technion and Faculty committees. His thinking and advice over a wide spectrum of issues have been much valued and always sought after. In his advisory capacity, Libai served main aeronautical industries like the I.A.I. (1971-1975) and the Ministry of Defense (1978-1984). In 1992 Avinoam was named the L. Shirley Tark Professor in Aircraft Structures.

Avinoam's distinguished academic career spans nearly three decades during which he established himself as a leading authority, of world recognition, in the nonlinear theory of elastic shells. His research contributions appear in first rate periodicals, covering a wide range of topics like shell buckling under non-uniform loads, structural testing, structural analysis and design and numerical analysis of shells. However, his main achievements are in the general (nonlinear) theory of shells, plates and membranes. His studies of invariant formulations of nonlinear shell theories have paved the path into accepting that field as an integral branch of continuum mechanics. In this aspect, Avinoam belongs to a small group of researchers who managed to develop the engineering theory of shells into a wide research field in nonlinear mechanics, employing advanced mathematical methods, with a variety of applications to modern aerospace structures.

The style of his papers is a model of scientific writing and mastery of a rare blend of research tools: from tensor analysis through asymptotic methods to pragmatic engineering thinking. Publications authored by Libai are often cited in a few languages, and there is little doubt that the bulk of his research output is of permanent value.

Avinoam's calm and confident personality, integrity of character, noble manners and his exceptional ability to explain profound ideas in simple words, have made him an outstanding teacher in both undergraduate and graduate classes.

Libai has been on many visits abroad, including sabbatical periods at Harvard University (1977), University of Virginia (1978, 1984, 1996), University of Oklahoma (1990) and University of Texas at Austin (1991). His collaboration with Jim Simmonds of the University of Virginia has been particularly fruitful. A series of co-authored papers has finally culminated in the book *The Nonlinear Theory of Elastic Shells, One Spatial Dimension* (1988), along with a recent (1998) follow up *The Nonlinear Behavior of Elastic Shells*. Both volumes have received much acclaim and are already regarded as classics within professional circles. The books are surely to remain an inspiration for generations to come.

This volume is presented to Professor Avinoam Libai on the occasion of his seventieth birthday by colleagues and friends who have followed and valued his work over the years. Many ideas originated by Avinoam are scattered through the pages and we are confident that his scientific tree has still a wealth of fruit to bear in the future.

The Editors

## PUBLICATIONS BY AVINOAM LIBAI

### Thesis

*Theory of Doubly Curved Shells*, Ph.D. Thesis, Purdue University, 1959.

### Papers

1. *On the Nonlinear Elastokinetics of Shells and Beams*, Journal of the Aerospace Sciences, 29, 1190-1195,1209, 1962.

2. *Invariant Stress and Deformation Functions for Doubly Curved Shells*, ASME Journal of Applied Mechanics, 89, 43-48,1967.

3. *The Nonlinear Membrane Shell with Application to Noncircular Cylinders*, International Journal for Solids and Structures, 8, 923-943, 1972.

4. *Buckling of a Circular Cylindrical Shell in Axial Compression and SS4 Boundary Conditions*, AIAA Journal, 10, 935-936, 1972, (with D. Durban).

5. *Pressurized Cylindrical Membranes with Flexible Supports*, Proceedings International Symposium on Pneumatic Structures, 1, Delft, 1972 (Int. Assoc. for Shell Structures).

6. *A Method for Approximate Stability Analysis and its Application to Circular Cylindrical Shells under Circumferentially Varying Edge Loads*, ASME Journal of Applied Mechanics, 971-976, December 1973, (with D. Durban).

7. *Buckling of Short Cylindrical Shells under Axial Compression*, AIAA Journal, 12, 909-914, July 1974, (with D. Durban).

8. *Transverse Vibrations of Compressed Annular Plates*, Journal of Sound and Vibration, 40, 149-153, 1975, (with A. Rosen).

9. *Thickness Influence on the Buckling of Circular Cylindrical Shells Subjected to Circumferentially Varying Axial Loads*, Israel Journal of Technology, 9-17, Nov. 1975, (with D. Durban).

10. *Stability and Behaviour of an Annular Plate under Uniform Compression*, Experimental Mechanics, 16, 461-467, 1976, (with A. Rosen).

11. *Dynamics and Failure of Cylindrical Shells Subjected to Axial Impact*, AIAA Journal, 15, 1977, (with G. Maymon).

12. *Buckling of Cylindrical Shells Subjected to Non-Uniform Axial Loads*, ASME Journal of Applied Mechanics", 44, 714-720, Dec. 1977, (with D. Durban).

13. *Exact Equations for the Inextensional Deformation of Cantilevered Plates*, ASME Journal of Applied Mechanics, 46, Sept. 1979, (with J.G. Simmonds).

14. *Alternate Exact Equations for the Inextensional Deformation of Arbitrary Quadrilateral and Triangular Plates*, ASME Journal of Applied Mechanics, 46, Dec. 1979, (with J.G. Simmonds).

15. *Optimization of a Square Panel Subjected to Compressive Edge Loads*, AIAA Journal, 17, Dec. 1979.

16. *Quasilinear Behavior of Eccentrically Stiffened Compressed Annular Plates Near the Buckling Point*, Israel Journal of Technology, 17, 1979. Also Collection of Papers, 21st Israel Annual Conference on Aviation and Astronautics, Feb. 1979, (with M. Feder and A. Rosen).

17. *Large Strain Constitutive Laws for the Cylindrical Deformation of Shells*, International Journal for Nonlinear Mechanics, 16, 91-103, 1981, (with J.G. Simmonds).

18. *On the Nonlinear Intrinsic Dynamics of Doubly Curved Shells*, ASME Journal of Applied Mechanics, 48, 909-914, Dec. 1981.

19. *Repeated Buckling Tests of Stiffened Thin Shear Panels*, Israel Journal of Technology 20, 220-231, 1982. Also Collection of Papers, 24th Israel Annual Conference on Aviation and Astronautics, Feb. 1982, (with J. Ari-Gur and J. Singer).

20. *Highly Nonlinear Cylindrical Deformations of Rings and Shells*, International Journal of Nonlinear Mechanics, 18, 181-198, 1983. Also TAE Report 467, Technion 1981, (with J.G. Simmonds).

21. *Nonlinear Elastic Shells Theory*, Advances in Applied Mechanics, 23, 271-371, 1983, (with J.G. Simmonds).

22. *Nonlinear Shell Dynamics - Intrinsic and Semi-Intrinsic Approaches*, ASME Journal of Applied Mechanics, 531-536, Sept. 1983.

23. *A Two Dimensional Model for Fluid Structure Interaction in Curved Pipes*, Nuclear Engineering and Design, 80, 1-10, 1984, (with E. Bar-On, Y. Berlinsky, Y. Kivity and D. Peretz).

24. *Intrinsic and Semi-Intrinsic Approaches to Large Deformation Shell Dynamics*, Flexible Shells, Theory and Applications, Euromech, 165, May 1983.

25. *Durability under Repeated Buckling of Stiffened Shear Panels*, Proc. of the 14th ICAS Conference, Toulouse, France, Sept. 1984. Also AIAA Journal of Aircraft, 24, 1987, (with T. Weller, M. Kollet and J. Singer).

26. *A Simplified Version of Reissner's Nonlinear Equations for a First Approximation Theory of Shells of Revolution*, Computational Mechanics, 2, 99-103, 1987, (with J.G. Simmonds).

27. *Asymptotic forms of a Simplified Version of the Nonlinear Reissner Equations for Clamped Elastic Spherical Caps under Outward Pressure*, Computational Mechanics, 2, 231-244, 1987, (with J.G. Simmonds).

28. *Engineering Education 2001*, in *Engineering Education*, 105-124, Nov. 1987. Also published as a separate book by the Neaman Press, 1987, (with Z. Tadmor, Z. Kohavi, P. Singer and D. Kohn.

29. *On the Axisymmetric Buckling of Shearable Shells and Plates of Revolution, with Emphasis on the Effects on Localized Supports*, in: *Buckling of Structures, Theory and Experiment*, Elsevier, Amsterdam, 221-240, 1988.

30. *The Transition Zone Near Wrinkles in Pulled Spherical Membranes*, Int. J. Solids Struct., 26, 927-939, 1990.

31. *A Curved Axisymmetric Shell Element for Nonlinear Dynamic Elastoplastic Problems, Part I - Formulation*, Computers and Structures, Vol. 42, No. 4, pp. 631-639, 1992, (with R. Ben Zvi and M. Perl).

32. *A Curved Axisymmetric Shell Element for Nonlinear Dynamic Axisymmetric Problems, Part II - Implementation and Results*, Computers and Structures, Vol. 42, No. 4, pp. 641-648, 1992, (with R. Ben Zvi and M. Perl).

33. *Equations for the Nonlinear Planar Deformation of Beams*, ASME, Journal of Applied Mechanics, 59, 1028-1030, 1992.

34. *A Mixed Variational Principle and its Application to the Nonlinear Bending Problem of Orthotropic Tubes - I. Development of General Theory and Reduction to Cylindrical Shells*, Int. J. Solids Struct., 31, 1003-1018, 1994, (with C.W. Bert).

35. *A Mixed Variational Principle and its Application to the Nonlinear Bending Problem of Orthotropic Tubes - II. Application to Nonlinear Bending of Circular Cylindrical Tubes*, Int. J. Solids Struct., 31, 1019-1033, 1994, (with C.W. Bert).

36. *Fully Inverse Dynamics of Very Flexible Beam Using a Finite Element Approach and Lagrangian Formulation*, Computers and Structures, Vol. 53, No. 5, pp. 1073-1084, 1994, (with D. Rubinstein and N. Galili).

37. *Incremental Stresses in Loaded Orthotropic Circular Membrane Tubes. - I: Theory*, Int. J. Solids Struct., 32, 1907-1925, 1995, (with D. Givoli).

38. *Incremental Stresses in Loaded Orthotropic Circular Membrane Tubes. - II: Numerical Solution*, Int. J. Solids Struct., 32 , 1927-1947, 1995, (with D. Givoli).

39. *Direct and Inverse Dynamics of a Very Flexible Beam*, Computer Methods in Appl. Mechanics and Engineering, 131, 241-261, 1996, (with D. Rubinstein and N. Galili).

40. *Intrinsic Equations for the Nonlinear Dynamics of Space Beams*, AIAA Journal, 34, 1657-1663, 1996.

**State of the Art Survey Paper**

1. *Nonlinear Membrane Theory*, in *Theoretical and Applied Mechanics 1992*, S. Bodner, J. Singer, A. Solan and Z. Hashin, Eds., 257-280, Elsevier, 1993 (ICTAM, 1992).

**Books**

1. *The Nonlinear Theory of Elastic Shells, One Spatial Dimension*, Academic Press, Boston, Feb. 1988, (with J.G. Simmonds).

2. *The Nonlinear Behavior of Elastic Shells*, Cambridge University Press, 1998, (with J.G. Simmonds).

**Book (editing)**

*Buckling of Structures, Theory and Experiment*. Elsevier, Amsterdam, 1988, (with I. Elishakoff, J. Arbocz and C.D. Babcock, Editors).

# BREATHING OSCILLATIONS OF ROTATING NONLINEARLY ELASTIC AND VISCOELASTIC RINGS

Stuart S. Antman
*Department of Mathematics,*
*Institute for Physical Science and Technology,*
*and Institute for Systems Research*
*University of Maryland*
*College Park, MD 20742-4015, U.S.A.*
ssa@math.umd.edu

## 1. INTRODUCTION

We study special unforced planar motions of nonlinearly elastic and viscoelastic rings, namely, motions in which the rings simultaneously rotate and oscillate radially. The most general class of rings that we consider can suffer flexure, shear, and both longitudinal and transverse extensions. For shearable rings we find that the shear deformation plays a surprising and critical role in such motions. For unshearable rings, this role is played by the shear force.

**Notation.** We denote vectors (which are elements of Euclidean 3-space) and vector-valued functions by boldface italic symbols $\boldsymbol{a}$, $\boldsymbol{b}$, etc. We denote $n$-tuples of real numbers by boldface sanserif symbols $\mathsf{p}$, $\mathsf{q}$, etc. We denote derivatives by subscripts. To keep a systematic notation, especially when treating Hamiltonians, we use the superposed dot not to indicate a time-derivative, but to identify an argument of a function that is occupied by a time derivative. In particular, we shall treat real-valued functions of the form $(x, \dot{x}) \mapsto f(x, \dot{x})$. The partial derivative of $f$ with respect to its second argument is denoted $f_{\dot{x}}$. If $x$ is a given function of the time $t$, then the time-derivative of the composite function $t \mapsto f(x(t), x_t(t))$ at $t$ is $f_x(x(t), x_t(t))x_t(t) + f_{\dot{x}}(x(t), x_t(t))x_{tt}(t)$.



*D. Durban et al. (eds.), Advances in the Mechanics of Plates and Shells,* 1–16.

## 2. EQUATIONS OF MOTION FOR A NONLINEARLY VISCOELASTIC RING

We outline the basic geometrically exact, direct, plane-strain theory of rods that can suffer flexure, shear, and both longitudinal and transverse extensions. (These rods are termed *beamshells* by Libai and Simmonds [9] and are thereby subsumed within the theory of shells. For the connection of this theory to two-dimensional theory of continuum mechanics, see [1, Sec. XIV.5].)

Let $\{\boldsymbol{i}, \boldsymbol{j}, \boldsymbol{k}\}$ be a fixed orthonormal basis for Euclidean 3-space. A *planar configuration* at time $t$ of a ring that can suffer flexure, shear, and both longitudinal and transverse extensions is specified by a continuously differentiable vector-valued function with values $\boldsymbol{r}(s,t)$ in the $\{\boldsymbol{i}, \boldsymbol{j}\}$-plane and two continuously differentiable functions with real values $\theta(s,t)$ and $\delta(s,t)$, with each of these functions having period $2\pi$ in $s$. We take the domain of $s$ to be $[0, 2\pi]$. We think of the body under study as having a natural reference configuration in the form of a circular annulus, and we interpret $s$ as the arc-length parameter of a suitable unit (concentric) circle within the annulus, called the base circle. Then $s$ identifies material sections of the annulus. See Figure 1.

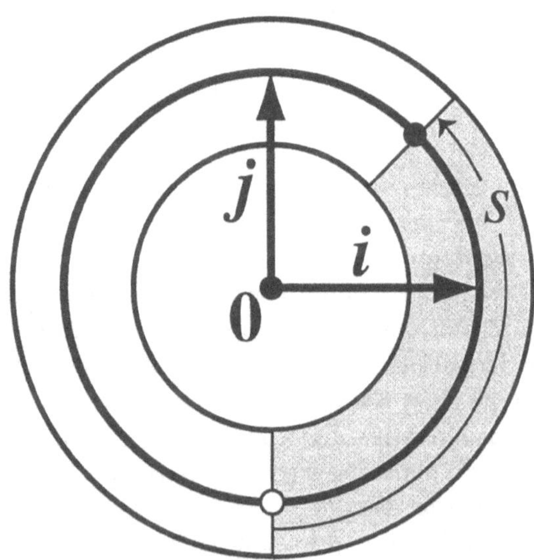


*Figure 1* Reference configuration of the ring (in its 2-dimensional interpretation). The base curve is the intermediate circle. The shaded region is a typical segment in which the arc-length parameter ranges from 0 to $s$. The material point on the base curve at the section $s$, shown with a black dot, has position $\sin s\boldsymbol{i} - \cos s\boldsymbol{j}$.

We interpret $\boldsymbol{r}(s,t)$ as the position at time $t$ of the material point on the base circle with coordinate $s$. We interpret $\theta(s,t)+\frac{\pi}{2}$ as characterizing the orientation at time $t$ of the material cross section $s$. We finally interpret $\delta(s,t)$ as characterizing the ratio of deformed to reference length of the section $s$.

We introduce the orthonormal basis

$$\boldsymbol{a}(\theta)=\cos\theta\boldsymbol{i}+\sin\theta\boldsymbol{j},\quad \boldsymbol{b}(\theta)=-\sin\theta\boldsymbol{i}+\cos\theta\boldsymbol{j}. \tag{2.1}$$

The *strains* are $(\nu,\eta,\mu,\delta,\omega)$ where

$$\nu\boldsymbol{a}+\eta\boldsymbol{b}:=\boldsymbol{r}_s,\quad \mu:=\theta_s,\quad \omega:=\delta_s. \tag{2.2}$$

See Figure 2.

Note that $\boldsymbol{r}_s$ is tangent to the deformed base curve, but it need not be a unit tangent. The stretch of the base curve (i.e., the local ratio of deformed to reference length of the base curve) is $|\boldsymbol{r}_s|=\sqrt{\nu^2+\eta^2}$. It is mathematically and mechanically convenient to decompose $\boldsymbol{r}_s$ with respect to the basis $\{\boldsymbol{a},\boldsymbol{b}\}$ and take $\nu,\eta$ as the corresponding strains, rather than taking the strains to be $|\boldsymbol{r}_s|$ and some shear angle. The strain $\eta\equiv\boldsymbol{b}\cdot\boldsymbol{r}_s$ measures shear. $\eta$ could easily be related to a shear angle, the use of which would greatly complicate the governing equations. The strain $\nu\equiv\boldsymbol{r}_s\cdot\boldsymbol{a}\equiv\boldsymbol{k}\cdot(\boldsymbol{r}_s\times\boldsymbol{b})$ measures a volume ratio, but it is convenient to interpret it as the main contributor to the longitudinal stretch $|\boldsymbol{r}_s|$.

For the special class of motions we shall consider, the rod-theoretic requirement that the deformation locally preserve orientation (which includes the requirement that the local ratio of actual to reference length of any fiber be positive) reduces to

$$\nu>0,\qquad \delta>0. \tag{2.3}$$

Let $N(s,t)\boldsymbol{a}(\theta(s,t))+H(s,t)\boldsymbol{b}(\theta(s,t))$ be the contact force and $M(s,t)$ be the contact couple (about $\boldsymbol{k}$) exerted across the material section at $s$. Let $\Delta$ and $\Omega$ be the generalized forces corresponding to the strains $\delta$ and $\omega$ through a principle of virtual power. (For mechanical interpretations of $\Delta$ and $\Omega$, see [1, Sec. XIV.5].) Then (provided that the base curve is appropriately located in the reference configuration) the equations of motion for the ring under no external loading have the form

$$\rho A\boldsymbol{r}_{tt}=(N\boldsymbol{a}+H\boldsymbol{b})_s, \tag{2.4a}$$

$$\begin{aligned}\rho J(\delta^2\theta_t)_t&=M_s+\boldsymbol{k}\cdot[\boldsymbol{r}_s\times(N\boldsymbol{a}+H\boldsymbol{b})]\\&\equiv M_s+\nu H-\eta N,\end{aligned} \tag{2.4b}$$

$$\rho J(\delta_{tt}-\delta{\theta_t}^2)=\Omega_s-\Delta, \tag{2.4c}$$

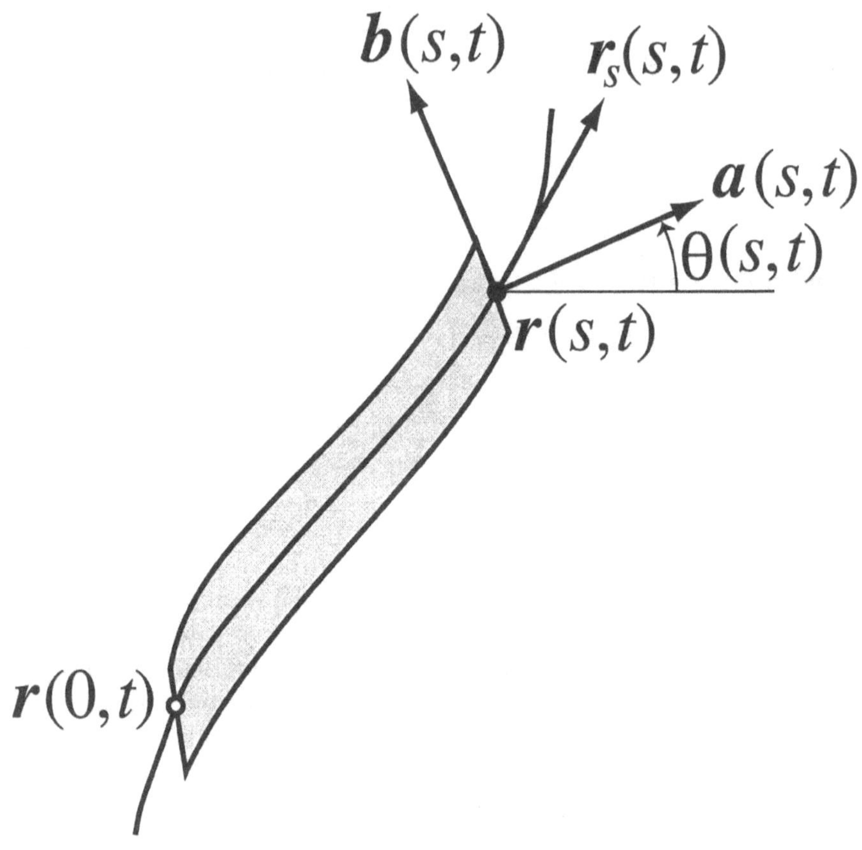


*Figure 2* The configuration at time $t$ of the material segment shaded gray in Figure 1.

where $\rho A$ and $\rho J$ are positive-valued functions of $s$, roughly corresponding to the mass and to the second mass moment of inertia of a cross section per unit reference length [1].

Like $\boldsymbol{r}_s$, the contact force is decomposed along the basis $\{\boldsymbol{a}, \boldsymbol{b}\}$. The components $N$ and $H$ (which are the natural duals of $\nu$ and $\eta$) are not the tension and (vertical) shear forces of elementary beam theory. (These latter components are taken with respect to the unit tangent and normal to the deformed base curve.)

When rods are given 2-dimensional interpretations, it is customary to take the base curve to be the curve of centroids of the reference configuration because this choice often simplifies the constitutive equations. If the base curve is curved (as in our problem), however, this simplification produces concomitant complications in the inertia terms. To avoid these complications, we are here making an alternative choice of the base curve (see [1]). Since we are treating very general constitutive equations, we would not notice any simplification in these equations. An inkling of the underlying issue can be gained from the observation that if the base curve for our problem were the circle of centroids and if the mass density of our 2-dimensional annular body were constant, then the mass of the material outside the base curve would exceed that on the inside.

We assume that the ring is uniform, so that its material properties are independent of $s$. Then $\rho A$ and $\rho J$ are constants. A uniform viscoelastic rod of strain-rate type has constitutive equations of the form

$$N = W_\nu(\nu, \eta, \mu, \delta, \omega) + N^{\mathrm{D}}(\nu, \eta, \mu, \delta, \omega, \nu_t, \eta_t, \mu_t, \delta_t, \omega_t), \quad \text{etc.}, \tag{2.5}$$

where

$$N^{\mathrm{D}}(\nu, \eta, \mu, \delta, \omega, 0, 0, 0, 0, 0) = 0, \quad \text{etc.}, \tag{2.6}$$

where $W$ is a given stored-energy function, assumed to be twice continuously differentiable, and $N^{\mathrm{D}}$, etc., are given continuously differentiable functions, accounting for the dissipative parts of the resultants $N$, etc. If $N^{\mathrm{D}} = 0$, etc., then the ring is hyperelastic. System (2.5), (2.6) is the most general set of constitutive equations for a 'visco-hyperelastic' rod of strain-rate type invariant under rigid motions having the kinematics just described.

We make the reasonable symmetry assumption [1, Prop. XIV.5.22]

$$\begin{gathered} W_\eta(\nu, 0, \mu, \delta, 0) = 0 = W_\omega(\nu, 0, \mu, \delta, 0), \\ H^{\mathrm{D}}(\nu, 0, \mu, \delta, 0, \dot\nu, 0, \dot\mu, \dot\delta, 0) = 0 = \Omega^{\mathrm{D}}(\nu, 0, \mu, \delta, 0, \dot\nu, 0, \dot\mu, \dot\delta, 0). \end{gathered} \tag{2.7}$$

We assume that the constitutive functions are such that (2.3) is never violated in regular motions, but do not pause to spell out specific assumptions that enforce this requirement (see [4]).

Let us mention some standard constitutive restrictions. The requirement that $W$ be a uniformly convex function of $\nu, \eta, \mu, \omega$ is a rod-theoretic analog of the Strong Ellipticity Condition of the three-dimensional theory. This condition implies the physically reasonable assumptions that $N$ is an increasing function of $\nu$ for fixed values of the other strains, that $H$ is an increasing function of $\eta$, that $M$ is an increasing function of $\mu$, etc. (Had we not used the special choices of strains and resultants associated with the basis $\{\boldsymbol{a}, \boldsymbol{b}\}$, then this condition would be so complicated that its simple mechanical consequences would not be apparent. This uniform convexity condition is not universally valid: It precludes several kinds of coexistent phases now under intensive study.)

A related condition for the dissipative response is that the $4 \times 4$ matrix of partial derivatives of $(N^{\mathrm{D}}, H^{\mathrm{D}}, M^{\mathrm{D}}, \Omega^{\mathrm{D}})$ with respect to the corresponding arguments $(\dot{\nu}, \dot{\eta}, \dot{\mu}, \dot{\omega})$ be uniformly positive-definite. This condition ensures that the resulting system is parabolic. A stronger condition, which is not unreasonable, is the uniform monotonicity condition that the $5 \times 5$ matrix of partial derivatives of $(N^{\mathrm{D}}, H^{\mathrm{D}}, M^{\mathrm{D}}, \Delta^{\mathrm{D}}, \Omega^{\mathrm{D}})$ with respect to their arguments $(\dot{\nu}, \dot{\eta}, \dot{\mu}, \dot{\delta}, \dot{\omega})$ be uniformly positive-definite. This condition ensures a certain uniform dissipativity. In Section 5, we discuss the effect of such conditions on the behavior of special kinds of solutions.

We get equations for a ring that cannot suffer transverse extension by constraining $\delta = 1$ (and $\omega = 0$), making this substitution into (2.5), (2.6) for $N, H, M$, and ignoring (2.4c), which now is just an equation for Lagrange multipliers $\Delta$ and $\Omega$. If we further require the ring to be unshearable, then we constrain $\eta = 0$, make this substitution in (2.5), (2.6) for $N, M$, and treat the Lagrange multiplier $H$ as a fundamental unknown. We do not ignore rotatory inertia, i.e., we do not set $\rho J = 0$, because doing so would change the type of the equations and the qualitative behavior of their solutions.

If we take the dot product of (2.4a) with $\boldsymbol{r}_t$, multiply (2.4b) by $\theta_t$, multiply (2.4c) by $\delta_t$, and integrate the sum of the resulting products by parts with respect to $s$, we obtain the energy equation

$$\frac{d}{dt}\int_0^{2\pi} \{\tfrac{1}{2}[\rho A \boldsymbol{r}_t \cdot \boldsymbol{r}_t + \delta^2 {\theta_t}^2 + {\delta_t}^2] + W(\nu, \eta, \mu, \delta, \omega)\}\, ds + 2\pi P = 0, \tag{2.8}$$

where

$$2\pi P := \int_0^{2\pi} (N^{\mathrm{D}} \nu_t + H^{\mathrm{D}} \eta_t + M^{\mathrm{D}} \mu_t + \Omega^{\mathrm{D}} \omega_t + \Delta^{\mathrm{D}} \delta_t)\, ds \geq 0 \tag{2.9}$$

is the stress power, which is non-negative if the matrix of partial derivatives of $(N^{\mathrm{D}}, H^{\mathrm{D}}, M^{\mathrm{D}}, \Delta^{\mathrm{D}}, \Omega^{\mathrm{D}})$ with respect to $(\dot{\nu}, \dot{\eta}, \dot{\mu}, \dot{\delta}, \dot{\omega})$ is uniformly positive-definite. For hyperelastic materials, for which $P = 0$, regular solutions of (2.4)–(2.6) conserve energy:

$$\int_0^{2\pi} \{\tfrac{1}{2}[\rho A \boldsymbol{r}_t \cdot \boldsymbol{r}_t + \rho J(\delta^2 {\theta_t}^2 + {\delta_t}^2)] + W\}\, ds = \text{const.} \tag{2.10}$$

## 3. SHEARLESS OSCILLATIONS

We study the nature of unforced motions in which the ring rotates and oscillates radially with no shearing, starting from some initial configuration in which the ring is inflated and rotating. Since conservation of angular momentum requires that the ring rotate faster when its radius is smaller, we seek solutions of the form

$$\theta(s,t) = s + \psi(t), \qquad \boldsymbol{r}(s,t) = -\bar{\nu}(t)\boldsymbol{b}(s+\psi(t)), \qquad \delta(s,t) = \bar{\delta}(t). \tag{3.1a,b,c}$$

We now drop the superposed bars. Assumption (3.1) implies that

$$\boldsymbol{r}_t = -\nu_t\boldsymbol{b} + \nu\psi_t\boldsymbol{a}, \qquad \boldsymbol{r}_{tt} = [2\nu_t\psi_t + \nu\psi_{tt}]\boldsymbol{a} + [-\nu_{tt} + \nu{\psi_t}^2]\boldsymbol{b}, \tag{3.2}$$

$$\eta = 0, \quad \mu = 1, \quad \delta_s = 0. \tag{3.3}$$

We set

$$\bar{N}(\nu,\delta,\nu_t,\delta_t) := W_\nu(\nu,0,1,\delta,0) + N^{\mathrm{D}}(\nu,0,1,\delta,0,\nu_t,0,0,\delta_t,0), \quad \text{etc.}$$

Under our symmetry assumption (2.7), $\bar{H} = 0 = \bar{\Omega}$, so that the substitution of (3.1) into (2.4)–(2.6) reduces these equations to

$$\rho A[\nu_{tt} - \nu{\psi_t}^2] + \bar{N}(\nu,\delta,\nu_t,\delta_t) = 0, \tag{3.4a}$$

$$(\nu^2\psi_t)_t = 0, \tag{3.4b}$$

$$(\delta^2\psi_t)_t = 0, \tag{3.4c}$$

$$\rho J[\delta_{tt} - \delta{\psi_t}^2] + \bar{\Delta}(\nu,\delta,\nu_t,\delta_t) = 0. \tag{3.4d}$$

Equations (3.4b,c) together with (2.3) say that provided that $\psi_t(0) \geq 0$, there are constants $\alpha, \beta$ of the same sign (determined by the initial conditions) such that

$$\psi_t = \frac{\alpha^2}{\nu^2} = \frac{\beta^2}{\delta^2}. \tag{3.5}$$

Thus $\psi_t$ either is identically zero or it never vanishes. In the former case, (3.4) reduces to the coupled pair of ordinary differential equations

$$\rho A\nu_{tt} + \bar{N}(\nu,\delta,\nu_t,\delta_t) = 0, \qquad \rho J\delta_{tt} + \bar{\Delta}(\nu,\delta,\nu_t,\delta_t) = 0, \tag{3.6}$$

which describe purely radial motions. In the latter case, the motion is overdetermined with $\nu$ and $\delta$ proportional; equations (3.4a,d), (3.5) say that this motion is not possible unless $\bar{N}$ and $\bar{\Delta}$ satisfy

$$\frac{\bar{N}(\nu,\beta\nu/\alpha,\nu_t,\beta\nu_t/\alpha)}{\rho A\alpha} = \frac{\bar{\Delta}(\nu,\beta\nu/\alpha,\nu_t,\beta\nu_t/\alpha)}{\rho J\beta}. \tag{3.7}$$

There is no reason to expect this condition to hold for given $\alpha, \beta$ (determined from initial conditions), much less for a range of these parameters. (For three-dimensional interpretations of these functions, which support this assertion, see [1, Section XIV.5]. Physically reasonable constitutive functions that account for a Poisson-ratio effect (in which a longitudinal tensile force produces a longitudinal extension and a smaller transverse contraction) cannot meet this condition except trivially.) Note that rotatory inertia, i.e., the positivity of $\rho J$, is crucial in leading to (3.7).

In summary, we have shown that generically the only solutions of the form (3.1) describe purely radial motions. For the purposes of this paper, these motions are degenerate.

## 4. OSCILLATIONS WITH SHEARING

To try to discover the source of these difficulties, we relax the requirement that (3.1) represent a shearless motion by seeking solutions of the more general form

$$\begin{aligned} \boldsymbol{r}(s,t) &= \bar{\eta}(t)\boldsymbol{a}(s+\psi(t)) - \bar{\nu}(t)\boldsymbol{b}(s+\psi(t)), \\ \theta(s,t) &= s+\psi(t), \quad \delta(s,t) = \bar{\delta}(t). \end{aligned} \tag{4.1}$$

Note that this $\boldsymbol{r}$ is perpendicular to $\boldsymbol{r}_s$, so that this motion represents a combination of breathing and rotation.

We drop the superposed bars from $\nu, \eta, \delta$, replace $W(\nu, \eta, 1, \delta, 0)$ by $W(\nu, \eta, \delta)$, replace $N^{\mathrm{D}}(\nu, \eta, 1, \delta, 0, \nu_t, \eta_t, 0, \delta_t, 0)$ by $N^{\mathrm{D}}(\nu, \eta, \delta, \nu_t, \eta_t, \delta_t)$ (we use analogous conventions elsewhere), and set

$$\bar{N}(\nu, \eta, \delta, \nu_t, \eta_t, \delta_t) := W_\nu(\nu, \eta, \delta) + N^{\mathrm{D}}(\nu, \eta, \delta, \nu_t, \eta_t, \delta_t), \quad \text{etc.} \tag{4.2}$$

Substituting (4.1) into (2.4)–(2.6) we obtain the following replacement for (3.4):

$$\rho A[\nu_{tt} - \nu\psi_t{}^2 - 2\eta_t\psi_t - \eta\psi_{tt}] + \bar{N}(\nu, \eta, \delta, \nu_t, \eta_t, \delta_t) = 0, \tag{4.3a}$$

$$\rho A[2\nu_t\psi_t + \nu\psi_{tt} + \eta_{tt} - \eta\psi_t{}^2] + \bar{H}(\nu, \eta, \delta, \nu_t, \eta_t, \delta_t) = 0, \tag{4.3b}$$

$$\rho J(\delta^2\psi_t)_t + \eta\bar{N}(\nu, \eta, \delta, \nu_t, \eta_t, \delta_t) - \nu\bar{H}(\nu, \eta, \delta, \nu_t, \eta_t, \delta_t) = 0, \tag{4.3c}$$

$$\rho J[\delta_{tt} - \delta\psi_t{}^2] + \bar{\Delta}(\nu, \eta, \delta, \nu_t, \eta_t, \delta_t) = 0. \tag{4.3d}$$

Unlike (3.4a,b,c), this system is not degenerate; as we show in the next section, where we discuss its integrals, it has a rich collection of solutions.

How is this system modified for an unshearable ring for which $\eta$ is constrained to be 0? In this case, $H$ is the corresponding Lagrange multiplier, which is a fundamental unknown not specified by a constitutive equation. The other constitutive functions clearly do not depend on $\eta$ or $\eta_t$. We retain (3.1), but do not adopt a mechanical symmetry condition requiring $H$ to vanish. In this case, (4.3) is replaced by

$$\rho A[\nu_{tt} - \nu \psi_t{}^2] + \bar{N}(\nu, \delta, \nu_t, \delta_t) = 0, \tag{4.4a}$$

$$\rho A(\nu^2 \psi_t)_t + \nu H = 0, \tag{4.4b}$$

$$\rho J(\delta^2 \psi_t)_t - \nu H = 0, \tag{4.4c}$$

$$\rho J[\delta_{tt} - \delta \psi_t{}^2] + \bar{\Delta}(\nu, \delta, \nu_t, \delta_t) = 0. \tag{4.4d}$$

Equations (4.4b,c) yield the conservation of angular momentum:

$$(\rho A \nu^2 + \rho J \delta^2)\psi_t = \text{const.} \tag{4.5}$$

The substitution of (4.5) into (4.4a,d) produces a pretty fourth-order system for $\nu$ and $\delta$. This system also has a rich collection of non-degenerate solutions. The substitution of these solutions into (4.4b) or (4.4c) typically yields a nonzero time-varying expression for $H$. Thus even for an unshearable ring, a nonzero shear force is essential for non-degenerate solutions.

The modifications of (4.3) and (4.4) necessary to accommodate the constraint $\delta = 1$ are immediate. When $\delta = 1$, the substitution of (4.5) into (4.4a) reduces the latter to an autonomous second-order ordinary differential equation, which can be readily analyzed by phase-plane methods.

## 5. PROPERTIES OF SOLUTIONS

Let us first study (4.3). Replacing $\bar{N}$ and $\bar{H}$ in (4.3c) with the expressions coming from (4.3a,b), we obtain

$$\begin{aligned} \rho J(\delta^2 \psi_t)_t - \rho A[\nu_{tt} - \nu \psi_t{}^2 - 2\eta_t \psi_t - \eta \psi_{tt}]\eta \\ + \rho A[2\nu_t \psi_t + \nu \psi_{tt} + \eta_{tt} - \eta \psi_t{}^2]\nu = 0, \end{aligned} \tag{5.1a}$$

which yields the conservation of angular momentum

$$[\rho A(\nu^2 + \eta^2) + \rho J \delta^2]\psi_t + \rho A(\nu \eta_t - \nu_t \eta) = \text{const.} \tag{5.1b}$$

We could solve this equation for $\psi_t$ and differentiate the solution to get $\psi_{tt}$. Substituting these expressions into (4.3a,b,d), we would obtain a complicated system of ordinary differential equations for $\nu, \eta, \delta$.

Let us introduce the kinetic energy function $K$ and the stress-power function $P$ by

$$K(\nu,\eta,\delta,\dot\nu,\dot\eta,\dot\delta,\dot\psi) := \tfrac12\rho A[(\dot\nu-\eta\dot\psi)^2+(\dot\eta+\nu\dot\psi)^2] + \tfrac12\rho J[\delta^2\dot\psi^2+\dot\delta^2], \tag{5.3}$$

$$P(\nu,\eta,\delta,\dot\nu,\dot\eta,\dot\delta) := N^{\mathrm D}(\nu,\eta,\delta,\dot\nu,\dot\eta,\dot\delta)\dot\nu + H^{\mathrm D}(\nu,\eta,\delta,\dot\nu,\dot\eta,\dot\delta)\dot\eta + \Delta^{\mathrm D}(\nu,\eta,\delta,\dot\nu,\dot\eta,\dot\delta)\dot\delta. \tag{5.4}$$

Then the energy equation (2.8) reduces to

$$\tfrac{d}{dt}[K(\nu,\eta,\delta,\nu_t,\eta_t,\delta_t,\psi_t)+W(\nu,\eta,\delta)]+P(\nu,\eta,\delta,\nu_t,\eta_t,\delta_t)=0. \tag{5.5}$$

To write (4.3) as a system of first-order equations, we denote by

$$\mathbf q := (\nu,\eta,\delta,\psi), \quad \dot{\mathbf q} := (\dot\nu,\dot\eta,\dot\delta,\dot\psi) \tag{5.6}$$

the quadruples of generalized coordinates and their derivatives, and we introduce the corresponding quadruple $\mathbf p = (p_1,p_2,p_3,p_4)$ of generalized momenta corresponding to $\mathbf q$ by $\mathbf p = \partial K(\mathbf q,\dot{\mathbf q})/\partial\dot{\mathbf q}$, i.e., by

$$\begin{aligned}
p_1 &= K_{\dot\nu} = \rho A(\dot\nu-\eta\dot\psi),\\
p_2 &= K_{\dot\eta} = \rho A(\dot\eta+\nu\dot\psi),\\
p_3 &= K_{\dot\delta} = \rho J\dot\delta,\\
p_4 &= K_{\dot\psi} = \rho A[-\eta(\dot\nu-\eta\dot\psi)+\nu(\dot\eta+\nu\dot\psi)]+\rho J\delta^2\dot\psi.
\end{aligned} \tag{5.7}$$

We solve (5.7) for $\dot{\mathbf q}$ in terms of $\mathbf q$ and $\mathbf p$, and denote the solution by $\mathbf g(\mathbf q,\mathbf p)$. Then we define the Hamiltonian function $E$ (the total energy) by

$$\begin{aligned}
E(\mathbf q,\mathbf p) :&= K(\nu,\eta,\delta,\mathbf g(\mathbf q,\mathbf p))+W(\nu,\eta,\delta)\\
&= \frac{p_1^2+p_2^2}{2\rho A}+\frac{p_3^2}{2\rho J}+\frac{(p_4+\eta p_1-\nu p_2)^2}{2\rho J\delta^2}+W(\nu,\eta,\delta).
\end{aligned} \tag{5.8}$$

We can now write (4.3) as the system of first-order equations:

$$\mathbf q_t = \frac{\partial E}{\partial\mathbf p}(\mathbf q,\mathbf p), \qquad \mathbf p_t = -\frac{\partial E}{\partial\mathbf q}(\mathbf q,\mathbf p)-\mathbf f^{\mathrm D}(\mathbf q,\mathbf p) \tag{5.9}$$

where

$$\mathbf f^{\mathrm D}(\mathbf q,\mathbf p) = (N^{\mathrm D},H^{\mathrm D},\Delta^{\mathrm D}) \tag{5.10}$$

and where the arguments of $N^{\mathrm D},H^{\mathrm D},\Delta^{\mathrm D}$ are

$$(\nu,\eta,\delta,g_1(\mathbf q,\mathbf p),g_2(\mathbf q,\mathbf p),g_3(\mathbf q,\mathbf p)).$$

In components, (5.9) has the form

$$\begin{aligned}
\nu_t &= \frac{p_1}{\rho A} + \frac{(p_4 + \eta p_1 - \nu p_2)\eta}{\rho J \delta^2}, \\
\eta_t &= \frac{p_2}{\rho A} - \frac{(p_4 + \eta p_1 - \nu p_2)\nu}{\rho J \delta^2}, \\
\delta_t &= \frac{p_3}{\rho J}, \\
\psi_t &= \frac{p_4 + \rho A(\eta p_1 - \nu p_2)}{\rho J \delta^2}, \\
p_{1t} &= \frac{(p_4 + \eta p_1 - \nu p_2)p_2}{\rho J \delta^2} - W_\nu(\nu, \eta, \delta) - N^{\mathrm{D}}, \\
p_{2t} &= -\frac{(p_4 + \eta p_1 - \nu p_2)p_1}{\rho J \delta^2} - W_\eta(\nu, \eta, \delta) - H^{\mathrm{D}}, \\
p_{3t} &= \frac{(p_4 + \eta p_1 - \nu p_2)^2}{\rho J \delta^3} - W_\delta(\nu, \eta, \delta) - \Delta^{\mathrm{D}}, \\
p_{4t} &= 0,
\end{aligned} \tag{5.11}$$

where the arguments of $N^{\mathrm{D}}, H^{\mathrm{D}}, \Delta^{\mathrm{D}}$ are those shown above. Of course, the last equation of (5.11), which is a consequence of the ignorability of the generalized coordinate $\psi$, is equivalent to the conservation (5.1b) of angular momentum. In terms of these new variables, the energy equation (5.5) becomes

$$\tfrac{d}{dt}E(\mathbf{q}, \mathbf{p}) + P(\nu, \eta, \delta, g_1(\mathbf{q}, \mathbf{p}), g_2(\mathbf{q}, \mathbf{p}), g_3(\mathbf{q}, \mathbf{p})) = 0. \tag{5.12}$$

Now let us assume that $P$ is bounded below by a constant:

$$P \geq -C. \tag{5.13}$$

(As our remarks in the paragraph following that containing (2.7) show, this is a very mild assumption: If the material is hyperelastic, then $C = 0$. The material is dissipative if

$$P > 0 \quad \text{except where} \quad \dot{\nu} = \dot{\eta} = \dot{\delta} = 0. \tag{5.14}$$

The material may be said to be uniformly dissipative if $(N^{\mathrm{D}}, H^{\mathrm{D}}, \Delta^{\mathrm{D}})$ is a monotone function of $(\dot{\nu}, \dot{\eta}, \dot{\delta})$. This condition and (2.6) ensure (5.14)). The energy equation (5.12) and inequality (5.14) imply that

$$E|_t \leq E|_0 + Ct. \tag{5.15}$$

Next, let us further assume that

$$W(\nu,\eta,\delta)\to\infty \quad \text{as} \quad \begin{cases}\nu^2+\eta^2+\delta^2\to\infty,\\ \nu\to 0,\\ \delta\to 0.\end{cases} \tag{5.16}$$

This property and the bound on $E$ imply that for any $t$, solutions of (4.3) satisfy $0<\nu(t),\delta(t)<\infty$, $|\eta(t)|,|\nu_t(t)|,|\eta_t(t)|,|\delta_t(t)|,|\psi_t(t)|<\infty$. This means that solutions neither blow up nor violate the bounds (2.3) in finite time. By the standard continuation theory for ordinary differential equations, it follows that solutions of (initial-value problems for) (4.3) (or of (5.11)) exist for all time. If $C=0$, then we get $E|_t\le E|_0$, so that $\nu$ and $\delta$ are confined to compact sets of $(0,\infty)$ and that $\eta,\nu_t,\eta_t,\delta_t,\psi_t$ are confined to bounded sets. In this case, there is a number $B$ such that $|\psi(t)-\psi(0)|\le Bt$.

Now we determine qualitative properties of solutions. Let (5.14) hold. Then $E$ is a Lyapunov function for (5.11). Let us further assume that (5.15) holds, so that all solutions are bounded for all time. The LaSalle Invariance Principle (see [7,Thm. X.1.3; 13]) then says that each solution approaches the largest invariant subset of $(\mathbf{q},\mathbf{p})$-space where the derivative of $E$ along solutions of (5.11) vanish. By (5.12), this derivative vanishes exactly where $P$ vanishes, i.e., where $\dot\nu=\dot\eta=\dot\delta=0$. To find this invariant set, we revert to the formulation (4.3) and seek solutions with $\nu_t=\eta_t=\delta_t=0$. These solutions satisfy

$$\rho A[-\nu{\psi_t}^2-\eta\psi_{tt}]+W_\nu(\nu,\eta,\delta)=0, \tag{5.17a}$$

$$\rho A[\nu\psi_{tt}-\eta{\psi_t}^2]+W_\eta(\nu,\eta,\delta)=0, \tag{5.17b}$$

$$\rho J\delta^2\psi_{tt}+\eta W_\nu(\nu,\eta,\delta)-\nu W_\eta(\nu,\eta,\delta)=0, \tag{5.17c}$$

$$-\rho J\delta{\psi_t}^2+W_\delta(\nu,\eta,\delta)=0. \tag{5.17d}$$

From these equations, or better yet from the conservation (5.1b) of angular momentum, we find that $\psi_t$ is the constant given by

$$[\rho A(\nu^2+\eta^2)+\rho J\delta^2]\psi_t=\alpha \tag{5.18}$$

where $\alpha$ is determined by initial conditions. In this case, the invariant set consists of those constant solutions $(\nu,\eta,\delta,\psi_t)$ of (5.18) and the following reduced form of (5.17):

$$\rho A \nu \psi_t{}^2 = W_\nu(\nu, \eta, \delta), \tag{5.19a}$$

$$\rho A \eta \psi_t{}^2 = W_\eta(\nu, \eta, \delta), \tag{5.19b}$$

$$\eta W_\nu(\nu, \eta, \delta) = \nu W_\eta(\nu, \eta, \delta), \tag{5.19c}$$

$$\rho J \delta \psi_t{}^2 = W_\delta(\nu, \eta, \delta). \tag{5.19d}$$

Note that (5.19c) is a consequence of (5.19a,b); we accordingly ignore it. Thus (5.18), (5.19a,b,d) form a system of four equations for the unknown real numbers $\psi_t, \nu, \eta, \delta$.

We use (2.7) to reduce (5.19b) to an identity by taking $\eta = 0$. If we substitute (5.18) into (5.19a,d), we get a pair of algebraic equations for $\nu$ and $\delta$. Now for fixed $\delta > 0$, the left-hand side of this version of (5.19a) behaves like $\nu$ for small $\nu$ and behaves like $\nu^{-3}$ for large $\nu$, whereas for reasonable $W$, the right-hand side approaches $-\infty$ as $\nu \to 0$ and approaches $\infty$ as $\nu \to \infty$. An analogous remark applies to (5.19d). An elementary degree-theoretic generalization of these observations shows that these coupled equations have solutions, not necessarily unique. For certain materials (having shear instabilities, which are compatible with the Strong Ellipticity Condition), the full system may admit solutions with $\eta \neq 0$.

Thus we have shown that for dissipative materials, solutions must approach a steady rotating solution. When there is more than one such solution, the limiting state is determined by the initial conditions. Associated with each stable steady state is a basin of attraction of initial conditions.

Now let us briefly discuss hyperelastic materials, for which $\mathbf{f}^{\mathrm{D}} = \mathbf{0}$ whence $P = 0$, so that (5.11) is a Hamiltonian system. The process of solving (5.1b) for $\psi_t$ and substituting the result into (4.3a,b,d) is equivalent to taking $p_4$ equal to a constant in (5.8) and (5.11). We accordingly redefine $\mathbf{q}$ and $\mathbf{p}$ to be the triples of their first three components. Of course, (5.5) or (5.9) or (5.11) yields the conservation of energy

$$E(\mathbf{q}, \mathbf{p}) = \text{const.} \tag{5.20}$$

Then for fixed $\mathbf{q}$, this new Hamiltonian $E$ is a (nonhomogeneous) positive-definite quadratic form in $\mathbf{p}$, and its dependence on $\mathbf{q}$ is largely dictated by $W$. We can, however, choose the constant angular momentum $p_4$ so that the Hessian matrix of the kinetic energy term of $E$ is not positive-definite, i.e., this term is not convex. This means that we can choose $p_4$ so that $E$ itself is not convex. Note further that the kinetic energy term is a quartic in $(\mathbf{q}, \mathbf{p})$.

Now the Strong Ellipticity Condition requires that $\nu \mapsto W(\nu, \eta, \delta)$ be convex for all $\eta, \delta$ and that $\eta \mapsto W(\nu, \eta, \delta)$ be convex for all $\nu, \delta$, but it does not require $W$ itself to be convex, although it is not unreasonable to require it to be so on large parts of its domain. On the other hand, much modern work on coexistent phases is based on the assumption that the stored-energy function has multiple local minima (so that the limited convexity required by the Strong Ellipticity Condition is completely violated).

The importance of these remarks inheres in the fact that there is now a very rich (but still incomplete) global theory on the behavior of solutions of Hamiltonian systems. This theory contains a host of theorems on the existence of solutions with a prescribed period and theorems on the number of qualitatively distinct solutions for a given constant value of the total energy $E$. Typical hypotheses for many of the available theorems for the first kind of problems involve the specification of the behavior of $E$ at infinity, in particular whether it is sub- or superquadratic. (The quartic growth of the kinetic energy term in our $E$ precludes the use of results based on subquadratic growth.) A typical hypothesis for the second kind of problem is that the region of $(\mathbf{q}, \mathbf{p})$-space enclosed by the level surface of energy be bounded and convex. Clearly, for our problem, this convexity depends crucially on the nature of $W$ and on the size of $p_4$. Because of the richness of material response we are allowing, a whole battery of theorems on periodic solutions would be necessary to delimit the variety of behavior possible. For specific theorems, see [6] and [11], and the references cited therein.

Of course, periodic solutions represent but a small range of the response available to Hamiltonian systems, which admit very complicated and chaotic solutions. For the theory and references, see [5, 10, 14]. Finally, it should be mentioned that while Hamiltonian systems are eminently appropriate for the description of celestial mechanics, the ubiquity of dissipation in terrestrial mechanics makes them somewhat artificial here.

The study of stability of solutions of (5.11), with or without dissipation, would be a purely academic exercise, because stability would be treated only within the special class of motions (4.1). The correct setting of stability of these solutions is within the class of solutions of the partial differential equations (2.4)–(2.6) (cf. [12]). Nevertheless, our result showing that the solutions of (5.11) for viscoelastic rings approach purely rotating states begs to be generalized to the partial differential equations (2.4)–(2.6) or even to the partial differential equations for a full two-dimensional theory.

# 6. COMMENTS

The system (2.4) for a hyperelastic ring for which $W$ is a uniformly convex function of $\nu, \eta, \mu, \omega$ is a quasilinear hyperbolic system susceptible to shocks. There are a host of numerical methods designed to handle such hyperbolic systems having but one independent variable [8]. Virtually all these methods had their source in gas dynamics. They consist in imposing some sort of mathematical viscosity on the system with a form suggested by the viscosity in fluids. It is shown in [3] (cf. [2]) that some (and maybe all) of these dissipative mechanisms do not correspond to constitutive equations invariant under rigid motion. (A mathematical dissipation that is properly invariant in the spatial (Eulerian) formulation is not invariant in the material (Lagrangian) formulation.) Consequently, the numerical treatment of hyperelastic structures undergoing rapid rotation may well lead to serious error.

The special problems treated above are each solutions to the corresponding system of partial differential equations for appropriate initial conditions. Since we know so much about these solutions, especially the solution that rotates without breathing, the initial-value problems that generate them are ideal settings for showing which numerical methods lack the requisite invariance and for testing the efficacy of new numerical methods that respect the invariance [3].

## Acknowledgments

The research reported here was supported in part by NSF Grant DMS 99 71823 and by ARO-MURI97 Grant No. DAAG55-97-1-0114 to the Center for Dynamics and Control of Smart Structures.

## References

[1] S. S. Antman, *Nonlinear Problems of Elasticity*, Springer-Verlag, 1995.

[2] S. S. Antman, Physically unacceptable viscous stresses, *Z. angew. Math. Mech.* **49** (1998), 980–988.

[3] S. S. Antman and Jian-Guo Liu, Errors in the numerical treatment of hyperbolic conservation laws caused by lack of invariance, in preparation.

[4] S. S. Antman and T. I. Seidman, Quasilinear hyperbolic-parabolic equations of one-dimensional viscoelasticity, *J. Diff. Eqs.* **123** (1995), 132–185.

[5] H. Dankowicz, *Chaotic Dynamics in Hamiltonian Systems*, World Scientific, 1997.

[6] I. Ekeland, *Convexity Methods in Hamiltonian Mechanics*, Springer, 1990.

[7] J. K. Hale, *Ordinary Differential Equations*, Wiley Interscience, 1969.

[8] R. LeVeque, *Numerical Methods for Conservation Laws*, 2nd ed., Birkhäuser, 1992.

[9] A. Libai and J. G. Simmonds, *The Nonlinear Theory of Elastic Shells*, 2nd edn. Cambridge Univ. Press, 1998.

[10] A. J. Lichtenberg and M. A. Lieberman, *Regular and Chaotic Dynamics*, 2nd edn., Springer-Verlag, 1992.

[11] J. Mawhin and M. Willem, *Critical Point Theory and Hamiltonian Systems*, Springer-Verlag, 1989.

[12] P. J. Rabier and J. T. Oden, *Bifurcation in Rotating Bodies*, Masson, 1989.

[13] N. Rouche, P. Habets, and M. Laloy, *Stability Theory by Liapunov's Direct Method*, Springer-Verlag, 1977.

[14] G. M. Zaslavskii, *Physics of Chaos in Hamiltonian Systems*, World Scientific, 1998.

# A CONSISTENT THEORY OF, AND A VARIATIONAL PRINCIPLE FOR, THICK ELASTIC SHELLS UNDERGOING FINITE ROTATIONS

S.N.ATLURI*, M.IURA** and Y.SUETAKE***

* *UCLA, 7704 Boelter Hall, Los Angeles, CA 90095-1600, USA.*
** *Tokyo Denki University, Hatoyama, Hiki, Saitama, Japan.*
*** *Ashikaga Institute of Technology, Ashikaga, Tochigi, Japan.*

## 1. Introduction

The concept of a finite rotation vector has been introduced by Simmonds and Danielson [15, 16] to develop a nonlinear shell theory. Initially, the objective of introducing the finite rotation vector was, to derive a simple form for the governing equations of nonlinear shell theory (see Atluri [3] and Pietraszkiewicz [13]). With the progress in computer-simulation methodologies , several new issues for the finite rotation shell theory have arisen. Some of the current issues are: (1)how to preserve the symmetry of the tangent stiffness of a shell-finite-element; (2)how to incorporate the drilling degrees of freedom, and (3) how to develop a consistent thick shell theory.

A discussion of the symmetry or unsymmetry of the tangent stiffness has been recently given in Makowski and Stumpf [9]. The Eulerian rotational variation has most often been used in the existing literature, since the moment equilibrium equation in the current configuration is conjugate with the Eulerian rotational variation. As a result, the tangent stiffness matrix of the shell-element becomes unsymmetric. Iura and Atluri [6, 7] have, much earlier, shown in the case of a beam theory, that the variation of the Lagrangian finite rotation vector, and its conjugate balance equation, leads always to a symmetric tangent stiffness matrix. It is advantageous, from a computational point of view, to use a tangent stiffness matrix which is always symmetric. The application of finite rotation vector in nonlinear solid mechanics, and attendant computational nuances, have been discussed in detail by Atluri and Cazzani [4].

Drilling degrees of freedom are especially important in shell structures, with faceted joints. As indicated by Zienkiewicz [19], often, an artificial stiffness for the drilling d.o.f. has been used in the finite element analysis. Cazzani and Atluri [5], and Iura and Atluri [8], on the basis of a mixed variational principle developed by Atluri [1-4], have constructed a linear membrane element with the drilling d.o.f.. Suetake, Iura and Atluri [17] have developed a generalized functional for shells, in which unsymmetric strain and stress measures are used, and in which the drilling d.o.f. can be naturally incorporated. Another approach has been given by Reissner [14], in which the generalized Piola stresses have been used to construct



*D. Durban et al. (eds.), Advances in the Mechanics of Plates and Shells,* 17–32.

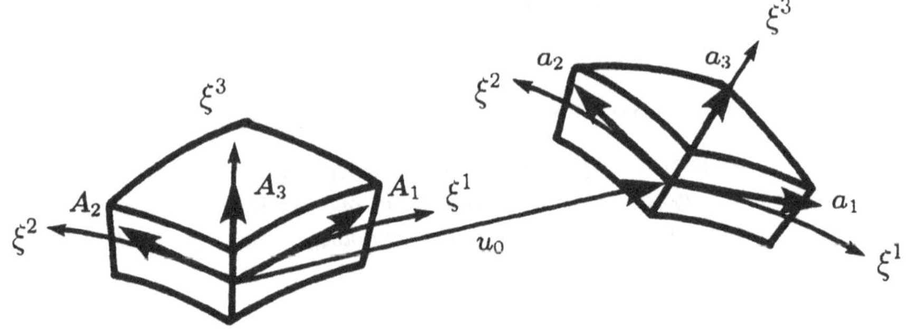


Figure 1: Shell Geometry and Notations

a functional for geometrically nonlinear elasticity.

Suetake, Iura and Atluri [17] have indicated that more attention should be paid to the assumptions of stress states in thick shells, in order to develop a consistent thick shell theory. The Reissner-Mindlin shell theory has often been used in an analysis of thick shells. Although the kinematic assumptions of this shell theory are clear, the assumptions for the stress states have not been fully discussed in literature. The angular momentum balance and director momentum balance conditions for the shell take different forms, depending on the assumptions for stress states . In this paper, the important role of the AMB condition is emphasized, especially in the considerations of mechanical power, and of the invariance of strain energy function.

A generalized variational functional is constructed on the basis of the present shell theory. The importance of the assumptions on the stress states is again emphasized in the variational principle. Reductions of this multifield variational principle are also discussed.

## 2. Preliminaries

We define the reference surface of the undeformed shell by two convected curvilinear coordinates $\xi^\alpha$ $(\alpha = 1, 2)$. Let $\xi^3$ denotes the through-thickness coordinate and $\boldsymbol{A}_3$ the associated base vector in the reference state (see Fig.1). Let $\boldsymbol{X}_0$ be the position vector of the shell mid-surface. The covariant base vectors are give by

$$\boldsymbol{A}_\alpha = \boldsymbol{X}_{0,\alpha}\,, \quad \boldsymbol{A}_3 = \frac{1}{2\sqrt{A}} e^{\alpha\beta} \boldsymbol{A}_\alpha \times \boldsymbol{A}_\beta, \tag{1}$$

where $(\cdot)_{,\alpha}$ denote the partial differentiation with respect to $\xi^\alpha$, $e^{\alpha\beta}$ the permutation tensor, $A = \det|\boldsymbol{A}_\alpha \cdot \boldsymbol{A}_\beta|$. The position vector at an arbitrary material point of the shell is expressed by

$$\boldsymbol{X} = \boldsymbol{X}_0 + \xi^3 \boldsymbol{A}_3. \tag{2}$$

The base vectors at an arbitrary material point are given by

$$\boldsymbol{G}_\alpha = \boldsymbol{A}_\alpha + \xi^3 \boldsymbol{A}_{3,\alpha} \quad \boldsymbol{G}_3 = \boldsymbol{A}_3. \tag{3}$$

For later convenience, we introduce the rotated triad defined by

$$\boldsymbol{k}_i = \mathbf{R}\boldsymbol{A}_i, \qquad (i = 1, 2, 3) \tag{4}$$

where $\mathbf{R}$ is the finite rotation tensor. The contravariant base vectors are defined by

$$\boldsymbol{A}^i \cdot \boldsymbol{A}_i = \delta^i_j, \quad \boldsymbol{G}^i \cdot \boldsymbol{G}_i = \delta^i_j, \quad \boldsymbol{k}^i \cdot \boldsymbol{k}_i = \delta^i_j, \tag{5}$$

where $\delta^i_j$ is Kronecker's delta.

We briefly describe the finite rotation tensor $\mathbf{R}$ (see Atluri and Cazzani [4] for more detailed discussion). Let $\boldsymbol{e}$ be a unit vector satisfying $\mathbf{R}\boldsymbol{e} = \boldsymbol{e}$, and $\theta$ the magnitude of rotation about the axis of rotation defined by $\boldsymbol{e}$. Then the finite rotation vector $\boldsymbol{\theta}$ is defined by $\boldsymbol{\theta} = \theta\boldsymbol{e}$. With the use of finite rotation vector $\boldsymbol{\theta}$, we obtain the relationships between $\mathbf{R}$ and $\boldsymbol{\theta}$, expressed as

$$\mathbf{R} = \mathbf{I} + \sin\theta(\boldsymbol{e} \times \mathbf{I}) + (1 - \cos\theta)\boldsymbol{e} \times (\boldsymbol{e} \times \mathbf{I}). \tag{6}$$

Since $\mathbf{R}\mathbf{R}^t = \mathbf{I}$, $\delta\mathbf{R}\mathbf{R}^t$ is the skewsymmetric tensor. Then there exists a variation $\delta\boldsymbol{\phi}$ satisfying the following relation:

$$\delta\boldsymbol{\phi} \times \mathbf{I} = \delta\mathbf{R}\mathbf{R}^t. \tag{7}$$

The variation $\delta\boldsymbol{\phi}$, which we refer to as the Eulerian variation, has often been used to construct the weak form of moment balance laws, since $\delta\boldsymbol{\phi}$ is conjugate with the moment equilibrium equation in the current configuration. The disadvantage of using $\delta\boldsymbol{\phi}$ is that it results in an unsymmetric tangent stiffness. Iura and Atluri [6, 7] have shown, for the beam problem, that the use of the finite rotation vector $\boldsymbol{\theta}$ always leads to the symmetric tangent stiffness. We call the variation $\delta\boldsymbol{\theta}$ as the Lagrangian variation of the finite rotation vector. The relation between $\delta\boldsymbol{\phi}$ and $\delta\boldsymbol{\theta}$ is given by $\delta\boldsymbol{\phi} = \boldsymbol{\Gamma}\delta\boldsymbol{\theta}$ where $\boldsymbol{\Gamma}$ is expressed as (see Atluri and Cazzani [4])

$$\boldsymbol{\Gamma} = \mathbf{I} + \frac{1 - \cos\theta}{\theta^2}(\boldsymbol{\theta} \times \mathbf{I}) + \frac{1}{\theta^2}(1 - \frac{\sin\theta}{\theta})\boldsymbol{\theta} \times (\boldsymbol{\theta} \times \mathbf{I}).$$

Let $\boldsymbol{a}_\alpha$ and $\boldsymbol{a}_3$ denote the base vectors of the deformed shell mid-surface, and $\boldsymbol{u}_0$ the displacement vector at the mid-surface (see Fig.1). The base vectors $\boldsymbol{a}_\alpha$ are expressed as

$$\boldsymbol{a}_\alpha = \boldsymbol{A}_\alpha + \boldsymbol{u}_{0,\alpha}\,. \tag{8}$$

The definition of $\boldsymbol{a}_3$ depends on the hypothesis used and shall be discussed later. The base vectors in the deformed shell domain are expressed as

$$\boldsymbol{g}_\alpha = \boldsymbol{a}_\alpha + \xi^3 \boldsymbol{a}_{3,\alpha}\,, \quad \boldsymbol{g}_3 = \boldsymbol{a}_3. \tag{9}$$

The balance laws for three dimensional continuum are written as (see Atluri [3] for a detailed discussion)

$$\text{(LMB)} \quad : \quad \nabla_0 \mathbf{t} + \rho_0 \boldsymbol{p} = 0, \tag{10}$$
$$\text{(AMB)} \quad : \quad \mathbf{F}\mathbf{t} = \mathbf{t}^t \mathbf{F}^t, \tag{11}$$

where $\mathbf{t}$ is the first Piola-Kirchhoff stress tensor, $\mathbf{F}$ the deformation gradient tensor, $\nabla$ the gradient operator, $\rho_0$ the mass density in the undeformed state, $\boldsymbol{p}$ the body force vector and $(\ )^t$ the transpose.

## 3. Linear Theory of Plates

In this section, we briefly describe a linear theory of membrane and bending problems for elastic plates. The Kirchhoff-Love hypothesis is used in the bending problem. Beginning with the LMB and AMB equations for three dimensional continuum, we obtain the LMB and AMB equations for the membrane problem, and the LMB, AMB and DMB equations for the bending problem. The drilling d.o.f. is introduced naturally in the membrane problem by using the unsymmetric membrane strains.

### 3.1 MEMBRANE PROBLEMS OF PLATES

#### 3.1.1 *Balance laws for membrane problem*

The LMB and AMB equations for 3-D continuum is written as

$$\boldsymbol{t}^1{}_{,1} + \boldsymbol{t}^2{}_{,2} + \boldsymbol{t}^3{}_{,3} + \rho_0 \boldsymbol{p} = 0, \tag{12}$$
$$\boldsymbol{G}_1 \times \boldsymbol{t}^1 + \boldsymbol{G}_2 \times \boldsymbol{t}^2 + \boldsymbol{G}_3 \times \boldsymbol{t}^3 = 0. \tag{13}$$

In the case of the membrane problem, the balance equations are obtained by integrating Eqs.(12) and (13) with respect to $\xi^3$. The LMB equations in the direction of $\xi^\alpha$ and the AMB equation about $\xi^3$ are written, respectively, as

$$\text{(LMB)} \quad : \quad N^{\alpha\beta}{}_{,\alpha} \boldsymbol{A}_\beta + P^\beta \boldsymbol{A}_\beta = 0, \tag{14}$$
$$\text{(AMB)} \quad : \quad \boldsymbol{A}_1 \times N^{12} \boldsymbol{A}_2 + \boldsymbol{A}_2 \times N^{21} \boldsymbol{A}_1 = 0, \quad \text{or} \quad (N^{12} - N^{21}) \boldsymbol{A}_3 = 0 \tag{15}$$

where

$$\boldsymbol{N}^\alpha = \int \boldsymbol{t}^\alpha d\xi^3, \quad \boldsymbol{P} = \int \rho_0 \boldsymbol{p} d\xi^3 + [\boldsymbol{t}^3]_{h^-}^{h^+}; \quad \boldsymbol{N}^\alpha = N^{\alpha m} \boldsymbol{A}_m, \quad \boldsymbol{P} = P^m \boldsymbol{A}_m. \tag{16}$$

in which $h^+$ and $h^-$ denote the coordinates of upper and lower surface of the plate, respectively.

3.1.2 *Kinematics for the membrane problem*
The deformed base vectors are written as

$$a_1 = (1 + u_{1,1})A_1 + u_{2,1} A_2, \tag{17}$$
$$a_2 = u_{1,2} A_1 + (1 + u_{2,2})A_2, \tag{18}$$

where $u_0 = u_1 A_1 + u_2 A_2$. Let $\theta_3$ be the rotational variable which transforms $A_\alpha$ into $k_\alpha$ such that

$$k_1 = A_1 + \theta_3 A_2, \quad k_2 = -\theta_3 A_1 + A_2. \tag{19}$$

The deformed vectors are written in the component forms, as

$$a_1 = c_{11} k_1 + c_{12} k_2, \quad a_2 = c_{21} k_1 + c_{22} k_2. \tag{20}$$

Since $c_{\alpha\beta} = a_\alpha \cdot k_\beta$, we have

$$c_{11} = 1 + u_{1,1}, \quad c_{12} = u_{2,1} - \theta_3, \quad c_{21} = u_{1,2} + \theta_3, \quad c_{22} = 1 + u_{2,2}, \tag{21}$$

where $c_{11}$ and $c_{22}$ denote the stretching tensors, and $c_{12}$ and $c_{21}$ the shearing tensors. As shown in Eq.(21), the tensors $c_{\alpha\beta}$ are not symmetric. The unsymmetric strain tensors $c_{\alpha\beta}$, accompanied with proper constitutive equations, have been used to develop a membrane finite element with drilling d.o.f. (see Cazzani and Atluri [5], and Iura and Atluri [8]).

When the components of the stress resultant vector are used, the AMB equation expressed by Eq.(15) takes a simple form such that $N^{12} = N^{21}$. The proper constitutive equations for $N^{12}$ and $N^{21}$ may be expressed by $N^{12} = E^0 c_{12}$ and $N^{21} = E^0 c_{21}$ where $E^0$ is the elastic modulus. Therefore, with the use of strain tensors, the AMB equation is rewritten as $c_{12} = c_{21}$. Then. from Eq.(21), we have

$$u_{2,1} - \theta_3 = u_{1,2} + \theta_3, \quad \text{or} \quad \theta_3 = \frac{1}{2}(u_{2,1} - u_{1,2}). \tag{22}$$

Note that the above equation is derived under the condition that the AMB equation holds. Substituting Eq.(22) into Eq.(21), we obtain the following conventional expression for the symmetric shearing strain:

$$c_{12} = c_{21} = \frac{1}{2}(u_{1,2} + u_{2,1}). \tag{23}$$

## 3.2 BENDING PROBLEMS OF PLATES

3.2.1 *Balance laws for bending problem*
In the case of bending problem, the balance equations consist of the LMB, AMB and DMB equations. The DMB equation is obtained by integrating the product of

Eq.(12) and $\xi^3$ with respect to $\xi^3$. The balance equations for the bending problem are written as

$$\begin{aligned} \text{(LMB)} &: \quad N^{\alpha 3}{}_{,\alpha} \boldsymbol{A}_3 + P^3 \boldsymbol{A}_3 = 0, && (24)\\ \text{(AMB)} &: \quad \boldsymbol{A}_\alpha \times N^{\alpha 3} \boldsymbol{A}_3 + \boldsymbol{A}_3 \times \boldsymbol{N}^3 + \boldsymbol{A}_{3,\alpha} \times \boldsymbol{H}^\alpha = 0, && (25)\\ \text{(DMB)} &: \quad \boldsymbol{H}^\alpha{}_{,\alpha} - \boldsymbol{N}^3 + \boldsymbol{T} = 0, && (26) \end{aligned}$$

where

$$\begin{aligned} \boldsymbol{N}^3 &= \int \boldsymbol{t}^3 d\xi^3, \quad \boldsymbol{H}^\alpha = \int \xi^3 \boldsymbol{t}^\alpha d\xi^3 = H^{\alpha m} \boldsymbol{A}_m, && (27)\\ \boldsymbol{T} &= \int \xi^3 \rho_0 \boldsymbol{p} d\xi^3 + [\xi^3 \boldsymbol{t}^3]_{h^-}^{h^+} = T^m \boldsymbol{A}_m, && (28) \end{aligned}$$

in which $\boldsymbol{H}^\alpha$ is the director moment vector. The elimination of $\boldsymbol{N}^3$ from the AMB and DMB equations gives the following moment equilibrium equation:

$$\boldsymbol{A}_\alpha \times N^{\alpha 3} \boldsymbol{A}_3 + \boldsymbol{M}^1{}_{,1} + \boldsymbol{M}^2{}_{,2} + \boldsymbol{m} = 0, \tag{29}$$

where the internal and external moment vectors are defined by

$$\boldsymbol{M}^\alpha = \boldsymbol{A}_3 \times \boldsymbol{H}^\alpha = \int \xi^3 \boldsymbol{A}_3 \times \boldsymbol{t}^\alpha d\xi^3, \quad \boldsymbol{m} = \boldsymbol{A}_3 \times \boldsymbol{T}. \tag{30}$$

When the shearing forces $N^{\alpha 3}$ are eliminated from Eqs.(24) and (28), we obtain the following LMB equation:

$$H^{11}{}_{,11} + H^{12}{}_{,12} + H^{21}{}_{,21} + H^{22}{}_{,22} + T^1{}_{,1} + T^2{}_{,2} + P^3 = 0 \tag{31}$$

3.2.2 *Kinematics for the bending problem*

The deformed in-plane base vectors are written as

$$\boldsymbol{a}_1 = \boldsymbol{A}_1 + u_{3,1} \boldsymbol{A}_3, \quad \boldsymbol{a}_2 = \boldsymbol{A}_2 + u_{3,2} \boldsymbol{A}_3, \tag{32}$$

where $\boldsymbol{u}_0 = u_3 \boldsymbol{A}_3$. Since the Kirchhoff-Love theory is used, the deformed normal vector is written as

$$\boldsymbol{a}_3 = \boldsymbol{a}_1 \times \boldsymbol{a}_2 = -u_{3,1} \boldsymbol{A}_1 - u_{3,2} \boldsymbol{A}_2 + \boldsymbol{A}_3. \tag{33}$$

Let $\theta_2$ and $\theta_3$ be the components of rotational variable which transforms $\boldsymbol{A}_3$ into $\boldsymbol{a}_3 (= \boldsymbol{k}_3)$ such that

$$\boldsymbol{a}_3 = \theta_2 \boldsymbol{A}_1 - \theta_1 \boldsymbol{A}_2 + \boldsymbol{A}_3. \tag{34}$$

The comparison of Eq.(33) and (34) leads to

$$\theta_1 = u_{3,2}, \quad \theta_2 = -u_{3,1}. \tag{35}$$

The mapping of base vectors $\boldsymbol{a}_i$ into $\boldsymbol{k}_i$ due to the rigid rotation is expressed as

$$\boldsymbol{k}_1 = \boldsymbol{A}_1 - \theta_2 \boldsymbol{A}_3, \quad \boldsymbol{k}_2 = \boldsymbol{A}_2 + \theta_1 \boldsymbol{A}_3, \quad \boldsymbol{k}_3 = \boldsymbol{a}_3. \tag{36}$$

The curvature tensor $b_{\alpha\beta}$ is defined by

$$\boldsymbol{a}_{3,\alpha} = b_{\alpha\beta} \boldsymbol{k}_\beta. \tag{37}$$

Since $b_{\alpha\beta} = \boldsymbol{a}_{3,\alpha} \cdot \boldsymbol{k}_\beta$, we have

$$b_{11} = -u_{3,11}\,, \quad b_{12} = b_{21} = -u_{3,12}\,, \quad b_{22} = -u_{3,22}\,. \tag{38}$$

## 4. Nonlinear Thick-Shell Theory

The plane-stress assumption has always been used in a thin shell theory. As far as a thick shell is concerned, the stresses along the shell thickness might not be negligible. The out-of-plane shearing stresses are taken into account in the Mindlin-Reissner shell theory, while the change of thickness is not taken into account. In this paper, we pay attention to the stress-states of the shell in addition to the kinematic assumptions. We develop a shell theory on the basis of the following four assumptions :

$$\begin{array}{lll} \text{(G1)} & : & \boldsymbol{a}_3 = \boldsymbol{k}_3 \ \text{ and } \ t^{3i} \neq 0, \\ \text{(G2)} & : & \boldsymbol{a}_3 = z\boldsymbol{k}_3 \ \text{ and } \ t^{3i} \neq 0, \\ \text{(GM1)} & : & \boldsymbol{a}_3 = \boldsymbol{k}_3 \ \text{ and } \ t^{3i} = 0, \\ \text{(GM2)} & : & \boldsymbol{a}_3 = z\boldsymbol{k}_3, \ \ t^{3\alpha} = 0 \ \text{ and } \ t^{33} \neq 0, \end{array}$$

where $\boldsymbol{k}_3 = \mathbf{R}\boldsymbol{A}_3$. The assumption such that $\boldsymbol{a}_3 = \boldsymbol{k}_3$ has been employed in the Mindlin-Reissner shell theory. The assumption such that $\boldsymbol{a}_3 = z\boldsymbol{k}_3$ allows the shell thickness to change. The assumptions (G1) and (GM1) correspond to the Mindlin-Reissner shell theory. The normal stresses $t^{3i}$ to the shell mod-surface are assumed to be zero in (GM1), while the kinematic assumption only is given in (G1). The change of shell thickness is allowed in (G2) and (GM2). The kinematic assumption only is used in (G2), while both kinematic and stress states assumptions are used in (GM2).

In the assumptions (GM1) and (GM2), we assume that $t^{3\alpha} = 0$. This assumption does not indicate that the out-of-plane shearing stress resultants vanish. It will be shown later that the shearing stress resultants $N^{\alpha 3}$ do not vanish even if $N^{3\alpha} = 0$ holds.

The time derivative of $\boldsymbol{a}_3$ corresponding each assumption is given by

$$\text{(G1) and (GM1)} \ : \ \dot{\boldsymbol{a}}_3 = \boldsymbol{W} \times \boldsymbol{a}_3, \tag{39}$$

$$\text{(G2) and (GM2)} \ : \ \dot{\boldsymbol{a}}_3 = \dot{z}\mathbf{R}\boldsymbol{A}_3 + \boldsymbol{W} \times \boldsymbol{a}_3, \tag{40}$$

where $\boldsymbol{W} \times \mathbf{I} = \dot{\mathbf{R}}\mathbf{R}^t$. We introduce the strain vectors $\boldsymbol{c}_\alpha$ and $\boldsymbol{b}_\alpha$ defined by

$$\boldsymbol{a}_\alpha = c_{\alpha i} \boldsymbol{k}_i = \mathbf{R}\boldsymbol{c}_\alpha, \qquad \boldsymbol{c}_\alpha = c_{\alpha i} \boldsymbol{A}^i, \tag{41}$$

$$\boldsymbol{a}_{3,\alpha} = b_{\alpha i} \boldsymbol{k}_i = \mathbf{R}\boldsymbol{b}_\alpha, \qquad \boldsymbol{b}_\alpha = b_{\alpha i} \boldsymbol{A}^i, \tag{42}$$

where $\boldsymbol{c}_\alpha$ and $\boldsymbol{b}_\alpha$ denote the stretching and bending strain vectors, respectively. The present strain vectors are associated with $\boldsymbol{A}^i$ so that they are the Lagrangian measures. Since the AMB equation is not embedded, the stretch tensor $c_{\alpha\beta}$ is no longer symmetric. With the help of the strain vectors, the deformation gradient tensor is written by

$$\text{(G1) and (GM1)} \quad : \quad \mathbf{F} = \mathbf{R}(\boldsymbol{c}_\alpha + \xi^3 \boldsymbol{b}_\alpha)\boldsymbol{G}^\alpha + \boldsymbol{k}_3 \boldsymbol{A}_3, \tag{43}$$

$$\text{(G2) and (GM2)} \quad : \quad \mathbf{F} = \mathbf{R}(\boldsymbol{c}_\alpha + \xi^3 \boldsymbol{b}_\alpha)\boldsymbol{G}^\alpha + z\boldsymbol{k}_3 \boldsymbol{A}_3. \tag{44}$$

## 4.1 BALANCE LAWS

The LMB and AMB equations for the shell are written, with the help of Eqs.(10) and(11), as (see Suetake, Iura and Atluri [17])

$$\text{(LMB)} \quad : \quad \frac{1}{\sqrt{G}}(\sqrt{G}\boldsymbol{t}^i)_{,i} + \rho_0 \boldsymbol{p} = 0, \tag{45}$$

$$\text{(AMB)} \quad : \quad \boldsymbol{g}_i \times \boldsymbol{t}^i = 0, \tag{46}$$

where $\boldsymbol{t}^i$ is the first Piola-Kirchhoff stress vector and $G = \det|\boldsymbol{G}_\alpha \cdot \boldsymbol{G}_\beta|$. For later use, we introduce the following stress resultant vectors:

$$\boldsymbol{N}^\alpha = \int \boldsymbol{t}^\alpha \mu_0 d\xi^3, \; \boldsymbol{N}^3 = \int \boldsymbol{t}^3 \mu_0 d\xi^3, \; \boldsymbol{N} = \boldsymbol{a}_3 \int t^{33} \mu_0 d\xi^3, \; \boldsymbol{H}^\alpha = \int \xi^3 \boldsymbol{t}^\alpha \mu_0 d\xi^3. \tag{47}$$

where $\mu_0 = \sqrt{G}/\sqrt{A}$. It should be noted that these vectors are the stress resultant vectors of the 1st Piola-Kirchhoff stress type.

The LMB equation is derived by integrating Eq.(45) with respect to $\xi^3$ and using the divergence theorem. The LMB equation under (G1) and (G2) is given by

$$\frac{1}{\sqrt{A}}(\sqrt{A}\boldsymbol{N}^\alpha)_{,\alpha} + \boldsymbol{P} = 0, \tag{48}$$

where the body force vector is defined by

$$\text{(G1) and (G2)} : \boldsymbol{P} = \int \rho_0 \boldsymbol{p} \mu_0 d\xi^3 + [\boldsymbol{t}^3 \mu_0]_{h^-}^{h^+}. \tag{49}$$

The LMB equation under (GM1) and (GM2) takes the same form as that of Eq.(48) except the definition of the external force which is expressed by

$$\text{(GM1)} \quad : \quad \boldsymbol{P} = \int \rho_0 \boldsymbol{p} \mu_0 d\xi^3, \tag{50}$$

$$\text{(GM2)} \quad : \quad \boldsymbol{P} = \int \rho_0 \boldsymbol{p} \mu_0 d\xi^3 + [\boldsymbol{a}_3 t^{33} \mu_0]_{h^-}^{h^+}. \tag{51}$$

The AMB equation is derived by integrating Eq.(46) with respect to $\xi^3$. The AMB equations under each assumption are written as

$$\text{(G1) and (G2)} \quad : \quad \boldsymbol{a}_\alpha \times \boldsymbol{N}^\alpha + \boldsymbol{a}_{3,\alpha} \times \boldsymbol{H}^\alpha + \boldsymbol{a}_3 \times \boldsymbol{N}^3 = 0, \tag{52}$$

$$(\text{GM1}) \quad : \quad \boldsymbol{a}_\alpha \times \boldsymbol{N}^\alpha + \boldsymbol{a}_{3,\alpha} \times \boldsymbol{H}^\alpha = 0, \tag{53}$$

$$(\text{GM2}) \quad : \quad \boldsymbol{a}_\alpha \times \boldsymbol{N}^\alpha + \boldsymbol{a}_{3,\alpha} \times \boldsymbol{H}^\alpha + \underline{\boldsymbol{a}_3 \times \boldsymbol{N}} = 0. \tag{54}$$

Note that $\boldsymbol{a}_3 = \mathbf{R}\boldsymbol{A}_3$ in (G1) and (GM1) while $\boldsymbol{a}_3 = z\mathbf{R}\boldsymbol{A}_3$ in (G2) and (GM2). The following points should be stressed herein:

- The AMB equation (54) under (GM2) includes the underlined term such that $\boldsymbol{a}_3 \times \boldsymbol{N}$. Since $\boldsymbol{N} = \boldsymbol{a}_3 \int t^{33}\mu_0 d\xi^3$, this term vanishes because of the definition of the cross product. As a result, the AMB equation under (GM2) is the same as that under (GM1).
- The AMB equation under (G1) and (G2) does not indicate that $N^{3\alpha} = N^{\alpha 3}$ where $\boldsymbol{N}^i = N^{ij}\boldsymbol{k}_j$.
- The AMB equation plays an important role in the mechanical power and the objectivity, which shall be discussed in detail, later in this paper.

The DMB equation for the shell is obtained by integrating the product of Eq.(45) and $\xi^3$ with respect to $\xi^3$. The DMB equations under each assumption are expressed by

$$\begin{aligned}
(\text{G1}) \text{ and } (\text{G2}) \quad &: \quad \frac{1}{\sqrt{A}}(\sqrt{A}\boldsymbol{H}^\alpha)_{,\alpha} - \boldsymbol{N}^3 + \boldsymbol{T} = 0, \\
&\quad \text{where } \boldsymbol{T} = \int \rho_0 \boldsymbol{p}\xi^3\mu_0 d\xi^3 + [\boldsymbol{t}^3\xi^3\mu_0]_{h^-}^{h^+},
\end{aligned} \tag{55}$$

$$\begin{aligned}
(\text{GM1}) \quad &: \quad \frac{1}{\sqrt{A}}(\sqrt{A}\boldsymbol{H}^\alpha)_{,\alpha} + \boldsymbol{T} = 0, \\
&\quad \text{where } \boldsymbol{T} = \int \rho_0 \boldsymbol{p}\xi^3\mu_0 d\xi^3,
\end{aligned} \tag{56}$$

$$\begin{aligned}
(\text{GM2}) \quad &: \quad \frac{1}{\sqrt{A}}(\sqrt{A}\boldsymbol{H}^\alpha)_{,\alpha} - \boldsymbol{N} + \boldsymbol{T} = 0, \\
&\quad \text{where } \boldsymbol{T} = \int \rho_0 \boldsymbol{p}\xi^3\mu_0 d\xi^3 + [t^{33}\boldsymbol{a}_3\xi^3\mu_0]_{h^-}^{h^+}.
\end{aligned} \tag{57}$$

The conventional AMB for the shell, or the moment equilibrium equation, is obtained by using AMB and DMB equations. Whichever assumption is employed, the moment equilibrium equation for the sell is given by

$$\boldsymbol{a}_\alpha \times \boldsymbol{N}^\alpha + \frac{1}{\sqrt{A}}(\sqrt{A}\boldsymbol{M}^\alpha)_{,\alpha} + \boldsymbol{m} = 0, \tag{58}$$

where

$$\boldsymbol{M}^\alpha = \boldsymbol{a}_3 \times \boldsymbol{H}^\alpha = \int \xi^3 \boldsymbol{a}_3 \times \boldsymbol{t}^\alpha \mu_0 d\xi^3, \quad \boldsymbol{m} = \boldsymbol{a}_3 \times \boldsymbol{T}. \tag{59}$$

The DMB equation is used to derive the moment equilibrium equation. Another aspect of the DMB condition is the definition of $\boldsymbol{N}^3$ or $\boldsymbol{N}$. In the case of (G1) and (G2), $\boldsymbol{N}^3$ is not connected directly with the associated strain vectors through the constitutive equations. In the case of (GM2), one equation is added to the constitutive equations while there are two additional variables $\boldsymbol{N}$ and $z$. Although the number of unknowns seems not to equal to the number of equations, the DMB equation is used to define $\boldsymbol{N}^3$ or $\boldsymbol{N}$. Therefore, the number of unknowns equals to the number of equations.

## 4.2 MECHANICAL POWER

The mechanical power for a three-dimensional continuum is expressed as (see Atluri [3])

$$\text{M.P.} = \int \mathbf{t} : \dot{\mathbf{F}} dv. \tag{60}$$

The expression of the M.P. leads to the conjugate relationships between strain and stress measures. The present strains defined by Eqs.(41) and (42) are the Lagrangian measures. We introduce, herein, the following Lagrangian stress resultant vectors:

$$\underline{\mathcal{N}}^\alpha = \mathbf{R}^t \boldsymbol{N}^\alpha = \boldsymbol{N}^\alpha \mathbf{R}, \quad \underline{\mathcal{N}} = \mathbf{R}^t \boldsymbol{N} = \boldsymbol{N}\mathbf{R}, \quad \underline{\mathcal{H}}^\alpha = \mathbf{R}^t \boldsymbol{H}^\alpha = \boldsymbol{H}^\alpha \mathbf{R}. \tag{61}$$

The above defined vectors are the stress resultant vectors of the Biot-Lure stress type (see Atluri [3]). According to Atluri and Murakawa [10, 11], the inversion of 1st Piola-Kirchhoff stress tensor and the associated strain tensor is not one-to-one. The symmetric part of the Biot-Lure stress tensor or the Jaumann stress tensor and the conjugated strain tensor leads to a one-to-one mapping.

With the use of Eqs.(60) and (61), we obtain the expression for the mechanical power written as

$$\begin{aligned}(\text{G1}) \quad : \quad \text{M.P.} = & \int (\underline{\mathcal{N}}^\alpha \cdot \dot{\boldsymbol{c}}_\alpha + \underline{\mathcal{H}} \cdot \dot{\boldsymbol{b}}_\alpha)\sqrt{A}d\xi^1 d\xi^2 \\ & + \int \boldsymbol{W} \times (\boldsymbol{a}_\alpha \times \boldsymbol{N}^\alpha + \boldsymbol{a}_{3,\alpha} \times \boldsymbol{H}^\alpha + \boldsymbol{a}_3 \times \boldsymbol{N}^3)\sqrt{A}d\xi^1 d\xi^2, \end{aligned} \tag{62}$$

$$\begin{aligned}(\text{G2}) \quad : \quad \text{M.P.} = & \int (\underline{\mathcal{N}}^\alpha \cdot \dot{\boldsymbol{c}}_\alpha + \underline{\mathcal{H}}^\alpha \cdot \dot{\boldsymbol{b}}_\alpha + \underline{\mathcal{N}} \cdot \dot{\boldsymbol{z}})\sqrt{A}d\xi^1 d\xi^2 \\ & + \int \boldsymbol{W} \times (\boldsymbol{a}_\alpha \times \boldsymbol{N}^\alpha + \boldsymbol{a}_{3,\alpha} \times \boldsymbol{H}^\alpha + \boldsymbol{a}_3 \times \boldsymbol{N}^3)\sqrt{A}d\xi^1 d\xi^2, \end{aligned} \tag{63}$$

$$\begin{aligned}(\text{GM1}) \quad : \quad \text{M.P.} = & \int (\underline{\mathcal{N}}^\alpha \cdot \dot{\boldsymbol{c}}_\alpha + \underline{\mathcal{H}}^\alpha \cdot \dot{\boldsymbol{b}}_\alpha)\sqrt{A}d\xi^1 d\xi^2 \\ & + \int \boldsymbol{W} \times (\boldsymbol{a}_\alpha \times \boldsymbol{N}^\alpha + \boldsymbol{a}_{3,\alpha} \times \boldsymbol{H}^\alpha)\sqrt{A}d\xi^1 d\xi^2, \end{aligned} \tag{64}$$

$$\begin{aligned}(\text{GM2}) \quad : \quad \text{M.P.} = & \int (\underline{\mathcal{N}}^\alpha \cdot \dot{\boldsymbol{c}}_\alpha + \underline{\mathcal{H}}^\alpha \cdot \dot{\boldsymbol{b}}_\alpha + \underline{\mathcal{N}} \cdot \dot{\boldsymbol{z}})\sqrt{A}d\xi^1 d\xi^2 \\ & + \int \boldsymbol{W} \times (\boldsymbol{a}_\alpha \times \boldsymbol{N}^\alpha + \boldsymbol{a}_{3,\alpha} \times \boldsymbol{H}^\alpha)\sqrt{A}d\xi^1 d\xi^2, \end{aligned} \tag{65}$$

where $\boldsymbol{z} = z\boldsymbol{A}_3$ is the Lagrangian stretching vector along $\xi^3$. In contrast to the three-dimensional continuum mechanics, the mechanical power of the shell consists of the two parts: one is associated with the expression for the conjugate relationships between stress and strain measures, and the other the expression for the AMB equation. When the AMB equation under each assumption holds, the second terms in Eqs.(62)–(65) vanish. Then we obtain the following conjugate relationships:

$$\begin{array}{lcl}\text{(G1) and (GM1)} & : & (\underline{\mathcal{N}}^{\alpha}, \underline{\mathcal{H}}^{\alpha}) \Longleftrightarrow (\boldsymbol{c}_{\alpha}, \boldsymbol{b}_{\alpha}), \\ \text{(G2) and (GM2)} & : & (\underline{\mathcal{N}}^{\alpha}, \underline{\mathcal{H}}^{\alpha}, \underline{\mathcal{N}}) \Longleftrightarrow (\boldsymbol{c}_{\alpha}, \boldsymbol{b}_{\alpha}, \boldsymbol{z}).\end{array}$$

The strain energy function may be expressed as

$$\begin{array}{lcl}\text{(G1) and (GM1)} & : & W_0 = W_0(\boldsymbol{c}_{\alpha}, \boldsymbol{b}_{\alpha}), \\ \text{(G2) and (GM2)} & : & W_0 = W_0(\boldsymbol{c}_{\alpha}, \boldsymbol{b}_{\alpha}, \boldsymbol{z}).\end{array}$$

The constitutive equations are written as

$$\frac{\partial W_0}{\partial \boldsymbol{c}_{\alpha}} = \underline{\mathcal{N}}^{\alpha}, \quad \frac{\partial W_0}{\partial \boldsymbol{b}_{\alpha}} = \underline{\mathcal{H}}^{\alpha}, \quad \frac{\partial W_0}{\partial \boldsymbol{z}} = \underline{\mathcal{N}}. \tag{66}$$

When the strain energy function is expressed in terms of the displacement and rotational variables, it is crucial to use the finite rotation vector as the rotational variable. As discussed before, the variation $\delta\boldsymbol{\phi}$ has often been used in the existing literature. The vector $\boldsymbol{\phi}$, however, does not exist so that the strain energy function can not be expressed in terms of $\boldsymbol{\phi}$.

## 4.3 OBJECTIVITY

We consider the objectivity or the invariance of strain energy function. Since the present strain vectors $\boldsymbol{c}_{\alpha}$, $\boldsymbol{b}_{\alpha}$ and $\boldsymbol{z}$ are associated with the reference base vectors $\boldsymbol{A}^i$, they are the Lagrangian measures. It is, therefore, clear that the present strain vectors are objective (see Ogden [12]). Let $\mathbf{Q}$ be a proper orthogonal tensor. The objective strain vector is transformed such that

$$\boldsymbol{c}_i^* = \mathbf{Q}\boldsymbol{c}_i, \quad \boldsymbol{b}_{\alpha}^* = \mathbf{Q}\boldsymbol{b}_{\alpha}, \quad \boldsymbol{A}_i^* = \mathbf{Q}\boldsymbol{A}_i, \tag{67}$$

where $(\;)^*$ denotes the value after the rigid rotation. The strain energy function $W_0^*$ after the rigid rotation is expressed as

$$\begin{array}{lcl}\text{(G1) and (GM1)} & : & W_0^* = W_0^*(\boldsymbol{c}_{\alpha}^*, \boldsymbol{b}_{\alpha}^*), \\ \text{(G2) and (GM2)} & : & W_0^* = W_0^*(\boldsymbol{c}_{\alpha}^*, \boldsymbol{b}_{\alpha}^*, \boldsymbol{z}^*).\end{array}$$

The variation of strain energy function under (G1) and (GM1) is written as

$$\begin{aligned}\delta W_0^* &= \frac{\partial W_0^*}{\partial \boldsymbol{c}_{\alpha}^*}\delta\boldsymbol{c}_{\alpha}^* + \frac{\partial W_0^*}{\partial \boldsymbol{b}_{\alpha}^*}\delta\boldsymbol{b}_{\alpha}^* \\ &= \delta W_0 + \delta\boldsymbol{\psi}\mathbf{R}^t(\boldsymbol{a}_{\alpha} \times \boldsymbol{N}^{\alpha} + \boldsymbol{a}_{3,\alpha} \times \boldsymbol{H}^{\alpha}),\end{aligned} \tag{68}$$

where

$$\delta W_0 = \frac{\partial W_0}{\partial \boldsymbol{c}_{\alpha}}\delta\boldsymbol{c}_{\alpha} + \frac{\partial W_0}{\partial \boldsymbol{b}_{\alpha}}\delta\boldsymbol{b}_{\alpha}, \quad \delta\boldsymbol{\psi} \times \mathbf{I} = \delta\boldsymbol{Q}\boldsymbol{Q}^t.$$

The variation of strain energy function under (G2) and (GM2) is written as

$$\begin{aligned}\delta W_0^* &= \frac{\partial W_0^*}{\partial \boldsymbol{c}_\alpha^*}\delta \boldsymbol{c}_\alpha^* + \frac{\partial W_0^*}{\partial \boldsymbol{b}_\alpha^*}\delta \boldsymbol{b}_\alpha^* + \frac{\partial W_0^*}{\partial \boldsymbol{z}^*}\delta \boldsymbol{z}^* \\ &= \delta W_0 + \delta\boldsymbol{\psi}\mathbf{R}^t(\boldsymbol{a}_\alpha \times \boldsymbol{N}^\alpha + \boldsymbol{a}_{3,\alpha} \times \boldsymbol{H}^\alpha),\end{aligned} \tag{69}$$

where

$$\delta W_0 = \frac{\partial W_0}{\partial \boldsymbol{c}_\alpha}\delta \boldsymbol{c}_\alpha + \frac{\partial W_0}{\partial \boldsymbol{b}_\alpha}\delta \boldsymbol{b}_\alpha + \frac{\partial W_0}{\partial \boldsymbol{z}}\delta \boldsymbol{z}.$$

Since the strain energy function should be invariant under the rigid rotation, we have $\delta W_0^* = \delta W_0$. Therefore, the invariance of strain energy function under any assumption asserts that

$$\boldsymbol{a}_\alpha \times \boldsymbol{N}^\alpha + \boldsymbol{a}_{3,\alpha} \times \boldsymbol{H}^\alpha = 0. \tag{70}$$

The above equation is not always satisfied under the present assumptions. The AMB equation under (G1) and (G2) is different from Eq.(70). Therefore, the invariance condition of strain energy function is not satisfied when the assumptions (G1) and (G2) are used. However, the AMB equation under (GM1) and (GM2) takes the same form as that of Eq.(70). This fact shows the importance of using the correct assumptions for the stress states .

## 5. Variational Principle

According to Atluri [1-4] and Suetake, Iura and Atluri [17], the generalized functional for the shell is given by

$$\begin{aligned}\mathcal{F}_1 = \int [W_0 + \hat{\underline{\mathcal{N}}}^\alpha \cdot \{\mathbf{R}^t(\boldsymbol{A}_\alpha + \boldsymbol{u}_{0,\alpha}) - \boldsymbol{c}_\alpha\} + \hat{\underline{\mathcal{H}}}^\alpha \cdot (\mathbf{R}^t\boldsymbol{a}_{3,\alpha} - \boldsymbol{b}_\alpha) \\ + \hat{\underline{\mathcal{N}}}^3 \cdot (\mathbf{R}^t\boldsymbol{a}_3 - \boldsymbol{c}_3) - \boldsymbol{P}\cdot\boldsymbol{u}_0 + \boldsymbol{T}\cdot(\boldsymbol{a}_3 - \boldsymbol{A}_3)]\sqrt{A}d\xi^1 d\xi^2 \\ - \int_{s_\sigma} \{\bar{\boldsymbol{N}}\cdot\boldsymbol{u}_0 + \bar{\boldsymbol{H}}\cdot(\boldsymbol{a}_3 - \boldsymbol{A}_3)\}\nu_{0\alpha}\sqrt{A}ds \\ - \int_{s_u} \{(\mathbf{R}\hat{\underline{\mathcal{N}}}^\alpha)\cdot(\boldsymbol{u}_0 - \bar{\boldsymbol{u}}_0) + (\mathbf{R}\hat{\underline{\mathcal{H}}}^\alpha)\cdot(\boldsymbol{a}_3 - \bar{\boldsymbol{n}})\}\nu_{0\alpha}\sqrt{A}ds,\end{aligned} \tag{71}$$

where $(\hat{\ })$ denotes the Lagrange multiplier, $(\bar{\ })$ the prescribed value at the boundary, and $W_0$ and $\boldsymbol{c}_3$ are given, under each assumption, by

$$\begin{aligned}&\text{(G1) and (GM1)}: \quad W_0 = W_0(\boldsymbol{c}_\alpha, \boldsymbol{b}_\alpha), \quad \boldsymbol{c}_3 = \boldsymbol{A}_3, \\ &\text{(G2) and (GM2)}: \quad W_0 = W_0(\boldsymbol{c}_\alpha, \boldsymbol{b}_\alpha, \boldsymbol{z}), \quad \boldsymbol{c}_3 = z\boldsymbol{A}_3.\end{aligned}$$

The variables subjected to variation in Eq.(71) are $\boldsymbol{u}_0, \mathbf{R}(\boldsymbol{\theta}), \boldsymbol{c}_\alpha, \boldsymbol{b}_\alpha, \boldsymbol{z}, \hat{\underline{\mathcal{N}}}^\alpha, \hat{\underline{\mathcal{H}}}^\alpha, \hat{\underline{\mathcal{N}}}^3$, and $\boldsymbol{a}_3$. The first variation of the functional $\mathcal{F}_1$ is written as

$$\delta\mathcal{F}_1 = \int [(\frac{\partial W_0}{\partial \boldsymbol{c}_\alpha} - \hat{\underline{\mathcal{N}}}^\alpha)\delta\boldsymbol{c}_\alpha + (\frac{\partial W_0}{\partial \boldsymbol{b}_\alpha} - \hat{\underline{\mathcal{H}}}^\alpha)\delta\boldsymbol{b}_\alpha + \underline{\text{c.e.}}$$

$$+\delta\underline{\hat{\mathcal{N}}}^{\alpha} \cdot \{\mathbf{R}^t(\boldsymbol{A}_\alpha + \boldsymbol{u}_{0,\alpha}) - \boldsymbol{c}_\alpha\} + \delta\underline{\hat{\mathcal{H}}}^{\alpha} \cdot (\mathbf{R}^t\boldsymbol{a}_{3,\alpha} - \boldsymbol{b}_\alpha) + +\delta\underline{\hat{\mathcal{N}}}^{3} \cdot (\mathbf{R}^t\boldsymbol{a}_3 - \boldsymbol{c}_3)$$

$$+(\boldsymbol{a}_\alpha \times \hat{\boldsymbol{N}}^{\alpha} + \boldsymbol{a}_{3,\alpha} \times \hat{\boldsymbol{H}}^{\alpha} + \boldsymbol{a}_3 \times \hat{\boldsymbol{N}}^{3}) \cdot \boldsymbol{\Gamma}\delta\boldsymbol{\theta}$$

$$-\{\frac{1}{\sqrt{A}}(\sqrt{A}\hat{\boldsymbol{N}}^{\alpha})_{,\alpha} + \boldsymbol{p}\} \cdot \delta\boldsymbol{u}_0 + \{\frac{1}{\sqrt{A}}(\sqrt{A}\hat{\boldsymbol{H}}^{\alpha})_{,\alpha} - \hat{\boldsymbol{N}}^{3} + \boldsymbol{T}\} \cdot \delta\boldsymbol{a}_3]\sqrt{A}d\xi^1 d\xi^2$$

$$+\int_{s_\sigma} [(\hat{\boldsymbol{N}}^{\alpha} - \bar{\boldsymbol{N}}^{\alpha}) \cdot \delta\boldsymbol{u}_0 + (\hat{\boldsymbol{H}}^{\alpha} - \bar{\boldsymbol{H}}^{\alpha}) \cdot \delta\boldsymbol{a}_3]\nu_{0\alpha}\sqrt{A}ds$$

$$-\int_{s_u} \{(\boldsymbol{u}_0 - \bar{\boldsymbol{u}}_0) \cdot \delta\hat{\boldsymbol{N}}^{\alpha} + (\boldsymbol{a}_3 - \bar{\boldsymbol{n}}) \cdot \delta\hat{\boldsymbol{H}}^{\alpha}\}\nu_{0\alpha}\sqrt{A}ds, \tag{72}$$

where the following notations are used:

$$\hat{\boldsymbol{N}}^{i} = \mathbf{R}\underline{\hat{\mathcal{N}}}^{i}, \quad \hat{\boldsymbol{H}}^{\alpha} = \mathbf{R}\underline{\hat{\mathcal{H}}}^{\alpha}, \quad \boldsymbol{a}_\alpha = \boldsymbol{A}_\alpha + \boldsymbol{u}_{0,\alpha}.$$

The underlined term $\underline{c.e.}$ in Eq.(72) is given by

$$\begin{aligned} &(\text{G1}) \text{ and } (\text{GM1}) &:& \quad \underline{c.e.} = 0, \\ &(\text{G2}) \text{ and } (\text{GM2}) &:& \quad \underline{c.e.} = (\frac{\partial W_0}{\partial \boldsymbol{z}} - \underline{\hat{\mathcal{N}}}^{3}) \cdot \delta z. \end{aligned}$$

Since $\boldsymbol{a}_3$ is subjected to variation, the DMB equation is recovered as the Euler equation. The assumptions (G1) and (GM2) gives the AMB and DMB equations which are the same as those derived in section 4.1. In the case of (G2), the constitutive equation gives the physical meaning of $\underline{\hat{\mathcal{N}}}^{3}$, which states that $\underline{\hat{\mathcal{N}}}^{3} = \underline{\mathcal{N}}$. The resulting AMB and DMB equations including $\underline{\mathcal{N}}$ is different from Eqs.(52) and (55), respectively. In the case of (GM1), $\underline{\hat{\mathcal{N}}}^{3}$ appears in the AMB and DMB equations which is not the case in Eqs. (53) and (56). It should be noted that, with the help of the AMB and DMB equations, the moment equilibrium equation (58) can be recovered irrespective of $\underline{\hat{\mathcal{N}}}^{3}$. Therefore, under any assumption used herein, the exact moment equilibrium equation is obtained from this functional.

In the Hu-Washizu variational principle, the stress and strain tensors, and the displacement and rotational vectors are subjected to variation (see Washizu [18]). The vector $\boldsymbol{a}_3$ is subjected to variation in $\mathcal{F}_1$. The vector $\boldsymbol{a}_3$ is not the strain vector so that $\boldsymbol{a}_3 = \mathbf{R}\boldsymbol{A}_3$ (under (G1) and (GM1)) or $\boldsymbol{a}_3 = z\mathbf{R}\boldsymbol{A}_3$ (under (G2) and (GM2)) is the subsidiary condition in the Hu-Washizu type functional, expressed as

$$\mathcal{F}_{HW} = \int [W_0 + \underline{\hat{\mathcal{N}}}^{\alpha} \cdot \{\mathbf{R}^t(\boldsymbol{A}_\alpha + \boldsymbol{u}_{0,\alpha}) - \boldsymbol{c}_\alpha\} + \underline{\hat{\mathcal{H}}}^{\alpha} \cdot (\mathbf{R}^t\boldsymbol{a}_{3,\alpha} - \boldsymbol{b}_\alpha)$$

$$+c.c. - \boldsymbol{P} \cdot \boldsymbol{u}_0 + \boldsymbol{T} \cdot (\boldsymbol{a}_3 - \boldsymbol{A}_3)]\sqrt{A}d\xi^1 d\xi^2$$

$$-\int_{s_\sigma} \{\bar{\boldsymbol{N}} \cdot \boldsymbol{u}_0 + \bar{\boldsymbol{H}} \cdot (\boldsymbol{a}_3 - \boldsymbol{A}_3)\}\nu_{0\alpha}\sqrt{A}ds$$

$$-\int_{s_u} \{(\mathbf{R}\hat{\underline{\mathcal{N}}}^{\alpha}) \cdot (\boldsymbol{u}_0 - \bar{\boldsymbol{u}}_0) + (\mathbf{R}\hat{\underline{\mathcal{H}}}^{\alpha}) \cdot (\boldsymbol{a}_3 - \bar{\boldsymbol{n}})\} \nu_{0\alpha} \sqrt{A} ds, \tag{73}$$

where $W_0$ and *c.c.* are given, under each assumption, by

$$\begin{aligned}
&\text{(G1) and (GM1)} \quad : \quad W_0 = W_0(\boldsymbol{c}_\alpha, \boldsymbol{b}_\alpha), \quad c.c. = 0, \\
&\text{(G2) and (GM2)} \quad : \quad W_0 = W_0(\boldsymbol{c}_\alpha, \boldsymbol{b}_\alpha, \boldsymbol{z}), \quad c.c. = \hat{\underline{\mathcal{N}}} \cdot (\mathbf{R}^t \boldsymbol{a}_3 - \boldsymbol{z}).
\end{aligned}$$

The Euler equations obtained from $\delta \mathcal{F}_{HW} = 0$ are the constitutive equations, the compatibility equations, the balance equations, and the boundary conditions. Since the constitutive equations and the compatibility equations are the same as those of Eq.(72), we show the balance equations and the boundary conditions as follows:

$$\begin{aligned}
\delta \mathcal{F}_{HW} = \int &[\text{Constitutive Equations} + \text{Compatibility equations} \\
&+\{\boldsymbol{a}_\alpha \times \hat{\boldsymbol{N}}^{\alpha} + \frac{1}{\sqrt{A}}(\sqrt{A}\boldsymbol{a}_3 \times \hat{\boldsymbol{H}}^{\alpha})_{,\alpha} + \boldsymbol{a}_3 \times \boldsymbol{T}\} \cdot \boldsymbol{\Gamma} \delta \boldsymbol{\theta} \\
&-\{\frac{1}{\sqrt{A}}(\sqrt{A}\hat{\boldsymbol{N}}^{\alpha})_{,\alpha} + \boldsymbol{p}\} \cdot \delta \boldsymbol{u}_0] \sqrt{A} d\xi^1 d\xi^2 \\
+\int_{s_\sigma} &[(\hat{\boldsymbol{N}}^{\alpha} - \bar{\boldsymbol{N}}) \cdot \delta \boldsymbol{u}_0 + (\boldsymbol{a}_3 \times \hat{\boldsymbol{H}}^{\alpha} - \boldsymbol{a}_3 \times \bar{\boldsymbol{H}}) \cdot \boldsymbol{\Gamma} \delta \boldsymbol{\theta}] \nu_{0\alpha} \sqrt{A} ds \\
-\int_{s_u} &\{(\boldsymbol{u}_0 - \bar{\boldsymbol{u}}_0) \cdot \delta \hat{\boldsymbol{N}}^{\alpha} + \{\boldsymbol{a}_3 - \bar{\boldsymbol{n}}\} \cdot \delta \hat{\boldsymbol{H}}^{\alpha}\} \nu_{0\alpha} \sqrt{A} ds.
\end{aligned} \tag{74}$$

Note that, in the case of Hu-Washizu type functional, the moment equilibrium equation is recovered in place of the AMB and DMB equations.

With the help of Legendre transformation, we may have the complementary function defined by

$$\text{(G1) and (GM1)} \quad : \quad B_0 = \underline{\mathcal{N}}^{\alpha} \cdot \boldsymbol{c}_\alpha + \underline{\mathcal{H}}^{\alpha} \cdot \boldsymbol{b}_\alpha - W_0, \tag{75}$$

$$\text{(G2) and (GM2)} \quad : \quad B_0 = \underline{\mathcal{N}}^{\alpha} \cdot \boldsymbol{c}_\alpha + \underline{\mathcal{H}}^{\alpha} \cdot \boldsymbol{b}_\alpha + \underline{\mathcal{N}} \cdot \boldsymbol{z} - W_0. \tag{76}$$

Substituting Eq.(75) or Eq.(76) into Eq.(73), we have the Hellinger-Reissner type functional expressed as

$$\begin{aligned}
\mathcal{F}_{HR} = \int &[\underline{\mathcal{N}}^{\alpha} \cdot \boldsymbol{c}_\alpha(\boldsymbol{u}_0, \boldsymbol{\theta}) + \underline{\mathcal{H}}^{\alpha} \cdot \boldsymbol{b}_\alpha(\boldsymbol{u}_0, \boldsymbol{\theta}) + c.c. - B_0 \\
&-\boldsymbol{P} \cdot \boldsymbol{u}_0 + \boldsymbol{T} \cdot (\boldsymbol{a}_3(\boldsymbol{\theta}) - \boldsymbol{A}_3)] \sqrt{A} d\xi^1 d\xi^2 \\
&-\int_{s_\sigma} \{\bar{\boldsymbol{N}} \cdot \boldsymbol{u}_0 + \bar{\boldsymbol{H}} \cdot (\boldsymbol{a}_3(\boldsymbol{\theta}) - \boldsymbol{A}_3)\} \nu_{0\alpha} \sqrt{A} ds \\
-\int_{s_u} &\{(\mathbf{R}\hat{\underline{\mathcal{N}}}^{\alpha}) \cdot (\boldsymbol{u}_0 - \bar{\boldsymbol{u}}_0) + (\mathbf{R}\hat{\underline{\mathcal{H}}}^{\alpha}) \cdot (\boldsymbol{a}_3(\boldsymbol{\theta}) - \bar{\boldsymbol{n}})\} \nu_{0\alpha} \sqrt{A} ds,
\end{aligned} \tag{77}$$

where $B_0$ and *c.c.* are given, under each assumption, by

$$\begin{aligned}
&\text{(G1) and (GM1)} \quad : \quad B_0 = B_0(\underline{\mathcal{N}}^\alpha, \underline{\mathcal{H}}^\alpha), \quad c.c. = 0, \\
&\text{(G2) and (GM2)} \quad : \quad B_0 = B_0(\underline{\mathcal{N}}^\alpha, \underline{\mathcal{H}}^\alpha, \underline{\mathcal{N}}), \quad c.c. = \underline{\mathcal{N}} \cdot \boldsymbol{z}.
\end{aligned}$$

The Euler equations are the constitutive equations, the LMB equation, the moment equilibrium equation and the boundary conditions.

The purely kinematic functional is obtained from Eq.(77) by eliminating the stress resultant vectors through the constitutive equations.

## 6. Conclusion

A nonlinear thick shell theory has been developed on the basis of kinematic and stress states assumptions. Balance laws of the shell are derived from the LMB and AMB equations of three dimensional continuum. The resulting AMB equation plays an important role in the mechanical power and the objectivity. When the AMB equation is satisfied, the mechanical power leads to the conjugate relationships between the stress and strain measures. The objectivity or invariance of strain energy function is not always satisfied even if the AMB equation is satisfied. The satisfaction of objectivity depends on the assumptions used. When both kinematic and stress states assumptions are used, the objectivity is satisfied. The use of kinematic assumption only is not enough to preserve the objectivity.

A generalized variational principle has been derived from the Atluri's variational principle. Unsymmetric stress resultant vectors are used so that the drilling degrees of freedom are introduced in a natural way. In contrast to the existing literature where the Eulerian rotational variation has been used, the rotational variation is expressed in terms of the variation of finite rotation vector. The use of the finite rotation vector enables us to construct the strain energy function. Therefore, the resulting tangent stiffness is always symmetric. The generalized functional leads to the LMB, AMB and DMB equations. The resulting AMB equations are not always the same as those derived from the balance laws of 3-D continuum.

## 7. References

1. Atluri, S.N. (1979) On rate principle for finite strain analysis of elastic and inelastic nonlinear solids, *in Recent Research on Mechanical Behavior of Solids*, University of Tokyo Press, pp.79-107.

2. Atluri, S.N. (1980) On some new general and complementary energy theorems for the rate problems in finite strain, classical elastoplasticity, *Journal of Structural Mechanics*, **8** (1), 61-92.

3. Atluri, S.N. (1983) Alternate stress and conjugate strain measures, and mixed variational formulations involving rigid rotations, for computational analysis of finitely deformed solids, with application to plates and shell - I Theory, *Computers and Structures*, **18**, 93-116.

4. Atluri, S.N. and Cazzani, A. (1995) Rotations in computational solid mechanics, *Archives of Computational Methods in Engineering*, **2**(1), 49-138.

5. Cazzani, A and Atluri, S.N. (1992) Four-nodded mixed finite elements, using unsymmetric stresses, for linear analysis of membranes, *Computational Mechanics*, **11**, 229-251.

6. Iura, M. and Atluri, S.N. (1988) Dynamic analysis of finitely stretched and rotated three-dimensional spaced-curved beams, *Computers and Structures*, **29** (5), 875-889.

7. Iura, M. and Atluri, S.N. (1989) On a consistent theory, and variational formulation of finitely stretched and rotated 3-D space-curved beams, *Computational Mechanics*, **4**, 73-88.

8. Iura, M. and Atluri, S.N. (1992) Formulation of a membrane finite element with drilling degrees of freedom, *Computational Mechanics*, **9**, 417-428.

9. Makowski, J. and Stumpf, H. (1995) On the "symmetry" of tangent operators in nonlinear mechanics, *Z. angew. Math. Mech.*, **75** (3), 189-198.

10. Murakawa, H. and Atluri, S.N. (1978) Finite elasticity solutions using hybrid finite elements based on a complementary energy principle, *Journal of Applied Mechanics*, ASME, **45**, 539-547.

11. Murakawa, H. and Atluri, S.N. (1979) Finite elasticity solutions using hybrid finite elements based on a complementary energy principle. Part 2: incompressible materials, *Journal of Applied Mechanics*, ASME, **46**, 71-77.

12. Ogden, R.W. (1984) *Non-linear Elastic Deformations*, Ellis Horwood Limited.

13. Pietraszkiewicz, W. (1979) *Finite Rotations and Lagrangian Description in the Nonlinear Theory of Shells*, Polish Scientific Publications.

14. Reissner, E. (1984) Formulation of variational theorems in geometrical nonlinear elasticity, *J. Eng. Mech.*, **110**(9), 1979-1988.

15. Simmonds, J.G. and Danielson, D.A. (1970) Nonlinear shell theory with a finite rotation vector I and II, *Proc. Kon. Ned. Ak. Wet.*, **B 73**, 460-478.

16. Simmonds, J.G. and Danielson, D.A. (1972) Nonlinear shell theory with finite rotation and stress-function vectors, *J. Appl. Mech.*, ASME, **39**, 1085-1090.

17. Suetake, Y., Iura, M. and Atluri, S.N. (1999) Shell theories with drilling degrees of freedom and geometrical and material assumptions, *Computer Modeling and Simulation in Engineering*, **4**(1), 42-49.

18. Washizu, K. (1982) *Variational Methods in Elasticity and Plasticity*, 3rd ed., Pergamon Press.

19. Zienkiewicz, O.C. (1977) *The Finite Element Method*, 3rd ed., McGraw-Hill.

# ON THE THEORY OF QUASI-SHALLOW SHELLS

E.L. AXELRAD

*137 Chapman Rd., Woodside, CA 94062, USA*

**Abstract** – The theory of small-strain unrestricted deformation of thin elastic shells is specialized to the 'Donnell-type' theory of quasi-shallow shells, which owes its consequent - intrinsic and dual - formulation to the idea first enunciated by Avinoam Libai. This theory is explored with respect to the accuracy, adequacy range and the physical meaning of its basic hypotheses.

## 1. Introduction

***Are the specialized branches of the shell theory still required?*** Will not the numerical solutions soon enable the general shell theory, or even the three-dimensional continuum theory, to become fully sufficient for any practical requirements? The past experience indicates clearly: in the foreseeable future the specialized theories shall remain useful, even indispensable. This situation is well recognized in Physics: "the more complicated is the system, the further simplified must be its theory" (Y.I. Frenkel). Moreover, vis-a-vis numerical data a structural engineer is often in a position depicted by A.Einstein: "Confronted with the individual results of empirical investigation, he has to remain in the state of helplessness, until the principles needed for deductive judgment become accessible to him." This role of specialized analysis is substantiated by the treatment of modern composite shell structures [23].

In the applications of shell theory, most basic results of practical value have been achieved not by the general theory, but by its ramifications simplified by *specialization*. The contribution of the membrane theory is memorable. And the break-through to problems encorporating the wall-bending the shell theory owes to the axisymmetric analysis, given by H.Reissner [2] (1912) and extended by his associate E. Schwerin in 1918 to the laterally loaded shells of revolution (which owes much to the F.Y.M. Wan work - cf. ref. in [16, 18]). The treatment of shells designed for large deformation has been started by E.Reissner's [20] (1950) nonlinear axisymmetric analysis.

H.Reissner founded in 1912 [2] also the *intrinsic formulation* of the theory - one free of any use of displacements as unknowns in the equations. This approach, based on the compatibility and equilibrium equations, proved the most effective in the general shell theory [4] and indispensable in the specialized branches, which serve virtually all the nonaxisymetric nonlinear problems. This includes the treatment of shells allowing



*D. Durban et al. (eds.), Advances in the Mechanics of Plates and Shells,* 33–48.

large deformation (flexibility); the influence of the large precritical deformation on buckling has been evaluated by means of the local-stability approach.[9, 18].

The Donnell-type theory [3, 6, 10, ...] has guided most investigations of buckling and postbuckling of elastic shells. It has achieved its unrivalled results thanks to its striking simplification, compared to the general shell theory.

***The Donnell-type theory*** has been founded in 1933-1934 [3] for cylinders. Simultaneously, a similar theory has been invented in USSR by H.M.Mushtary.

In 1944 V.S. Vlassov extended the theory to non-cylinder shells and introduced the Airy function for the stress resultants. (This was preceded in 1939 by the work of S.Feinberg, also in Russian.) However, the bending strains were still determined in terms of displacements. Moreover, this was done by means of linear strain-displacement relations, where only the terms with the normal component of deflection had been retained. That is, the nonlinear problems were treated on the basis of the drastically abridged linear relations. (These were relations introduced in 1874 in the first ever work on shells [1] and constantly criticized till the middle of 1950-s.)

The insufficiency of the part of the Donnell-type theory, which treated the strain, particularly its illegitimacy for large rotations, has been fist certified in 1963 by L.H. Donnell himself. In an unpublished lecture “General shell displacement-strain relations” (given 1964 also in Leningrad) L.H. Donnell investigated the strain caused by unlimited displacements. But this result could not be incorporated into the theory. The unreliable strain-displacement relations remained an indispensable part of the Donnell-type theory. - For nearly 30 years after the inception of the theory in 1934, there was no alternative.

Despite its illegitimate part, the theory did not incur any widely perceived grave errors in applications. This ‘skating on thin ice’ [10] was, in actual fact, made possible by the concentration on a quite specific class of problems - those of buckling with small wave-length pattern. Such modes do, indeed, involve nearly exclusively displacements of the kind shown later to justify the strain-displacement relations employed by Donnell [3]. - These modes encompass intensively varyable, predominantly normal, deflection and no large rotations.

***The idea enunciated by A.Libai*** in 1962 [6] has opened the way to the consequent formulation of the Donnell-type theory. - The bending strain has been expressed, in terms of a curvature function $W$, dual to the Airy function $F$. This idea of A.Libai allowed to free the theory of its compromise and to make its statement fully intrinsic and dual. The general solution of all problems, linear and nonlinear, has been formulated in terms of merely two resolving functions.- $W$ and $F$ - determined by means of two simultaneous equations, dual in their linear terms [10].

The term "theory of *quasi-shallow* shells”, or “a theory for shells of small Gaussian curvature" [7], [10], gained wide acceptance. The theory has been later recognized to be adequately representing ‘strongly varying deformation’ [18]. As the theory proved not to be unconditionally applicable to shells which may be described as “quasi-shallow" or "of small Gaussian curvature", it will be further referred to as Donnell-type theory or as Donnell-Mushtary-Vlassov-Koiter- theory, short: *DMVK*- theory.

***The task of what follows is threefold***: 1) To state the basic assumptions of the Donnell-type theory in a somewhat more consequent form, one not dependent on the values of stress or strain. 2) To display the physical sense of these assumptions. 3) To estimate the inherent error of the theory and to delineate the domain of problems, for which this theory is adequate in accuracy to the general one.

The error of the Donnell-type theory proves to depend on both the shape and the stress state of the shell. Specifically, both turn out to be represented by intensities of variation along the reference surface These intensities are measured: a) by intervals $L_\alpha$ of variation of the stress state; b) by principal curvature radiuses $R_\alpha$, which are, in fact, the intervals of variation of unit normal vector $\boldsymbol{n}$ of the surface. For the Donnell-type theory the error estimate is $L_\alpha^2/R_\alpha^2$. This means: the theory does not involve additional errors for those stress states which vary with both surface coordinate much more intensely than the unit normal vector $\boldsymbol{n}$ - when $L_\alpha^2/R_\alpha^2 \leq h/|R_\alpha|$.

To compare, recall the other two specialized shell theories. The *membrane* theory is, in certain respect, complementary to the *DMVK*-theory. - It is adequate for stress states varying with the two surface coordinates less intensely than $\boldsymbol{n}$. The third specialized theory, which covers stress states assuring *flexibility* of a shell - large deformability by small strain, is adequate for $(L_2/L_1)^2 << 1$. It serves the domain between those of the first two [24].

***A review of the field equations of the general theory*** of thin shells precedes in what follows (Sect.2-4) the discussion of the Donnell-type theory. This part of the paper has to make it self-contained. Its other purpose is to state the shell theory entirely in terms of resultants defined physically - representing the strain and stress without any requirement for the specially defined "best" or "best modified" resultants.

This formulation of the general theory originates in a vectorial treatment of the local deformation. It starts with the metric and the curvature of the reference surface determined by local-geometry quasi-vectors $\boldsymbol{a}_\alpha$, $\boldsymbol{b}_\alpha$. The strain and curvature-change are defined and determined by subtracting the *initial* local-geometry vectors, moved with the tangent plane during the deformation, from such vectors of *deformed* shape. The essential point is: *both* the initial and the deformed local shape variables are decomposend in one and the same local reference basis, which moves with each point of the surface but does not deform . (Such basis is due to Alumae *(PMM,* 1956, p.136ff*)*, and, in a later context, to Simmonds and Danielson .) The so defined membrane strain $E_{\alpha\beta}$ is equal to $(a_{\alpha\beta}{}^* - a_{\alpha\beta})/2$. However, the vectorialy defined bending-strain $\rho_{\alpha\beta}$ is *not* identical to the "natural choice" $b_{\alpha\beta}{}^* - b_{\alpha\beta}$. The symmetric part of the tensor $\rho_{\alpha\beta}$ proves to be nothing else but the "*best modified*" tensor of changes of curvature. The $\rho_{\alpha\beta}$ resolves the "difficulty in defining a finite bending strain tensor " [7]. - The variations $\delta E_{\alpha\beta}$ and $\delta\rho_{\alpha\beta}$ prove to be the virtual-work conjugates of the actual (not of any modified) stress resultants $n^{\alpha\beta}$ and $m^{\alpha\beta}$.

The *nonlinear* equations of compatibility and equilibrium display a simple *duality*.

## 2. Surface shape and strain. Compatibility

***The reference surface*** is set at the middle of the wall thickness - the optimum choice for homogeneous isotropic shells (cf., e.g., [18]). The radius vector $\boldsymbol{r}$, from a fixed pole to a generic point of the surface, is determined by Gaussian coordinates $x^\alpha$. - All Greek-letter indices take the values 1, 2. The tangential base vectors are defined as $\boldsymbol{a}_\alpha = \boldsymbol{r},_\alpha$. A comma preceding a subscript $\alpha$ denotes a partial (*not* a covariant) differentiation with respect to $x^\alpha$.

The contravariant base $\boldsymbol{a}^\alpha$ is defined by the orthogonality $\boldsymbol{a}_1 \bullet \boldsymbol{a}^2 = \boldsymbol{a}^1 \bullet \boldsymbol{a}_2 = 0$ and by $\boldsymbol{a}_\alpha \bullet \boldsymbol{a}^\alpha = 1$. Components of the metric tensor are: $a_{\alpha\beta} = \boldsymbol{a}_\alpha \bullet \boldsymbol{a}_\beta$, $a^{\alpha\beta} = \boldsymbol{a}^\alpha \bullet \boldsymbol{a}^\beta$
The unit normal vector of the surface is $\boldsymbol{a}_3 \equiv \boldsymbol{n} = \boldsymbol{a}_1 \times \boldsymbol{a}_2 / \sqrt{a}$, $a = |\boldsymbol{a}_1 \times \boldsymbol{a}_2|^2$.

The local shape of the reference surface, is displayed by the rotation of a tangent plane, when shifted along the surface. The position of the plane is indicated by the normal vector $\boldsymbol{n}(x^\alpha)$. This leads to the description of the curvature of the surface by vectors $\boldsymbol{b}_\alpha$ [19]:

$$\boldsymbol{b}_\alpha \equiv b_{\alpha\beta}\boldsymbol{a}^\beta = \boldsymbol{n},_\alpha , \quad b_{\alpha\beta} = \boldsymbol{n},_\alpha \bullet \boldsymbol{a}_\beta , \quad b_\alpha^\beta = \boldsymbol{b}_\alpha \bullet \boldsymbol{a}^\beta = b_{\alpha\lambda}a^{\lambda\beta} . \tag{2.1}$$

The Einstein summation convention is employed. The $b_{\alpha\beta}$ defined in (2.1) is equal to that of Sanders [7]. (The tensors $\boldsymbol{a}^\alpha \boldsymbol{a}_\alpha$ and $\boldsymbol{a}^\alpha \boldsymbol{b}_\alpha$ - can, following [21, 22], be defined in the coordinate-free form - as $\nabla \boldsymbol{r}$ and $\nabla \boldsymbol{n}$.)

***The local shape is determined***, besides $\boldsymbol{b}_\alpha$, by *curvature vectors* $\boldsymbol{k}_\alpha$ [14] which, in contrast to $\boldsymbol{b}_\alpha$, include $\boldsymbol{n}$-components. The $\boldsymbol{k}_\alpha dx^\alpha$ is defined as the angle between the tangent planes at the points $x^\alpha$ and $x^\alpha + dx^\alpha$ with a distance $\boldsymbol{a}_\alpha dx^\alpha$ between them. Thus, $\boldsymbol{k}_\alpha$ determines the derivative, with respect to $x^\alpha$, of any unit vector $\boldsymbol{v}$ bound to the tangent plane:

$$\boldsymbol{k}_\alpha = \boldsymbol{n} \times \boldsymbol{b}_\alpha + l_\alpha \boldsymbol{n} , \quad [\boldsymbol{n} \quad \boldsymbol{v}],_\alpha = \boldsymbol{k}_\alpha \times [\boldsymbol{n} \quad \boldsymbol{v}]. \tag{2.2}$$

Three coordinates $x^\alpha$, $z$ label a material point; they are not changed by deformation. Other variables take with the deformation of the surface new values which will be denoted by an *asterisk* superscript - $\boldsymbol{n}^*$, $\boldsymbol{b}_\alpha^*$, $l_\alpha^*$, $\boldsymbol{a}_\alpha^* = \boldsymbol{r}^*,_\alpha$. With the new values of variables, the relations (2.2) determine $\boldsymbol{k}_\alpha^*$ and the derivatives for the deformed, current, state:

$$\boldsymbol{k}_\alpha^* = \boldsymbol{n}^* \times \boldsymbol{b}_\alpha^* + l_\alpha^* \boldsymbol{n}^* , \quad [\boldsymbol{n}^* \quad \boldsymbol{v}^*],_\alpha = \boldsymbol{k}_\alpha^* \times [\boldsymbol{n}^* \quad \boldsymbol{v}^*] .$$

For a continuous surface, the radius vector $\boldsymbol{r}^*$ and any unit vector $\boldsymbol{v}^*$, bound to the tangent plane, are continuous functions of $x^\alpha$. - For any current shape (also for the initial one), the *continuity* conditions: $\boldsymbol{v}^*,_{12} = \boldsymbol{v}^*,_{21}$, $\boldsymbol{r}^*,_{12} = \boldsymbol{r}^*,_{21}$ lead to

$$\boldsymbol{k}_1^*,_2 - \boldsymbol{k}_2^*,_1 + \boldsymbol{k}_1^* \times \boldsymbol{k}_2^* = 0 , \quad \boldsymbol{a}_1^*,_2 = \boldsymbol{a}_2^*,_1 . \tag{2.3}$$

These equations are, of course, valid also for the initial shape - for the $\boldsymbol{k}_\alpha$ and $\boldsymbol{a}_\alpha$. The first of eqs. (2.3) is equivalent to the three scalar relations of Gauss and Codazzi .

***The initial-curvature vectors*** $\boldsymbol{b}_\alpha$ ***and*** $\boldsymbol{k}_\alpha$ are unchanged, with respect to the local reference basis $\boldsymbol{a}_\alpha$. During a deformation of the surface the $\boldsymbol{b}_\alpha$ and $\boldsymbol{k}_\alpha$ move, rotate, together with this basis - with the tangent plane.

Positions, attained by the vectors $\boldsymbol{a}_\alpha$, $\boldsymbol{b}_\alpha$ and $\boldsymbol{k}_\alpha$ in the course of deformation, may be denoted by a special index (as in [24]). However, different shapes of the surface and the corresponding positions of the local basis $\boldsymbol{a}_\alpha$, of $\boldsymbol{b}_\alpha$ and $\boldsymbol{k}_\alpha$, rarely appear in an analysis simultaneously with their initial position. ( An exception occurs in the sequel - in the derivation of the eq.(2.11), where the local basis $\boldsymbol{a}_\alpha$ of a deformed surface appears together with this basis in the undeformed-state position .)

The vector $\boldsymbol{a}_3 \equiv \boldsymbol{a}^3 \equiv \boldsymbol{n}$ moves into the normal vector of the deformed surface $\boldsymbol{n}^*$. The vectors $\boldsymbol{a}_\alpha$, $\boldsymbol{n}^*$ constitute a local basis, which in the course of deformation *rotates* with the tangent plane without being deformed. In terms of the rotated local basis there are expansions:

$$\boldsymbol{k}_\alpha = \boldsymbol{n}^* \times \boldsymbol{b}_\alpha + l_\alpha \boldsymbol{n}^*, \qquad \boldsymbol{b}_\alpha = b_{\alpha\beta}\boldsymbol{a}^\beta, \quad \boldsymbol{a}_\alpha = a_{\alpha\beta}\,\boldsymbol{a}^\beta \; . \tag{2.4}$$

As the shear strain changes the angle between $x^\alpha$ -lines, these lines do not stay tangent to the $\boldsymbol{a}_\alpha$. The basis vectors tangent to $x^\alpha$ -lines on the deformed surface are $\boldsymbol{a}_\alpha^* = \boldsymbol{r},_\alpha^*$. All variables describing the deformed shape can, of course, be decomposed in the deformed basis $\boldsymbol{a}_\alpha^*$, $\boldsymbol{a}^{\beta*}$. But, ordinarily, the undeformable local basis $\boldsymbol{a}_\alpha$, $\boldsymbol{a}^\beta$ proves preferable. Components of the characteristics of deformed geometry in the local basis $\boldsymbol{a}_\alpha$ will be denoted by a *prime* superscript :

$$\boldsymbol{a}_\alpha^* \equiv a_{\alpha\beta}^* \boldsymbol{a}^{\beta *} \equiv a_{\alpha\beta}'\boldsymbol{a}^\beta \;; \qquad \boldsymbol{b}_\alpha^* \equiv b_{\alpha\beta}^* \boldsymbol{a}^{\beta *} \equiv b_{\alpha\beta}'\boldsymbol{a}^\beta \,, \qquad b_\alpha^{\beta\prime} = a^{\lambda\beta}\, b_{\alpha\lambda}' \; . \tag{2.5}$$

A pleasing feature of the basis $\boldsymbol{a}_\alpha$, displayed in (2.5): the raising and lowering of indices is done with the metric of the *initial* geometry - with $a_{\alpha\beta}$, $a^{\alpha\beta}$ .

***The strain of the surface*** will be characterized by vectors $\boldsymbol{E}_\alpha$, $\boldsymbol{\rho}_\alpha$ and $\boldsymbol{K}_\alpha$. These are defined as the difference between the $\boldsymbol{a}_\alpha^*$, $\boldsymbol{b}_\alpha^*$ and $\boldsymbol{k}_\alpha^*$ of the current shape, and the respective vectors $\boldsymbol{a}_\alpha$, $\boldsymbol{b}_\alpha$ and $\boldsymbol{k}_\alpha$ (which are recalled to be the initial $\boldsymbol{a}_\alpha$, $\boldsymbol{b}_\alpha$ and $\boldsymbol{k}_\alpha$, moved with the tangent plane, but not otherwise changed by the deformation). The strain components are defined by the following relations, where *all* vectors - $\boldsymbol{a}_\alpha^*$ and $\boldsymbol{a}_\alpha$, $\boldsymbol{b}_\alpha^*$ and $\boldsymbol{b}_\alpha$, - are decomposed in *the same* basis, $\boldsymbol{a}_\alpha$:

$$\boldsymbol{E}_\alpha \equiv E_{\alpha\beta}\boldsymbol{a}^\beta = \boldsymbol{a}_\alpha^* - \boldsymbol{a}_\alpha \,, \qquad E_{\alpha\beta} = a_{\alpha\beta}' - a_{\alpha\beta} \;; \tag{2.6}$$

$$\boldsymbol{\rho}_\alpha \equiv \rho_{\alpha\beta}\boldsymbol{a}^\beta = \boldsymbol{b}_\alpha^* - \boldsymbol{b}_\alpha \,, \qquad \rho_{\alpha\beta} = b_{\alpha\beta}' - b_{\alpha\beta} \; . \tag{2.7}$$

$$\boldsymbol{K}_\alpha = \boldsymbol{k}_\alpha^* - \boldsymbol{k}_\alpha = \boldsymbol{n}^* \times \boldsymbol{\rho}_\alpha + \lambda_\alpha \boldsymbol{n}^* \,, \qquad \lambda_\alpha = l_\alpha^* - l_\alpha . \tag{2.8}$$

The components $E_{\alpha\beta}$ represent the extension and the shear strain. This is displayed clearer by ortogonal $x^\alpha$. The curvature-change $\boldsymbol{K}_\alpha$ will be useful besides the $\boldsymbol{\rho}_\alpha$. A position of the $\boldsymbol{a}_\alpha$ inside the tangent plane of the deformed surface specifies a partition of the shear angle into the two angles between $\boldsymbol{a}_\alpha$ and $\boldsymbol{a}_\alpha^*$. The *symmetricity* condition $E_{\alpha\beta} = E_{\beta\alpha}$ is with (2.6) represented by $a_{\alpha\beta}' = a_{\beta\alpha}'$ . This choice makes $E_{\alpha\beta}$ equal to the standard strain tensor $(a_{\alpha\beta}{}^* - a_{\alpha\beta})/2$.

However, the $\rho_{\alpha\beta} = b_{\alpha\beta}' - b_{\alpha\beta}$ is different from the tensor $b_{\alpha\beta}{}^* - b_{\alpha\beta}$ , known as the "obvious choice" for the bending strain. The relation between the two is given by (2.4)-(2.7) :

$$b_{\alpha\beta}{}^* - b_{\alpha\beta} = \rho_{\alpha\beta} + \boldsymbol{b}_\alpha{}^* \cdot \boldsymbol{E}_\beta \, , \quad \boldsymbol{b}_\alpha{}^* \cdot \boldsymbol{E}_\beta = b_{\beta\lambda}{}' E_\alpha{}^\lambda \, . \tag{2.9}$$

***Compatibility equations for the strain vectors $\boldsymbol{E}_\alpha$, $\boldsymbol{K}_\alpha$*** follow from (2.3) with (2.6) and (2.8). The transformation of (2.3) requires derivatives of the local basis $\boldsymbol{a}_\alpha$ and curvature vectors $\boldsymbol{k}_\alpha$, which have moved in the course of surface deformation.

To differentiate with the vectors $\boldsymbol{a}_\alpha$ and $\boldsymbol{k}_\alpha$ in their current-state orientation, the *initial*-state directions of these vectors are *in this subsection* indexed by $^0$ - written as $\boldsymbol{a}_\alpha{}^0$, $\boldsymbol{k}_\alpha{}^0$. A rotation of a vector bound to the tangent plane, during the deformation of the surface, will be represented by the index "$_{rotated}$". With this, the relation between the vectors in their current and initial states is: $(\boldsymbol{a}_\alpha \;\; \boldsymbol{k}_\alpha) = (\boldsymbol{a}_\alpha{}^0 \;\; \boldsymbol{k}_\alpha{}^0)_{rotated}$ .

In the initial state, the $\boldsymbol{a}_\alpha{}^0(x^\alpha + dx^\alpha)$ constitute with $\boldsymbol{a}_\alpha{}^0(x^\alpha)$ the angle $\boldsymbol{k}_\alpha{}^0 dx^\alpha$ . The angle between the *rotated* local-basis vector $\boldsymbol{a}_\alpha(x^\alpha + dx^\alpha)$ and the $\boldsymbol{a}_\alpha(x^\alpha)$ - the angle between the tangent planes of the *deformed* surface - is $\boldsymbol{k}_\alpha{}^* dx^\alpha$ .

With $\boldsymbol{k}_\alpha{}^* = \boldsymbol{k}_\alpha + \boldsymbol{K}_\alpha$ from (2.8), with the expansion of $\boldsymbol{k}_\alpha$ from (2.4) and (2.2), the above considerations lead to formulas for the derivatives of the local vectors $\boldsymbol{a}_\alpha$ and $\boldsymbol{k}_\alpha$ in terms of the derivatives of these vectors on the initial-shape surface $\boldsymbol{a}_\alpha{}^0$, $\boldsymbol{k}_\alpha{}^0$:

$$\boldsymbol{a}_{\alpha,\beta} = (\boldsymbol{a}_\alpha{}^0{}_{,\beta})_{rotated} + \boldsymbol{K}_\beta \times \boldsymbol{a}_\alpha \, , \quad \boldsymbol{k}_{\alpha,\beta} = (\boldsymbol{k}_\alpha{}^0{}_{,\beta})_{rotated} + \boldsymbol{K}_\beta \times \boldsymbol{k}_\alpha \, . \tag{2.10}$$

These relations read: the derivative of a rotated vector is equal to the rotated derivative of this vector on the initial-state surface plus a term reflecting the bending strain.

Formulas (2.10) are useful also for determining the covariant derivatives.

Insert into (2.3) the $\boldsymbol{a}_\alpha{}^* = \boldsymbol{a}_\alpha + \boldsymbol{E}_\alpha$ and $\boldsymbol{k}_\alpha{}^* = \boldsymbol{k}_\alpha + \boldsymbol{K}_\alpha$ and their derivatives, from (2.7), (2.8) and (2.10). Take into account, the equations (2.3), written for the initial $a_\alpha$, $k_\alpha$ (specified above as $\boldsymbol{a}_\alpha{}^0$, $\boldsymbol{k}_\alpha{}^0$). This leads, finally, to the equations:

$$\boldsymbol{K}_{1,2} - \boldsymbol{K}_{2,1} + \boldsymbol{K}_1 \times \boldsymbol{K}_2 + \boldsymbol{q}_c = 0 \, , \quad \boldsymbol{E}_{1,2} - \boldsymbol{E}_{2,1} - \boldsymbol{a}_1 \times \boldsymbol{K}_2 + \boldsymbol{a}_2 \times \boldsymbol{K}_1 + \boldsymbol{m}_c = 0 \, . \tag{2.11}$$

The "load" terms $\boldsymbol{q}_c$ and $\boldsymbol{m}_c$ serve in (2.11) to complement the analogy with the equations of equilibrium, they may represent temperature expansions or be zero.

The nonlinear vector compatibility equations (2.11) have initially been obtained in a different way [14] . They are an extension of the linear equations of E.Reissner (1974 - [20], p.353). The recent work of A.Libai and J.G.Simmonds [22], made these equations to a nearly self-evident consequence of the new form of the theory.

Written for a small increment of deformation, equations (2.11) do not contain the nonlinear term $\boldsymbol{K}_1 \times \boldsymbol{K}_2$. The nonlinearity is then represented only implicitly. - The equations (2.11), are referred to the *current* shape. The derivatives take this into account as exemplified in (2.10).

***The component form of the compatibility equations*** (2.11) is obtained with the decompositions of $E_\alpha$ and $K_\alpha$ in the standard way. These six equations, written without terms of relative magnitude of the strain and for $q_c, m_c = 0$, are

$$\varepsilon^{\alpha\beta}(\varepsilon^{\delta\lambda}\rho_{\alpha\lambda;\beta} + b_\alpha^{\delta\prime}\lambda_\beta) = 0, \tag{2.12}$$

$$\varepsilon^{\alpha\beta}\lambda_{\alpha;\beta} + \varepsilon^{\alpha\beta}\varepsilon^{\sigma\lambda}\rho_{\alpha\lambda}(b_{\beta\sigma} + \rho_{\beta\sigma}/2) = 0, \tag{2.13}$$

$$\lambda_\beta = -\varepsilon^{\sigma\kappa}E_{\sigma\beta;\kappa}, \tag{2.14}$$

$$\varepsilon^{\alpha\beta}(\rho_{\beta\alpha} + E_\alpha^{\lambda}b_{\beta\lambda}') = 0. \tag{2.15}$$

Here $\varepsilon^{\alpha\beta}$ is the Levi-Civitta tensor: $\varepsilon^{\alpha\alpha} = 0$, $\varepsilon^{12} = -\varepsilon^{21} = 1/\sqrt{a}$.

The semicolon subscript denotes the *covariant derivative*, which in all cases concerns here components with respect to the rotated local basis $\boldsymbol{a}_\alpha$ and $\boldsymbol{a}^\beta$, not to the deformed basis $\boldsymbol{a}_\alpha^*$, $\boldsymbol{a}^{\beta*}$.

For small strain, the deformed curvature components $b_\alpha^{\delta\prime}$ and $b_{\beta\lambda}'$ can be replaced in (2.12), (2.15) by the initial-geometry components $b_\alpha^\delta$ and $b_{\beta\lambda}$.

The expression in brackets in (2.15) coinsides with $b_{\alpha\beta}^* - b_{\alpha\beta}$, as given by (2.9). The non-differential equation (2.15) determines the skew part of the bending-strain $\rho_{\alpha\beta}$ and, thus, renders its *symmetric* part ${}'\rho_{\alpha\beta}$.

## 3. Stress resultants. Equilibrium. Duality

Tractions acting in the normal sections $x^\alpha = const$ of the shell are represented by resultants reduced to the reference surface. The force and moment resultants, acting on the element $a_\beta\, dx^\beta$ of this section, are defined and determined by

$$[\boldsymbol{T}^\alpha \quad \boldsymbol{G}^\alpha]\, dx^\beta = [\boldsymbol{n}^\alpha \quad \boldsymbol{m}^\alpha]\sqrt{a}\, dx^\beta \qquad (\beta \neq \alpha). \tag{3.1}$$

Consider an element of the shell, bounded by sections $x^\alpha$, $x^\alpha + dx^\alpha = const$ and encompassing the reference-surface area $dA = \sqrt{a}\, dx^1 dx^2$. Denote by $\boldsymbol{q}dA$, $\boldsymbol{m}dA$ the force and moment resultants of the load acting on the element. The equilibrium equations for the element are readily obtained in the form

$$\boldsymbol{T}^\alpha{}_{,\alpha} + \boldsymbol{q}\sqrt{a} = 0, \quad \boldsymbol{G}^\alpha{}_{,\alpha} + \boldsymbol{a}_\alpha^* \times \boldsymbol{T}^\alpha + \boldsymbol{m}\sqrt{a} = 0. \tag{3.2}$$

The nonlinearity, resulting here from the deformed shape of the shell, is taken into account in the derivatives and in $\boldsymbol{a}_\alpha^*$. For small strain, $\boldsymbol{a}_\alpha^* = \boldsymbol{a}_\alpha + \boldsymbol{E}_\alpha$ can be replaced in (3.2) by $\boldsymbol{a}_\alpha$.

The components of the resultants are defined and denoted by:

$$\boldsymbol{n}^\alpha = n^{\alpha j}\boldsymbol{a}_j, \quad \boldsymbol{m}^\alpha = m^{\alpha\lambda}\boldsymbol{n}^* \times \boldsymbol{a}_\lambda, \quad \boldsymbol{q} = q^\alpha \boldsymbol{a}_\alpha + q\boldsymbol{n}^* \qquad (j = 1, 2, 3). \tag{3.3}$$

The component equations following from (3.1) - (3.3) are (for $\boldsymbol{m} = 0$):

$$n^{\alpha\beta}{}_{;\alpha} + n^{\alpha 3} b^{\beta}{}_{\alpha}{}' + q^{\beta} = 0 \quad , \tag{3.4}$$

$$n^{\alpha 3}{}_{;\alpha} - n^{\alpha\beta}\, b_{\alpha\beta}{}' + q = 0 \quad , \tag{3.5}$$

$$n^{\alpha 3} = m^{\lambda\alpha}{}_{;\lambda} \quad , \tag{3.6}$$

$$\varepsilon^{\alpha\beta}\,(n^{\alpha\beta} - b_{\lambda}{}^{\beta}{}' m^{\lambda\alpha}) = 0 \ . \tag{3.7}$$

The non-differential equations (3.7) determine the skew and the symmetric part of the tensor $n^{\alpha\beta}$. With the accuracy of the small-strain theory, it can be set in (3.7) $b_{\lambda}{}^{\alpha}{}' = b_{\lambda}{}^{\alpha}$.

The ***duality*** of the equations of compatibility (2.11) and of equilibrium (3.2) is made transparent by their simplicity. It renders each equation from the dual one by the replacement of variables

$$[\boldsymbol{T}^1 \quad \boldsymbol{T}^2 \quad \boldsymbol{q}\sqrt{a}\,] \Leftrightarrow [-\boldsymbol{K}_2 \quad \boldsymbol{K}_1 \quad (\boldsymbol{K}_1{\times}\boldsymbol{K}_2 + \boldsymbol{q}_c)] \ .$$

$$[\boldsymbol{G}^1 \quad \boldsymbol{G}^2 \quad \boldsymbol{m}\sqrt{a}] \Leftrightarrow [-\boldsymbol{E}_2 \quad \boldsymbol{E}_1 \qquad \boldsymbol{m}_c \quad ] \ . \tag{3.8}$$

The static-geometric analogy (3.8) extends to the nonlinear theory the duality relations, which the theory owes to Lur'e [4]. This duality is perturbed merely by the term $\boldsymbol{K}_1{\times}\boldsymbol{K}_2$ of the compatibility equation (2.11). And in the incremental, step-by-step, solutions, usual for nonlinear problems, the term $\boldsymbol{K}_1{\times}\boldsymbol{K}_2$ can be made negligible.

***The virtual work of the inner forces*** pro unit aeria of the reference surfaces ($\delta W$) can be expressed either in terms of the vector resultants of strain and stress or of the symmetric parts $'\rho_{\alpha\beta} = (\rho_{\alpha\beta} + \rho_{\beta\alpha})/2$, $'n^{\alpha\beta} = (n^{\alpha\beta} + n^{\beta\alpha})/2$ of their respective components. As discussed in [24]:

$$\delta W = \boldsymbol{n}^{\alpha}{\bullet}\delta \boldsymbol{E}_{\alpha} + \boldsymbol{m}^{\alpha}{\bullet}\delta \boldsymbol{\rho}_{\alpha} = n^{\alpha\beta}\delta E_{\alpha\beta} + m^{\alpha\beta}\delta \rho_{\alpha\beta} = {}'n^{\alpha\beta}\delta E_{\alpha\beta} + m^{\alpha\beta}\delta({}'\rho_{\alpha\beta}) \ . \tag{3.9}$$

We recall here, that $m^{\alpha\beta} \cong m^{\beta\alpha}$ [19]. With the definition (2.9) of the bending strain $\rho_{\alpha\beta}$, the $\delta W(n^{\alpha\beta},...)$ follows for small strain directly from the $\delta W$ of [7]. - The $\rho_{\alpha\beta}$ eliminates the "difficulty in defining the finite bending strain because the coefficient of $M_0{}^{\alpha\beta}$ in this expression is not the exact variation of anything" [7]. For small strain the $M_0{}^{\alpha\beta}$ of [7] is identical to $\boldsymbol{m}^{\alpha\beta}$ - the mentioned coefficient is $\delta({}'\rho_{\alpha\beta})$.

It is abundantly clear: the need for "modified stress tensors" had been caused solely by the use of $b_{\alpha\beta}{}^* - b_{\alpha\beta} \equiv \boldsymbol{b}_{\alpha}{}^*\cdot\boldsymbol{a}_{\beta}{}^* - \boldsymbol{b}_{\alpha}\cdot\boldsymbol{a}_{\beta}$. The need disappears with the use of bending strain $\rho_{\alpha\beta} \equiv b_{\alpha\beta}{}' - b_{\alpha\beta} \equiv \boldsymbol{b}_{\alpha}{}^*\cdot\boldsymbol{a}_{\beta} - \boldsymbol{b}_{\alpha}\cdot\boldsymbol{a}_{\beta}$, defined by the same basis for $b_{\alpha\beta}{}'$ and $b_{\alpha\beta}$ - - by the $\boldsymbol{a}_{\beta}$..

With the virtual work (3.9) as the starting point, the principles of virtual stress and of virtual strain are formulated directly [24]. The above equations of compatibility and of equilibrium follow from the respective principles, which, vice versa, are obtainable from these equations.

## 4. Shell-volume strain. Elasticity.

***Two basic hypotheses (of Kirchhoff-Love)*** determine the strain and stress in the volume of a thin shell in terms of the deformation of its reference surface:
(a) In the *analysis of strain*, particles comprising a straight line, normal to the middle surface, can be assumed to constitute such a normal and to retain the distances between them also after a deformation. (b) In the *stress-strain relations*, the stresses on the sections parallel to the plane tangent to the reference surface can be disregarded.

These assumptions are known [12] to introduce errors which have the relative magnitude order not exceeding that of the quantities:

$$h/R, \quad h^2/L^2, \; h^2/d^2, \quad \eta \qquad (4.1)$$

Here $\eta$ and $1/R$ are the maximum absolute values of the principal strain and of the normal-section curvature; $d$ denotes the distance to the shell edge; $L$ is defined [5, 10, 12] as the minimum of *intervals of variation* $L_\alpha$ of any function $\boldsymbol{F}(x^\alpha)$, which is substantial in the description of the stress and deformation:

$$L = \min L_\alpha, \quad 1/L_\alpha \sim |\partial \boldsymbol{F}/a_\alpha \partial x^\alpha| / (\max|\boldsymbol{F}|) \quad (a_\alpha = |\boldsymbol{a}_\alpha|). \qquad (4.2)$$

The sign $\sim$ indicates the equality of the magnitude orders.

The physical meaning of the $L_\alpha$ is clarified by a simple case $F = \sin(x^1/c)$, $a_1 = 1$. The definition (4.2) gives in this case the intervals of variation: $L_1 = c$, $L_2 = \infty$.

The hypothesis (a) renders for the radius vector $\boldsymbol{R}^*$ of a point in the deformed *volume* ($\boldsymbol{R}^*$ has no direct relation to the $1/R$), as well as for the deformed and for the rotated local bases ($\boldsymbol{g}_\alpha^*$, $\boldsymbol{g}_\alpha$) of the three-dimensional metric the formulas:

$$\boldsymbol{R}^* = \boldsymbol{r}^* + z\boldsymbol{n}^*, \qquad \boldsymbol{g}_\alpha^* \equiv \boldsymbol{R}^*{}_{,\alpha} = \boldsymbol{a}_\alpha^* + z\boldsymbol{b}_\alpha^*, \qquad \boldsymbol{g}_\alpha = \boldsymbol{a}_\alpha + z\boldsymbol{b}_\alpha. \qquad (4.3)$$

***The strain in the shell volume*** is defined and denoted by $\boldsymbol{\gamma}_\alpha$ and $\gamma_{\alpha\beta}$, determined by the vector version of the standard Cauchy strain formula and the definitions (2.6), (2.7) of $\boldsymbol{E}_\alpha$, $\boldsymbol{\rho}_\alpha$:

$$\boldsymbol{\gamma}_\alpha \equiv \gamma_{\alpha\beta}\boldsymbol{a}^\beta = \boldsymbol{g}_\alpha^* - \boldsymbol{g}_\alpha = \boldsymbol{E}_\alpha + z\boldsymbol{\rho}_\alpha, \qquad \gamma_{\alpha\beta} = E_{\alpha\beta} + z\rho_{\alpha\beta}, \qquad (4.4)$$

$$\gamma_{\alpha\beta} = (g_{\alpha\beta}^* - g_{\alpha\beta})/2 + \Delta_{\alpha\beta}, \qquad g_{\alpha\beta}^* = \boldsymbol{g}_\alpha^* \cdot \boldsymbol{g}_\beta^*, \qquad (4.5)$$

$$2\Delta_{\alpha\beta} = z(\boldsymbol{E}_\alpha \cdot \boldsymbol{b}_\beta^* + \boldsymbol{E}_\beta \cdot \boldsymbol{b}_\alpha^*) + z^2(\boldsymbol{\rho}_\alpha \cdot \boldsymbol{b}_\beta^* + \boldsymbol{\rho}_\beta \cdot \boldsymbol{b}_\alpha^*) + (\boldsymbol{g}_\alpha^* - \boldsymbol{g}_\alpha)(\boldsymbol{g}_\beta^* - \boldsymbol{g}_\beta).$$

The term $\Delta_{\alpha\beta}$ can in (4.5) be neglected. Its magnitude is of the order of $\gamma_{\alpha\beta}\,\eta/R$, it does not exceed the error (4.1) of the basic hypotheses. Thus, the vectorial definition (4.4) of the strain $\gamma_{\alpha\beta}$ proves equivalent to the standard one - to $\gamma_{\alpha\beta} = (g_{\alpha\beta}^* - g_{\alpha\beta})/2$. (If the $z$-term of $\Delta_{\alpha\beta}$ is retained, the formula (4.5) determines the $\gamma_{\alpha\beta}$ not as in (4.4) by $\rho_{\alpha\beta} = b_{\alpha\beta}' - b_{\alpha\beta}$, but, with equivalent accuracy, by the $b_{\alpha\beta}^* - b_{\alpha\beta}$.)

***The strain energy*** of a shell made of Hookean elastic isotropic material can, as a consequence of the assumption (b), be determined solely in terms of the components $E_{\alpha\beta}$ and $\rho_{\alpha\beta}$ of the strain vectors. The starting point for the derivation is provided by the (based on the assumption) formula from [5] for the energy, per unit area of the reference surface

$$V = \int G \{ g^{\alpha\lambda} g^{\beta\mu} + [\nu/(1-\nu)] g^{\alpha\beta} g^{\lambda\mu} \} \gamma_{\alpha\beta} \gamma_{\lambda\mu} \sqrt{g}\, dz \,, \quad 2G = E/(1+\nu), \qquad (4.6)$$

here the integral extends over the shell thickness, $E$ and $\nu$ denote the Young modulus of elasticity and the Poisson's ratio.

The definitions of $g_{\alpha\beta}$, $g^{\alpha\beta}$ and $g = det \mid g_{\alpha\beta} \mid$ let each of them to decompose into a main term, identical, respectively, to $a_{\alpha\beta}$, $a^{\alpha\beta}$ and $a$, plus an additive of relative magnitude of $z/R$, which is less than the the error (4.1) of the thin-shell theory. With this in (4.6), the elastic-energy $V$ also decomposes into a main term and an additive of the negligible order of $h/R$. With this, formula (4.6) renders for homogeneous material and the corresponding bounds of integration $z = -h/2$ and $h/2$ :

$$V = G\{a^{\alpha\lambda} a^{\beta\mu} + [\nu/(1-\nu)] a^{\alpha\beta} a^{\lambda\mu}\} ( E_{\alpha\beta} E_{\lambda\mu} h + {}'\rho_{\alpha\beta} {}'\rho_{\lambda\mu} h^3/12) . \qquad (4.7)$$

Thus, the elastic energy is directly determined by the components of the strain vectors $\boldsymbol{E}_\alpha$ and $\rho_\alpha$ . It depends solely on the symmetric components: on $E_{\alpha\beta} = E_{\beta\alpha}$ of the membrane strain and on the ${}'\rho_{\alpha\beta} = (\rho_{\alpha\beta} + \rho_{\beta\alpha})/2$ of the bending strain (For nonhomogeneous and anisotropic materials the elastic energy may have to retain also mixed terms - products of the membrane and bending strains. These terms can be minimized by a choice of the reference surface [18].)

***The elasticity relations*** follow from the equality of the virtual work $\delta W$ of the inner forces to the corresponding variation of the elastic energy $\delta V$. With (3.9) and (4.7), this renders

$$
\begin{aligned}
{}'n^{\alpha\beta} &\equiv (n^{\alpha\beta} + n^{\beta\alpha})/2 = \partial V/\partial E_{\alpha\beta} = B[(1-\nu) E^{\alpha\beta} + \nu a^{\alpha\beta} E_\lambda{}^\lambda], \quad B = Eh/(1-\nu^2) \\
m^{\alpha\beta} &= \partial V/\partial ({}'\rho_{\alpha\beta}) = D[ (1-\nu)\, {}'\rho_{\alpha\beta} + \nu a^{\alpha\beta} ({}'\rho_\alpha{}^\alpha)], \quad D = Eh^3/[12(1-\nu^2)] . \qquad (4.8)
\end{aligned}
$$

## 5. The standard theory of 'quasi-shallow' shells.

***The relevant key work*** [10] ***justifies the Donnell-type theory*** by two assumptions: The *maximum absolute* values of wall-bending strain $h\rho/2$ and of middle-surface extension $\eta$ are assumed to have *not too different* orders of magnitude

$$h/R \ll h\rho/\eta \ll \min(R/h, 1/h\rho) \,. \qquad (5.1)$$

The undeformed middle surface is assumed to be 'quasi-shallow' - to have small Gaussian curvature K , compared with the square of the minimum interval of variation $L$ (defined in (4.2)):

$$KL^2 \ll 1 \,. \qquad (5.2)$$

***The basic simplifications of the Donnell-type theory***, justified in [10] by the assumption (5.1), are:

(i) In the tangential-forces equilibrium (3.4), all terms depending, directly or through $n^{\alpha 3}$, on the moments $m^{\alpha\beta}$ are disregarded. (ii) In the compatibility equations (2.12), all terms depending, directly or through $\lambda_\beta$, on the $E_{\alpha\beta}$ are neglected. (iii) The stress and strain components $n^{\alpha\beta}$, $\rho_{\alpha\beta}$ are set equal to their symmetric parts $(n^{\alpha\beta}+n^{\beta\alpha})/2$, $(\rho_{\alpha\beta}+\rho_{\beta\alpha})/2$. This means: the $m^{\alpha\beta}$ - terms in the equilibrium equations (3.7) and the $E_{\alpha\beta}$ - terms in the dual eqs. (2.15) are dropped.

After these simplifications, the equations (2.12), (2.15) and (3.4), (3.7) become:

$$\rho_{2\lambda,1} - \rho_{1\lambda;2} \cong 0 , \qquad \rho_{\alpha\beta} \cong {}'\rho_{\alpha\beta} = (\rho_{\alpha\beta}+\rho_{\beta\alpha})/2 ; \tag{5.3}$$

$$n^{1\lambda}{}_{,1} + n^{2\lambda}{}_{;2} + q^{\lambda} \cong 0 , \qquad n^{\alpha\beta} \cong {}'n^{\alpha\beta} = (n^{\alpha\beta}+n^{\beta\alpha})/2 . \tag{5.4}$$

***The assumption*** (5.2) ***justifies the interchange of the sequence of covariant surface differentiation.*** This interchange directly makes the expressions

$$\rho_{\alpha\beta} = W_{;\alpha\beta} , \qquad n^{\alpha\beta} = \varepsilon^{\alpha\lambda}\varepsilon^{\beta\mu} F_{;\lambda\mu} + P^{\alpha\beta} . \tag{5.5}$$

to a solution of the simplified equations of equilibrium and compatibility (5.3), (5.4). The $W(x^\alpha)$ and $F(x^\alpha)$ are three times continuously differentiable functions - the curvature function, first proposed by A. Libai [6], and the Airy function. The term $P^{\alpha\beta}$ denotes a particular solution of the equations (5.4).

***The system of equations for the functions*** $W$ ***and*** $F$, making these functions to a *general-solution* of the Donnell-type theory, follows with (5.5) from the remaining compatibility and equilibrium equations (2.13) and (3.5). The derivation of this system consists of three steps:

(a) The $\lambda_\beta$ and the $n^{\alpha 3}$ are expressed in terms of $E_{\alpha\beta}$ and $m^{\lambda\alpha}$, by means of (2.14) and (3.6). In turn, the $E_{\alpha\beta}$ and $m^{\lambda\alpha}$ are represented, respectively, in terms of $n^{\alpha\beta}$ and $\rho_{\alpha\beta}$ by means of the elasticity equations (4.8) and the relations $n^{\alpha\beta} \cong {}'n^{\alpha\beta}$, $\rho_{\alpha\beta} \cong {}'\rho_{\alpha\beta}$ from (5.4), (5.3). (b) Inserting (5.5) leads to :

$$Eh\varepsilon^{\alpha\beta}\lambda_\beta = -a^{\alpha\nu}(\nabla^2 F)_{;\nu} \qquad n^{\alpha 3} = Da^{\alpha\nu}(\nabla^2 W)_{;\nu} . \tag{5.6}$$

© With (5.6) and (5.5), the equations (2.13) and (3.5) render the system for the $F$ and $W$

$$\nabla^4 F + Eh\varepsilon^{\alpha\lambda}\varepsilon^{\beta\mu}(b_{\alpha\beta} + W_{;\alpha\beta}/2)\, W_{;\lambda\mu} = 0 , \tag{5.7}$$

$$D\nabla^4 W - \varepsilon^{\alpha\lambda}\varepsilon^{\beta\mu}(b_{\alpha\beta} + W_{;\alpha\beta})\, F_{;\lambda\mu} + q = 0 . \tag{5.8}$$

Here $\nabla^4 = \nabla^2\nabla^2$, with $\nabla^2$ denoting the two-dimensional Laplacian operator. The $P^{\alpha\beta}$ -terms are not written out for the sake of simplicity.

***Three questions are to be explored further*** : (1) Is the accuracy of the Donnell-type theory dependent solely on the *maximum* absolute values of the bending and membrane strain $(h\rho, \eta)$, or on the values of their specific components? (2) Can the basic hypotheses and the error estimates be stated in any terms, which allow to assess, whether the theory is adequate for a problem or not, *before* its solution has been obtained and evaluated? (3) Can the theory be characterized by any physical criterion, *besides* that of shallow shape or curvature restriction?

## 6. Accuracy of the Donnell-type theory

***The errors of the theory*** involve the terms dropped, first, to obtain the simplified equations of compatibility and equilibrium (5.3) and (5.4) and, second, to satisfy these equations and the remaining ones (2.13) and (3.5) by the solution (5.5).

The relative error $\delta_\lambda$ of the compatibility equations (5.3) is equal to the relation of the terms of (2.12) dropped in (5.3) to one of the terms retained there:

$$\delta_\lambda = |\varepsilon^{\alpha\beta} b_\alpha{}^\lambda \lambda_\beta| \,/\max|\,\varepsilon^{\mu\nu}\varepsilon^{\lambda\pi}\rho_{\alpha\delta;\,\beta}| \quad . \qquad (6.1)$$

The relative error $\Delta_\lambda$ of the equilibrium equations (5.4), has an estimate equal to the sum of terms of (3.4), dropped in (5.4), divided by one of the terms retained there. The formula for the $\Delta_\lambda$ can easily be written out, it is dual to the expression (6.1) of $\delta_\lambda$.

The application of the estimate (6.1) requires, besides the values of the strain resultants $\lambda_\beta$ and $\rho_{\alpha\delta}$, an assessment of the covariant derivatives $\rho_{\alpha\delta;\,\beta}$. Such derivatives with respect to *any* of the coordinates $x^\alpha$ are commonly estimated by means of the *minimum* interval of variation $L$ defined as in (4.2). Further, the factors $b_\gamma{}^\alpha$ are assessed by the absolute maximum $1/R$ of normal -section curvature $1/R_\alpha$. This approach [10] leads, for instance, to :

$$|\rho_{\alpha\lambda;\beta\gamma}| \sim |\rho_{\alpha\lambda}|/L^2 \;, \qquad |b_\gamma{}^\alpha E_{\alpha\lambda;\beta}| \sim |E_{\alpha\lambda}|/(LR)\,. \qquad (6.2)$$

***Closer estimates*** than (6.1) can be obtained, when the terms with $b_1{}^2$, $b_2{}^1$ may be dispensed with - when the $x^\alpha$ -lines are the *curvature lines*, or follow these lines approximately. For such coordinates the relative magnitude of the terms dropped in the full equations (2.12 ) and (3.4) to obtain (5.3) and (5.4) can (generalizing on [18]) be estimated by

$$\delta_\lambda \sim \max|(E_{\alpha\beta}/\rho_{\alpha\beta} h)(h/R_\gamma)| \qquad (\gamma\neq\beta), \qquad (6.3)$$

$$\Delta_\lambda \sim \max|(m^{\alpha\beta}/n^{\alpha\beta})/R_\beta|\,, \qquad (6.4)$$

where $R_\gamma$ and $R_\beta$ denote the principal radii of curvature of the reference surface.

The estimates (6.3) and (6.4) indicate an answer to the question (1) of the Sect.5. - The error of the Donnell-type theory depends *not* on the *maximum* absolute values of wall-bending strain, middle-surface extension and curvature ( $h\rho/2$ , $\eta$ and $1/R$). It is determined by the relations of *specific* components: $E_{\alpha\beta}/\rho_{\alpha\beta}$, $m^{\alpha\beta}/n^{\alpha\beta}$ and on $1/R_\alpha$.
- The components of strain and of stress in the shell volume have to be of *mixed* nature. The part of a component (e.g., of $\gamma_{\alpha\beta}$ ), which corresponds to membrane resultants, and its part determined by bending and torsion, must not differ too much in their magnitudes. Thus, the first basic assumption of the theory has to be stated more specifically than in (5.1). The actual approximation of the theory depends on the

restriction of relations of bending and membrane parts of the *same* component. - The validity of the theory depends on the conditions for the $\delta_\lambda$, $\Delta_\lambda$ defined in (6.3), (6.4):

$$\delta_\lambda\,,\;\; \Delta_\lambda \;<< 1 \qquad , \qquad\qquad (6.5)$$

However, the error estimates according to (6.3), (6.4) have a serious drawback. - They depend on the values of the stress and strain resultants. This means, the applicability of the theory and its accuracy, can be checked for a problem only, after a specific solution of this problem has been obtained and the resultants fully evaluated. The way to avoid, or at least to minimize, this requirement and also to obtain the clarification of the above questions (2) and (3) is suggested by the general solution (5.5).

***The error estimates not directly depending on the resultants of stress and strain*** will be obtained by eliminating all the resultants ($E_{\alpha\beta}$, $\rho_{\alpha\beta}$, $m^{\alpha\beta}$, $n^{\alpha\beta}$) from the formulas (6.3), (6.4) of $\delta_\lambda$ and $\Delta_\lambda$. This elimination hinges on the use of the general solution $W$, $F$; it consists of two steps. First, the estimates $\delta_\lambda$ and $\Delta_\lambda$-will be, expressed in terms of the W and F. Second, the dependence of the estimates on the functions W, F will be represented by the dependence on the intensity of variation ($1/L_\alpha$) of the stress state.

Both steps entail an assessment of the *derivatives* of $W$ and $F$ with respect to coordinates $x^\alpha$. It is indispensable, thereby, to take into account that the stress state can vary with the coordinates $x^1$ and $x^2$ with *different* intensities. - The derivative with respect to a coordinate $x^\alpha$ has to be estimated not by the $L = \min L_\alpha$, as in the standard relations (6.2), but by the interval of variation $L_\alpha$ with respect to the *specific* coordinate $x^\alpha$. In the following the *covariant* derivatives will be estimated by $L_\alpha$ - according to the formula:

$$|\, f_{\beta\delta;\alpha}| \sim |f_{\beta\delta}|\, a_\alpha / L_\alpha \,. \qquad\qquad (6.7)$$

With this, the stress and strain resultants in the estimates (6.3), (6.4) of $\delta_\lambda$, $\Delta_\lambda$ can be assessed in terms of W, F - not in terms of their derivatives.

To eliminate from the estimates $\delta_\lambda$, $\Delta_\lambda$ the functions $F$ and $W$, it remains to estimate the relation between them. This is done with the help of the equations (5.7) and (5.8). These provide relations between the $\nabla^4 F$ or $\nabla^4 W$ and the $W$ or , respectively, $F$ in the second term of the relevant equation. As $\nabla^4 F$ and $\nabla^4 W$ are invariants, the second terms of (5.7) and (5.8) must also be invariants. - They may be estimated for lines-of-curvature coordinates $x^\alpha$. The relevant order-of-magnitude estimates, following with (6.7) from (5.7) and (5.8), respectively, are:

$$|\,\nabla^4 F| \sim Eh\,|W|\,/\,|\,R_\alpha L_\beta^2\,|_{\min}\,, \quad D|\,\nabla^4 W| \sim |F|\,/\,|\,R_\alpha L_\beta^2\,|_{\min} \qquad (\alpha\neq\beta)\,. \quad (6.8)$$

Here $L_\beta$ denote the intervals of variation, defined by (6.7), for the case, when the $x^\beta$ -lines coinside with the curvature lines ; $|R_\alpha L_\beta^2|_{\min}$ is the smaller of the quantities $|R_1 L_2^2|$ and $|R_2 L_1^2|$.

The elimination of the stress and strain components from the error estimates (6.3), (6.4) can now be completed. - With $E_{\alpha\beta}$ and $n^{\alpha\beta}$ represented in in terms of $F$, the

$\rho_{\alpha\beta}$ and $m^{\alpha\beta}$ - in terms of $W$, the estimates (6.3), (6.4) depend on $|F/W|$ and $|W/F|$. This and (6.7), (6.8) results in the estimates of the errors $\Delta_\lambda$, $\delta_\lambda$, which depend only on the intensities of variation ($1/L_\alpha$) and on the local geometry ($R_\alpha$):

$$\Delta_\lambda \sim \delta_\lambda \sim (L_1 L_2)^2/(R_\lambda |R_\alpha L_\beta^2|_{\min}) \qquad (\alpha \neq \beta) \; . \tag{6.9}$$

A remarkable consequence of the duality - the estimate of the error in the equilibrium ($\Delta_\lambda$) and that in the compatibility ($\delta_\lambda$) have *equal* orders of magnitude.

***Another cause of inaccuracy*** in the Donnell-type theory is the approximation of the components $\rho_{\alpha\beta}$ and $n^{\alpha\beta}$ by the *symmetric* parts $'\rho_{\alpha\beta}$ and $'n^{\alpha\beta}$ of the respective tensors - the relations (5.3), (5.4). This amounts to an approximation in the equations (2.15) and (3.7). The estimates of the corresponding relative errors, determined similarly to (6.9), are:

$$|(\rho_{\alpha\beta}/'\rho_{\alpha\beta}) - 1| \sim |(n^{\alpha\beta}/'n^{\alpha\beta}) - 1| \sim L_\alpha L^3/(R\,|R_\alpha L_\beta^2|_{\min}) \; . \tag{6.10}$$

Finally, the error of the theory is caused by the interchange of the sequence of covariant surface differentiation, required to satisfy (5.3), (5.4) by (5.5). The estimate of this error can be found to be somewhat lower than in (5.2), where the intervals $L_\alpha \geq L$ are replaced by $L$. However, this error, just as the one assessed in (6.10), is *under* those ($\Delta_\lambda$, $\delta_\lambda$) of the other basic simplifications, estimated in (6.9). - These errors are of no significance for the overall accuracy of the Donnell-type theory.

The characteristic feature of the error estimates (6.9), (6.10) : they do not directly depend on stress and strain variables. - The accuracy of the theory and its adequacy for a problem can in certain cases be assessed even before a solution for the problem has been obtained and evaluated (cf. an example in §7).

## 7. Physical significance of the assumptions and the domain of the theory.

***The physical meaning of the assumptions shaping the theory*** is made more transparent by the simpler estimates, which follow from (6.9). - After a rather direct analysis these error estimates can be reduced to a telling form:

$$\Delta_\alpha \sim \delta_\alpha \sim L_\alpha^2/R_\alpha^2 \; . \tag{7.1}$$

Hence, in what concerns the accuracy of the Donnell-type theory, any stress state and local shape are characterized by merely two dimensionless parameters - by the relations of the intervals of variation $L_\alpha$ along the lines-of-curvature $x^\alpha$ to the corresponding principal curvature radiuses $R_\alpha$.

The above estimates show the theory to be adequate for stress states which vary with both coordinates $x^\alpha$ much more intensely than the unit normal vector $\boldsymbol{n}$. Indeed, the $1/L_\alpha$ is the intensity of variation of the stress state, the curvature $1/R_\alpha$ represent the intensity of variation of the $\boldsymbol{n}$.

It has to be noted, finally, that the above estimates rate *partial* errors of specific equations, which are *dual* one to the other. - These errors may, to some extent, compensate each other.

***A simplified theory is adequate*** to the general one, when the error of its additional assumptions does not exceed the error of the thin-shell theory, displayed in (4.1), that is, when $\Delta_\alpha \sim \delta_\alpha \leq h/R$. Thus, the estimate (7.1) defines problems, for which the Donnell-type theory is adequate, by the conditions

$$L_\alpha / R_\alpha \leq \sqrt{(h/R)} \tag{7.2}$$

In other words, the theory does not involve any additional inaccuracy by treating stress states which vary along the reference surface at least $\sqrt{(R/h)}$ - times more intensely, than the unit normal vector $\boldsymbol{n}$ (which indicates the direction of the tangent plane). It is a theory specialized for the strongly variable stress states [18].

The relations (7.1), (7.2) make it quite plain: the applicability of the theory of 'quasi-shallow' shells depends, not alone on the shell shape, however shallow it may be.

***An example***. To demonstrate the reach of the theory and the use of its error-estimate, consider a most simple limit case, discussed by L.H.Donnell himself [3].

Consider an infinitely long cylinder ($1/R_1=0$, $K=0$) with closed circular cross section, buckling under radial pressure. As easily perceivable, the buckling mode is

$$W = C\sin(x_2/L_2), \quad L_2 = R_2/n\,, \quad n = 2, 3, \dots .$$

The estimate (7.1) of the error is for this mode $L_2^2/R_2^2 = 1/2^2 = 25\%$.

The classical formula gives the critical pressure: $p = 3D/R_2^3$ [16]. The Donnell-type *theory* renders (e.g., [16]) $p = 4D/R^3$, which is 33% above the correct $p$. - As predicted by estimate (7.1), this case is, out of reach of the theory.

## References


1. Aron, H.: Das Gleichgewicht und die Bewegung einer unendlich dünnen, beliebig gekrümmten elastischen Schale. *Journ. f. reine u. angew. Mathem.* **78**(1874), 136-173.
2. Reissner, H.: Spannungen in Kugelschalen. *Festschrift H.Müller-Breslau.* Leipzig,1912.
3. Donnell, L.H.: A new theory for the buckling of thin cylinders under axial compression and bending Trans. ASME, **56** (1934).
4. Lur'e, A.I. General theory of thin elastic shells. (In Russian.) *Prikl. Mat. Mekh.* (*PMM*) **4** (1940), 7-34.
5. Koiter, W.T.: A consistent first approximation in the general theory of thin elastic shells. *Proc. IUTAM Symp.*, Delft. North-Holland, Amsterdam,1960.
6. Libai, A.: On the nonlinear elastokinetics of shells and beams. *J. Aerosp. Sci.* **29** (1962), 1190-1195.
7. Sanders, J.L.: Nonlinear theories for thin shells, *Quart. Appl. Math.***21**(1963), 21-36.
8. Budiansky, B. and Sanders, J.L.: On the "best" first-order linear shell theory. *Progress in Applied Mechanics, the Prager Anniv. Volume*, pp.129-140, New York, 1963.

9. Axelrad, E.L.: Refinement of buckling-load analysis for tube flexure by way of considering precritical deformation (in Russian). *Izv. AN SSSR, Mekhanika* **4**(1965), 133-139.

10. Koiter, W.T. : On the nonlinear theory of thin elastic shells. *Proc. Kon. Neth. Akad. Wet.* B **69** (1966),1-54.

11. Axelrad, E.L.: On different definitions of measures of shell-curvature change and on compatibility equations. *Mech. of Solids* (translated) **2** (1967), 105-108.

12. Koiter, W.T. and Simmonds, J.G.: Foundations of shell theory. *Proc.XIII Congr. on Theoret. and Appl. Mech.* (E.Becker, G.K.Mikhailov, eds.). Springer, 1973, 150-176.

13. Axelrad, E.L.: Flexible shells. In: *XV Intern. Congr. of Theoret. and Appl. Mech.* (F.P.J.Rimrott and B.Tabarrok, eds.) Amsterdam, North-Holland, 1980, pp.45-56

14. Axelrad, E.L.: On vector description of arbitrary deformation of shells. *Int. J. Solids and Struct.*, **17** (1981), 301-304.

15. Libai A.and Simmonds, J.G.: *Nonlinear Elastic Shell Theory.* Acad.Press, San Diego, 1983.

16. Axelrad, E.L.: Schalentheorie. Teubner, Stuttgart, 1983

17. Simmonds, J.G. : The non-linear thermodynamical theory of shells. In *Flexible Shells* (E.L.Axelrad and F.A.Emmerling, eds.). Springer, Berlin, 1984.

18. Axelrad, E.L. : *Theory of Flexible Shells.* North-Holland, Amsterdam, 1987.

19. Axelrad, E.L. and Emmerling, F.A.: On variational principles and consistency of elasticity relations of thin shells. *Int. J. Non-Linear Mech.* **25** (1990) 27-44.

20. Reissner, E.: *Selected Works in Applied Mechanics and Mathematics*, Jones and Bartlett, Boston, 1996.

21. Simmonds, J.G.: Some comments on the status of the shell theory at the end of the 20th century: complaints and correctives. *Amer. Inst. of Aer. and Astron.* 1997.

22. Libai, A. and Simmonds, J.G. : *The Nonlinear Theory of Elastic Shells.* 2nd ed., Cambridge University Press.Cambridge, 1998.

23. Librescu, L. and Hause, T.: Recent developements in the modeling and behaviour of advanced sandwich constructions: a survey. *Composite Structures,* **48** (2000) 1-17

24. Axelrad, E.L.: Shell theory and its specialized branches, *Int. J. Solids and Struct* (in press).

# EXPERIMENTS FOR MEASURING INTERFACE FRACTURE PROPERTIES

LESLIE BANKS-SILLS
*The Dreszer Fracture Mechanics Laboratory*
*Department of Solid Mechanics, Materials and Structures*
*The Fleischman Faculty of Engineering*
*Tel Aviv University*
*69978 Ramat Aviv, Israel*

Abstract

In this investigation, experiments carried out to measure interface fracture properties of joined materials are reviewed. In particular, results from tests carried out by means of the Brazilian disk specimen and two material pairs are presented. An energy based fracture criterion is exploited which agrees well with the experimental results. Scatter in test results for interface toughness is discussed.

## 1. Introduction

There have been many studies investigating the fracture properties of a crack along an interface between two materials. Two approaches have been taken. In the majority of studies, two materials are bonded by an interlayer. There have been few studies in which the two materials are joined without an interlayer. Investigations in the literature will be described in Section 2.

For completeness, relevant concepts related to interface fracture are presented. These may be found in other sources, as well. In two dimensions and referring to Fig. 1, the in-plane stresses in the neighborhood of a crack tip at an interface are given by

$$\sigma_{\alpha\beta} = \frac{1}{\sqrt{2\pi r}} \left[ \mathrm{Re}(Kr^{i\epsilon})\, \Sigma^{(1)}_{\alpha\beta}(\theta,\epsilon) \;+\; \mathrm{Im}(Kr^{i\epsilon})\, \Sigma^{(2)}_{\alpha\beta}(\theta,\epsilon) \right] \tag{1}$$

where $\alpha, \beta = x, y$, $i = \sqrt{-1}$, the complex stress intensity factor

$$K = K_1 + iK_2 \tag{2}$$



*D. Durban et al. (eds.), Advances in the Mechanics of Plates and Shells,* 49–66.

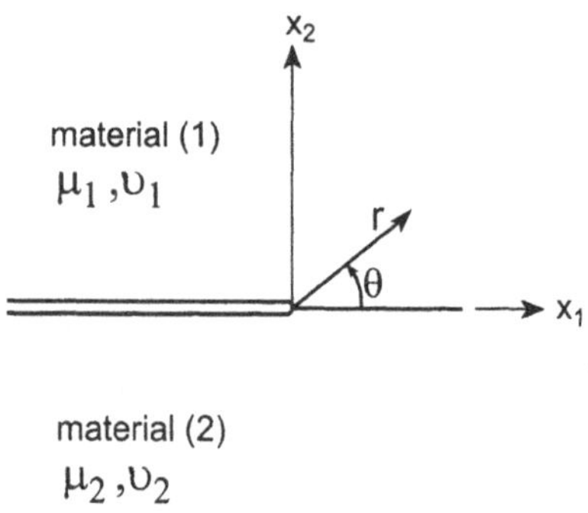


*Figure 1.* Crack tip coordinates.

and the oscillatory parameter

$$\epsilon = \frac{1}{2\pi} \ln \left( \frac{\kappa_1 \mu_2 + \mu_1}{\kappa_2 \mu_1 + \mu_2} \right) . \tag{3}$$

In (3), $\mu_i$ are the shear moduli of the upper and lower materials, respectively, $\kappa_i = 3 - 4\nu_i$ for plane strain and $(3 - \nu_i)/(1 + \nu_i)$ for generalized plane stress, and $\nu_i$ are Poisson's ratio. The stress functions $\Sigma^{(1)}_{\alpha\beta}$ and $\Sigma^{(2)}_{\alpha\beta}$ are given in polar coordinates by Rice, *et al.* (1990) and in Cartesian coordinates by Deng (1993).

The complex stress intensity factor in (2) may be written in non-dimensional form as

$$\tilde{K} = \frac{K L^{i\epsilon}}{\sigma \sqrt{\pi L}} \tag{4}$$

where $L$ is an arbitrary length parameter and $\sigma$ is the applied stress. The non-dimensional complex stress intensity factor may also be expressed as

$$\tilde{K} = \left| \tilde{K} \right| e^{i\psi} \tag{5}$$

so that the phase angle

$$\psi = \arctan \left[ \frac{\mathrm{Im}(\mathrm{K\,L}^{i\epsilon})}{\mathrm{Re}(\mathrm{K\,L}^{i\epsilon})} \right] = \arctan \left[ \frac{\sigma_{12}}{\sigma_{22}} \right] \Bigg|_{\theta=0, r=L} . \tag{6}$$

The interface energy release rate $\mathcal{G}_i$ is related to the stress intensity factors by

$$\mathcal{G}_i = \frac{1}{H} (K_1^2 + K_2^2) \tag{7}$$

where

$$\frac{1}{H} = \frac{1/\bar{E}_1 + 1/\bar{E}_2}{2\cosh^2 \pi\epsilon} \tag{8}$$

$\bar{E}_i = E_i/(1-\nu_i^2)$ for plane strain conditions and $E_i$ for generalized plane stress. Note that the subscript $i$ in (7) represents interface and $\mathcal{G}_i$ has units of force per length.

Inherently for any interface both $K_1$ and $K_2$ must be prescribed or equivalently $\mathcal{G}_i$ and $\psi$. In describing an interface crack propagation criterion, one may prescribe a relation between $K_1$ and $K_2$ or what is commonly done, the critical energy release rate $\mathcal{G}_{ic}$ is given as a function of the phase angle $\psi$.

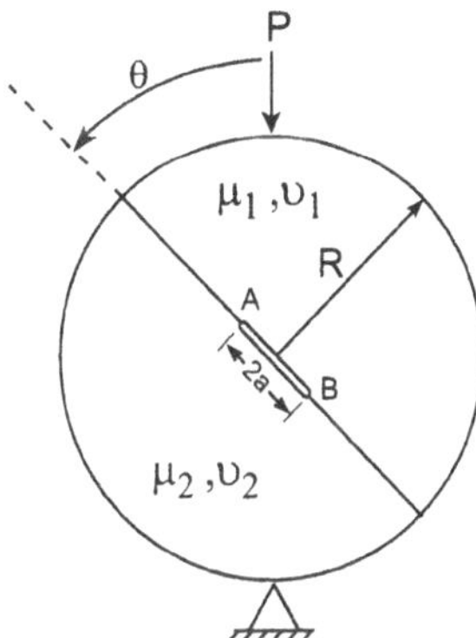


*Figure 2.* Bimaterial Brazilian disk specimen.

In Section 3, test results are presented which were obtained by means of the bimaterial Brazilian disk specimen (see Fig. 2). The Brazilian disk specimen shown in Fig. 2 was chosen for measuring the interface fracture toughness $\mathcal{G}_{ic}$ because it leads to a wide range of mixed mode values. Two material pairs were selected for these tests: (a) glass and epoxy and (b) two ceramic clays (K-142 and K-144). In each case, a mathematical interface is achieved since there is no apparent interlayer between the materials. An energy based fracture criterion is employed and shown to describe well the experimental results. The scatter in the experimental results is examined.

## 2. Experimental Results

In this section, experimental techniques and results presented in the literature will be considered. The majority of experiments have been carried out on sandwich specimens. A smaller number of tests have been performed on bimaterial specimens without an interlayer.

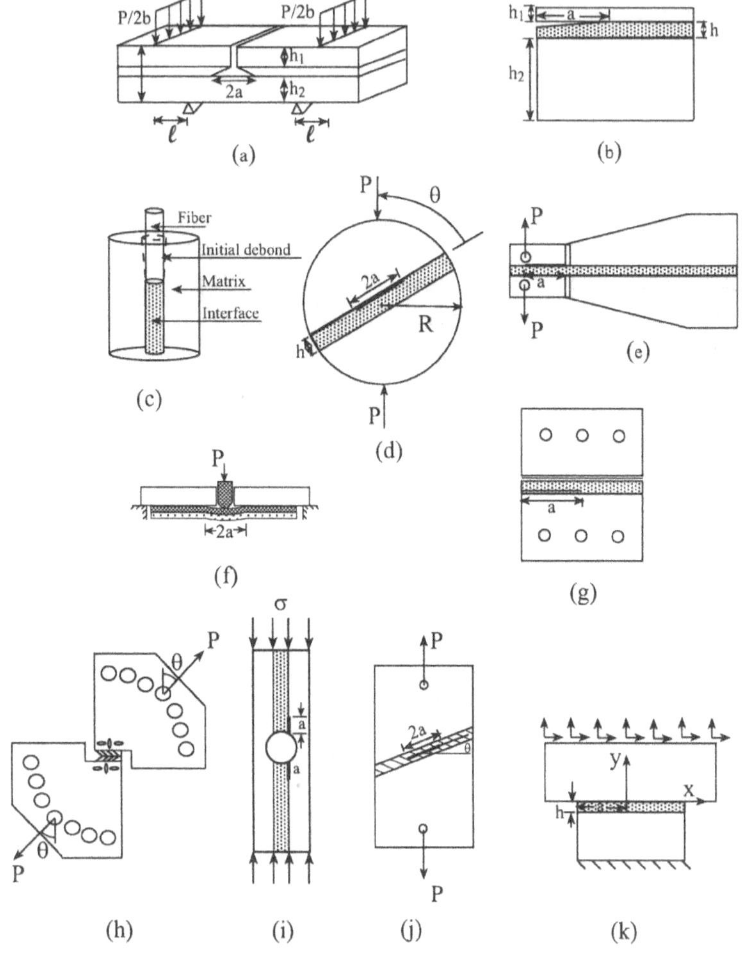


*Figure 3.* Specimens containing bonds employed to measure interface fracture toughness.

## 2.1. SANDWICH SPECIMENS

The specimens exhibited in Fig. 3 are those for which two materials have been joined by means of an interlayer of another material. In some cases the adherends are of the same material; for others, they are of different materials.

Charalambides, *et al.* (1989) presented a notched four point bend specimen containing symmetric interface cracks (the so-called Santa Barbara specimen, see Fig. 3a). A strip of aluminum and PMMA were bonded together using epoxy. Although the epoxy interlayer should not have been neglected, the experiments were considered to measure interface toughness between the aluminum and PMMA. The specimen (without the epoxy interlayer) was analyzed by means of the finite element method. For a constant ratio of $h_1/h_2$, the energy release rate, non-dimensionalized complex stress intensity factor and phase angle are constant for a wide range of crack lengths. Thus, this specimen produces a steady state value for $\mathcal{G}_i$. The phase angle is not very sensitive to variation of the thickness ratio $h_1/h_2$. Tests were carried out with $1.19 \leq \psi \leq 1.29$. It was concluded that testing of other specimens is necessary to obtain a wider range of mode mixity. It should be noted that the epoxy layer does not greatly influence the value of the interface energy release rate, but it does change the phase angle which was calculated by means of the finite element method. For example, for the geometries considered, an interlayer thickness of 25 $\mu$m, changes $\psi$ by about $5^\circ$ (Suo and Hutchinson, 1989).

In a study by Cao and Evans (1989), the three specimens illustrated in Figs. 3a through 3c were employed to measure the critical interface energy release rate $\mathcal{G}_{ic}$ as a function of the phase angle $\psi$. Approximate analytic expressions for the interface energy release rate were developed for each specimen. Whereas the phase angle $\psi$ was obtained for long crack lengths by means of the finite element method. The flexure and asymmetric specimens in Figs. 3a and 3b, respectively, were made from aluminum glued to glass. The axisymmetric fiber/matrix specimen was fabricated from a glass fiber glued to a glass matrix. The glue was a thermoplastic adhesive. Results were given for the crack propagating between the glass and glue and between the aluminum and glue. It was seen that the critical interface energy release rate as a function of phase angle was much greater for the glass/adhesive interface than the aluminum adhesive interface.

Wang and Suo (1990) employed a Brazilian disk sandwich specimen illustrated in Fig. 3d composed of aluminum, brass, steel and perspex substrates bonded by an epoxy adhesive. It was hypothesized that residual stresses in sandwich specimens do not influence the measured toughness values unless they are excessively high. This is something which should be examined further. For a sandwich specimen in which the layer is thin compared to other in-plane geometric length scales, the interface energy release rate is that of the same homogeneous specimen. The phase angle, however, changes as

$$\psi = \arctan\left(\frac{K_{\mathrm{II}}}{K_{\mathrm{I}}}\right) + \omega + \epsilon\left(\frac{L}{h}\right) \tag{9}$$

where $K_{\mathrm{I}}$ and $K_{\mathrm{II}}$ are, respectively, the mode I and II stress intensity factors of the homogeneous specimen, $\omega$ is tabulated by Suo and Hutchinson (1989) and depends on the interlayer/substrate material mismatch, $\epsilon$ is the material parameter given in (3), $L$ is the length parameter associated with

$\psi$ and $h$ is interlayer thickness. All systems exhibited the typical increase in critical interface energy release rate $\mathcal{G}_{ic}$ with increasing phase angle $\psi$. There was much scatter in the results. It was observed that large phase angles caused the epoxy to debond from the upper adherend. When the remote loading was predominately mode I ($\theta = 0$), the crack tended to kink out of the interface and propagate cohesively (within the interlayer).

Reimanis, *et al.* (1991) studied the micromechanical behavior of a metal/ceramic interface in which a gold foil was bonded between two single crystal alumina (sapphire) layers. The gold foil of thickness between 10 to 100 $\mu$m was diffusion bonded to the alumina creating a mathematical interface between the gold and sapphire. The flexure specimen in Fig. 3a was employed with a mode mixity $\psi \sim 0.79$. Tests were performed in a dry nitrogen environment. The critical interface energy release rate, as well as the resistance $\mathcal{G}_{iR}$ were determined as a function of gold interlayer thickness. The behavior of the gold interlayer was carefully examined. The mechanism of fracture was plastic void growth from pre-existing interface pores accompanied by brittle interface debonding. Bridging ligaments account for the increase of resistance as the crack propagates. Crack propagation was along one of the gold/sapphire interfaces.

In 1991, Kinloch, *et al.* carried out tests on three specimen types. These included a single edge notched (SEN) specimen fabricated from an epoxy adhesive, a bimaterial single edge notch (SBM) specimen fabricated from epoxy and aluminum (see Fig. 4a) and a tapered double cantilever beam (TDCB) specimen (see Fig. 3e) in which two aluminum substrates are glued by a 0.5 mm thick epoxy interlayer. Critical energy release rate measurements were made for the epoxy, post cured at five temperatures. The results showed that the critical energy release rate was essentially the same for the SEN specimen made only from epoxy, the TDCB specimen (the crack grew cohesively–within the epoxy layer) and the SBM specimen when the crack grew out of the interface and into the epoxy. In all three cases, it appears that the bulk fracture toughness of the epoxy was measured. It may be noted that the crack was growing by means of mode I deformation in the TDCB specimen. The residual stresses in the SBM specimens were observed by photoelastic means. They were seen to increase with post-cure temperature. Approximate values for the energy release rate due to residual stresses were calculated and superposed with those values due to applied force. The phase angle $\psi$ was not determined. It was seen that the residual stresses greatly affect the critical interface energy release rate values obtained with the SBM specimens.

Akisanya and Fleck (1992) investigated the effect of applied mode mixity $\phi = \arctan(K_{\mathrm{II}}^{\infty}/K_{\mathrm{I}}^{\infty})$ on crack path within a brittle epoxy layer between two aluminum alloy substrates. Two specimens were employed in the testing: a symmetric ($h_1 = h_2$) and asymmetric double cantilever beam (DCB) specimen illustrated in Fig. 3b and a Brazilian disk specimen illustrated in Fig. 3d. For the DCB specimens $\phi$ varied between -0.42 and 0.42 radians; for the Brazilian disk specimens $0.4 < \phi \leq 1.34$. For both specimens, the starter crack was placed along the upper interface of the joint. For the DCB specimens, the interfacial mode of growth is the most frequently encountered for small $\phi$ and becomes the only mode observed at large absolute values of applied mode mixity. The local mode mixity for the crack tip which propagated in the Brazilian disk specimens varied from $-0.05 \leq \psi \leq 1.52$. The crack was interfacial when $0 \leq \psi \leq 0.35$. For the DCB specimens, three joint thicknesses (0.2, 0.4 and 0.6 mm) were studied. For these specimens, four different cure regimes were employed inducing different residual stress levels into the epoxy layer. It was concluded in this study, that the residual stresses did not affect the toughness results for interface crack propagation. For both specimen types, the interface toughness increased with increasing $|\psi|$. For serrated or alternating propagation, the toughness results increased slightly with increasing residual stresses. The probability of serrated crack growth increased with increased layer thickness. Cohesive propagation occurred when the remote mode II stress intensity factor was small. In this case, the measured toughness was comparable to that of the bulk toughness. This crack path did not seem to be affected by layer thickness or residual stress level. The local cracking pattern of the adhesive joint greatly affects the measured toughness. For the serrated and alternating crack paths, the toughness is more than twice that of the interfacial toughness. Duer, *et al.* (1996) raised some doubts about the calibration equations for $\mathcal{G}_i$ and the phase angle $\psi$ which the authors obtained from Bao, *et al.* (1992) and Suo and Hutchinson (1989) for the DCB specimens. On careful reading, it appears that Akisanya and Fleck did interpret these formulas correctly and that their results are sound. The usual behavior of $\mathcal{G}_{ic}$ increasing with increase of $|\psi|$ is observed. The results presented by Duer, *et al.* (1996) exhibit no dependence on phase angle.

Liechti and Liang (1992) employed the strip blister specimen exhibited in Fig. 3f for two types of tests. In one there is an interlayer of epoxy between an upper layer of glass and a lower layer of aluminum. In the second, the lower layer is comprised of epoxy only (see Fig. 4c). The surface preparation of the adherends was such that debonding occurred between the glass and epoxy in all cases. Curing took place at room temperature for seven days in order to minimize residual stresses. The interface energy release rate $\mathcal{G}_i$ was determined from beam theory. Finite element studies with an $M$-integral post processor were carried out to obtain the phase angle $\psi$. The energy release rate was also obtained and compared well with the expression from beam theory. For the sandwich specimen, the phase angle remained nearly constant as crack length increased, although it increased as the thickness of the epoxy interlayer decreased (from about -0.70 radians to -0.64 radians). The values of $\mathcal{G}_{ic}$ decreased for thinner epoxy interlayers. The results showed a much more brittle behavior. For the two thickest layers of 0.61 mm and 0.356 mm, an $R$-curve behavior was observed. Behavior of the plastic zones may qualitatively explain the increase of toughness with increasing interlayer thickness and crack length for the thicker layers. It may be that the aluminum layer is disturbing the assumed asymptotic interfacial behavior between the glass and epoxy for the thinner interlayers. One of the conclusions here is that different sandwich specimens may produce different results. But they yield very different values as compared to bimaterial specimens, so that one must be careful in using data from these tests. Finally as mentioned by the authors, for design purposes, it would appear that the sandwich layer thickness in a testing environment should match that of the engineering application.

Thurston and Zehnder (1993) tested silica/copper sandwich specimens exhibited in Fig. 3g by means of the loading frame shown in Fig. 3h. A 0.127 mm layer of copper foil was hot pressed between two pieces of fused silica. The crack was interfacial between the silica and copper. Crack propagation was unstable in all cases. For positive phase angles, the crack kinked into the ceramic. For moderately negative phase angles, the high toughness of the metal tended to promote interface fracture. For large negative phase angles, fracture was observed to occur in the ceramic before the interfacial toughness was reached. Since an interlayer was considered here, the interface energy release rate was obtained from that of a cracked, homogeneous specimen of the same geometry and loading. The phase angle was determined from the relation between global and local stress intensity factors (see Suo and Hutchinson, 1989). In addition, a plastic analysis was presented, to account for the plasticity of the copper layer. In the tests, small scale yielding occurred for 12 out of 20 specimens. Critical interface energy release rate values were plotted versus the plastic phase angle. It was observed that $\mathcal{G}_{ic}$ increased with increase of the absolute value of phase angle. This phase angle varied between -0.31 and 0.66 radians. The difference between this phase angle and $\psi$ for $L$ equal to the layer thickness is no more that 0.05 radians. Much scatter in the results is observed. Of course, there were two types of fracture behavior: interface and kinking into the ceramic.

For the same specimen geometry and loading (Figs. 3g and 3h) in which nickel foil was hot pressed between alumina adherends, Thurston and Zehnder (1996) carried out tests. The effect of residual stresses was considered here. Since there is an order of magnitude difference between the thermal expansion coefficients of the two materials, residual stresses in the interlayer will result. Numerical analyses were carried out which showed that residual stresses exist which change the $\mathcal{G}_{ic}$ value by as much as 20% and the phase angle by as much as 40%. There is scatter in the test data. A maximum hoop stress criterion matches the data fairly well. As expected the toughness increases with increasing mode mixity. In this paper the authors demonstrate an important point. The length parameter $L$ in (6) should be chosen within the $K$-dominant region when employing test data for failure prediction.

Wang (1995) employed the Brazilian disk sandwich specimen exhibited in Fig. 3d; the interlayer was taken to be copper and the semi-circular adherends were alumina. A non-planar notch was induced by means of a graphite layer about 20 nm thick. The non-planar notch enabled both cracks to have the same mixity. Cracks grew along the interface between the copper and alumina. Specimen calibration (i.e. determination of the interface energy release rate and phase angle) was carried out as in Wang and Suo (1990). The interface fracture toughness is seen to increase as $|\psi|$ increases. There were few experimental results; had there been more, it appears that there would be large scatter. It was observed that there was no large scale plastic deformation of the copper. It was assumed that residual stresses are unimportant for this testing situation. Wang does note that residual stresses can have an effect on the $K$ solutions which in turn will influence the toughness results. Plasticity and surface roughness were assumed to cause the increase of $\mathcal{G}_{ic}$ with $|\psi|$.

Test were carried out by Turner, *et al.* (1995) on the specimen in Fig. 3i. It was composed of either glass adherends bonded by a thermoset resin or sapphire diffusion bonded with a 25 $\mu$m thick layer of platinum or a 10 $\mu$m thick layer of gold. The phase angle for all tests was between about 0.02 and 0.09 rad. For the glass thermoset, $\mathcal{G}_{ic}$ was found as $24 \pm 4$ N/m, whereas the critical energy release rate of the glass is about 8 N/m. For the sapphire/gold specimens $\mathcal{G}_{ic} = 10 \pm 2$ N/m and for the sapphire/platinum, it is found to be about 52 N/m. The toughness of sapphire is between 10 and 20 N/m. The specimen employed allows measurement of fracture toughness for interfaces which are tougher than the adherends.

Experiments were conducted on scarf joints (see Fig. 3j) by Wang (1997) in which a brittle epoxy was used to bond two steel adherends. The bond was formed at six different angles between zero and $\pi/2$. The thickness of the adhesive layer was between 0.15 and 0.2 mm. An artificial crack was formed by a thin piece of Teflon tape. It was observed that the crack ran within the adhesive. It is rather unusual for such a wide range of mixity. Perhaps the T-stress term is negative for this specimen and all loading angles (see Fleck, *et al*, 1991). A fracture criterion was considered for the case of a crack within an interlayer.

Another sandwich specimen (see Fig. 3k) was employed by Swadener and Liechti (1998). Here the upper material was glass and the lower material aluminum with a thin epoxy bond between them. The bond thickness was between 0.13 and 0.40 mm. The residual stresses in the epoxy were observed by a polariscope to be very small. A combination of displacements in the horizontal and vertical directions was applied to the specimen so that the phase angle ranged between -0.87 and 1.48 radians. An analytical expression for the interface energy release rate was found. For the phase angle, finite element analyses were carried out. Finite element analyses were employed to determine the plastic work. It was observed that this contribution to the critical interface toughness is asymmetric.

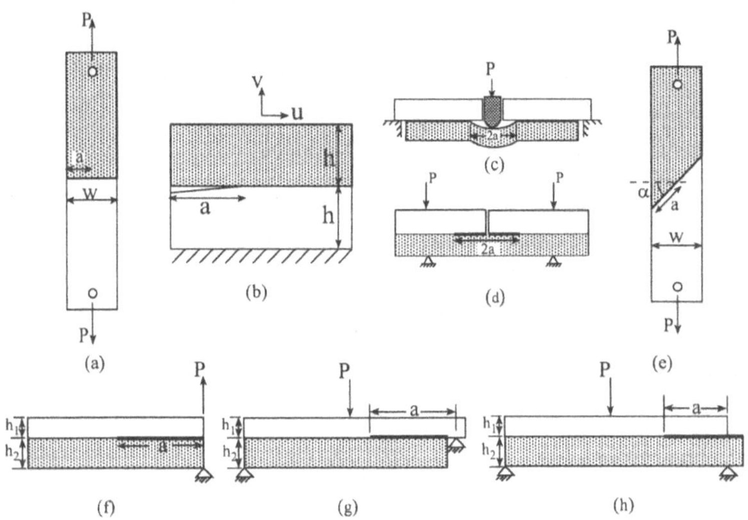


*Figure 4.* Bimaterial specimens employed to measure interface fracture toughness.

Another approach has been taken in the literature in order to study interface fracture. In this case, two materials are bonded together without an interlayer. As mentioned earlier, Kinloch, *et al.* (1991) employed the specimen illustrated in Fig. 4a to measure interface toughness between an aluminum alloy and epoxy. They considered the effect of residual stresses by post-curing the specimens at temperatures of 20°, 45°, 80°, 120° and 150° C. For the lowest temperature, the crack immediately diverted into the epoxy, and the measured fracture toughness is in close agreement with that reported earlier for bulk and cohesive crack growth in a specimen with an epoxy interlayer. For the other temperatures, the crack propagated along the interlayer and the critical interface energy release rate was much lower. Photoelastic studies showed an increase in residual stress with increase of post-cure temperature. The effect upon $\mathcal{G}_i$ of the residual stresses was calculated. When superposed with the interface energy release rate resulting from the applied load, this critical value was not a function of crack length for the two highest curing temperatures. The dependence of $\mathcal{G}_{ic}$ upon the phase angle $\psi$ was not noted.

In two studies by Liechti and Chai (1991, 1992), experiments on glass/epoxy specimens illustrated in Fig. 4b were reported. Displacements in the horizontal and vertical directions are imposed to produce a wide range of mode mixity. The specimens were cured at room temperature. Photoelastic observation showed no evidence of residual stresses. The energy release rate and mode mixities were obtained by means of finite element analyses. Tests were carried out with $-1.05 < \psi < 1.57$. The results were obtained from four specimens and showed little scatter indicating consistent adhesion between specimens. The critical interface energy release rate increased from about 3 N/m to more than 35 N/m as $|\psi|$ increased. The plastic zone within the epoxy was determined by means of finite element analyses and seen to be within the realm of small scale yielding. Larger plastic zones were observed as $|\psi|$ increased. An estimate of the contribution of plasticity and viscoelasticity effects to the fracture toughness increased with increasing $|\psi|$, but did not quantitatively account for the increase in toughness. The surface roughness model of Evans and Hutchinson (1989) did not seem to correlate with the toughness data.

Liechti and Liang (1992) employed the bimaterial strip blister specimen in Fig. 4c in which the upper material was glass and the lower was epoxy. Details of specimen preparation and calibration are described earlier. In these tests, the phase angle changed from about -1.06 to -0.93 radians, remaining constant as crack length increased. The critical interface energy release rate values $\mathcal{G}_{ic}$ increased with crack length demonstrating an $R$-curve behavior. The authors could not seem to attribute it to plasticity. It seems to occur with this type of specimen as also observed for a circular blister specimen employed by Liechti and Hanson (1988), as well as the sandwich specimens also tested here.

Cazzato and Faber (1997) employed the specimen shown in Fig. 4d to determine the toughness of a glass/alumina interface. With the specimen geometry chosen, the phase angle varied between 0.77 and 0.84 radians. The glass was melted onto the alumina to create a direct bond between the materials. The thermal mismatch between the glass and alumina is sufficiently low so as to prevent failure of the specimens during processing. Precracks grew from a machined notch in the top of the glass layer. Five types of alumina were employed to examine the effect of alumina purity on interface toughness. Five different surface roughnesses were induced on the ceramic layers. Testing took place in three different testing environments of nitrogen gas, liquid water and ambient air. Residual stresses were accounted for. For the lowest purity alumina, the crack did not propagate along the interface but crossed it, breaking the alumina. For the other alumina types, the interface fracture toughness increased with decreasing purity. There was much scatter in the results. It would appear that an alumina with a moderate amount of glassy phase is the ideal for enhancing debonding. Interface roughness did not seem to affect the toughness values. It did however, cause the crack to propagate near the interface within the glass. One type of alumina bimaterial specimen was tested in the different environments. It was found that interface toughness was lowest in water and highest in nitrogen. The liquid environment lowered the toughness by about 30% from the air environment. The limited phase angle range achieved here indicates the need for a specimen which allows for a wider range of this quantity.

The specimen exhibited in Fig. 4e was employed by Ikeda, *et al.* (1998) with the angle $\alpha$ taken as $0^{\circ}$, $\pm 45^{\circ}$ and $\pm 60^{\circ}$. The upper material was either aluminum or methacrylic resin and the lower material was epoxy. The amplitude of the stress intensity factor and the phase angle were presented, each as a function of crack length. For each material pair, graphs of $\hat{K}_1$ vs. $\hat{K}_2$ with a length parameter $L$ chosen were presented. It was observed that for the proper choice of $L$, the experimental values could be fit between two bounding ellipses.

Sundararaman and Sitaraman (1999) employed three bend type specimens exhibited in Figs. 4f through 4h to measure the interface fracture toughness between aluminum and epoxy. The application is for electronic packaging. It was assumed that residual stresses are negligible. For the specimen in Fig. 4f, the critical interface energy release rate was related to the change in energy obtained for two different crack lengths. The phase angle value for tests with several crack lengths was found from an analytic/numerically based expression as 0.41 radians. For the specimens in Figs. 4g and 4h, a compliance calibration method was employed to relate the critical interface energy release rate to the critical load and crack length at fracture. Two experiments were carried out with the specimen in Fig. 4g; one with the aluminum as the upper material and one in which it was the lower. In the first case, $\psi$ was negative with a value of -0.76; in the second case it was determined as 0.83. For the specimen in Fig. 4h, the phase angle of the test was found to be 1.23 radians. Comparison was made between $\mathcal{G}_{ic}$ values obtained experimentally and those found from analytic approximations, as well as finite element analyses.

It may be seen that it is difficult to obtain a wide range of mode mixity from one specimen. In the next section, results obtained for bimaterial Brazilian disk specimens are reported. This specimen produces a wide range of mode mixity.

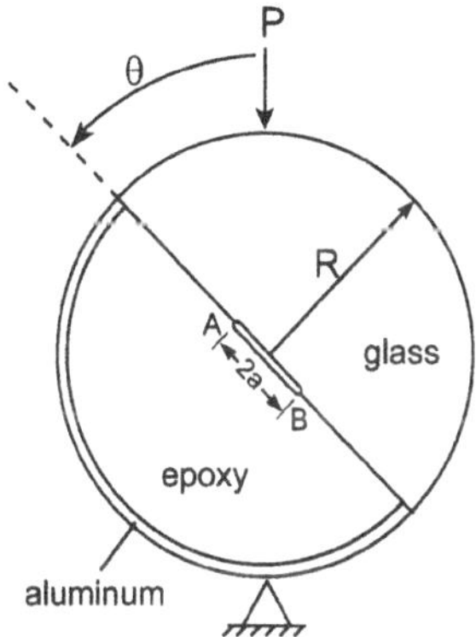


*Figure 5.* Bimaterial Brazilian disk specimen composed of glass and epoxy.

## 3. Bimaterial Brazilian Disk Specimens

Critical interface energy release rates were obtained by Banks-Sills, *et al.* (1999a, 1999b) with the Brazilian disk specimen exhibited in Fig. 2 for two material pairs: glass/epoxy and two ceramic clays (K-142/K-144). Material parameters for each pair are presented in Table 1.

TABLE 1. Some properties of materials studied.

| material | E (GPa) | $\nu$ | $\alpha(10^{-6}/^{\circ}C)$ |
|---|---|---|---|
| glass | 73.0 | 0.22 | 8.0 |
| epoxy | 2.9 | 0.29 | 73.0 |
| aluminum | 70.0 | 0.33 | 23.5 |
| K-142 | 19.5 | 0.29 | 6.01 |
| K-144 | 23.3 | 0.20 | 5.38 |

As illustrated in Fig. 5, the glass/epoxy specimens have a thin aluminum arc about the epoxy in order to reduce tensile residual stresses at the interface edges. It may be noted that the oscillatory parameter $\epsilon$ for the glass/epoxy pair is -0.088 and for the ceramic clay pair is -0.00563.

Finite element analyses (Bathe, 1995) were carried out on both specimen pairs in order to obtain calibration equations relating the stress intensity factors, the applied load, specimen geometry and material properties. In addition, the effect of residual stresses upon the stress intensity factors was also accounted for. For details, see Banks-Sills, *et al.* (1999a, 1999b).

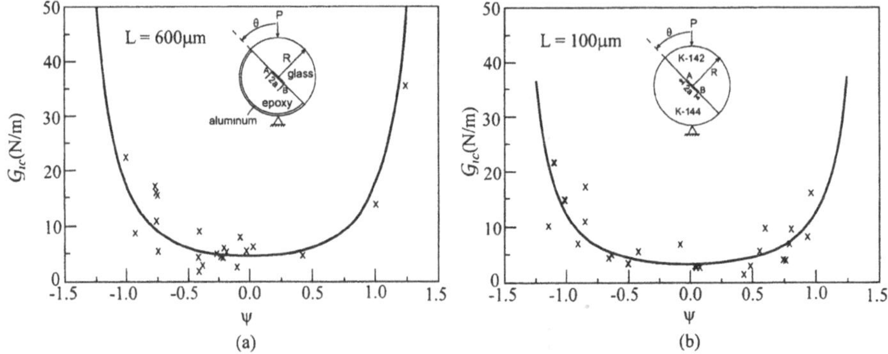


*Figure 6.* Interface fracture toughness as a function of phase angle $\psi$ for (a) glass/epoxy and (b) bimaterial ceramic clay specimens.

Results for the critical interface energy release rate $\mathcal{G}_{ic}$ as a function of phase angle $\psi$ for each material pair are exhibited in Fig. 6. Experimental details may be found in Banks-Sills, *et al.* (1999a, 1999b). The usual behavior is observed for both material pairs in which $\mathcal{G}_{ic}$ increases as $|\psi|$ increases. In addition, an energy based crack growth criterion is exhibited in Fig. 6 as a solid line. For each material pair, the length parameter $L$ in eq. (6) was chosen so that the test results are approximately centered with respect to $\psi = 0$. For the glass/epoxy pair, $L = 600$ $\mu$m, whereas for the ceramic clay pair $L = 100$ $\mu$m. As noted by Thurston and Zehnder (1996), the length parameter $L$ should be chosen within the $K$-dominant region when employing this data for failure prediction. Since for the ceramic clay pair, the value of $\epsilon$ is very small (-0.00563), the test results may be centered with $100\,\mu\text{m} \leq \text{L} \leq 1000\,\mu\text{m}$. This change in $L$ shifts the graph by less than $1^o$ or 0.013 radians. The shift is obtained from

$$\psi_2 = \psi_1 + \epsilon \ln\left(\frac{L_2}{L_1}\right) \tag{10}$$

where $\psi_1 = 0$. Since both materials are very brittle, it may be assumed that the length $L = 100$ $\mu$m is within the $K$-dominance region. For the glass/epoxy pair, an estimate of the plastic zone size, as well as finite element analyses to determine the $K$-dominance region (see Banks-Sills, *et al.*, 1999c) showed $L = 600$ $\mu$m to be within this zone.

Several fracture criteria presented by Banks-Sills and Ashkenazi (1999c) were compared with experimental data from tests on both material pairs. One of the energy release rate fracture criteria presented there is employed in this study. This criterion is given by

$$\mathcal{G}_{ic} = \mathcal{G}_1\left[1 + \tan^2\psi\right] \tag{11}$$

where $\mathcal{G}_1 \equiv \hat{K}_1^2/H$, $\hat{K}_1 \equiv \mathrm{Re}(\mathrm{KL}^{i\epsilon})$ and $\hat{K}_2 \equiv \mathrm{Im}(\mathrm{KL}^{i\epsilon})$. In fact, $\mathcal{G}_1$ is the value of $\mathcal{G}_{ic}$ when $\psi = 0$. Before applying this criterion, the test values are centered about $\psi = 0$. For each material pair, $\mathcal{G}_1$ is determined as explained in Banks-Sills and Ashkenazi (1999c). For the glass/epoxy pair, $\mathcal{G}_1 = 5.1$ N/m; for the ceramic clay pair $\mathcal{G}_1 = 3.7$ N/m.

For all experimental results in which there is a wide mixity range the interface toughness $\mathcal{G}_{ic}$ increases as $|\psi|$ increases. In experiments by Liechti and Chai (1992) on glass/epoxy bimaterial specimens, analyses were carried out to assess the effect of plastic and viscoelastic deformation upon values of $\mathcal{G}_{ic}$. It was shown that these do not contribute sufficiently to explain the steep increase in the toughness values. However, employing a cohesive zone to model the plastic zone in front of the crack tip in a sandwich specimen, Swadener and Liechti (1998) showed that plastic deformation is sufficient to increase $\mathcal{G}_{ic}$ when shear deformation increases.

It may be observed in Fig. 6 that there is much scatter in the test data for both data sets. For both bimaterial and sandwich specimens employed in other investigations in which a wide range of mixity was achieved, scatter may also be observed. Only tests carried out by Liechti and Chai (1992) exhibited little scatter. In that study, only four specimens were employed to obtain the data. This minimized the difference in surface finish and joining conditions from specimen to specimen.

In this study, finite difference calculations were carried out on ceramic clay specimens to assess error induced in $\mathcal{G}_i$ and $\psi$ for a variation in the applied loading angle $\theta$. For several of the load angles, these values were calculated for a difference of $\pm 0.5^\circ$. These are presented in Table 2. It may be observed that the percentages obtained for $\mathcal{G}_i$ are rather small. For $\theta = 2^\circ$, $\psi$ is approximately -0.5 radians. If the load angle is $1.5^\circ$, $\psi$ will be about -0.45. This change will hardly be discerned in Fig. 6b. Hence, it appears that the scatter resulting from load angle error is small.

TABLE 2. Percent change in $\mathcal{G}_{ic}$ and $\psi$ for loading angle change of $\pm 0.5^\circ$.

| $\theta$ | $\mathcal{G}_{ic}$ | $\psi$ |
|---|---|---|
| 15° | -2.5%/2.5% | -2.1%/2.1% |
| 10° | -2.8%/2.8% | -3.2%/3.2% |
| 2° | -3.6%/2.1% | -9.0%/6.8% |
| -15° | -2.2%/2.1% | -3.6%/3.7% |

Next, for the ceramic clay specimens, the deviations of the observed $\mathcal{G}_{ic}$ values (test data) from the theoretical curve in eq. (11) were examined statistically. The $\chi^2$ goodness of fit test was applied to the observed deviations assuming they are normally distributed experimental errors. The $\chi^2$ statistic is computed to be 4.0 which is less than $\chi^2_{0.5} = 5.99$ with two degrees of freedom. Thus, the normal distribution provides a satisfactory fit to the deviations. Furthermore, the correlation between the errors and the $\psi$ values was $|R| = 0.09$. This is considerably lower than a value of 0.45 at the 1% level of significance. Hence, it may be assumed that the errors and $\psi$ values are independent.

## 4. Summary and Discussion

Many studies have been reviewed on the subject of fracture toughness of interface crack propagation. Two types of specimens are studied: a bimaterial specimen and a sandwich specimen. Interface behavior between the two specimen types, as well as sandwich specimens with bonds of different thicknesses, may be different and the test should be tailored to the application.

In some early investigations, the stress intensity factor phase angle $\psi$ was not necessarily noted. More recent studies have presented results for a wide range of mode mixity. For those studies, the critical interface energy release rate $\mathcal{G}_{ic}$ increases with increasing $|\psi|$. In one study with sandwich specimens this increase in toughness was quantitatively attributed to increased plastic deformation with greater phase angle (see Swadener and Liechti, 1998).

With sandwich specimens, it was observed that crack path may change; that is, a crack along the interface between the adherend and joining layer may not remain along the interface. An excellent study examining the factors which contribute to these changes was presented by Fleck, *et al.* (1991). According to that study, the crack path depends upon the sign of $K_{\rm II}$ and the T-stress.

The residual stresses created by the material mismatch has not been taken into consideration in all studies. It has been seen in some studies, that when they exist, they can significantly influence the critical quantities. Hence, it is concluded here that this factor should be addressed.

In many studies in which a wide range of mode mixity is attained there is much scatter in the $\mathcal{G}_{ic}$ values. For a series of tests on ceramic clay Brazilian disk specimens, statistical analyses were carried out on the data. It was found that the deviations of the experimental data points from the $\mathcal{G}_{ic}$ criterion very nearly satisfy the standard Gaussian distribution of errors.

Finally, because of space limitations, there are many other studies which have not been mentioned. Some of them include Dalgleish, *et al.* (1989), Charalambides, *et al.* (1992), Chen, *et al.* (1999), Jiao, *et al.* (1998), Smith, *et al.* (1993), Swadener, *et al.* (1999), as well as others.

## Acknowledgment

I would like to thank Mr. Avraham Doroguy for providing finite difference results employed in this investigation.

## References

Akisanya, A.R. and Fleck, N.A. (1992) Brittle fracture of adhesive joints, Int. J. Fract. **58**, 93–114.

Banks-Sills, L., Travitzky, N., Ashkenazi, D. and Eliasi, R. (1999a) A methodology for measuring interface fracture properties of composite materials, Int. J. Fract. **99**, 143–160.

Banks-Sills, L., Travitzky, N. and Ashkenazi, D. (1999b) Interface fracture properties of a bimaterial ceramic composite, submitted for publication.

Banks-Sills, L. and Ashkenazi, D. (1999c) A note on fracture criteria for interface fracture, to appear: Int. J. Fract.

Bao, G., Ho, S., Suo, Z. and Fan, B. (1992) The role of material othotropy in fracture specimens for composites, Int. J. Solids Struct. **29**, 1105–1116.

Bathe, K.J. (1995) *ADINA - Automatic Dynamic Incremental Nonlinear Analysis System*, Version 6.2, Adina Engineering, Inc. USA.

Cao, H.C. and Evans, A.G. (1989) An experimental study of the fracture resistance of bimaterial interfaces, Mech. Mater. **7**, 295–304.

Cazzato A. and Faber K.T. (1997) Fracture energy of glass-alumina interfaces via the bimaterial bend test, J. Am. Ceram. Soc. **80**, 181–188.

Charalambides, M., Kinloch, A.J., Wang, Y. and Williams, J.G. (1992) On the analysis of mixed-mode failure, Int. J. Fract. **54**, 269–291.

Charalambides, P.G., Lund, J., Evans, A.G. and McMeeking, R.M. (1989) A test specimen for determining the fracture resistance of bimaterial interfaces, J. of Appl. Mech. **56**, 77–82.

Chen, Z., Cotterell, B. and Chen, W.T. (1999) Characterizing the interfacial fracture toughness for microelectronic packaging, Surf. Interface Anal. **28**, 146–149.

Dalgleish, B.J., Trumble, K.P. and Evans, A.G., (1989) The strength and fracture of alumina bonded with aluminum alloys, Acta metall. mater. **37**, 1923–1931.

Deng, X. (1993) General crack-tip fields for stationary and steadily growing interface cracks in anisotropic bimaterials, J. Appl. Mech. **60**, 183–189.

Duer, R., Katevatis, D., Kinloch, A.J. and Williams, J.G. (1996) Comments on mixed-mode fracture in adhesive joints, Int. J. Fract. **75**, 157–162.

Evans, A.G. and Hutchinson, J.W. (1989) Effects of the non-planarity on the mixed-mode fracture resistance of bimaterial interfaces, Acta Metall. Mater. **37**, 909–916.

Fleck, N.A., Hutchinson, J.W. and Suo, Z. (1991) Crack path selection in a brittle adhesive layer, Int. J. Solids Struc. **27**, 1683–1703.

Ikeda, T. Miyazaki, N. and Soda, T. (1998) Mixed mode fracture criterion of interface crack between dissimilar materials, Eng. Fract. Mech. **59**, 725–735.

Jiao, J., Gurumurthy, C.K, Kramer, E.J., Sha, Y., Hui, C.Y. and Borgesen, P. (1998) Measurement of interfacial fracture toughness under combined mechanical and thermal stresses, J. Elec. Pack. **120**, 349–353.

Kinloch, A.J., Thrusabanjong, E. and Williams, J.G. (1991) Fracture at bimaterial interfaces: the role of residual stresses, J. Mater. Sci. **26**, 6260–6270.

Liechti, K.M. and Hanson, E.C. (1988) Nonlinear effects in mixed-mode interfacial delamination, Int. J. Fract. **36**, 199–217.

Liechti, K.M. and Chai, Y.S. (1991) Biaxial loading experiments for determining interfacial fracture toughness, J. Appl. Mech. **58**, 680–687.

Liechti, K.M. and Chai, Y.S. (1992) Asymmetric shielding in interfacial fracture under in-plane shear, J. Appl. Mech. **59**, 295–304.

Liechti, K.M. and Liang, Y-M. (1992) The interfacial fracture characteristics of bimaterial and sandwich blister specimens, Int. J. Fract. **55**, 95–114.

Rice, J.R., Suo, Z. and Wang, J.-S. (1990) Mechanics and thermodynamics of brittle interface failure in bimaterial systems, in: Metal-Ceramic Interfaces, (eds) Rühle, M., Evans, A.G., Ashby, M.F. and Hirth, J.P., Pergamon Press, Oxford, 269–294.

Reimanis, I.E., Dalgleish, B.J. and Evans, A.G. (1991) The fracture resistance of a model metal/ceramic interface, Acta Metall. Mater. **39**, 3133–3141.

Smith, J.W., Kramer, E.J., Xiao, F., Hui, C-Y, Reichert, W. and Brown, H.R. (1993) Measurement of the fracture toughness of polymer-non-polymer interfaces. J. Mat. Sci. **28**, 4234–4244.

Sundararaman, V and Sitaraman, S.K. (1999) Determination of fracture toughness for metal/polymer interfaces. J. Elec. Pack. **121**, 275–281.

Suo, Z. and Hutchinson, J.W. (1989) Sandwich test specimens for measuring interface crack toughness. Mat. Sci. and Eng. **A107**, 135–143.

Swadener, J.G. and Liechti, K.M. (1998) Asymmetric shielding mechanisms in the mixed-mode fracture of a glass/epoxy interface, J. Appl. Mech. **65**, 25–29.

Swadener, J.G., Liechti, K.M. and de Lozanne, A.L. (1999) The intrinsic toughness and adhesion mechanisms of a glass/epoxy interface, J. Mech. Phys. Solids **47**, 223–258.

Thurston, M.E. and Zehnder, A.T. (1993) Experimental determination of silica/copper interfacial toughness, Acta Metall. Mater. **41**, 2985–2992.

Thurston, M.E. and Zehnder, A.T. (1996) Nickel-alumina interfacial fracture toughness: experiments and analysis of residual stress effects, Int. J. Fract. **76**, 221–241.

Turner, M.R., Dalgleish, B.J., He, M.Y. and Evans, A.G. (1995) A fracture resistance measurement method for bimaterial interfaces having large debond energy, Acta. Metall. Mater. **43**, 3459–3465.

Wang, C.H. (1997) Fracture of interfacial cracks under combined loading, Eng. Fract. Mech. **56**, 77–86.

Wang, J.-S. and Suo, Z. (1990) Experimental determination of interfacial toughness curves using Brazil-nut-sandwiches, Acta Metall. Mater. **38**, 1279–1290.

Wang, J.-S. (1995) Interfacial fracture toughness of a copper/alumina system and the effect of loading phase angle, Mech. Mater. **20**, 251–259.

# SOME REFINEMENTS IN ANALYSIS OF THICK-WALLED TUBES IN AXIAL BENDING

CHARLES W. BERT
*School of Aerospace and Mechanical Engineering*
*The University of Oklahoma*
*Norman, Oklahoma 73019-1052 USA*

## Abstract

This paper investigates analytically certain refinements in the nonlinear analysis of tubes undergoing axial bending, with emphasis on thick-walled effects: axial moment of inertia, circumferential bending and membrane action, and higher-order terms in angular position.

## 1. Introduction

There have been dozens of analyses devoted to elastic tubes subjected to axial beam bending with cross-sectional deformation. However, most of them have been based on the same assumptions as those used by Brazier (1927), namely:

(1) Deflections are uniform along the length, i.e., the shell is infinitely long.
(2) Circumferential extension of the shell middle surface is neglected.
(3) Axial shell bending is neglected.
(4) The thin-walled shell assumption is made.
(5) The shell is constructed of homogeneous and isotropic material.
(6) The shell material is linearly elastic.
(7) Negligible local shell buckling occurs.
(8) Cross-sectional deflection is symmetric about both the horizontal and vertical axes of the cross section.
(9) The cross-sectional deflection is of the form $\cos 2\theta$.
(10) Shear deformation due to bending is omitted.



*D. Durban et al. (eds.), Advances in the Mechanics of Plates and Shells,* 67–78.

Hypothesis no. 1 was removed in analytical investigations by Axelrad (1965), Libai and Bert (1994), and Tatting et al. (1997) and in finite element analyses (FEA) by Fuchs and Hyer (1992) and Stockwell and Cooper (1992).

Apparently, the only investigations removing hypotheses no. 2 and no. 3 have been the FEA by Fuchs and Hyer (1992) and Stockwell and Cooper (1992). No investigations known to the author have removed hypothesis no. 4.

The assumption of homogeneous isotropic material (hypothesis no. 5) was lifted by Hayashi (1949), Kedward (1978), Fuchs and Hyer (1992), Stockwell and Cooper (1992), Libai and Bert (1994), Li (1996), Tatting et al. (1996, 1997), and Harursampath and Hodges (1999).

The hypothesis (no. 6) of linear elastic material has been removed in numerous analytical and numerical investigations involving the plastic range. Apparently, the first analysis to include simultaneously cross-sectional ovalization and local (short-wavelength) buckling (hypothesis no. 7) was the recent FEA by Knight et al. (1995) and analytical study by Tatting et al. (1996).

The only investigations that removed hypothesis no. 8 (doubly symmetric cross-sectional deflection) were the FEA by Prinja and Chitkara (1986), Fuchs and Hyer (1992) and the recent analytical study by Libai and Bert (1999). The aforementioned FEA also lifted hypothesis no. 9 ($\cos 2\theta$ form). Finally, hypothesis no. 10 (no in-surface shear warping deformation) was removed by Libai and Bert (1994) and Tatting et al. (1997), who showed that it was negligible except in the case of very short shells.

The purpose of the present investigation is to analytically remove hypotheses 2, 3, 4, and 9. The objective is an attempt to improve upon the previous analytical investigations especially for the case of relatively thick-walled shells.

## 2. Strain Energy of Axial-Stress Action (Arbitrary Wall Thickness)

The strain energy per unit length of shell resulting from axial-stress action may be expressed as

$$U_x = \iint_V \int_0^{\varepsilon_x} \sigma_x d\varepsilon_x dV \tag{1}$$

where $V \equiv$ volume per unit length, $\varepsilon_x \equiv$ axial normal strain, and $\sigma_x \equiv$ axial normal stress component.

The axial bending moment is given by

$$M_x = \iint_A (r\sin\theta + \eta)\sigma_x \, dA \tag{2}$$

where $A \equiv$ cross-sectional area, $r$ and $\theta$ are plane polar coordinates in the cross-sectional plane. The axial normal stress is given by

$$\sigma_x = E_x \varepsilon_x \quad (3)$$

Here $E_x$ is the axial elastic modulus and the axial normal strain is assumed to be small and to obey the hypothesis that plane sections remain plane. Then

$$\varepsilon_x = C(r\sin\theta + \eta) \quad (4)$$

where $C$ is the axial bending curvature and $\eta$ is the resultant vertical deflection component given by

$$\eta = w\sin\theta + v\cos\theta \quad (5)$$

Here $w$ and $v$ are the respective radial and circumferential displacement components. It is noted that all previous investigations used hypotheses 3 and 4 and thus used the mean radius (a) instead of the general radius ($r$) in eqn. (4).

The differential area referred to in eqn. (2) is expressed as follows:

$$dA = r\ dr\ d\theta \quad (6)$$

Again it is to be emphasized that previous investigations used only the following approximate expression:

$$dA = at\, d\theta \quad (6a)$$

Here $t \equiv$ tube wall thickness.

Substituting eqns. (3) – (6) into eqn. (2), one obtains

$$M_x = E_x C \int_o^{2\pi} \int_{R_i}^{R_o} (r\sin\theta + w\sin\theta + v\cos\theta)^2\ r\, dr\, d\theta \quad (7)$$

where $R_i$ and $R_o$ are the respective inside and outside radii of the tube. By definition, the centroidal area moment of inertia of the deformed cross section is

$$I = \int_o^{2\pi} \int_{R_i}^{R_o} (r\ \sin\theta + w\ \sin\theta + v\ \cos\theta)^2\ r\ dr\, d\theta \quad (8)$$

From eqns. (7) and (8), it is clear that

$$M_x = E_x I C \quad (9)$$

and eqn. (1) becomes

$$U_x = \frac{1}{2} E_x I C^2 \quad (10)$$

Up to now, all of the equations presented have been exact within the hypotheses used. Now, however, in the spirit of Brazier (1927) and Calladine (1983), the following one-term assumptions are made for each of the deflection components:

$$w = a\xi\cos 2\theta, \quad v = a\gamma\sin 2\theta \tag{11}$$

Here, $\xi$ and $\gamma$ are dimensionless coefficients to be determined later by the principle of virtual work.

Inserting eqns. (11) into the integral (8) and performing all of the integrations exactly yields

$$I = \pi\left[\frac{1}{4}\left(R_o^4 - R_i^4\right) + \frac{1}{3}\left(R_o^3 - R_i^3\right)a\left(\gamma - \xi\right) + \frac{1}{4}a^2\left(R_o^2 - R_i^2\right)\left(\xi^2 + \gamma^2\right)\right] \tag{12}$$

It is noted that the first term on the right side of eqn. (12) is the moment of inertia of the undeformed cross section. For purposes of comparison, the classical thin-walled version of eqn. (12) is

$$I \approx \pi a^3 t\left[1 + \left(\gamma - \xi\right) + \frac{1}{2}\left(\xi^2 + \gamma^2\right)\right] \tag{13}$$

Both eqns. (12) and (13) can be written in the following general form

$$I = I_o + I_1\left(\gamma - \xi\right) + I_2\left(\xi^2 + \gamma^2\right) \tag{14}$$

It can be shown that the approximate $I_2$ is identical to the exact one and that the approximate values of $I_o$ and $I_1$ are always higher than the exact as shown in Table 1. It can be concluded that the approximate expressions are adequate for $a/t \geq 3$.

TABLE 1. Comparison of ratios of the exact moment-of-inertia terms to the approximate ones as a function of $a/t$ or $R_i/R_o$

| $a/t$ | $R_i/R_o$ | $\overline{I_o}$ | $\overline{I_1}$ |
|---|---|---|---|
| 1 | 0.3333 | 1.250 | 1.083 |
| 2 | 0.6000 | 1.063 | 1.021 |
| 3 | 0.7143 | 1.028 | 1.009 |
| 5 | 0.8182 | 1.010 | 1.003 |
| 10 | 0.9048 | 1.003 | 1.001 |
| 20 | 0.9512 | 1.001 | 1.000 |

## 3. Strain Energy of Circumferential-Bending-Stress Action (Arbitrary Wall Thickness and Curved-Beam Effect)

The strain energy per unit length of the shell resulting from circumferential-bending-stress action may be expressed

$$U_{b\theta} = \iint_V \int_o^{\varepsilon_{b\theta}} \sigma_{b\theta}\, d\varepsilon_{b\theta}\, dV \tag{15}$$

where $\varepsilon_{b\theta}$ and $\sigma_{b\theta}$ are the circumferential bending strain and stress, respectively.

Generalized Hooke's law for a thin orthotropic shell may be written as

$$\sigma_{b\theta} = (E_\theta/\lambda)(\varepsilon_{b\theta} + v_{x\theta}\,\varepsilon_x) \tag{16}$$

where $E_\theta \equiv$ circumferential elastic modulus, $v_{v\theta} \equiv$ Poisson's ratio associated with uniaxial stress in the circumferential direction, and $\lambda \equiv 1 - v_{\theta x}\, v_{x\theta}$. Using $dA$ from eqn. (6) for $dV$ and $\sigma_\theta$ from eqn. (16), one can rewrite eqn. (15) as follows:

$$U_{b\theta} = \frac{E_\theta}{2\lambda} \int_o^{2\pi} \int_{R_i}^{R_o} \left(\varepsilon_{b\theta}^2 + v_{x\theta}\,\varepsilon_{b\theta}\,\varepsilon_x\right) r\, dr\, d\theta \tag{17}$$

At this point, previous analyses have invoked the following simplifications:

(1) Neglect of the Poisson coupling term $v_{x\theta}\,\varepsilon_{b\theta}\,\varepsilon_x$

(2) Use of eqn. (6a) instead of eqn. (6) for the cross-sectional area

(3) Use of straight-beam theory instead of ring or curved-beam theory

(4) Neglect of the middle-surface circumferential extensional strain

Here we make none of those simplifications except the last one which is covered in the next section. Using the Winkler-Bach curved-beam theory as presented by Boresi et al. (1993), the circumferential bending strain can be expressed as

$$\varepsilon_{b\theta} = \left[(R_n/r) - 1\right]\beta \tag{18}$$

where $R_n \equiv$ radius of the <u>neutral</u> surface and $\beta \equiv$ circumferential rotation in the cross-sectional plane. Thus, eqn. (17) becomes

$$U_{b\theta} = \frac{E_\theta}{2\lambda} \int_o^{2\pi} \int_{R_i}^{R_o} \left\{\left[(R_n/r) - 1\right]^2 \beta^2 + v_{x\theta}\left[(R_n/r) - 1\right]\beta C(r\sin\theta + w\sin\theta + v\cos\theta)\right\} r\, dr\, d\theta \tag{19}$$

Eqn. (19) is exact within the hypotheses used. Now we introduce assumed modes for $w$ and $v$ as given in eqns. (11) and the following consistent mode for $\beta$ into eqn. (17)

$$\beta = 3\xi \cos 2\theta \tag{20}$$

to obtain the following result:

$$U_{b\theta} = (9\pi/2)(E_\theta/\lambda)\left[R_n^2 \ln(R_o/R_i) - 2R_n(R_o - R_i) + \frac{R_o^2 - R_1^2}{2}\right]\xi^2 \tag{21}$$

It is noted that the Poisson effect ( $\nu_{x\theta}$ term) has vanished. All that remains is to evaluate $R_n$ for a rectangular cross section of unit length and inside and outside radii $R_i$ and $R_o$, respectively. From Boresi et al. (1993), we have

$$R_n = (R_o - R_i)/\ln(R_o/R_i) \tag{22}$$

Substituting eqn. (22) into eqn. (21) and simplifying yields

$$U_{b\theta} = (9\pi/4)(E_\theta/\lambda)\left[R_o^2 - R_i^2 - 2(R_o - R_i)^2/\ln(R_o/R_i)\right]\xi^2 \tag{23}$$

It can be shown that for the thin-walled case (see Calladine, 1983, for instance)

$$U_{b\theta} = (3\pi/8)(E_\theta/\lambda)(t^3/a)\xi^2 \tag{24}$$

Table 2 shows a comparison of $U_{b\theta}$ according to eqns. (23) and (24) as a function of $a/t$. Again, for $a/t \geq 3$, the error in using the approximate expression is less than 5%.

TABLE 2. Ratio of the exact circumferential bending strain energy to the approximate one as a function of $a/t$ or $R_i/R_o$

| $a/t$ | $R_i/R_o$ | $(U_{b\theta})_e/(U_{b\theta})_a$ |
|---|---|---|
| 1 | 0.33333 | 1.0770 |
| 2 | 0.60000 | 1.0178 |
| 3 | 0.71429 | 1.0100 |
| 5 | 0.81818 | 0.9800 |
| 10 | 0.90476 | 0.9591 |
| 20 | 0.95122 | 0.9532 |

## 4. Strain Energy of Circumferential-Membrane-Stress Action (Arbitrary Wall Thickness)

Here two different models are presented: one based on the thin-walled assumption and the other based on arbitrary wall thickness.

In general, the strain energy per unit length of tube due to circumferential extension is

$$U_{e\theta} = \frac{1}{2}\iint_V \sigma_{e\theta}\,\varepsilon_{e\theta}\,dV \tag{25}$$

Also,

$$\sigma_{e\theta} = \frac{E_\theta}{\lambda}\varepsilon_{e\theta} \tag{26}$$

For the thin-walled case

$$dV = at\,d\theta;\;\; \varepsilon_{e\theta} = (1/a)\left(w + \frac{\partial v}{\partial \theta}\right) \tag{27}$$

Thus,

$$U_{e\theta} = (E_\theta/2\lambda)(t/a)\int_o^{2\pi}\left(w + \frac{\partial v}{\partial \theta}\right)^2 d\theta \tag{28}$$

Substituting eqns. (11) into eqn. (28), one obtains

$$U_{e\theta} = (\pi/2)(E_\theta/2\lambda)(at)(\xi + 2\lambda)^2 \tag{29}$$

For the case of arbitrary wall thickness,

$$dV = r\;dr\;d\theta;\;\; \varepsilon_{e\theta} = (1/r)\left(w + \frac{\partial v}{\partial \theta}\right) \tag{30}$$

Then substituting eqns. (11), (26), and (30) into eqn. (25) one obtains

$$U_{e\theta} = (E_\theta/2\lambda)a^2(\xi + 2\lambda)^2\int_{R_i}^{R_o}(1/r)\cos^2 2\theta\,dr\,d\theta \tag{31}$$

Performing the integration yields

$$U_{e\theta} = (\pi/2)(E_\theta/\lambda)(\xi + 2\lambda)^2 a^2 \ln(R_o/R_i) \tag{32}$$

Table 3 shows a comparison of $U_{e\theta}$ as a function of $a/t$. Again, for $a/t \geq 3$, the error in using the thin-walled approximation is less than 1%.

TABLE 3. Ratio of the exact circumferential extensional strain energy to the approximate one as a function of $a/t$ or $R_i/R_o$

| $a/t$ | $R_i/R_o$ | $(U_{e\theta})_e/(U_{e\theta})_a$ |
|---|---|---|
| 1 | 0.33333 | 1.0986 |
| 2 | 0.60000 | 1.0217 |
| 3 | 0.71429 | 1.0094 |
| 5 | 0.81818 | 1.0034 |
| 10 | 0.90476 | 1.0009 |
| 20 | 0.95122 | 1.0002 |

## 5. Determination of the Dimensionless Deflection Parameters

The total strain energy is the sum of strain energies due to axial bending, circumferential bending, and circumferential extension:

$$U = U_x + U_{b\theta} + U_{e\theta} \tag{33}$$

Thus,

$$\begin{aligned} U &= (\pi/2)E_x C^2\left[I_o + I_1(\gamma-\xi) + I_2(\xi^2+\gamma^2)\right] \\ &+(9\pi/4)(E_\theta/\lambda)\left[R_o^2 - R_i^2 - 2(R_o-R_i)^2/\ln(R_o/R_i)\right]\xi^2 \\ &+(\pi/2)(E_\theta/\lambda)\left[a^2\ln(R_o/R_i)\right](\xi+2\gamma)^2 \end{aligned} \tag{34}$$

Nondimensionalizing, we introduce

$$\begin{aligned} &\overline{U} \equiv U\Big/\left(\frac{\pi}{2}E_x a^2\right), \quad \Delta_o \equiv \left(R_o^4 - R_i^4\right)/4a^4, \\ &\Delta_1 \equiv \left(R_o^3 - R_i^3\right)/3a^3, \quad \Delta_2 \equiv \left(R_o^2 - R_i^2\right)/4a^2, \\ &\Delta_3 \equiv (9/2)\left[\left(R_o^2 - R_i^2\right)/\left(a^2\right) - 2(R_o-R_i)^2/a^2\ln(R_o/R_i)\right] \\ &\Delta_4 \equiv \ln(R_o/R_i), \quad \overline{E} \equiv E_\theta/\lambda E_x \end{aligned}$$

Then

$$\begin{aligned} \overline{U} &= \left[\Delta_o + \Delta_1(\gamma-\xi) + \Delta_2(\xi^2+\gamma^2)\right](aC)^2 \\ &+\overline{E}\left[\Delta_3\xi^2 + \Delta_4(\xi+2\gamma)^2\right] \end{aligned} \tag{35}$$

By use of the principle of virtual work which leads to the principle of stationary potential energy, we set

$$\partial\overline{U}/\partial\xi = 0, \quad \partial\overline{U}/\partial f\gamma = 0 \tag{36}$$

to obtain the appropriate values for $\xi$ and $\gamma$, which are solved to obtain the following results:

$$\xi = \frac{B_1(A_{22}+A_{12})}{A_{11}A_{22}-A_{12}^2}, \quad \gamma = \frac{-B_1(A_{11}+A_{12})}{A_{11}A_{22}-A_{12}^2} \tag{37}$$

where

$$A_{11} \equiv \Delta_2 (aC)^2 + (\Delta_3 + \Delta_4)\bar{E}, \quad A_{12} \equiv 2\Delta_4 \bar{E}$$
$$A_{22} \equiv \Delta_2 (aC)^2 + 4\Delta_4 \bar{E}, \quad B_1 \equiv (\Delta_1/2)(aC)^2 \tag{38}$$

For purposes of comparison of several mathematical models, the calculations are carried out for a series of eight thick-walled nylon tubes tested by Luo (1992). The pertinent data and results are listed in Table 4. It can be concluded that the Karam (1994) theory always predicts significantly smaller values of $\xi$ than the Brazier (1927) theory. However, the improved thick-walled extensible theory presented here offers very little change from the values predicted by Karam. Also, it is noted that the ratio of $\gamma/\xi$ predicted by the present theory ranged from –0.5041 to –0.5160 in comparison with the inextensional value of ½.

TABLE 4. Comparison of dimensionless displacements predicted by various theories for the eight tubes tested by Luo (1992)

| Test No. | $R_o$ (in) | $R_i$ (in) | Dimensionless Curvature, $Ca$ | Brazier* $\xi_B$ | Karam* $\xi_K$ | Present $\xi$ | Present $\gamma/\xi$ |
|---|---|---|---|---|---|---|---|
| 1 | 5 | 4 | 0.1139 | 0.2095 | 0.1783 | 0.1773 | -0.5067 |
| 2 | 5 | 4 | 0.09091 | 0.1335 | 0.1201 | 0.1195 | -0.5064 |
| 3 | 5 | 3.5 | 0.1083 | 0.07510 | 0.07067 | 0.07065 | -0.5160 |
| 4 | 5 | 3.5 | 0.08629 | 0.04767 | 0.04585 | 0.04589 | -0.5158 |
| 5 | 6.25 | 5.25 | 0.1411 | 0.5250 | 0.3652 | 0.3653 | -0.5045 |
| 6 | 6.25 | 5.25 | 0.1133 | 0.3386 | 0.2641 | 0.2635 | -0.5041 |
| 7 | 6.25 | 4.75 | 0.1358 | 0.1977 | 0.1697 | 0.1701 | -0.5098 |
| 8 | 6.25 | 4.75 | 0.1089 | 0.1272 | 0.1150 | 0.1151 | -0.5067 |

* In the Brazier (1927) and Karam (1994) theories, the *a priori* assumption of circumferential inextensibility requires that $\gamma/\xi = -1/2$.

Luo (1992) measured the dimensionless ovalization factor $P$ defined as follows:

$$P = D_{max} / D_{min}$$

In terms of the dimensionless deflection factor $\xi$, $P$ can be expressed as

$$P = (1+\xi)/(1-\xi) \tag{39}$$

The results are compared in Table 5. Clearly, the improved thick-walled theory presented here gives predictions closest to the experimental values. However, it should be cautioned that in all of Luo's tests, some material was beyond the yield strength.

TABLE 5. Comparison of ovalization factor P for the experiments conducted by Luo (1992); see Table 4 for data

| Test | $R_\xi/R_o$ | $Ca$ | P | | | |
|---|---|---|---|---|---|---|
| | | | Exp. | Predicted | | |
| | | | Luo | Brazier | Karam | Present |
| 1 | 0.80 | 0.1139 | 1.65 | 1.53* | 1.434 | 1.431 |
| 2 | 0.80 | 0.90901 | 1.28 | 1.31 | 1.273* | 1.271 |
| 3 | 0.70 | 0.1083 | 1.13 | 1.16 | 1.152* | 1.152* |
| 4 | 0.70 | 0.08629 | 1.08 | 1.10 | 1.096* | 1.096* |
| 5 | 0.84 | 0.1411 | 1.96 | 3.21 | 2.151* | 2.151* |
| 6 | 0.84 | 0.1133 | 1.415 | 2.02 | 1.718 | 1.716* |
| 7 | 0.76 | 0.1358 | 1.445 | 1.49 | 1.409 | 1.410* |
| 8 | 0.76 | 0.1089 | 1.195 | 1.29 | 1.260* | 1.260* |

* An asterisk denotes the prediction which is closest to the measured value.

## 6. Effect of Higher-Order Terms in $\theta$

All of the previous analytical investigations of the Brazier problem assumed that the radial deflection depended on $\theta$ as $\cos 2\theta$. To investigate the effect of higher-order terms in $\theta$, the following forms were used:

$$w = a(\xi_1 \cos 2\theta + \xi_2 \cos 4\theta)$$

$$v = a\left(-\frac{1}{2}\xi_1 \sin 2\theta - \frac{1}{4}\xi_2 \sin 4\theta\right)$$

in a circumferentially inextensible Calladine-type analysis. The results, for the same eight tubes as previously mentioned are summarized in Table 6. It is noted that although $\xi_2$ is only approximately 2.6% of $\xi_1$, the effect on $\xi_1$ itself is significant, as can be seen by comparing with the results of Karam's analysis in Table 4. In this case, the ovalization factor is given by

$$P = (1 + \xi_1 + \xi_2)/(1 - \xi_1 - \xi_2)$$

Results are presented in the last column of Table 6.

TABLE 6. Effect of a second $\theta$-dependent term in an inextensional analysis

| Test No. | One $\theta$-dependent term $\xi$ (Karam, 1994) | Two $\theta$-dependent terms[†] $\xi_1$ | $\xi_2/\xi_1$ | $P$ |
|---|---|---|---|---|
| 1 | 0.1783 | 0.2528 | -0.0262 | 1.653* |
| 2 | 0.1201 | 0.1498 | -0.0264 | 1.342 |
| 3 | 0.07067 | 0.07997 | -0.0265 | 1.169 |
| 4 | 0.04585 | 0.04959 | -0.0266 | 1.101 |
| 5 | 0.3652 | 0.9239 | -0.0254 | 19.08 |
| 6 | 0.2641 | 0.4684 | -0.0259 | 2.678 |
| 7 | 0.1697 | 0.2359 | -0.0262 | 1.596 |
| 8 | 0.1150 | 0.1419 | -0.0264 | 1.321 |

* An asterisk indicates the only $P$ value that is closer to Luo's experimental results than that predicted by Karam (1994); see Table 5.

† $P = (1+\xi_1+\xi_2)/(1-\xi_1-\xi_2)$

In summary, it can be concluded that the use of higher-order terms in $\theta$ offer little or no improvement in ovalization factor.

## Acknowledgments

This paper is dedicated in memory of the late Professor Eric Reissner of the University of California, San Diego, (UCSD). The author acknowledges very gratefully some very helpful discussions with him during the initiation of the present work in the period January through March 1996. The author also acknowledges helpful discussions with Professor Avinoam Libai of the Technion-Israel Institute of Technology, with whom he has collaborated in research for nearly a decade.

Finally, the author acknowledges financial support in the form of a sabbatical leave from the University of Oklahoma and subsistence support from the UCSD Institute for Mechanics and Materials, of which the late Professor Richard Skalak was the director.

## References

Axelrad, E. (1965) Refinement of the upper critical loading of pipe bending taking account of the geometrical nonlinearity, *Izvestiya Akad. Nauk SSR*, OTN, Mekh. i Mash. no. 4, 133-139.

Boresi, A.P., Schmidt, R.J., and Sidebottom, O.M. (1993) *Advanced Mechanics of Materials*, 5th ed., John Wiley & Sons, New York.

Brazier, L.G. (1927) On the flexure of thin cylindrical shells and other "thin" sections, *Proc., Royal Society, London*, Ser. A **116**, 104-114.

Calladine, C.R. (1983) *Theory of Shell Structures*, Cambridge University Press, Cambridge, UK, sec. 16.

Fuchs, H.P. and Hyer, M.W. (1992) Bending response of thin-walled laminated composite cylinders, *Composite Structures* **22**, 87-107.

Harursampath, D. and Hodges, D.H. (1999) Asymptotic analysis of the non-linear behavior of long anisotropic tubes, *International Journal of Non-Linear Mechanics* **34**, 1003-1018.

Hayashi, T. (1949) On the elastic instability of orthogonal anisotropic cylindrical shells, especially the buckling loads due to compression, bending and torsion, *Journal of the Society of Naval Architects, Japan* no. 81, 85-98.

Karam, G.N. (1994) On the ovalisation in bending of nylon and plastic tubes, *International Journal of Pressure Vessels and Piping* **58**, 147-149.

Kedward, K.T. (1978) Nonlinear collapse of thin-walled composite cylinders under flexural loading, *Proc., 2nd. Internat. Conference on Composite Materials*, Metallurgical Society of AIME, Warrendale, PA, 353-365.

Knight, Jr., N.F., Macy, S.C., and McCleary, S.L. (1995) Assessment of structural analysis technology for static collapse of elastic cylindrical shells, *Finite Elements in Analysis and Design* **18**, 403-431.

Li, L.Y. (1996) Bending instability of composite tubes, *Journal of Aerospace Engrg.* **9**, 58-61.

Libai, A. and Bert, C.W. (1994) A mixed variational principle and its application to the nonlinear bending of orthotropic tubes – II. Application to nonlinear bending of circular cylindrical tubes, *International Journal of Solids and Structures* **31**, 1019-1033.

Libai, A and Bert, C.W. (1999) On the problem of a possible nonsymmetric deformation component in the nonlinear pure bending of orthotropic straight tubes, *Proc., 2nd. International Conference on Nonlinear Problems in Aviation and Aerospace*, Dayton Beach, FL, 1998; European Conference Publications, Cambridge, UK, **2**, 431-438.

Luo, D.Z. (1992) Effects of outside diameter, wall thickness and bend radius on ovality of nylon tubes during bending operation, *International Journal of Pressure Vessels and Piping*, **52**, 145-148.

Prinja, N.K. and Chitkara, N.R. (1986) Finite-element analyses of post-collapse plastic bending of thick pipes, *Nuc. Engrg. and Design* **91**, 1-12.

Stockwell, A.E. and Cooper, P.A. (1992) Collapse of composite tubes under end moments, *Proc., 33rd. Structures, Structural Dynamics, and Materials Conference*, Dallas, TX, Pt. 4, 1841-1850 (AIAA Paper 92-2389).

Tatting, B.F., Gürdal, Z., and Vasilev, V.V. (1996) Nonlinear response of long orthotropic tubes under bending including the Brazier effect, *AIAA Journal* **34**, 1934-1940.

Tatting, B.F., Gürdal, Z., and Vasilev, V.V. (1997) The Brazier effect for finite length composite cylinders under bending, *International Journal of Solids and Structures* **34**, 1419-1440.

# OPTIMIZATION OF PANELS WITH RIVETED Z-SHAPED STIFFENERS VIA PANDA2

D. BUSHNELL
*Senior Consulting Scientist, Retired*
*Lockheed Martin Advanced Technology*
*3251 Hanover Street. Palo Alto, California*

## 1. Abstract

The PANDA2 computer program has been modified to permit minimum weight design of imperfect panels with riveted Z-shaped stiffeners for service in a load regime in which the panel is in its locally postbuckled state. Perfect and imperfect panels optimized with PANDA2 are evaluated via nonlinear STAGS analyses. The agreement between predictions by PANDA2 and STAGS is sufficient to qualify PANDA2 as a preliminary design tool for panels with riveted Z-shaped stringers. Optimum designs for panels with Z-shaped stringers are compared to those with J-shaped and T-shaped stringers.

## 2. Introduction

In the late 1970's van der Neut [1] obtained approximate buckling load factors for the overall buckling of uniformly axially compressed flat panels with either bonded or riveted Z-shaped stringers. He checked his results by comparing with predictions from the VIPASA code by Wittrick and Williams [2,3].

Riks [4] performed an analysis with use of the STAGSC1 program [5]. He included a study of sensitivity of the load factor corresponding to overall buckling to initial bowing imperfections, finding unstable postbuckling behavior (imperfection sensitivity) caused by deformation of the stringer cross section in the overall buckling mode. In "classical" wide column buckling of a panel stiffened with T-shaped stringers, for example, the T-stringer cross section remains undeformed as it translates normal to the skin surface in the wide column buckling mode. This is not so with Z-stiffened panels. In that case, because the stringers have a nonsymmetric cross section, they undergo significant sidesway as the panel skin essentially translates normal to the undeformed panel skin in the overall buckling mode. Hence, the load at which a Z-stiffened panel collapses under uniform axial compression is sensitive to an initial overall bowing imperfection even if the local buckling load factor significantly exceeds that corresponding



*D. Durban et al. (eds.), Advances in the Mechanics of Plates and Shells,* 79–102.

to general ("wide column") instability.

Local and overall bifurcation buckling of panels with Z-shaped stringers can also be determined with the BUCLASP code [6] and with the newer successors to BUCLASP and VIPASA: the PANDA2 [7], POSTOP [8], VICONOPT [9], and PASCO [10] codes. PASCO, VICONOPT, PANDA2 and POSTOP are capable of obtaining optimum designs of such panels and PANDA2 and POSTOP can do so including the effect of local postbuckling [11] of the panel skin and/or parts of the stringers. The authors of VICONOPT [9] are currently working on a postbuckling capability [9]. One of the PANDA2 processors, called STAGSMODEL [12] automatically sets up a finite element model of a panel previously optimized with PANDA2. The [PANDA2, STAGSMODEL, STAGS] combination has been used many times to optimize and evaluate optimum designs of panels under combined loads for service in the postbuckling regime [11-15]. Other works are briefly surveyed in [16,17]. This paper is a condensed version of [16], which is a condensed version of [17].

## 3. Method of Analysis

The purpose of the work reported here is to enhance the capability of PANDA2 [7] by inclusion of Z-shaped stiffeners riveted to the panel skin. A discretized panel skin-stringer single module is constructed as shown in Figs. 1 - 4. The stringer spacing is called "b" and the rivet line, considered to be continuous in the axial direction and located at the midwidth of the attached flange (Segment 2 of the single module shown in Fig. 1), is located at b/2, the midwidth of the entire module. The toe and heel of the attached flange are free to separate from or to "penetrate" the panel skin in the local buckling mode and in the post-local buckling regime: no intermittant contact conditions are imposed in the PANDA2 model. At the rivet line compatibility conditions are imposed between panel skin and attached flange. Eccentricity of the reference surface of the attached flange with respect to the reference surface of the panel skin is accounted for.

The sketches in Figs. 1(a) and 1(b) are displayed on the computer screen as a guide for the user during his/her interactive input session. Figures 2 - 4 show the discretization of the cross section of the single module. Symmetry conditions are imposed at Node 1 of Segment 1 and at Node 11 of Segment 5 [Fig. 1(b)]. Figure 2, a local buckling mode of a nonoptimized cross section, demonstrates that the attached flange of the Z-stiffener can bend differently from the panel skin to which it is attached along the rivet line. In contrast, T and J stiffeners are assumed to be bonded to the panel skin.

Local buckling is determined by preventing the root of the web of the Z-stiffener [Node 1 of Segment 3] from moving in a direction normal to the panel skin, as shown in Figs. 2 and 3. The location of this nodal point is slightly above the dashed horizontal line at the ordinate value of zero ($z=0$) because Segment 2, which represents the reference surface of the attached flange of the Z-stiffener, is located in this example at a distance, $(t_1 + t_2)/2$, above the reference surface of the panel skin, which is at $z=0$. The quantity $t_1$ is the thickness of the panel skin and $t_2$ is the thickness of the attached flange.

Overall buckling is predicted from both a single-module discretized "wide column" model, such as shown in Fig. 4, and from a model in which the stiffeners are "smeared

(a) MODULE WITH Z-SHAPED STIFFENER...

```
                                   !<-- w -->!
                                   ___________
                                   !   ^     .
               Segment No. 3 -----> !   !      .
                                   !   !      .Seg. No. 4
                  Seg. No. 2-.     !   h
                             .     !   !
             Seg. No. 1-.      .   !   !         .-Seg. No. 5
                       .     R  .  !   !        .(same as Seg.1)
                      .  _____I_____!  V        .
----------------------- V ----------------------
                        !    E     !
                        !    T     !
                        !<---b2--->!
!<---- Module width  =  stiffener spacing, b ---->!
```

(b) EXPLODED VIEW, SHOWING LAYERS and (SEGMENT, NODE) NUMBERS

```
                         Layer No. 1___
                                        .
                                         .
            (Segment,Node)=(4,1)____________(4,11)
                                           .
                          (3,11)        .__Layer No. j
                             !
            Layer No. 1 -----> ! <----------- Layer No. k
                             !
            Layer No. 1-.    !
  Layer No. 1-.          .     !         .-Layer No. 1
               .         .   !          .
              .        R  . (3,1)        .
             .  (2,1)______I______(2,11)       .
--------------------- V -----------------------
(1,1)      .      (1,11) E (5,1)           .   (5,11)
              .          T                   .
               .                              .
            Layer No. m                       Layer No. m
```

*Figure 1.* Cross section dimensions, segment numbering and nodal point numbering of riveted Z-stiffened panel module. These sketches are presented to the PANDA2 user during interactive input and in the output.

out" (averaged) over the panel in the manner of Baruch and Singer [18]. In the "wide column" model the root of the web of the Z-stiffener [Node 1 of Segment 3 in Fig. 1(b)] is permitted to move, as demonstrated in Fig. 4.

Discretization of the cross section of the single module model is via the finite difference energy method, as described in [19]. There are a number of nodal points in each of the segments of the module cross section, as shown in Fig. 1(b). Variation along the axis of the panel (normal to the plane of the paper) is assumed to be trigonometric. The critical number of axial halfwaves and critical slope of the buckling nodal lines in the plane of the panel skin (for an anisotropic panel and/or a panel in which in-plane shear loading $N_{xy}$ is present) are determined in the analysis described in detail in [11].

Local buckling modes typically have a form such as shown in Fig. 3, which is assumed to be sinusoidal in the axial direction with a computed critical number of axial halfwaves and a computed slope of the local buckling nodal lines [11] in the critical local buckling mode. The local buckling nodal lines are assumed to be straight.

The nonlinear local postbuckling analysis [11] is analogous to that of Koiter [20]. In PANDA2 the axial wavelength and the slope of the nodal lines of the postbuckled pattern are permitted to change as the applied load is increased above that corresponding to initial local bifurcation buckling. Also, the post-locally buckled panel skin is permitted to "flatten" in the region midway between stringers, as described by Koiter [20], who introduced a "flattening parameter", $a$, into his analysis [11]. Details of the nonlinear post-local buckling analysis and predictions from PANDA2 and STAGS [21] appear in [11].

The "wide column" buckling mode has a form such as shown in Fig. 4. There is assumed to be one-half wave in the axial direction in the "wide column" mode. The "softening" effect which influences the effective overall axial, hoop, and in-plane shear membrane stiffness components of a locally imperfect and/or locally postbuckled panel module (See Fig. 15 of [12], for example) is accounted for in the computation of the bending-torsional, "wide column" and overall buckling load factors.

## 4. Implementation of Z-shaped Stringers into PANDA2

The file that contains prompting phrases and "help" paragraphs for the user was modified to include Z-shaped stiffeners. Special sketches for Z-shaped modules were introduced into the PANDA2 input prompts and output as depicted in Figs. 1 (a,b).

In order to maintain conservativeness of optimized designs, no allowance is made in PANDA2 for clamping Z-shaped stringer-stiffened panels along the two axially loaded ends as opposed to simple support there. In the computation of "effective" axial length (described for clamped panels in [7] and in ITEMS 3, 79a, 105e, 106, and 113d,r of the PANDA2 documentation file, ...panda2/doc/panda2.news [22]), a panel with Z-shaped stringers is ALWAYS treated as if it were simply supported along its two axially loaded ends even if the user indicates clamping there. This strategy was introduced during the testing phase of implementation of the "Z" capability. Comparison of results from PANDA2 and STAGS [21] revealed that this strategy is required to maintain conservativeness of optimum designs of imperfect (bowed) panels generated by PANDA2.

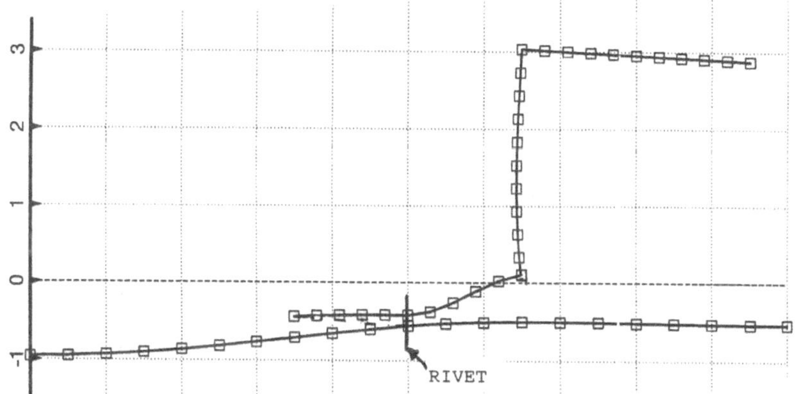


*Figure 2.* Local buckling of a module with a very weak attached flange.

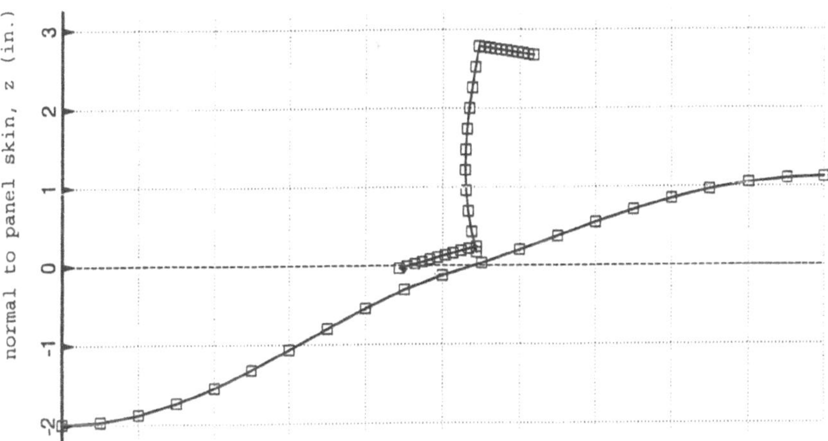


*Figure 3.* Local buckling of a module of an optimized panel.

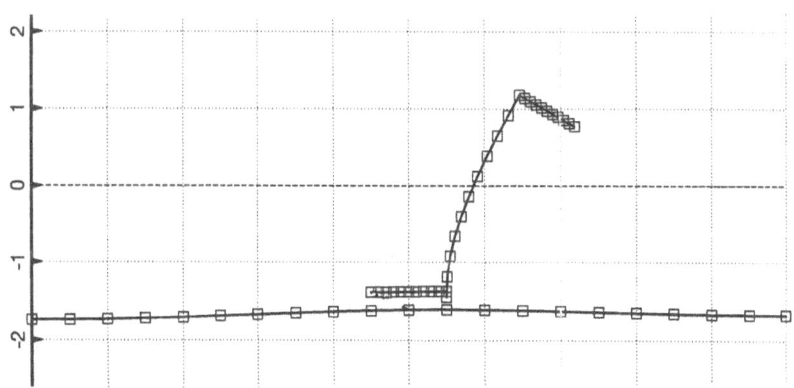


*Figure 4.* Wide column (general) buckling of a module of an optimized panel.

It is thought that this new strategy, now introduced into PANDA2 also for J-shaped stringer stiffened panels, compensates for the softening effect on effective axial stiffness of sidesway of the "unbalanced" stringers as an initially axially bowed panel with Z or J stringers deforms further under application of axial compression [4]. (By "unbalanced" is meant here a stringer which is not symmetrical with respect to a plane normal to the panel skin and containing the line of attachment of the stringer to the panel skin). Unfortunately, in its nonlinear Koiter-type postbuckling analysis [11], PANDA2 accounts for deformation of the single module cross section ONLY in the LOCAL buckling mode, such as that shown in Fig. 3, not also in the "wide column" (general) buckling mode, such as that shown in Fig. 4. (Note, however, that in one of the branches in which the maximum stresses are computed, PANDA2 does include this effect in an approximate manner, as described in [23]).

It is emphasized here that the neglect of further stiffener sidesway in PANDA2's analysis of the overall collapse of Z or J stiffened panels with overall initial buckling modal imperfections may well lead to the generation of unconservative designs in the case of imperfect Z or J stiffened panels that are actually simply supported along the two axially loaded ends. In such cases, the user should introduce factors of safety for inter-ring and overall buckling that are greater than unity, perhaps in the range 1.2 - 1.5.

The sequence of execution of PANDA2 modules called BEGIN, DECIDE, MAINSETUP, PANDAOPT, CHOOSEPLOT, DIPLOT, etc. is described in [7] and [22]; the use of the processor called SUPEROPT (for obtaining global optimum designs) is discussed in [14]; and the "automatic" generation of STAGS finite element models of panels previously optimized by PANDA2 via the PANDA2 processor called STAGSMODEL is demonstrated in [12]. There also exists a PANDA2 processor called PANEL [7] that generates an input file for BOSOR4 [19] for a panel previously optimized by PANDA2. (PANEL is valid only for panels with insignificant in-plane shear loading). The BOSOR4 model of the panel generated by PANEL is of "annular" form, as described in [24].

Optimization in PANDA2 is performed via the ADS routines written by Vanderplaats and his colleagues [25, 26].

## 5. Summary of Numerical Results

Optimization is performed for flat aluminum, elastic panels with Z-shaped, J-shaped, and T-shaped stringers. There are no transverse stiffeners (no rings). The panels are subjected to uniform axial compression, $N_x$ = -2000 lb/in, and are clamped along the two axially loaded edges. All properties and decision variables and their lower and upper bounds are listed in Table 1. A typical PANDA2 runstream for optimization is listed in Table 2. Results for this study are summarized in Tables 3 and 4. Full details appear in [17].

Results from ten cases are summarized in Table 3 (columns 1 - 10 in Table 3): eight of the cases for a panel with riveted Z-shaped stringers, one for a panel with bonded J-shaped stringers and one for a panel with bonded T-shaped stringers. The first five cases (columns 1 - 5 in Table 3) are for perfect panels and the second five cases

TABLE 1. Panel to be optimized by PANDA2 (Z-stiffened, aluminum)

| | |
|---|---|
| Overall length of the panel | L1 = 50 in. |
| Overall width of the panel | L2 = 50 in. |
| Young's modulus, | E = 10 msi |
| Poisson ratio | nu = 0.3 |
| Weight density | rho = 0.1 lb/in**3 |
| Maximum allowable effective stress | SIGBAR = 45 ksi |
| Boundary conditions: | clamped along two axially loaded edges<br>simple support along two unloaded edges |
| Applied load | Nx = -2000 lb/in |

| DECISION VARIABLES (inches) | LOWER BOUND | INITIAL VALUE | UPPER BOUND |
|---|---|---|---|
| b = stringer spacing | 5.0 | 10.0 | 10.0 |
| b2 = width of attached flange | 1.0 | 3.0 | 5.0 |
| h = height of web | 0.5 | 3.0 | 5.0 |
| w = width of outstanding flange | 0.5 | 3.0 | 5.0 |
| t1 = thickness of panel skin | 0.01 | 0.15 | 0.5 |
| t2 = thickness of attached flange | 0.01 | 0.05 | 0.5 |
| t3 = thickness of web | 0.01 | 0.10 | 0.5 |
| t4 = thickness of outstanding flange | 0.01 | 0.10 | 0.5 |

TABLE 2. Sequence of PANDA2 commands used to obtain an optimum design

| Command | Function performed by command |
|---|---|
| BEGIN | (Provide starting design; input data listed in Table 2 in [17]) |
| SETUP | (Set up matrix templates. no input data required. See Ref. [7]) |
| DECIDE | (Choose decision variables and bounds. Table 7 in [17]) |
| MAINSETUP | (Choose loading, analysis type, model type... Table 8 in [17]) |
| PANDAOPT | ("batch" execution of PANDA2 mainprocessor. See Ref. [7]) |
| PANDAOPT | " |
| PANDAOPT | " |
| PANDAOPT | " |
| CHOOSEPLOT | (Choose which decision variables, margins to plot. ) |
| DIPLOT | (Obtain plots. See Refs. [12,13]) |
| SUPEROPT | (Attempt to find global optimum design. See Ref. [14]) |
| CHOOSEPLOT | |
| DIPLOT | |
| SUPEROPT | |
| CHOOSEPLOT | |
| DIPLOT | |
| SUPEROPT | |
| . | |
| . | |
| . | |

TABLE 3. Summary of results for various optimum designs from PANDA2, BOSOR4 and STAGS (Dimensions in inches)

| PART 1: CHARACTERISTICS OF CASE | | | | | | | | | | |
|---|---|---|---|---|---|---|---|---|---|---|
| Was SUPEROPT used?--> | no | yes | yes | yes | yes | yes | yes | no | yes | yes |
| Local postbuckling?-> | no | no | no | yes | yes | yes | yes | yes | yes | yes |
| Stringer spacing----> | 10in. | 10in. | var. | var. | var. | var. | var. | var. | var. | var. |
| Bowing Imperfection-> | 0 | 0 | 0 | 0 | 0 | -0.1 | -0.1 | +0.1 | -0.1 | -0.1 |
| Stringer type-------> | Z | Z | Z | Z | Z | Z | Z | Z | J | T |
| Modejump constraint-> | - | - | - | OFF | ON | OFF | ON | ON | ON | ON |
| Table number [17]---> | 9 | 14 | 15 | 16 | 18 | 19 | 20 | 21 | 22 | 23 |
| Column number-------> | 1 | 2 | 3 | 4 | 5 | 6 | 7 | 8 | 9 | 10 |
| **PART 2: OPTIMUM DESIGN OF 50 in. x 50 in. PANEL** | | | | | | | | | | |
| panel weight (lbs) | 51.23 | 49.33 | 36.15 | 30.07 | 32.21 | 33.49 | 35.22 | 35.22 | 31.08 | 29.58 |
| WIDTHS (in.): | | | | | | | | | | |
| stringer spacing,b | 10.00 | 10.00 | 5.00 | 5.94 | 5.00 | 6.25 | 5.19 | 5.19 | 7.31 | 6.06 |
| attached flange, b2 | 1.00 | 1.00 | 1.00 | 1.14 | 1.00 | 1.06 | 1.00 | 1.00 | 2.06 | 2.02 |
| stringer height, h | 2.00 | 2.62 | 2.17 | 2.35 | 2.29 | 2.30 | 2.19 | 2.19 | 2.22 | 1.69 |
| outstandng flnge,w | 1.42 | 0.723 | 0.500 | 0.907 | 0.539 | 1.08 | 0.979 | 0.979 | 1.49 | 1.07 |
| THICKNESSES (in.): | | | | | | | | | | |
| skin thickness, t1 | 0.1713 | .1504 | .0887 | .0680 | .0831 | .0779 | .0887 | .0887 | .0588 | .0640 |
| attched flange, t2 | 0.0726 | .3089 | .1503 | .0905 | .0644 | .0891 | .0568 | .0568 | .0780 | .0903 |
| web thickness, t3 | 0.0835 | .0438 | .0424 | .0682 | .0542 | .0759 | .0604 | .0604 | .0589 | .0502 |
| outstndng flnge,t4 | 0.0680 | .0634 | .0740 | .0521 | .0746 | .0754 | .0835 | .0835 | .1255 | .0578 |
| **PART 3: LOCAL BUCKLING LOAD FACTOR OF THE OPTIMIZED DESIGN (PANDA2 prediction)** | | | | | | | | | | |
| | 1.105 | 1.095 | 1.117 | 0.364 | 0.639 | 0.389 | 0.614 | 0.614 | 0.299 | 0.567 |
| **PART 4: LOCAL BUCKLING LOAD FACTOR OF THE OPTIMIZED DESIGN (BOSOR4 prediction)** | | | | | | | | | | |
| | 1.114 | 1.119 | 1.131 | | | | | | | |
| **PART 5: LOCAL BUCKLING LOAD FACTOR OF THE OPTIMIZED DESIGN (STAGS prediction)** | | | | | | | | | | |
| | 1.142 | 1.130 | 1.104 | 0.372 | 0.642 | 0.467 | 0.728 | 0.745 | 0.356 | 0.664 |
| **PART 6: WIDE COLUMN BUCKLING LOAD FACTOR (PANDA2 prediction)** | | | | | | | | | | |
| | 1.116 | 1.143 | 1.134 | 1.147 | 1.088 | 1.310 | 1.277 | 1.282 | 1.145 | 2.110 |
| **PART 7: WIDE COLUMN BUCKLING LOAD FACTOR (BOSOR4 prediction)** | | | | | | | | | | |
| | 1.125 | 1.161 | 1.185 | | | | | | | |
| **PART 8: LATERAL-TORSIONAL BUCKLING LOAD FACTOR (PANDA2 prediction)** | | | | | | | | | | |
| | 1.100 | 1.134 | 1.385 | 1.067 | 1.099 | 1.053 | 1.048 | 1.048 | 1.090 | 1.095 |
| **PART 9: COLLAPSE LOAD FACTOR (STAGS prediction)** | | | | | | | | | | |
| | 2.30 | 1.72 | 1.62 | 1.25 | 1.35 | 1.39 | 1.32 | 1.42 | >1.7 | 1.72 |
| **PART 10: LOAD FACTOR FOR CRITICAL EFFECTIVE STRESS (PANDA2 prediction)** | | | | | | | | | | |
| | 4.61 | 4.43 | 3.25 | 1.016 | 1.262 | 0.964 | 1.069 | 1.069 | 0.996 | 0.994 |
| **PART 11: LOAD FACTOR FOR CRITICAL EFFECTIVE STRESS (STAGS prediction)** | | | | | | | | | | |
| | | | | 0.91 | 1.1 | 1.15 | 1.20 | 1.20 | 1.05 | 1.07 |
| **PART 12: VALUES OF MARGINS AT THE OPTIMUM DESIGN ACCORDING TO PANDA2** | | | | | | | | | | |
| Column number-------> | 1 | 2 | 3 | 4 | 5 | 6 | 7 | 8 | 9 | 10 |
| local buckling | 0.005 | -.001 | -.016 | ---- | ---- | ---- | ---- | ---- | ---- | ---- |
| bendng-tors buckl. | 0.016 | 0.039 | 0.034 | -.001 | 0.261 | 0.020 | 0.011 | 0.011 | ---- | -.006 |
| effective stress | 3.61 | 3.43 | 2.250 | 0.016 | 0.262 | -.036 | 0.069 | 0.069 | -.004 | -.006 |
| mode jumping | 0.675 | 0.729 | 0.593 | ---- | -.002 | ---- | -.028 | -.028 | 0.482 | 0.627 |
| wide column buckl. | 0.060 | 0.144 | 0.082 | 0.043 | -.011 | 0.191 | 0.161 | 0.165 | 0.041 | 0.920 |
| laterl-tors buckl. | 0.000 | 0.031 | 0.259 | -.030 | -.001 | -.043 | -.047 | -.047 | -.009 | -.005 |
| stringer seg. 2 | 24.3 | 190.0 | 60.7 | 11.6 | 8.70 | 11.2 | 5.86 | 5.86 | ---- | ---- |
| stringer seg. 3 | 5.32 | -0.002 | -.003 | 0.402 | -.001 | 0.665 | 0.290 | 0.290 | 0.169 | 0.238 |
| stringer seg. 4 | 0.876 | 2.40 | 5.65 | 0.067 | 3.54 | -.030 | 0.497 | 0.497 | 1.07 | 0.630 |
| stringer seg. 3+4 | 1.37 | -0.003 | 0.217 | -.002 | -.002 | -.028 | -.019 | -.019 | -.007 | -.007 |
| general buckling | 0.026 | -.001 | -.001 | 0.150 | 0.103 | 0.369 | 0.381 | 0.427 | 0.843 | 1.50 |
| stringer rolling | 7.61 | 0.870 | 0.431 | 1.55 | 0.288 | 0.947 | 0.621 | 0.621 | 0.738 | 0.008 |
| hiwave str rolling | 1.39 | 1.27 | 2.33 | 0.554 | 1.46 | 0.159 | 0.340 | 0.340 | 0.213 | 0.452 |
| str. web buckling | 5.31 | -0.002 | -.003 | 0.462 | 0.041 | 0.408 | 0.072 | 0.072 | 0.034 | -.010 |

TABLE 4. Comparison of predicions from PANDA2 and STAGS for the maximum effective stresses in various panels optimized by PANDA2 and loaded well into their local postbuckling regimes. The panel is loaded by the design load, axial compression, $N_x$ = -2000 lb/in.

| Location in single module cross section | Max. Effective Stress (psi) from PANDA2 | from STAGS | Figure in [17] |
|---|---|---|---|
| PART 1: Perfect "Z" panel optimized with the modejump constraint turned OFF (4)* | | | |
| At x=25 in. midway between stringers ** | 26391. | 35200. | 90 |
| At x=25 in. in panel skin at rivet line | 44204. | 33500. | 91 |
| At x=29.75 in. in panel skin at rivet line | 44204. | 40000. | 92 |
| At x=25 in. in attached flange at rivet | 35322. | 33000. | 93 |
| At x=25 in. in attached flange next to web | 37458. | 36500. | 94 |
| At x=25 in. in web at web root | 39073. | 39000. | 95 |
| PART 2: Perfect "Z" panel optimized with the modejump constraint turned ON (5) | | | |
| At x=27.25 in. midway between stringers | 26385. | 35000. | 113 |
| At x=22.25 in. in panel skin at rivet line | 30164. | 22300. | 114 |
| At x=22.25 in. in attached flange at rivet | 28939. | not obtained | |
| At x=22.25 in. attached flange next to web | 34982. | 30500. | 115 |
| At x=22.25 in. in web at web root | 35663. | 30000. | 116 |
| PART 3: Imperfect "Z" panel optimized with the modejump constraint turned OFF(6) | | | |
| At x=26.25 in. midway between stringers | 29668. | 33000. | 138 |
| At x=25.75 in. in panel skin at rivet line | 44788. | 36000. | 139 |
| At x=25.75 in. in attached flange at rivet | 40511. | 25000. | 140 |
| At x=20.75 in. attached flange next to web | 45450. | 33000. | 141 |
| At x=20.75 in. in web at web root | 46698. | 32500. | 142 |
| PART 4: Imperfect "Z" panel optimized with the modejump constraint turned ON (7) | | | |
| At x=21.75 in. midway between stringers | 28199. | 33000. | 159 |
| At x=21.75 in. in panel skin at rivet line | 33234. | 21000. | 160 |
| At x=25.75 in. in panel skin at rivet line | 33234. | 26000. | 161 |
| At x=21.75 in. in attached flange at rivet | 30869. | 25000. | 162 |
| At x=25.75 in. in attached flange at rivet | 30869. | 20000. | 163 |
| At x=21.75 in. attached flange next to web | 42092. | 28000. | 164 |
| At x=25.75 in. attached flange next to web | 42092. | 24000. | 165 |
| At x=21.75 in. in web at web root | 39033. | 25000. | 166 |
| At x=25.75 in. in web at web root | 39033. | 24000. | 167 |
| PART 5: Imperfect "J" panel optimized with the modejump constraint turned ON (9) | | | |
| At x=23.75 in. in panel skin next to base | 45627. | 42000. | 205 |
| At x=20.75 in. in base next to panel skin | 45788. | 33000. | 206 |
| At x=21.75 in. in base at web root | 34904. | 30500. | 207 |
| At x=23.75 in. in web at web root | 35776. | 30000. | 208 |
| PART 6: Imperfect "T" panel optimized with the modejump constraint turned ON(10) | | | |
| At x=27.25 in. in skin midway betw. stiff. | 28794. | 40600. | 228 |
| At x=23.25 in. in panel skin next to base | 45169. | 43500. | 229 |
| At x=22.25 in. in base next to panel skin | 44743. | 32000. | 230 |
| At x=25.00 in. in base at web root | 31335. | not shown | |
| At x=25.00 in. in web at web root | 28521. | not shown | |

NOTES: * (n), where n = column number in Table 3

** x=25 in. corresponds to the panel midlength.

(columns 6 - 10) are for panels with an initial general buckling modal imperfection with amplitude equal to plus or minus 0.1 in. The first three cases (columns 1 - 3) are for a perfect panel in which local postbuckling is NOT permitted. In the first two cases the stringer spacing b is held constant at 10.0 inches, and in the remaining 8 cases the stringer spacing is one of the decision variables. The effect of the "modejump constraint" (See [15]) is explored for a perfect panel in Cases 4 and 5 and for an imperfect panel in Cases 6 and 7. For the optimum designs obtained by PANDA2 comparisons are made with predictions from BOSOR4 [19] and STAGS [21]. The units used in this study are inches and lbs.

Table 4 lists comparisons from PANDA2 and STAGS for the maximum effective stresses in the optimized designs corresponding to Cases 4, 5, 6, 7, 9, and 10. In all these cases the axial load, $N_x$ = -2000 lb/in, corresponds to the panel being loaded well beyond local buckling. (See PART 3 of Table 3).

Design margins corresponding to the optimum design for each of the 10 cases are listed in PART 12 of Table 3. In PANDA2 buckling margins are defined as follows:

buckling margin = (buckling load factor)/(factor of safety) -1.

Stress margins in PANDA2 are defined as follows

stress margin = (allowable stress)/[(actual stress)(factor of safety)] -1.

In this paper only results corresponding to Column 4 of Table 3 will be discussed. For more information see [16, 17].

## 6. Discussion of Results Listed in Column 4 of Table 3

Column 4 of Table 3, PART 1 of Table 4, and Figs. 5 - 14 pertain to this section. Local buckling is permitted and the "modejump" constraint [15] is turned OFF, that is, mode jumping IS PERMITTED in the optimized design.

### 6.1 OPTIMIZATION

Figure 5 shows the evolution of the panel weight during execution of the one SUPEROPT [14] performed in this case. The "modejump" constraint was turned OFF during optimization, that is, incipient mode jumping, a phenomenon modelled in PANDA2 as described in detail in [15], was ignored during optimization. Each "spike" in Fig. 5 represents a new starting design in the SUPEROPT cycle, a starting design generated automatically via the PANDA2 processor called AUTOCHANGE [14]. As explained in [14], the starting designs are generated by random changes in the vector of decision variables consistent with the lower and upper bounds and inequality constraint conditions provided by the PANDA2 user in DECIDE.

### 6.2 RESULTS FROM PANDA2 FOR THE OPTIMUM DESIGN

The optimum weight of the panel, 30.07 lbs (Col. 4, PART 2 of Table 3), is about 20 per cent less than that (36.15 lbs) obtained for the optimum design in which local postbuckling is not permitted. The local buckling load factor of the new optimum design, 0.364 (Col. 4, PART 3 of Table 3), indicates that at the design load the optimized panel is loaded well into the local postbuckling regime. The critical number of axial halfwaves in the local buckling mode is 10 [17]. As seen from Col. 4, PART 12 of Table 3 the following design margins are critical or almost so: bending-torsion buckling (-.001), maximum effective stress (0.016), wide column buckling (0.043), lateral-torsional buckling (-0.030), buckling of stringer segment 4 (outstanding flange, 0.067), buckling of stringer segments 3 and 4 together (web and outstanding flange, -.002), and general buckling (0.150). (In PANDA2 very small negative margins are permitted for feasible designs).

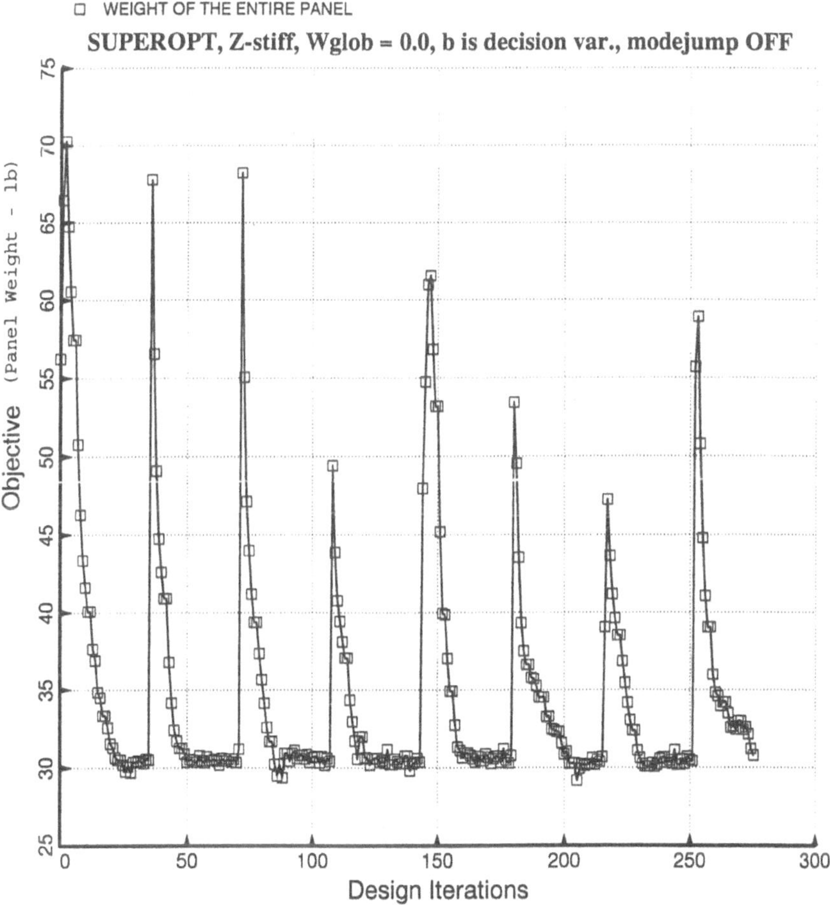


*Figure 5.* Evolution of objective during execution of SUPEROPT, a processor by means of which PANDA2 finds a global optimum design.

The margin called "bending-torsion buckling" is computed in exactly the same way as the margin for local buckling, that is, redistribution of stress resultants, $N_x$, $N_y$, $N_{xy}$, in the panel skin and stringer segments that occurs for loads in excess of the local buckling load is not accounted for in the computation of the "bending-torsion buckling" constraint condition. In contrast, the margin called "lateral-torsional buckling" is computed accounting for this stress redistribution, which is calculated in the "Koiter" (postbuckling) branch of PANDA2 [11]. In this particular case "bending-torsion buckling" is predicted by PANDA2 to occur with two axial halfwaves over the 50-inch length of the panel and "lateral-torsional buckling" is predicted to occur with one axial halfwave. Margins 3 and 10 in PART 16 of Table 16 in [17] appear as follows:

3 -7.31E-04 Bending-torsion buck.(bypassed low-m mode); M=2; FS=1.1 (FS = "factor of safety")

10 -2.98E-02 (m=1 lateral-torsional buckling load factor)/(FS)-1; FS=1.1

The margins called "wide column buckling" and "general buckling" are computed with the stress redistribution accounted for. The "wide column buckling" margin is computed from a single discretized module model such as shown in Fig. 4. The "general buckling" margin is computed from a model in which the effect of the stringers is "smeared" (averaged) over the panel width in the manner of Baruch and Singer [18]. The buckling margins called "stringer seg. n" are computed from a PANDA-type (closed form) model [27] in which stringer segment n is assumed to be simply supported along its line of intersection with other segments of the panel module if Segment n represents an "internal" segment (such as the web of the Z) and in which one of the longitudinal edges is considered to be free if Segment n represents an "end" segment (such as the outstanding flange of the Z). The effect of stress redistribution during local postbuckling is accounted for in the computation of these margins.

Values for "local buckling" and "mode jumping" margins are not given in Col. 4, PART 12 of Table 3 because the PANDA2 user has indicated in the PANDA2 input data for MAINSETUP that these phenomena should not constrain the design in this particular case. The "local buckling" margin becomes non-critical because the PANDA2 user has set the factor of safety for local buckling equal to 0.3. (See Margin No. 1 listed in PART 16 of Table 16 in [17]).

Figure 6 shows how the single discretized, optimized panel module deforms, according to PANDA2, as the panel is loaded beyond the local buckling load. As the panel is loaded further and further into the postbuckled regime, one can see the "flattening" of the postbuckled profile, especially in the panel skin on the left-hand side of the module cross section. Figure 7 shows how the local buckles grow and how the postbuckled panel bows with increasing axial compression, $N_x$. Since there is essentially zero initial local buckling modal initial imperfection (Wloc = 0.1E-06 in Table 17 of [17]), there is an abrupt change in behavior at the local buckling load factor, $N_x(cr)$ = -2000 x 0.364 = -728 lb/in. The "corner" becomes rounded off when there exists a local buckling modal initial imperfection with a significant amplitude (See Fig. 126 of [17], for example).

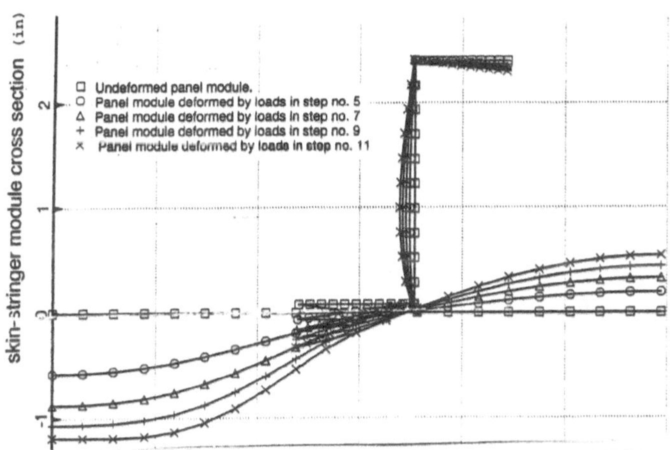


*Figure 6.* Postbuckling deformation of module cross section from PANDA2.

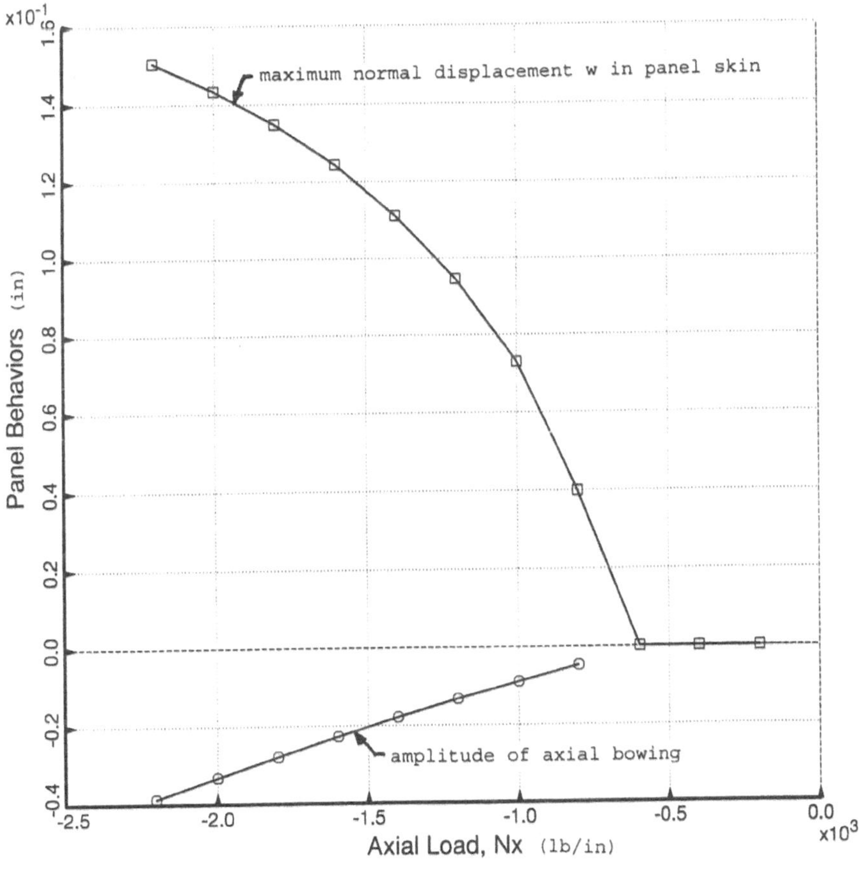

*Figure 7.* Postbuckling deflection *w* of panel skin and overall bowing from PANDA2.

## 6.3 COMPARISON OF RESULTS FROM PANDA2 AND STAGS

The PANDA2 processor called STAGSMODEL [12] makes it relatively easy to generate input data for STAGS [21] corresponding to a panel previously optimized by PANDA2. In this way, the quality of the optimum design obtained by PANDA2 can be evaluated by comparison with predictions from a general-purpose finite element computer program that does not use the many "tricks" and approximations employed in PANDA2 in order to save computer time, which must be efficiently used during optimization cycles.

At present, STAGSMODEL works only for panels that are clamped along the two axially loaded edges and only for panels that do not contain any transverse stiffeners (rings). STAGSMODEL produces the two STAGS input files, *.bin and *.inp (in which "*" denotes a user-selected name for the case), for what in STAGS jargon is called an "element unit" (no "shell units"). Therefore, the *.inp file is often very large (more than a megabyte). Typical input data for STAGSMODEL are listed in Table 12 of [17].

In this particular case the STAGS finite element model, shown in Fig. 8, consists of only a single module. Therefore, the STAGS model is analogous to the PANDA2 model displayed in Fig. 6. A one-module STAGS model is acceptable in this case because the panel is not subjected to in-plane shear loading and the walls of the panel skin and Z-stiffener are not anisotropic. (However, see the discussion below about the maximum postbuckling stress midway between stringers.)

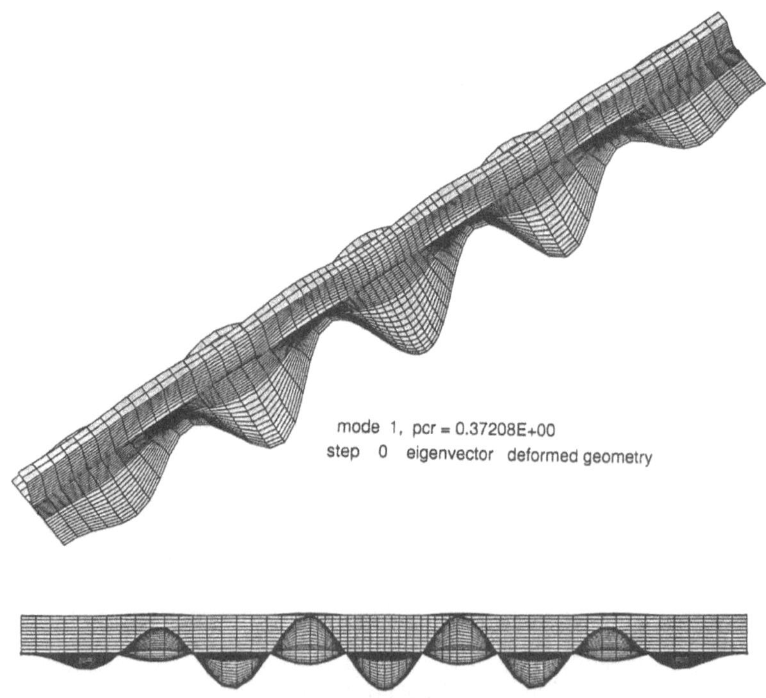


*Figure 8.* Bifurcation buckling mode from single-module STAGS model.

In the STAGS models generated during this study the two longitudinal edges are free to approach eachother and to undergo in-plane warping. This freedom of displacement along these edges generally leads to conservative results because the local buckles can become deeper in the postbuckling regime than would be the case if in-plane warping of the two longitudinal edges were prevented.

From linear analysis STAGS predicts a critical bifurcation buckling eigenvalue, pcr = 0.37208 (Fig. 8), which is in very good agreement with PANDA2's prediction of 0.364 (Col. 4, PART 3 of Table 3). As seen in Fig. 8(b) the local buckling mode from STAGS has 9 axial halfwaves, whereas PANDA2 predicts 10 axial halfwaves. The difference arises from the different boundary conditions used in the STAGS (clamped) and PANDA2 (simple support) at the two axially loaded ends of the panel module for the analysis of local buckling. In the central region of the panel the axial wavelengths of the local buckles as predicted by STAGS and PANDA2 are in very good agreement.

Next, it is necessary to find at what load STAGS predicts the optimized panel to collapse under uniform axial end shortening. The method of doing this is described in some detail in [12]. A nonlinear collapse analysis is performed with STAGS with use of the one-module model shown in Fig. 8. The STAGS model includes an initial imperfection in the form of the local buckling mode depicted in Fig. 8. This initial imperfection is required to avoid almost singular behavior in the neighborhood of the local buckling load at a load factor of about 0.372. With a small imperfection in the form of the local buckling mode there will be a smooth transition from prebuckling state to locally postbuckled state as the panel is loaded into its post-local-buckling regime. (See Fig. 6 of [11] for an example).

NOTE: In this paper for all the STAGS results a load factor, PA = 1.0, corresponds to the design load, $N_x$ = -2000 lb/in.

Figure 9 shows the load-end-shortening curve obtained via STAGS. There are several groups of load steps for which unloading occurs, as is demonstrated in Fig. 17 of [16]. Often in a case such as this the unloading represents “Riks reversal” (see Fig. 17 of [15]), in which the Riks path [28] doubles back on itself, converging to the same states determined in previous load steps. Superficially that appears to be the case here, since all the points in Fig. 9 appear to lie on the same fundamental curve. However the points do NOT all lie on the same fundamental curve, as will be shown next.

Figures 10(a) and 10(d) show the deformed state of the panel module at a load factor very near unity (PA = 1.00832), that is, very near the design load, PA=1.0, at Load Step 23. While the skin is deformed in a pattern similar to that shown in Fig. 8(b) (local buckling mode shape), the stringer undergoes a long-axial-wave sidesway similar to the deformation patterns corresponding to Margin Numbers 3 and 10 in PART 16 of Table 16 of [17]:

3 -7.31E-04 Bending-torsion buck.(bypassed low-m mode);M=2; FS=1.1
10 -2.98E-02 (m=1 lateral-torsional buckling load factor)/(FS)-1; FS=1.1

Figures 10(b-e) display edge-on views of the locally postbuckled panel module at four load steps, Steps 10, 15, 23, and 76. Note that the deformations shown in Figs. 10(a-d) (Load Steps 10, 15, 23) represent essentially growth of the local buckling lobes as the loading is increased. However, the equilibrium state of the panel at Load Step 76 is quite different from that at Load Step 23, even though the load factor PA and the end shortening *u* are, for all practical purposes, the same at these two load steps. At Load Step 76 the local buckling lobes along the one edge of the panel module have shifted relative to those along the other edge and an additional buckle has appeared (Fig. 11).

Figures 11(a,b), which are "fringe" plots of the normal displacement *w* in the panel skin as viewed from the surface of the panel skin opposite to the surface to which the stringer is attached, show more clearly the relative positions of the local buckling lobes on either side of the stringer at Load Steps 23 and 76. Note that at Load Step 76 (Fig. 11(b)) an additional buckle has appeared along the bottom half of the single panel module relative to the number of buckles apparent there at Load Step 23 (Fig. 10(a)). This change in equilibrium state represents a mode "jump". Often a mode "jump" can be captured only by means of a nonlinear transient STAGS run "sandwiched" between nonlinear static STAGS runs [15]. In this case, however, it turns out that the static Riks procedure [28] is capable of capturing the change in state of the panel between Load Step 23 and Load Step 76. As will be seen later, this mode "jump" has a significant influence on the maximum effective stress generated in the panel skin.

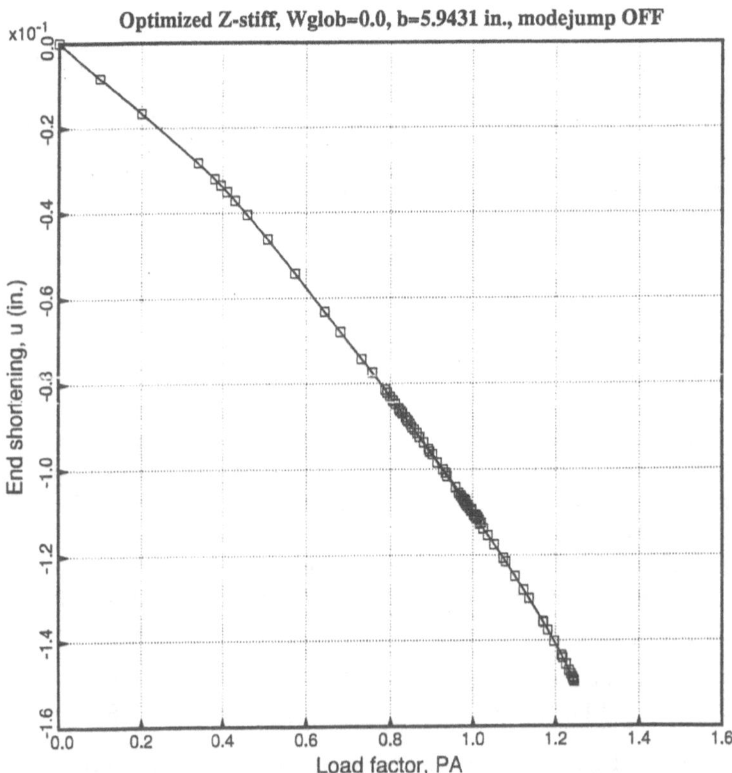


*Figure 9.* Load-end-shortening curve from the single-module STAGS model.

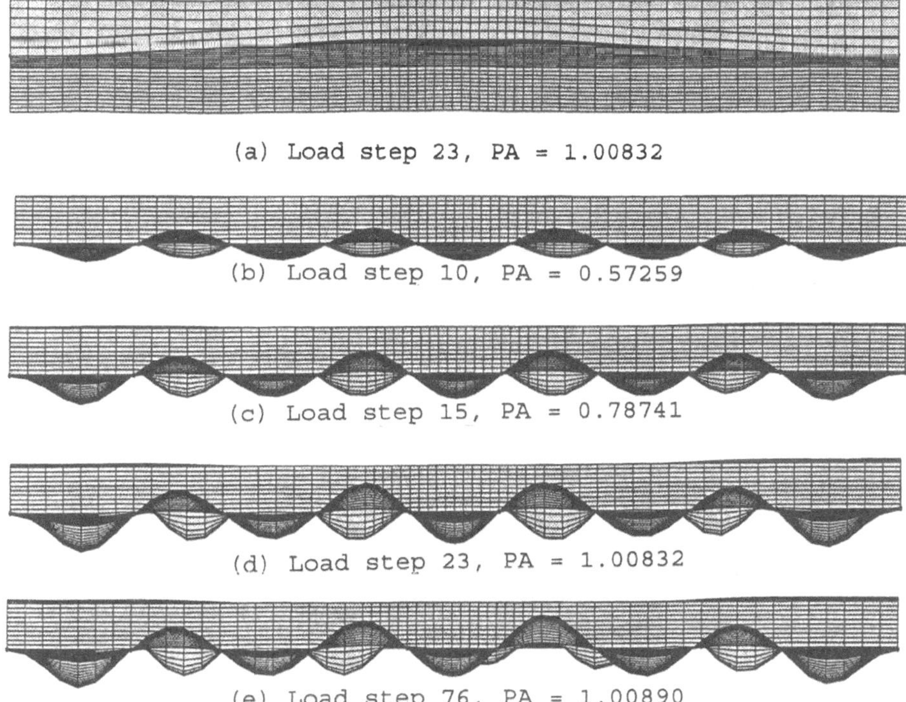


*Figure 10.* Edge-on views (b-e) of postbuckling deflection *w*, single-module STAGS model.

(a) Load step 23, PA = 1.00832

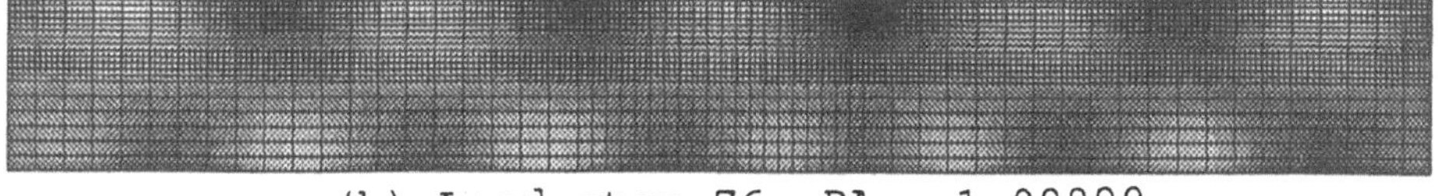

(b) Load step 76, PA = 1.00890

*Figure 11.* Fringe plots of postbuckling deflection *w* in skin of single-module STAGS model.

Figure 12 demonstrates the local buckling and postbuckling phenomena at a location along the axis of the panel, x = 30 inches from one end. The depth of the buckles is equal to the difference between the normal displacement *w* at either of the two longitudinal edges (curves with triangles and squares) and *w* in the panel skin under the stringer web (curve with circles). Up to Load Step 23 the buckle depth as a function of load obtained by STAGS agrees reasonably well with the prediction by PANDA2 shown in Fig. 7 as the curve with squares. After Load Step 23 it is clear that the buckle represented by the curves with triangles in Fig. 12 shifts along the panel axis, so that what was a maximum *w(x)* at Load Step 23 becomes a node (*w(x)*=0) for Load Step 58 in Fig. 12. The different axial positions of the local buckles along the longitudinal edge of the panel module represented by the curves with triangles in Fig. 12 are displayed for Load Steps 23 and 76 in the fringe plots of Figs. 11(a,b) corresponding to the bottommost of the two longitudinal edges.

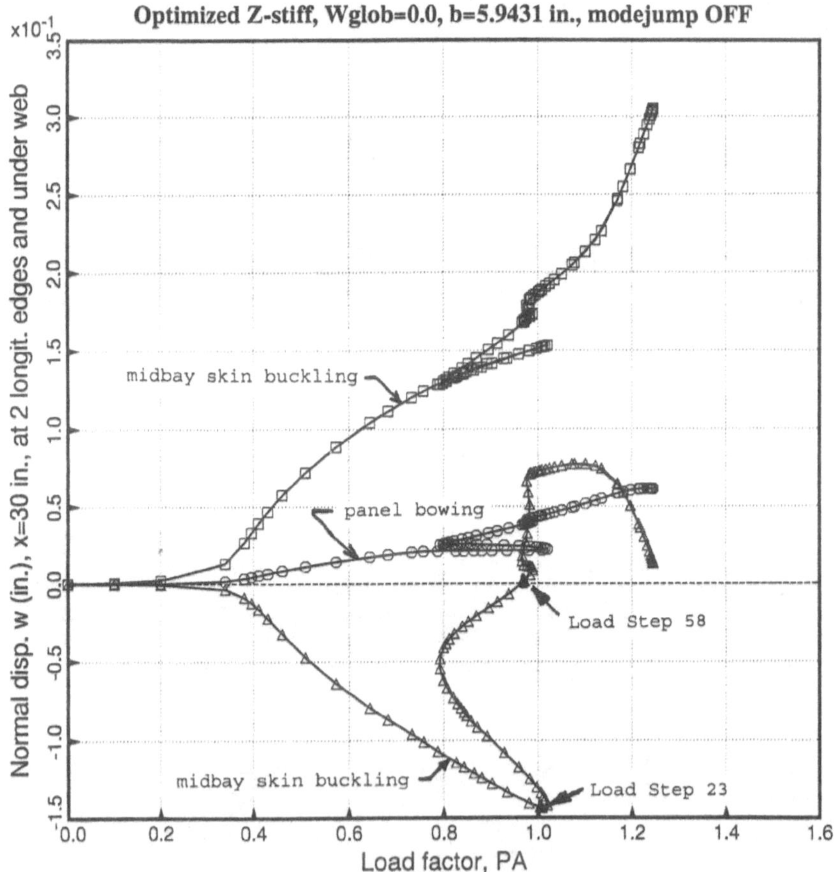


*Figure 12.* Normal deflection *w* of the panel skin at the web root and midway between Z-stiffeners from the single-module STAGS model.

Figure 13 shows sidesway of the stringer at the panel end (curve with squares) and at the panel midlength (curve with circles). Sidesway at the panel ends, a phenomenon that cannot be predicted by the PANDA2 analysis, begins immediately after local buckling. Sidesway at the panel midlength begins at a somewhat higher load. This also cannot be predicted by the PANDA2 analysis, which is based on the assumption that in the post-local buckling regime the panel module cross section deforms as shown in Fig. 6. According to PANDA2, sidesway of the stringer will exist in this case of a perfect panel only at loads in excess of the design load because the following margins listed in Part 16 of Table 16 in [17],

3 -7.31E-04 Bending-torsion buck.(bypassed low-m mode);M=2; FS=1.1
10 -2.98E-02 (m=1 lateral-torsional buckling load factor)/(FS)-1; FS=1.1

become significantly negative only for loads in excess of 1.1 times the design load, $N_x =$ -2000 lb/in.

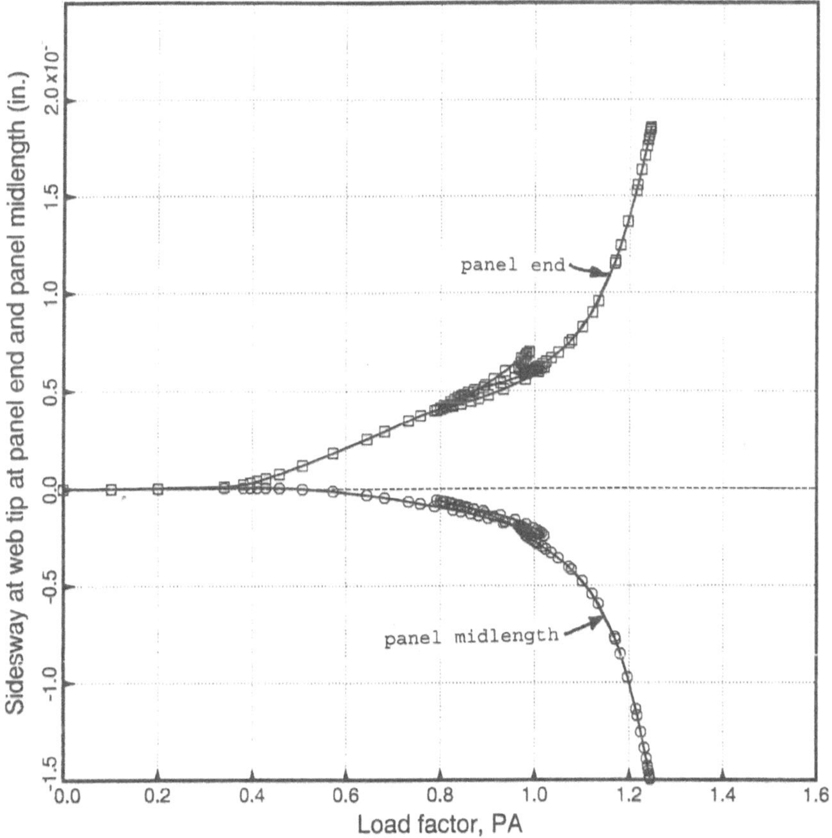


*Figure 13.* Sidesway of web tip at panel end and midlength from single-module STAGS model.

Figures 90 - 95 of [17] show extreme fiber effective (von Mises) stress vs load factor PA where fringe plots (not included because they are inadequate in black-and-white) indicate that these stresses are the highest of any in the single panel module and at which PANDA2 predicts that maxima occur in the single discretized module model. The values of these stresses at the design load, PA = 1.0, before the mode "jump" (at Load Step 23) can be compared with the values predicted by PANDA2. This comparison appears in PART 1 of Table 4.

Figure 14 shows the effective stress in the panel skin at the rivet line. This is the location of the maximum effective stress anywhere in the panel module cross section at the design load, PA = 1.0. Before the mode jump at Load Step No. 23 the maximum effective stress according to PANDA2 is 44204 psi (PART 1 of Table 4), a value that is in very good agreement with the STAGS prediction: 40000 psi. During the mode jump (Load Steps 24 - 76) the maximum effective stress grows very steeply well beyond the critical value of 45 ksi. **Mode jumping should be prevented in optimization runs.** This is done in the case listed as Column 5 of Table 3 and PART 2 of Table 4. Results obtained with the "mode-jump-prevention" switch turned ON are described in [16, 17].

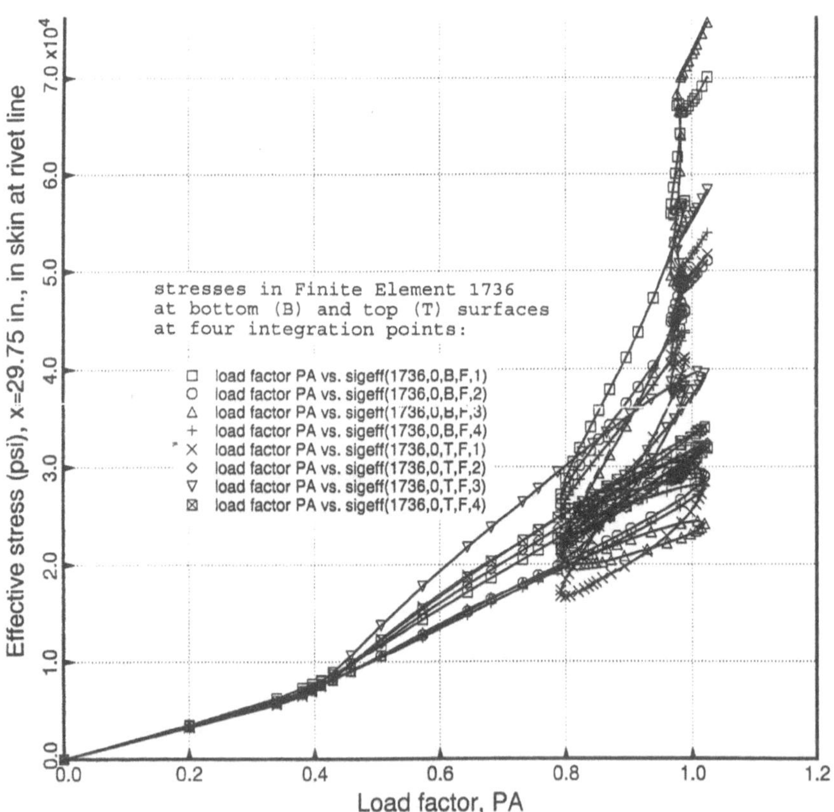


*Figure 14.* Maximum effective stress in panel skin from single-module STAGS model.

Compared to that predicted with the STAGS single module model, PANDA2 rather grossly underestimates the maximum effective stress midway between stringers (discussed in [16] in connection with Fig. 27 of [16]). However, the effective stress at this location is not critical. Also, the maximum effective stress midway between stringers predicted by STAGS is higher for a one-module STAGS model (Fig. 8) than for a STAGS model with three or more modules (Fig. 15, top). For example, for the optimum design determined with the "modejump prevention switch" turned ON (Col. 5 of Table 3 and PART 2 of Table 4) PANDA2 predicts the maximum effective stress midway between stringers to be 26385 psi and the one-module STAGS model predicts about 35000 psi. However, for a three-module STAGS model of this case, shown in Fig. 15, STAGS predicts a maximum effective stress midway between stringers of about 27500 psi at the design load, PA = 1.0. This is in very good agreement with the prediction of PANDA2. The lower maximum stress from the three-module STAGS model is caused by the outer modules acting to constrain the amplitude of the local skin buckles in the middle module. Compared to results from the one-module STAGS module, PANDA2 overestimates the critical effective stress anywhere in the panel by about 10 per cent.

The effective stresses from the STAGS model listed in PART 1 of Table 4 correspond to Load Step 23, that is, at the design load, PA = 1.0, but before mode jumping occurs. Figure 14 demonstrates the extremely harmful effect of mode jumping. The maximum effective stress in the panel skin at the rivet line becomes unacceptably high at loads below the design load as the postbuckling pattern changes between Load Step 23 and Load Step 76. Whereas PANDA2 predicts "stress failure" (defined here as the maximum effective stress reaching the value 45 ksi) at a load factor of 1.016 (Col. 4, PART 10 of Table 3), the STAGS model predicts "stress failure" at a load factor of approximately 0.91 (Col. 4, PART 11 of Table 3). PANDA2 yields unconservative stress constraints in this case if optimization is performed with the "modejump prevention switch" turned OFF.

Therefore, optimum designs should be obtained with the "modejump prevention switch" in PANDA2 turned ON. The results for optimization with the "modejump prevention switch" turned ON (Column 5 of Table 3) and subsequent analysis of the optimized panel are discussed in [16, 17].

## 7.0 Conclusions

The agreement between PANDA2 and STAGS appears to be sufficient to qualify PANDA2 as a preliminary design tool for panels with riveted Z-shaped stringers for service in the locally postbuckled regime. With proper (conservative) user input, such as specification that the mode jump constraint be turned ON during optimization cycles, PANDA2 errs on the conservative side, but does not appear to be overly conservative. Further work should include a similar study performed for laminated composite panels, for cylindrical panels, and for panels with both stringers and rings.

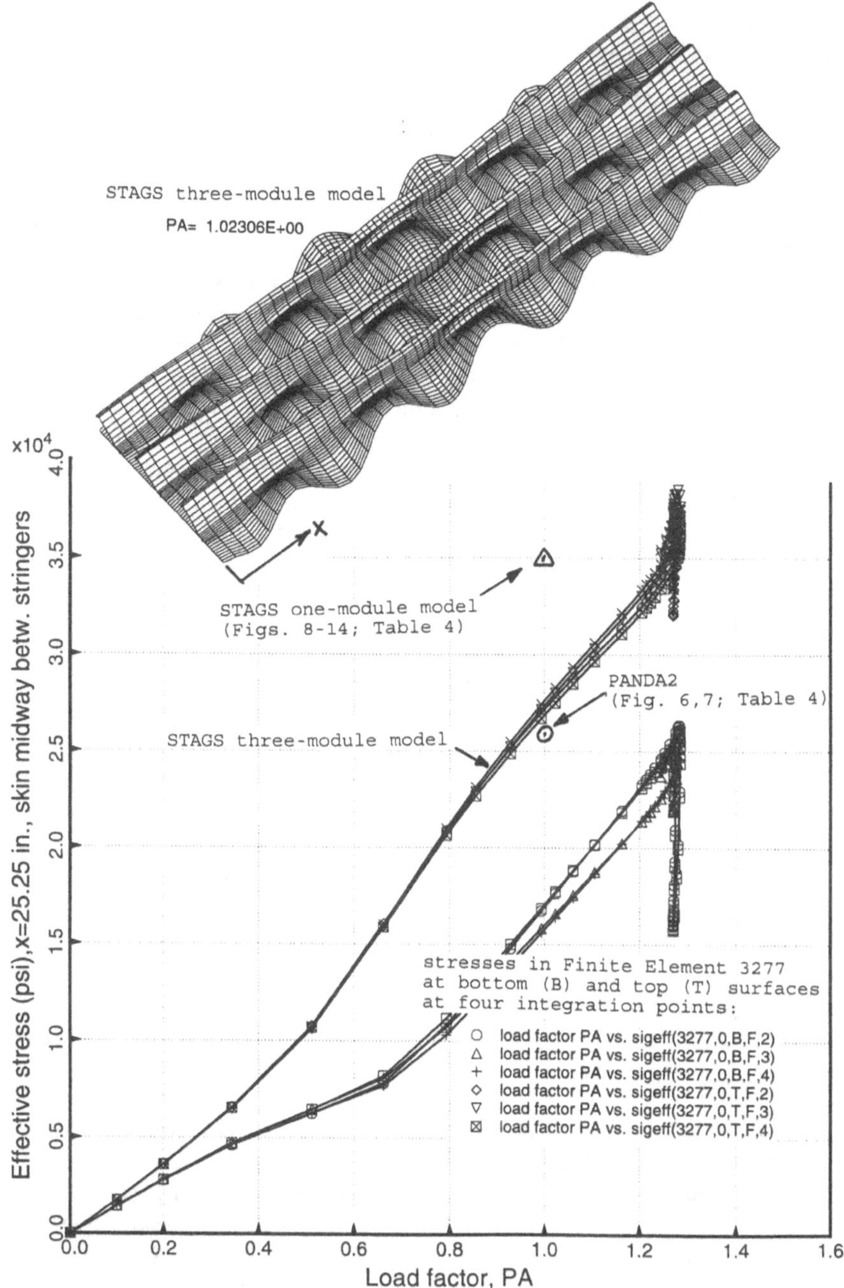


*Figure 15* Comparison of maximum midbay effective stress in panel skin predicted at the design load, PA = 1.0 from PANDA2 and STAGS three-module and one-module finite element models.

## 8.0 Acknowledgments

The author is grateful to Dr. Charles Rankin and Mr. Frank Brogan, developers of the STAGS program, for their helpful suggestions on the use of STAGS.

## 9.0 References

[1] van der Neut, A.: Overall buckling of Z-stiffened panels in compression, Delft University of Technology, Dept. of Aerospace Engineering, Report LR303, August 1980.

[2] Wittrick, W. H.: A unified approach to the initial buckling of stiffened panels in compression, *Aeronautical Quarterly* **19** (1968), 265-283.

[3] Williams, F. W. and Wittrick, W. H.: Computational procedures for a matrix analysis of the stability and vibration of thin flat-walled structures in compression, *International Journal of Mechanical Sciences* **11** (1969), 979-998.

[4] Riks, E.: A buckling analysis of a Z-stiffened compression panel simply supported on equidistant ribs, National Aerospace Laboratory NLR, The Netherlands, Report NLR TR 78145 L, December, 1978.

[5] Almroth, B. O. and Brogan, F. A.: The STAGS Computer Code, NASA CR-2950, NASA Langley Research Center, Hampton, VA (1978).

[6] Viswanathan, A. V. and Tamekuni, M.: Elastic buckling analysis for composite stiffened panels and other structures subjected to biaxial inplane loads, NASA CR-2216, September 1973.

[7] Bushnell, D.: PANDA2-Program for minimum weight design of stiffened, composite, locally buckled panels, *Computers and Structures* **25** (1987), 469-605.

[8] Dickson, J. N., Biggers, S. B. and Wang, J. T. S.: Preliminary design procedure for composite panels with open-section stiffeners loaded in the post-buckling range, in: *Advances in Composite Materials*, A. R. Bunsell, et al,editors, Pergamon Press Ltd., Oxford, England, 1980, pp 812-825. Also see, Dickson , J. N. and Biggers, S. B.: POSTOP: Postbuckled open- stiffened optimum panels, theory and capability, NASA Langley Research Center, Hampton, Va., NASA Contractor Report from NASA Contract NAS1 - 15949, May 1982.

[9] Butler, R. and Williams, F. W.: Optimum design features of VICONOPT, an exact buckling program for prismatic assemblies of anisotropic plates, AIAA Paper 90-1068-CP, *Proceedings 31st AIAA/ASME Structures, Structural Dynamics, and Materials Meeting* (1990), 1289-1299. Also see Williams, F. W., Kennedy, D., Anderson, M.S.: Analysis features of VICONOPT, an exact buckling and vibration program for prismatic assemblies of anisotropic plates, AIAA Paper 90-0970-CP, *Proceedings 31st AIAA/ASME Structures, Structural Dynamics, and Materials Meeting*, (1990), 920-929. Also see Kennedy, D., Powell, S. and Williams, F.: Local postbuckling analysis for perfect and imperfect longitudinally compressed plates and panels, AIAA-98-1770, *39th AIAA/ASME Structures, Structural Dynamics, and Materials Meeting*, 1998.

[10] Anderson, M. S. and Stroud, W. J.: General panel sizing computer code and its application to composite structural panels, *AIAA Journal*, **17** (1979), 892-897. Also see Stroud, W. J. and Anderson, M. S.: PASCO: Structural panel analysis and sizing code, capability and analytical foundations, NASA TM-80181, NASA Langley Research Center, Hampton, Va., 1981. Also see Stroud, W. J., Greene, W. H. and Anderson, M. S.: Buckling loads of stiffened panels subjected to combined longitudinal compression and shear: Results obtained with PASCO, EAL, and STAGS computer programs, NASA TP 2215, NASA Langley Research Center, Hampton, Va., January 1984.

[11] Bushnell, D.: Optimization of composite, stiffened, imperfect panels under combined loads for service in the postbuckling regime, *Computer Methods in Applied Mechanics and Engineering*, **103** (1993), 43-114.

[12] Bushnell, D. and Bushnell, W. D.: Minimum-weight design of a stiffened panel via PANDA2 and evaluation of the optimized panel via STAGS, *Computers and Structures* **50** (1994), 569-602.

[13] Bushnell, D. and Bushnell, W. D.: Optimum design of composite stiffened panels under combined loading, *Computers and Structures* **55** (1995), 819-856.

[14] Bushnell, D.: Recent enhancements to PANDA2 *37th AIAA Structures, Structural Dynamics and Materials Conference*, 1996.

[15] Bushnell, D., Rankin, C. C. and Riks, E.: Optimization of stiffened panels in which mode jumping is accounted for, AIAA Paper 97-1141, *Proceedings 38th Structures, Structural Dynamics and Materials Conference* (1997), 2133-2162.

[16] Bushnell, D.: Optimization of panels with riveted Z-shaped stiffeners via PANDA2, AIAA Paper 98-1990, *Proceedings 39th Structures, Structural Dynamics and Materials Conference* (1998), 2357-2388.

[17] Bushnell, D.: Optimization of panels with riveted Z-shaped stiffeners via PANDA2, LMCO-P489100, Lockheed-Martin Missiles and Space Co., Palo Alto, CA . October 1997.

[18] Baruch, M. and Singer, J.: Effect of eccentricity of stiffeners on the general instability of stiffened cylindrical shells under hydrostatic pressure, *Journal of Mechanical Engineering Science*, **5** (1963), 23-27.

[19] Bushnell, D.: BOSOR4: Program for stress, buckling, and vibration of complex shells of revolution, *Structural Mechanics Software Series* - Vol. 1, (N. Perrone and W. Pilkey, editors), University Press of Virginia, Charlottesville, 1977, pp. 11-131. See also *Computers and Structures* **4** (1974), 399-435; *AIAA J* **9** (1971), 2004-2013; *Structural Analysis Systems*, Vol. 2, A. Niku-Lari, editor, Pergamon Press, Oxford, 1986, pp. 25-54; and *Computers and Structures*, **18** (1984), 471-536.

[20] Koiter, W. T.: Het Schuifplooiveld by Grote Overshrijdingen van de Knikspanning, Nationaal Luchtvaart Laboratorium, The Netherlands, Report X295, November 1946 (in Dutch).

[21] Rankin, C. C., Stehlin, P. and Brogan, F. A.: Enhancements to the STAGS computer code, NASA CR 4000, NASA Langley Research Center, Hampton, VA (1986).

[22] Bushnell, D. ..../panda2/doc/panda2.news, a continually updated file distributed with PANDA2 that contains a log of all significant modifications to PANDA2 from 1987 through the present.

[23] Bushnell, D.: Approximate method for the optimum design of ring and stringer stiffened cylindrical panels and shells with local, inter-ring and general buckling modal imperfections, *Computers and Structures* **59** (1996), 489-527.

[24] Bushnell, D.: Stress, buckling and vibration of prismatic shells, *AIAA Journal* **9** (1971), 2004-2013.

[25] Vanderplaats, G. N.: ADS--a FORTRAN program for automated design synthesis, Version 2.01, Engineering Design Optimization, Inc, Santa Barbara, CA, January, 1987.

[26] Vanderplaats, G. N. and Sugimoto, H., A general-purpose optimization program for engineering design, *Computers and Structures* **24** (1986), 13-21.

[27] Bushnell, D.: Theoretical basis of the PANDA computer program for preliminary design of stiffened panels under combined in-plane loads, *Computers and Structures* **27** (1987), 541-563.

[28] Riks, E., Brogan, F. A., and Rankin, C. C.: Aspects of the stability analysis of shells in *Static and Dynamic Stability of Shells* (W. B. Kratzig and E. Onate, editors), Springer Series in Computational Mechanics, Springer-Verlag, Heidelberg (1990).

# NONDESTRUCTIVE TESTING OF THIN-WALLED CERAMIC MATRIX COMPOSITES WITH MATRIX CRACKS USING THERMOGRAPHY

LARRY W. BYRD[1] and VICTOR BIRMAN[2]

[1]Air Force Research Laboratory
AFRL/VAST, Bldg. 65, 2790 D Street
Wright-Patterson Air Force Base, Ohio 45433-7402, USA

[2]University of Missouri-Rolla
Engineering Education Center
8001 Natural Bridge Road
St. Louis, Missouri 63121, USA

## ABSTRACT

Three problems related to a nondestructive detection of matrix cracks in thin-walled ceramic matrix composite components are considered. First, the necessary background for the analysis of a cross-ply component with matrix cracks undergoing small-amplitude forced vibrations is prepared by specifying the stiffness that is affected by the presence of cracks. The second problem solved in the paper is the steady-state surface temperature rise over the ambient temperature in a vibrating component with cracks. This temperature depends on local stresses what explains the necessity in the analysis of forced vibrations. Finally, an exact solution of the problem of heat transfer in a component with matrix cracks in the central transverse layer is considered, including the transient period. The principal conclusion obtained based on numerical examples is that thermography is a feasible and reliable method capable of detecting matrix cracks in ceramic matrix composites.

## 1. INTRODUCTION

Ceramic matrix composites (CMC) have found numerous applications in situations where a component is subject to thermomechanical loading. Examples of these applications are internal chamber walls and nozzles of rocket motors, intake ramps for hypersonic propulsion systems, thermal protection blankets for re-entry vehicles, brake disks in transport systems, etc. (Birman and Byrd, 1999, 2000a).

The advantages of CMC related to their ability to endure high temperatures and to withstand initial damage without immediate failure are diminished due to their tendency to matrix cracking. While there is an experimental evidence of post-processing matrix



D. Durban et al. (eds.), Advances in the Mechanics of Plates and Shells, 103–118.

cracking, typical cracks develop in brittle matrices during lifetime of a CMC component. Although these cracks result in a reduction of strength and stiffness, the most dangerous effect is related to conduction of oxygen via the cracks to the fiber/matrix interface. The subsequent oxidation results in an abrupt embrittlement of the material.

The necessity to detect matrix cracks resulted in interest to nondestructive evaluation of CMC components. In particular, thermography has been considered as one of the candidates for nondestructive testing of these materials (Cho et al., 1991; Camden et al., 1998; Byrd and Birman, 1999). The application of this method necessitates us to excite forced vibrations. Then rapidly changing local stresses result in a local elevated temperature in the vicinity of the crack tip. In the case of bridging cracks perpendicular to the fibers, friction along the fiber/matrix interface results in a release of significant amount of heat (Cho et al., 1991; Byrd and Birman, 1999). An additional source of local elevated temperature is related to energy dissipation that occurs in the course of partial opening and closing of a crack during a cycle of motion. Thermoelastic coupling (Dunn, 1997) is another cause for a local elevated temperature due to the local stress concentration around the crack tip (this coupling is not considered in the paper).

Typical CMC components consist of a number of layers (or yarns, in the case of a woven material). Initial cracks develop in the transverse layers that are oriented in the direction perpendicular to the direction of tensile loads. As the load increases, the density of cracks in transverse layers increases as well. At an even higher load, the cracks begin to penetrate into longitudinal layers (Kuo and Chou, 1995; Domergue et al., 1996). While this scenario has been observed in cross-ply CMC, the general sequence of cracking remains the same in angle-ply materials, i.e. first cracks develop in the layers oriented in the direction that is close to perpendicular to the load direction. Therefore, a development of the nondestructive technique capable of detecting initial matrix cracks in transverse layers, prior to their propagation into longitudinal layers, is an important task. The feasibility of this task and related issues of heat transfer are discussed in this paper.

The paper includes three parts:

- Stiffness of CMC with matrix cracks is necessary to determine the stresses in a component subject to forced vibrations;
- Steady-state surface temperature due to the energy dissipation during vibrations can be used to detect the presence of cracks;
- Exact solution of the transient heat transfer problem is presented.

It is shown in the paper that once transient temperature variations become negligible, the results obtained using the exact solution coincide with those generated in the second section.

## 2. STIFFNESS OF CROSS-PLY CMC COMPONENTS WITH MATRIX CRACKS UNDERGOING SMALL-AMPLITUDE FORCED VIBRATIONS

The stiffness of CMC components with matrix cracks varies dependent on time and coordinates. This is related to the effect of cracks on stiffness of individual layers. In particular, the cracks in a transverse layer result in a reduced stiffness in the load direction. However, if the corresponding layer is subject to compression, the cracks may close and the stiffness of the layer in the load direction is recovered. The cracks in longitudinal layers reduce the stiffness in tension, as was shown by Pryce and Smith (1993) in the case of cracks developed under load and by Byrd and Birman (1999) for the cracks that appeared during post-processing cooling and affected residual stresses. However, similar to the case of cracks in transverse layers, the stiffness may recover under compressive stresses. Note that while these observations refer to the stiffness in the load direction, in-plane shear is also affected by the presence of cracks, even if they are closed, due to a possible sliding of the faces of a closed crack relative to each other. Therefore, the analysis of a thin-walled CMC component with matrix cracks represents a complex issue, that may be treated as a generalization of the problem of vibrations of bi-modular materials previously considered by Bert and Kumar (1982) and others. This analysis is presented in the forthcoming paper of the authors (Birman and Byrd, 2001).

The complexity of the analysis may be reduced in the case where the amplitude of forced vibrations excited during nondestructive tests is very small. Then, if the crack closing occurs at a certain compressive strain, this strain is not reached during the cycle of motion. Accordingly, the entire stress-strain relationship during the cycle corresponds to that for the material with open matrix cracks. This situation is illustrated in Figs. 1 and 2. The former figure illustrates matrix cracks in longitudinal and transverse layers and the orientation of the load that caused these cracks. Note that although the stiffness of materials with either bridging or tunneling cracks is a function of the applied stress, the problem discussed in this paper deals with CMC structures with preexisting matrix cracks. In this case, the stiffness appears to be independent of the stress, as long as the stress does not exceed the maximum value that caused cracking (the crack spacing that affects the stiffness should be evaluated at this maximum stress). Accordingly, the stress-strain relationship is linear, as is shown in Fig. 2.

In the case of matrix cracks limited to transverse layers of a cross-ply component, the stiffness of the laminate can be evaluated using the shear lag analysis or the energy considerations (see the recent paper of Kashtalyan and Soutis, 1999, for a list of work in this area). For example, Han and Hahn (1989) presented the results for the composite average engineering constants in the presence of matrix cracks. Therefore, the solution of a vibration problem can be obtained using standard methods of the theory of vibration with correspondingly adjusted stiffness coefficients, accounting for the presence of matrix cracks.

The limits of validity of this simple solution are available by comparing the crack opening displacement to maximum compressive deformations reached at the corresponding location during the cycle. In the case of a regular system of cracks contained within a transverse layer, the strain necessary to initiate the process of the crack closing can be obtained as a ratio of the crack opening displacement to the spacing of cracks. If the applied compressive strain reaches the crack closing value, the linear stress-strain relationship becomes unacceptable.

The value of the crack opening displacement may be available from experiments. An analytical estimate may be obtained using the observation that a crack in a transverse layer is parallel to the fibers. Therefore, the compression of the crack that may result in a complete or partial closing takes place in the plane perpendicular to the fibers where the material of the layer is transversely isotropic. Accordingly, the crack opening displacement can be estimated using isotropic models.

In the case where the cracks exist in both longitudinal as well as transverse layers, an estimate of the modulus of elasticity in the load direction becomes quite complicated. A possible approach is based on the fact that the cracks begin to appear in longitudinal layers only after they have reached saturation in transverse layers. However, it was found that even as the cracks reach the saturation in transverse layers, a reduction of the average modulus of elasticity of the composite material is rather small (Birman and Byrd, 2001). Using the average composite modulus corresponding to saturation of the cracks in transverse layers it is possible to calculate the modulus of the latter layers by the rule of mixtures. Similarly, other engineering constants of the transverse layers corresponding to the crack saturation can be evaluated from the corresponding micromechanical equations and the solution for these average material constants suggested by Han and Hahn (1989).

Now it is possible to estimate engineering constants of longitudinal layers with matrix cracks under a prescribed stress. In particular, the longitudinal modulus of elasticity $E_x$ can be evaluated as follows. The stress in the transverse layer corresponding to the crack saturation is available from Han et al. (1988). It is assumed that additional loading does not result in an increase of the stress in transverse layers. Therefore, the stress in longitudinal layers can be calculated as a function of the applied composite stress. Subsequently, the modulus of elasticity of the longitudinal layers is determined as a function of the stress in these layers.

As suggested in the previous work of the authors, it is possible to assume that the stiffness of the material in the planes parallel to the crack plane remains without change (Birman and Byrd, 2000b). Accordingly, in the longitudinal layer in the state of plane stress, the only engineering constants affected by open bridging cracks are the Poisson ratio $\nu_{yx}$ and the shear modulus $G_{xy}$. The values of these constants can be obtained as (Birman and Byrd, 2000b):

$$\nu_{yx}' = \nu_{xy}E_y/E_x' \qquad G_{xy}' = 2G_{xy}/(1 + E_x/E_x') \tag{1}$$

where the quantities with a prime denote engineering constants affected by matrix cracking. Although these equations may become inaccurate due to the presence of transverse layers, they should provide satisfactory qualitative results. A more accurate approach to the evaluation of the shear modulus, similar to the method suggested in the previous paragraph for the longitudinal modulus, could also be considered.

Note that the assumption that matrix cracks remain open during the entire cycle of motion does not necessarily introduce a noticeable error, even if it is violated in practice. This is because the surface temperature elevation due to the presence of cracks is dependent on the stresses, as shown in the next section. While deformations increase due to an underestimation of the stiffness, the stresses are proportional both to the stiffness as well as to deformations. Therefore, the overall effect of underestimating the stiffness on the stresses (and on temperature) is lower than the corresponding effect on deformations.

The previous discussion illustrating that the stresses can be estimated by assumption that the compressive strain remains below the closing value should not be applied to free vibration problems concerned with an estimate of natural frequencies. The frequency is proportional to the stiffness of the structure and the simplifying assumption introduced above could yield noticeable mistakes.

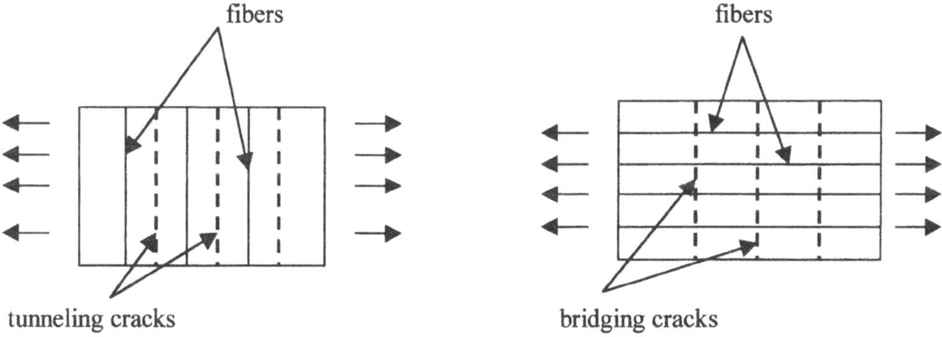


Fig. 1. Transverse layers with tunneling matrix cracks and longitudinal layers with bridging matrix cracks.

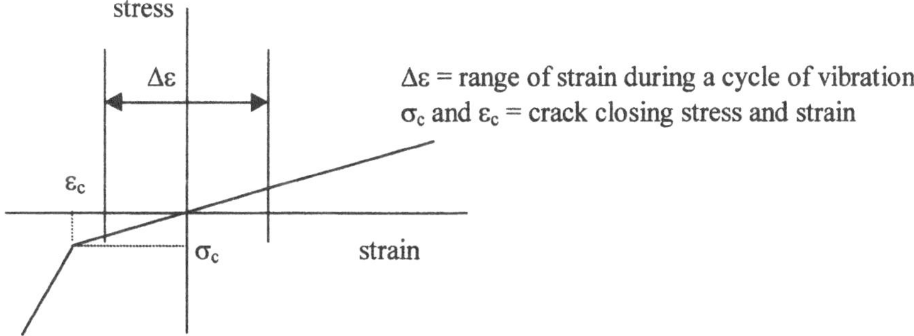


Fig. 2. Stress-strain relationship for transverse or longitudinal layers of a CMC composite with preexisting matrix cracks. Note: the stress-strain relationship may be nonlinear in the vicinity of $\sigma_c$ due to a gradual crack closing.

## 3. ESTIMATE OF SURFACE TEMPERATURE DUE TO ENERGY DISSIPATION AROUND CRACKS IN TRANSVERSE LAYERS DURING VIBRATIONS

In this section we illustrate that the surface temperature elevation in the vicinity of cracks in transverse layers of CMC components is sufficiently high to be detected by thermography. This elevated temperature can further be used to nondestructively predict the presence of internal damage.

Consider a transverse layer with a family of identical equally spaced tunneling matrix cracks (Fig. 1). Such cracks usually have a significant extent in the direction perpendicular to the plane of drawing in Fig. 1, so that it is possible to assume the state of plane strain and treat the cracks as mode I fracture. The analysis is conducted by assumption that kinetic energy is negligible, i.e. we consider a relatively slow quasi-static process. The cracks are present in transverse layers only, so that frictional energy dissipation that occurs along the fiber-matrix interface in the case of bridging cracks in longitudinal layers is absent. Accordingly, rate of the external work performed in the vicinity of the crack tip during vibrations (dW/dA) is equal to the local changes in the strain energy (dU/dA), the local energy of dissipation ($dQ_d/dA = Q'$) and the energy necessary to create the new surface area (dS/dA). Here A is the surface area of the crack. Using the definitions

$$G = dW/dA - dU/dA \qquad S = 2\gamma A \tag{2}$$

where G is the strain energy release rate and $\gamma$ is the fracture surface energy density of the material, one obtains the rate of the energy of dissipation (thermal energy) around the tip of the crack as

$$Q' = G - 2\gamma \tag{3}$$

The energy dissipation (thermal energy) for a crack that extends a distance B in the direction perpendicular to the plane of drawing in Fig. 1 is

$$Q_d = 2B\int(G - 2\gamma)da \tag{4}$$

where a is a half-length of the crack.

The energy release rate can be expressed in terms of the mode I stress intensity factor by (Kanninen, and Popelar, 1985).

$$G_I = K_I^2(b_{11}b_{22}/2)^{1/2}[(b_{22}/b_{11})^{1/2} + (2b_{12} + b_{66})/(2b_{11})]^{1/2} = K_I^2/E \tag{5}$$

where $K_I$ is the stress intensity factor, and $b_{ij}$ are the elements of the matrix of compliances in the plane strain constitutive relationships.

The stress intensity factor is the same function of the applied stress $\sigma$ and the crack half-length as in isotropic materials, i.e.

$$K_I = f\sigma(\pi a)^{1/2} \tag{6}$$

where f is the correction factor accounting for the surface effect. Williams (1989) indicated that the isotropic finite width correction factors may be adequate for composite materials, except for the cases where "extreme anisotropy is involved." For isotropic materials, the value of this factor is usually in the range between 1.0 and 1.2 (Bannantine et al., 1990). Therefore, if the location of the crack relative to the surface is not specified, the conservative approach would be to assume f = 1.0.

The values of the fracture surface energy density for CMC have not been reported, to the best knowledge of the authors. However, it is possible to estimate this energy using the solution of Chou (1992) for the critical strain corresponding to the matrix crack initiation in transverse layers of a cross-ply material:

$$\varepsilon = \{2bE_L\gamma\varphi^{1/2}/[(t_l + t_t/2)E_l E_c]\}^{1/2} \tag{7}$$

where $t_l$ is the thickness of longitudinal layers, $t_t$ is the thickness of transverse layers, and $E_L$, $E_T$ and $E_C$ are the longitudinal, and transverse moduli of the layers and the composite modulus, respectively. The factor $\varphi$ is based on the shear-lag analysis and reflects an additional stress transferred to longitudinal layers as a result of cracking of the transverse layers. This factor is given by

$$\varphi = E_c G_{LT}(t_l + t_t/2)/[E_L E_T t_l (t_t/2)^2] \tag{8}$$

where $G_{LT}$ is the shear modulus.

Note that the fracture surface energy density calculated from equation (7) is independent of the crack length (and time). Then evaluating the integral in equation (4) and taking a derivative with respect to time one obtains the expression for the heat transfer rate as a function of applied stresses and the crack length:

$$Q = dQ_d/dt = 2\pi Ba^2\sigma(d\sigma/dt)/E + C \tag{9}$$

where t is time and C is a constant of integration that should be chosen to ensure that Q is non-negative.

Note that equation (9) is obtained by assumption that the effect of the kinetic energy on the energy balance in the vicinity of the crack tip is negligible. Variations of the crack length during the cycle are also assumed small and their effect is neglected as compared to that of the rate of change of the applied stress.

If we assume that the thermal energy change within the material is equal to the heat flow through the surface, the balance condition is

$$2Q/s = qB \tag{10}$$

where q is a rate of heat gain or loss through the surface, s is the crack spacing, and the factor 2 accounts for two tips of the crack. Equation (10) should be modified if thermal capacitance is accounted for, but the present analysis refers to thin components with a regular system of cracks that are subject to low-frequency vibrations. In this case, equation (10) is sufficiently accurate.

If the composite is thin and the crack spacing is small, the effects of in-plane heat conduction and thermal resistance in the thickness direction are negligible. Then the rate of heat loss or gain through the surface of the element is

$$q = h_l(T_{sl} - T_{al}) + h_u(T_{su} - T_{au}) + \varepsilon_0\beta_0[(T_{sl}^4 - T_{al}^4) + (T_{su}^4 - T_{au}^4)] \tag{11}$$

where h is the heat transfer coefficient, $T_s$ is the surface temperature, $T_a$ is the ambient air temperature, $\varepsilon_0$ is the emissivity, $\beta_0$ is the Stefan-Boltzman constant, and subscripts "u" and "l" identify the upper and lower surfaces, respectively. Note that in many situations radiation from the surface becomes comparable to convection only at high elevations, so the term in the square brackets is often negligible.

If the ambient temperature on both sides of the component is the same, the closed-form expression for the elevation of the surface temperature over the ambient air temperature is available by assumption that radiation can be neglected:

$$\Delta T = [2\pi\sigma a^2/(shE)]d\sigma/dt + C \tag{12}$$

If the component is subject to a periodic in-plane stress $\sigma = \sigma_m \sin\omega t$, where $\omega$ is the frequency, the surface temperature elevation obtained subject to the requirement that $\Delta T$ remains non-negative during the entire cycle of motion is given by

$$\Delta T = \pi\sigma_m^2 a^2 \omega(1 + \sin 2\omega t)/(shE) \tag{13}$$

As follows from equation (13), a difference between the surface and ambient air temperature increases proportionally to the frequency and to the square of the applied stress. Note that the exact solution of the heat transfer problem presented in the next section illustrates that after the transient phase temperature reaches a steady-state constant value. Therefore, in the following discussion a reference is made to the mean temperature elevation $\Delta T_{ave} = \pi\sigma_m^2 a^2 \omega/(shE)$.

## 3.1. NUMERICAL EXAMPLES

Consider the SiC/CAS material that was experimentally studied by Beyerle et al. (1992). Details and discussion on the properties used in the analysis can be found in the paper of Birman and Byrd (2000c). The mean elevated surface temperature was calculated using the half-length of the crack equal to 0.18 mm, and the applied stress amplitude and frequency equal to 1 MPa and 10 Hz, respectively. The result of calculations yields

$\Delta T_{ave} = 0.038°C$. On the other hand, using the data employed to generate numerical results in the next section, one obtains the mean temperature elevation $\Delta T_{ave} = 3.39°C$ and 1.70°C, for h = 10 and 20 W/m$^2$C, respectively. Therefore, the feasibility of using thermography to detect initial failure in cross-ply and plain weave woven CMC is supported by these results.

## 4. HEAT TRANSFER PROBLEM IN CROSS-PLY CMC WITH MATRIX CRACKS IN A TRANSVERSE LAYER

The present section illustrates the solution of the problem of heat transfer for a symmetrically laminated cross-ply CMC laminate with a regular system of matrix cracks in the central transverse layer. Due to linearity of the problem, the elevated surface temperature associated with cracks in other layers can be obtained by superposition, as long as the cracks remain symmetric relative to the middle plane of the component.

The regular system of cracks analyzed in this section is shown in Fig. 3 for the case in which these cracks are present in the central transverse layer. Due to symmetry, it is possible to analyze a representative cell that is also shown in Fig. 3. The spacing between the cracks being quite small (less than 0.3 mm, according to reported experimental data), it is possible to neglect the effect of heat conduction in the x-direction and to treat the heat transfer problem as a one-dimensional dynamic process, according to Fig. 3.

Regularly spaced cracks in the central transverse layer (s = spacing)

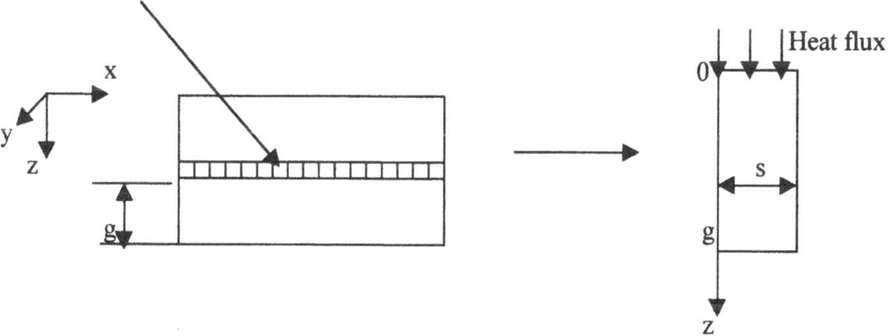


Fig. 3. Composite with matrix cracks in the central transverse layer and a representative one-dimensional cell used in the analysis.

Based on the analysis in the previous section, the heat flux into the component subject to dynamic stresses with the frequency $\omega$ is a sinusoidal function of time varying with the frequency $2\omega$. Therefore, the heat transfer problem for a representative cell can be modeled as shown in Fig. 4. The mathematical formulation of this problem is given as follows:

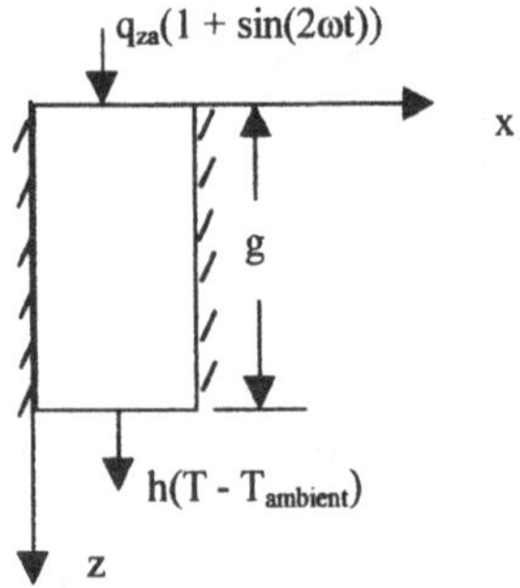


Fig. 4. Representative cell.

$$\frac{\partial^2 T_*}{\partial z^2} = \frac{1}{\alpha}\frac{\partial T_*}{\partial t} \tag{14}$$

$$-k\frac{\partial T_*}{\partial z} = q_{za}(1+\sin 2\omega t) \qquad at\ z=0 \tag{14a}$$

$$k\frac{\partial T_*}{\partial z} + hT_* = 0 \qquad at\ z=g \tag{14b}$$

$$T_* = 0 \qquad at\ t=0 \tag{14c}$$

where $T_* = (T - T_{ambient})$, and k and α are the thermal conductivity and diffusivity of the composite in the z-direction.

It is necessary to specify the conductivity in the z-direction. The conductivity of the i-th layer with isotropic fibers oriented in either x or y-direction can be obtained as

$$k_{zi} = (V_{fi}/k_f + V_{mi}/k_m)^{-1} \tag{15}$$

where $V_{fi}$ and $V_{mi}$ are volume fractions of the fibers and matrix, and $k_f$ and $k_m$ are the conductivites of the fibers and matrix, respectively.

The average conductivity of the laminate with an equal number of equal-thickness layers oriented in the x and y directions is

$$k = [0.5(1/k_{zm} + 1/k_{zn})]^{-1} \tag{16}$$

where subscripts m and n identify the layers oriented in the x and y directions, respectively.

Now the thermal diffusivity can be calculated as $\alpha = \rho C_p/k$ where ρ is the mass density and $C_p$ is a specific heat.

The problem is solved using Duhamel's theorem (Ozisik, 1993) for heat transfer within the component subject to a time dependent boundary condition at z = 0. According to this theorem, the temperature distribution is the function of the z-coordinate and time that can be represented by

$$T_{*}(z,t) = \int_{\tau=0}^{t} \Phi(z,t-\tau)\frac{df(\tau)}{d\tau}d\tau + f(0)\Phi(z,t) \tag{17}$$

where

$$f(t) = q_{za}(1+\sin 2\omega t) \tag{18}$$

and $\Phi(z,t)$ is the solution to an auxiliary problem defined as:

$$\frac{\partial^2\Phi}{\partial z^2} = \frac{1}{\alpha}\frac{\partial\Phi}{\partial t} \tag{19}$$

$$-k\frac{\partial\Phi}{\partial z} = 1 \qquad at\ z=0 \tag{19a}$$

$$k\frac{\partial\Phi}{\partial z} + h\Phi = 0 \qquad at\ z=g \tag{19b}$$

$$\Phi = 0 \qquad at\ t=0 \tag{19c}$$

The solution of the latter problem is

$$\Phi(z,t) = \Phi_s(z) + \Phi_t(z,t) \tag{20}$$

where $\Phi_s$ and $\Phi_t$ are the solutions of the following steady state and transient problems:

$$\frac{\partial^2\Phi_s}{\partial z^2} = 0 \tag{21}$$

$$-k\frac{\partial\Phi_s}{\partial z} = 1 \qquad at\ z=0 \tag{21a}$$

$$k\frac{\partial\Phi_s}{\partial z} + h\Phi_s = 0 \qquad at\ z=g \tag{21b}$$

$$\frac{\partial^2\Phi_t}{\partial z^2} = \frac{1}{\alpha}\frac{\partial\Phi_t}{\partial t} \tag{22}$$

$$-k\frac{\partial\Phi_t}{\partial z} = 0 \qquad at\ z=0 \tag{22a}$$

$$k\frac{\partial\Phi_t}{\partial z} + h\Phi_t = 0 \qquad at\ z=g \tag{22b}$$

$$\Phi_t = -\Phi_s \qquad at\ t=0 \tag{22c}$$

The solution of the problem given by equations (21) is

$$\Phi_s(z) = -\frac{z}{k} + \frac{1}{h} + \frac{g}{k} = c_1 z + c_2 \tag{23}$$

The solution $\Phi_t$ can be obtained by using a separation of variables technique (Ozisik, 1993) as:

$$\Phi_t(z,t) = \sum_{n=1}^{\infty} C_n \cos(\lambda_n z) e^{-\alpha \lambda_n^2 t} \tag{24}$$

The eignevalues $\lambda_n$ are found from the boundary condition at z = g (equation 22b) as the roots of the following equation:

$$\lambda_n \tan(\lambda_n g) = h/k \equiv H \tag{25}$$

The constants $C_n$ in equation (24) specified from the initial condition and the orthogonality of the eigenfunctions are given by

$$C_n = \frac{1}{N(\lambda_n)} \int_0^g -(c_1 z + c_2) \cos(\lambda_n z) dz = \frac{c_1}{\lambda_n^2 N(\lambda_n)} \tag{26}$$

where the normalization integral $N(\lambda_n)$ is

$$N(\lambda_n) = \frac{g\left(\lambda_n^2 + H^2\right) + H}{2\left(\lambda_n^2 + H^2\right)} \tag{27}$$

Substituting the results from equations (23-27) into equation (17) and simplifying yields the following expression for the $T_*$.

$$T_*(z,t) = \\ q_{za}\left\{(c_1 z + c_2)(1 + \sin 2\omega t) + \sum_{n=1}^{\infty} C_n \cos(\lambda_n z)\left[F_{t,n} e^{-\alpha \lambda_n^2 t} + F_{p,n} \sin(2\omega t + \phi_n)\right]\right\} \tag{28}$$

where

$$F_{t,n} = 1 - \frac{2\omega \alpha \lambda_n^2}{\alpha^2 \lambda_n^4 + 4\omega^2}; \qquad F_{p,n} = \frac{2\omega}{\sqrt{\alpha^2 \lambda_n^4 + 4\omega^2}}; \qquad \phi_n = \tan^{-1}\left(\frac{\alpha \lambda_n^2}{2\omega}\right) \tag{29}$$

## 4.1. NUMERICAL EXAMPLES AND DISCUSSION

The following examples were considered for the case where k = 1.25 W/mC, h = 10.0 W/m$^2$C (the value h = 20 W/m$^2$C is also considered in Fig. 5), E = 93.2 GPa, a = 0.0001 m, s = 0.00025 m, and g = 0.001 m. The value of $q_{za}$ = 33.887 W/m$^2$ was calculated based on the stress amplitude equal to $\sigma_m$ = 20 MPa. The frequency of vibrations was taken equal to $\omega$ = 62.832 rad/s.

The rise of the elevation of the mean surface temperature over the ambient air temperature is shown as a function of time in Fig. 5. As follows from this figure, the transient temperature increases rapidly (within 10-15 minutes, in this example) when a component is subject to periodic stresses. Once the transient temperature variations have decayed, the mean surface temperature rise, $T_*(g,\infty)$, is given as $T_*(g,\infty) = q_{za}/h$. This agrees with a simple energy balance over the entire region shown in Fig. 4 because the average rate of heat generation into the control volume is $q_{za}$ and this heat must be dissipated by convection at the surface. Note that the simple solution obtained in the previous section yields the steady-state mean temperature equal to 3.39°C and 1.70°C for h = 10 and 20 W/m$^2$C, respectively, i.e. exactly the results obtained by the present analysis. As follows from Fig. 5, decreasing h increases time corresponding to a noticeable transient portion of the response. Increasing the dimension g (i.e., a thicker specimen) also results in a longer time of transient response.

The periodic oscillations in $T_*$ are much smaller than the mean surface temperature rise, as shown in Fig. 6. These oscillations are too small to be measured reliably with current differential thermography techniques.

It is also interesting to note that increasing $\omega$ increases the heat flow rate $q_{za}$ and thus the mean temperature rise over the ambient. Therefore, higher-frequency vibrations can be appropriate for a thermographic nondestructive evaluation of CMC components.

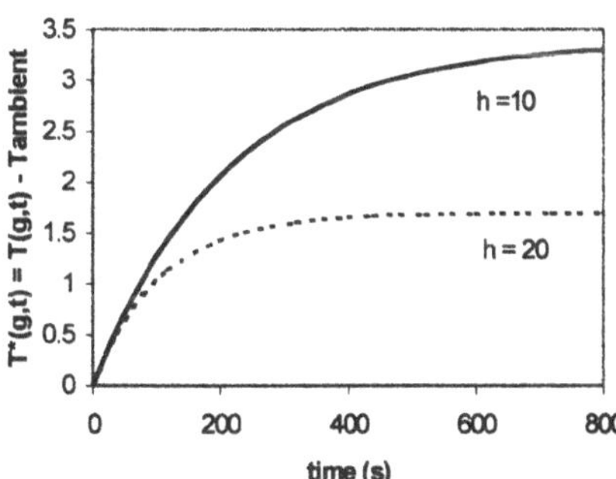


Fig. 5. Temperature rise as a function of time.

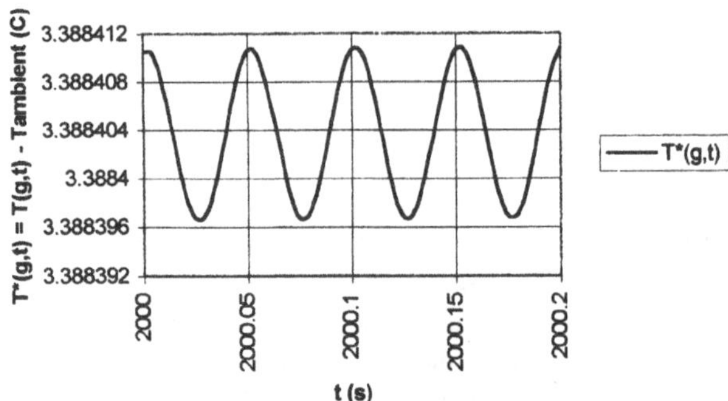


Fig. 6. Periodic temperature oscillations.

## 5. CONCLUSIONS

The paper presents the solution of several problems related to nondestructive testing of thin-walled ceramic matrix composite components using thermography. In particular, the stiffness of cross-ply CMC components with matrix cracks in longitudinal and transverse layers undergoing small-amplitude forced vibrations is discussed. The method for an estimate of the stiffness of a component is suggested for the case where the cracks remain open during the entire cycle of motion.

The problem of heat generation in a vibrating component with a system of regular matrix cracks in transverse layers is solved by assumption that kinetic energy is negligible. Based on the energy analysis, the closed-form expression for the surface temperature rise is obtained for the steady-state heat transfer problem.

The problem of heat transfer within a cross-ply CMC component with regular matrix cracks in the central transverse layer is also considered. The latter solution illustrates that the time interval corresponding to transient thermal response is relatively short. After transient variations of temperature have been reduced to a negligible level, the predicted mean surface temperature rise corresponds to the solution obtained based on the approximate energy-based analysis. At the same time, periodic temperature oscillations superimposed on the steady-state mean temperature are negligible and they should not affect the measurements.

Numerical examples presented in the paper illustrate that the mean surface temperature rise in thin-walled CMC components with matrix cracks subject to small-amplitude forced vibrations is sufficient to recommend thermography for damage detection. This method should be even more sensitive to damage, if the cracks propagate into longitudinal or angle-ply layers resulting in frictional heating and an even higher surface temperature.

*Acknowledgement:* This research has been supported by the Air Force Office of Scientific Research through the contract F49620-93-C-0063. The program manager is Dr. Brian Sanders.

**References**

Balantine, J.A., Comer, J.J. and Handrock, J.L. (1990) *Fundamentals of Metal Fatigue Analysis*, Prentice Hall, Englewood Cliffs, New Jersey.

Beyerle, D., Spearing, S.M. and Evans, A.G. (1992) Damage mechanisms and the mechanical properties of a laminated 0/90 ceramic matrix composite, *Journal of the American Ceramic Society*, **75**, 3321-3330.

Birman, V. and Byrd, L.W. (1999) *Review of Developments in Fracture and Fatigue of Ceramic Matrix Composites*, Report AFRL-VA-WP-TR-1999-3023, Wright-Patterson Air Force Base, Ohio.

Birman, V. and Byrd, L.W. (2000a) Review of fracture and fatigue in ceramic matrix composites, *Applied Mechanics Reviews*, **53**, 147-174.

Birman, V. and Byrd, L.W. (2000b) Selected issues of mechanics of ceramic matrix composites, *Composite Structures*. In press.

Birman, V. and Byrd, L.W. (2000c) Application of thermography to detection of matrix cracks in transverse layers and yarns of ceramic matrix composites, *International Journal of Fracture*. In press.

Birman, V. and Byrd, L.W. (2001) Matrix cracking in transverse layers of cross-ply beam subjected to bending and its effect on vibration frequencies, *Composites: Part B*. In press.

Bert, C.W. and Kumar,M. (1982) Vibration of cylindrical shells of bimodulus composite materials, *Journal of Sound and Vibration* **81**, 107-121.

Byrd, L.W. and Birman, V. (1999) Detection of cracks in ceramic matrix composites based on surface temperature, *Composite Structures*, **48**, 71-77.

Camden, M.P., Byrd, L.W. and Simmons L.W. (1998) Monitoring crack propagation in ceramic matrix composites in fully reversed bending using a differential thermography system, *Book of Abstracts, The Fifth International Conference on Composites Engineering,* Las Vegas, Nevada.

Cho, C., Holmes, J.W. and Barber, J.R. (1991) Estimation of interfacial shear in ceramic matrix composites from frictional heating measurements, *Journal of the American Ceramic Society* **74**, 2802-2808.

Chou, T-W. (1992) *Microstructural Design of Fiber Composites*, Cambridge University Press, Cambridge.

Domergue, J-M., Heredia, F.E. and Evans, A.G. (1996) Hysteresis loops and the inelastic deformation of 0/90 ceramic matrix composites, *Journal of the American Ceramic Society* **79**, 161-170.

Dunn, S.A. (1997) Using nonlinearities for improved stress analysis by thermoelastic techniques, *Applied Mechanics Reviews*, **50**, 499-513.

Han, Y.M., Han, H.T. and Croman, R.B. (1988) A simplified analysis of transverse ply cracking in cross-ply laminates, *Composites Science and Technology*, **31**, 165-177.

Han, Y.M. and Hahn, H.T. (1989) Ply cracking and property degradation of symmetric balanced laminates under general in-plane loading, *Composites Science and Technology*, **35**, 377-397.

Kanninen, M.F. and Popelar, C.H. (1985). *Advanced Fracture Mechanics*, Oxford University Press, Oxford.

Kashtalyan, M. and Soutis, C. (1999) Stiffness degradation in cross-ply laminates damaged by transverse cracking and splitting, *Composites: Part A*, **31**, 335-351.

Kuo, W-S and Chou, T-W. (1995) Multiple cracking of unidirectional and cross-ply ceramic matrix composites, *Journal of the American Ceramic Society* **78**, 745-755.

Özisik, M.N. (1993) *Heat Conduction*, 2nd ed., John Wiley & Sons, Inc., New York.

Pryce, A.W. and Smith, P.A. (1993) Matrix cracking in unidirectional ceramic matrix composites under quasi-static and cyclic loading, *Acta Metallurgica et Materialia* **41**, 1269-1281.

Williams, J.G. (1989) Fracture mechanics of anisotropic materials, *Applications of Fracture Mechanics to Composite Materials*, Ed. Frederich, K., Elsevier, Amsterdam, 3-38.

# A SHELL-BUCKLING PARADOX RESOLVED

C.R. CALLADINE
*Department of Engineering*
*University of Cambridge*
*Cambridge CB2 1PZ, U.K.*

## 1. Introduction

This paper is concerned with the buckling of uniform thin-walled cylindrical shells under uniform axial compressive loading.

The classical, linearised theory of buckling (e.g. Timoshenko and Gere, 1961) predicts that

$$\sigma_{c\ell}/E \approx 0.6t/R. \tag{1}$$

Here $\sigma_{c\ell}$ is the (compressive) buckling stress,
$E$ is the Young modulus of elasticity of the material,
$t$ is the thickness of the shell wall, and
$R$ is the radius of the shell.

Now it has been known since the 1930s that such shells actually buckle at loads considerably below those given by equation (1); and over the following decade or so

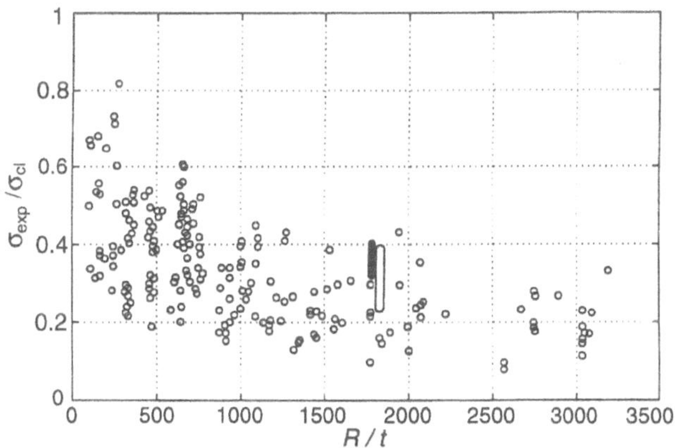


*Figure 1.* Experimental buckling loads for cylindrical shells, normalised with respect to the classical buckling prediction (1) and plotted *versus* the radius/thickness ratio. Circles are data from Brush and Almroth (1975). The vertical bars at R/t ≈ 1800 are data from Lancaster *et al.* (2000), and are referred to in section 11: the longer bar corresponds to cases where imperfections were deliberately introduced, while the shorter bar relates to cases where there were no such imperfections.



*D. Durban et al. (eds.), Advances in the Mechanics of Plates and Shells,* 119–134.

many experimental studies of the buckling strength of thin cylindrical shells have been made. By the mid-1950s it was possible to assemble data from many such studies. Thus, Fig. 1 shows a plot of almost 200 experimentally measured buckling strengths normalised with respect to the classical prediction (1), against $R/t$.

It is immediately obvious from this plot (a) that for any given value of $R/t$ there is considerable *scatter* in the buckling performance; and (b) that the measured strengths become smaller in comparison with the classical prediction as $R/t$ increases.

A replot of the data of Fig. 1 on double logarithmic scales is given in Fig. 2 (a). It is now clear that the experimental data lie in a well-defined band corresponding to the buckling-stress level being proportional to $(t/R)^{1.5}$, rather than to $(t/R)$ as in the classical theory. The plot also shows the best-fitting straight line, and parallel lines at ± 1 and 2 standard deviations from the mean. Practically all of the data-points lie within two standard deviations; and these two practical upper and lower bounds on the data correspond to the mean stress multiplied by 2 and 0.5 respectively.

The paradox that I shall consider in the present paper is concerned with the clear difference between the empirical observation

$$\sigma_{\text{mean}}/E \approx 5(t/R)^{1.5}, \tag{2}$$

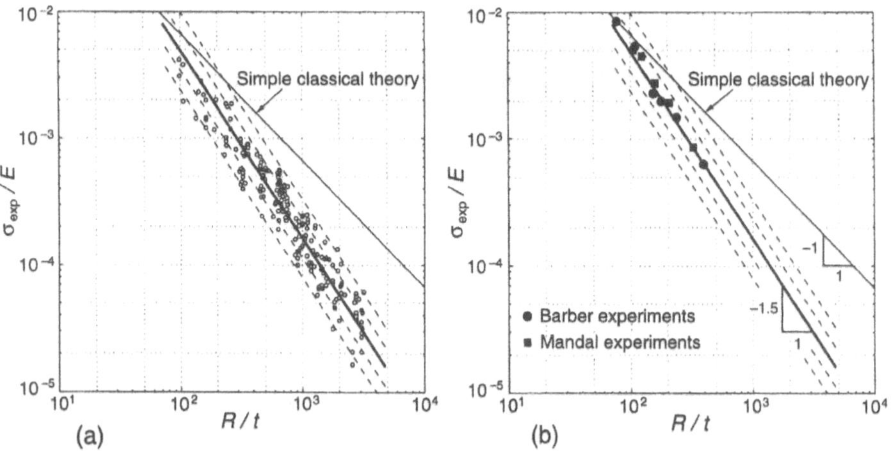


*Figure 2.* Double-logarithmic plot of experimental data on thin cylindrical shells, as in Fig. 1, with buckling stress normalised with respect to the Young modulus $E$, *versus* radius/thickness ratio. The heavy best-fitting line has a slope of −1.5; and lighter parallel lines at 1 and 2 standard deviations from the mean are also shown. The classical theory (1) is represented by a line of slope −1. (b) Self-weight buckling data from experiments of Calladine and Barber (1970) and Mandal and Calladine (2000) on open-topped silicone-rubber shells, plotted as the normalised self-weight vertical stress at the base of the shell *versus* radius/thickness ratio. The five lines from (a) are also shown.

where $\sigma_{mean}$ is the mean experimental buckling stress, and the classical prediction given in equation (1). How can we explain the empirical observations in terms of rational mechanics?

## 2. Remarks on previous work in this area

Th. von Kármán (von Kármán, Dunn and Tsien, 1940; von Kármán and Tsien, 1941) grappled with this as an example of a *non-linear* problem. He described an empirical relationship similar to (2), on the basis of the much smaller experimental data-set available in 1940, but with an exponent of 1.4 rather than 1.5. He set up his famous non-linear governing equations; but he did not provide a convincing explanation of (2).

Koiter (1945) made a seminal early study in this field. He took an asymptotic approach to the non-linearities, and focussed attention on two questions, as follows.
(a) Why are the experimentally observed buckling loads significantly lower than the predictions of the classical theory?
(b) Why do the experimental observations have so much scatter?
His explanation was in terms of *imperfection-sensitive buckling loads* and the *unavoidable presence of small imperfections*; and this has been abundantly fruitful in the general field of the mechanics of buckling and stability. But in Koiter's approach to the present problem one asks, essentially, why the experimental observations do not agree with (1). That is actually a different question from the one that I take to be more central, *viz.* why does (2) have an exponent of 1.5 rather than 1.0 ? This, then, is my main task in the present paper.

## 3. Method of investigation

Most papers on the buckling of thin-shell structures start with the classical theory, and use it as a foundation for their argument. But in my view the classical theory is not a good starting point if we wish to explain the exponent 1.5 in (2) : and I believe that we need to adopt an altogether different approach if we are to make progress.

The argument that I shall present in order to resolve the paradox connects together a number of ideas and phenomena which are not particularly new or difficult in themselves. Some of these connections occurred unexpectedly, as a consequence of chance encounters and conversations – which is of course a widespread, if largely unacknowledged, procedure in scientific discovery. And there were many blind alleys along the way. Thus, the point reached at the end of the paper was not attained by the application of some carefully formulated program or method, but by a haphazard chain of surprising connections. For this reason I am giving a somewhat anecdotal account of the development of the main ideas.

## 4 Experiments on self-weight buckling of open-topped cylindrical shells

My students and I, together with R.J. Denston, our laboratory technician, developed a simple method of casting uniform, thin silicone rubber shells in a rotating mould (Calladine and Barber, 1970; Mandal and Calladine, 2000). The shells were cast with a thick disc closure at one end, and were open at the other. When placed on a flat, horizontal table with the open end upwards, the shells were incapable of standing upright under their own weight. But if horizontal rings were successively cut off the top, eventually a height, $L_{cr}$, was reached at which the shell could just stand up under its own weight.

Barber's shells were of radius ~90 mm, and an experimenter's two hands were adequate to support the shell in a cylindrical shape in an attempt to stand it up. Mandal's shells were cast in a larger mould of radius ~120 mm; and it was necessary to have the hands of two experimenters active in the process of finding whether a shell of given height could be made to stand unaided.

Figure 2(b) shows the results of the experiments of Barber and Mandal. The values of $R/t$ ranged from 75 to 370. The observed critical heights $L_{cr}$ have been normalised with respect to the weight of material per unit volume ($\rho g$) and the Young modulus of elasticity ($E$) of the material. The relevant material parameter $E/\rho g$ was obtained from simple assays of the self-weight deflection of a horizontal beam of rectangular cross-section, cast from the same batch of rubber solution that had been used for the shell.

The thickest shells, with the smallest values of $R/t$, collapsed like the falling of a heavy curtain, with axisymmetric buckles at the base; and their critical heights almost reached the classical prediction. But for most of the shells the buckling involved a falling inwards at the top, like the collapse of a wall.

We were surprised that the results from all of the specimens lay on, or close to, a single straight line in the log-log plot of Fig. 2(b): they did not display the expected range of scatter that is familiar in the testing of cylindrical shells by external forces applied through conventional end-fixtures (Figs. 1, 2(a)). And we did not understand why that should be.

Now the experiments recorded in Figs 1 and 2(a) (which I shall refer to as "ordinary" experiments) were conducted on cylindrical shells with the same sort of end-fitting at both ends; and the self-weight of the shells was usually a tiny fraction of the imposed axial load. In our experiments, *per contra*, the top edge was free and the loading was entirely by self-weight. In one sense, therefore, the two sets of experiments are not comparable. Nevertheless, it seems reasonable to attempt a comparison by taking as a characteristic stress in our experiments the self-weight stress $L_{cr}\rho g$ at the base of the shell, and plotting this as $\sigma_{\text{exp}}$ in Fig. 2(b). For the sake of clarity the mass of points in Fig. 2(a) has not been transferred to (b), but only the mean line and the parallel lines at $\pm 1$ and 2 standard deviations from the mean.

The remarkable outcome of this exercise, which was a great surprise to us when we first did it, was that *the self-weight buckling data lay very close to the mean line of Fig. 2(a).* Now it is true that our range of $R/t$ values was not as extensive as that of the data in Figs 1 and 2(a). Nevertheless, the close coincidence of these two sets of experimental data suggests strongly that the two buckling situations are indeed closely related. Evidently the lack of scatter in the self-weight buckling data is attributable somehow to the different loading and boundary conditions; and I shall return to that point later, in section 11.

I shall therefore take as a working hypothesis that the buckling phenomena in "ordinary" buckling experiments are closely related to those of self-weight buckling. This leads directly to the obvious remark that if we can understand the self-weight buckling phenomena, we shall have made progress in understanding the paradoxical buckling of "ordinary" cylindrical shells under axial loading.

## 5. Computational study of self-weight buckling

Our next task, therefore, was to analyse the buckling behaviour of an open-topped shell under its own weight. We (Mandal and Calladine, 2000; Zhu, Mandal and Calladine, 2000) found this to be a fairly straightforward task. We used the standard ABAQUS finite-element package (Hibbitt *et al.*, 1995) to analyse the finite displacements of a particular experimental shell in Mandal's series, having $R$ = 120 mm, $t$ = 0.58 mm and $L = L_{cr}$ = 230 mm – the experimentally observed value. An advantage of the computer over a laboratory experiment, of course, is that the actual gravitational acceleration $g$ can be multiplied at will by a "load factor" $G$: so $G$ = 1.0 corresponds to the experimental "just stable" critical condition. Once a suitable imperfection pattern had been found (see Mandal, 1997 or Mandal and Calladine, 2000, for details) the computations ran smoothly, with the help of the "Riks" algorithm for following descending loads. Figure 3 shows a plot of $G$ against radial displacement at two different material points P and Q, whose locations on the shell is indicated in Fig. 4(a). There was hardly any sign of buckling until $G$ reached a value of around 1.8, corresponding to the self-weight vertical stress at the base reaching almost the classical buckling value. But then the load fell very sharply, and deflections only began to increase significantly when the load factor had fallen to $G \approx 1.3$. Thereafter, the radial displacement at the two chosen points increased steadily while $G$ fell slowly to a value of around 1.0.

We can describe this behaviour as a "post-buckling plateau", although as a plateau it is not absolutely flat. The "plateau" extends to radial displacements of around 10 wall thicknesses, with relatively little change in load-factor $G$. Further computations showed that the *initial* buckling load was only sensitive to the amplitude of the initial geometric imperfection if that imperfection featured a "dimple" near the base of the kind to be described below. But in any case the level of the post-buckling "plateau" was insensitive to the pattern or amplitude of any initial geometric imperfection.

Of particular interest is the fact that the post-buckling "plateau" occurs at around the experimentally-observed critical height $L_{cr}$. And computations on geometries corresponding to other experimental specimens at critical height showed essentially the same feature.

Thus we may conclude that, somehow, the self-weight experimental buckling assay is picking out the almost-stable behaviour of the post-buckling "plateau". The way in which the "plateau" is actually reached depends evidently on the pattern of assumed initial geometric imperfections. But it is not hard to imagine that the process of supporting the shell by the fingers of at least two hands, in attempts to get it to stand up, will introduce a wide range of imperfection patterns.

## 6. The post-buckling "plateau" mode

Our next task is to investigate the special post-buckling mode that grows at an almost-constant "plateau" load-factor, in the hope of finding a simple way of modelling and understanding its main features.

Mandal and Calladine (2000) have presented contour plots of the radial deflection of the shell in the "plateau" mode. But for present purposes the sketch of Fig. 4(a) is adequate. The key point is that there is an inward-directed "dimple" near the base, while vertically above the dimple the shell wall "leans outwards". Figure 5 shows a series of profiles of the leading generator as the deflection increases; and it is clear that

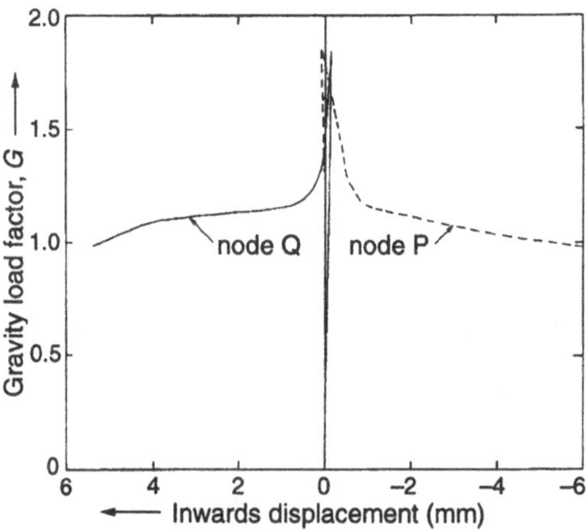


*Figure 3.* Plot of gravity load factor $G$ against normal deflection at two points on the central generator of an open-topped shell, in an ABAQUS computation: see Mandal and Calladine (2000) for details. Material points P and Q on the symmetry meridian are identified in Fig. 4(a).

as the dimple becomes deeper – i.e. as its current centre deflects further inwards – it also extends further up the shell; and indeed the dimple also becomes more extensive around the circumference. And when the dimple enlarges in this way, the upper part of the shell tilts outwards at an increasing angle.

There is a simple first-order kinematic interpretation of this well-defined post-buckling mode. If we suppose that the leading generator within the dimple deforms *inextensionally,* then it follows that its inwards deflection is accompanied by a small vertically-downwards movement of the shell wall at the upper boundary of the dimple. Then, if we consider the moiety of the shell above this level as an open-topped storage-tank standing on a level foundation, but which is "sinking" in the small region over the dimple, we can see the outward tilting of the shell wall above the dimple as an elementary example of inextensional deformation of the shell wall : cf Kamyab and Palmer (1989) or Calladine (1983 § 6.5.1).

## 7. Focus on the dimple

In an unrelated computational assay, Guggenberger, Greiner and Rotter (1999) have made an interesting and relevant computational study of the development of dimples in

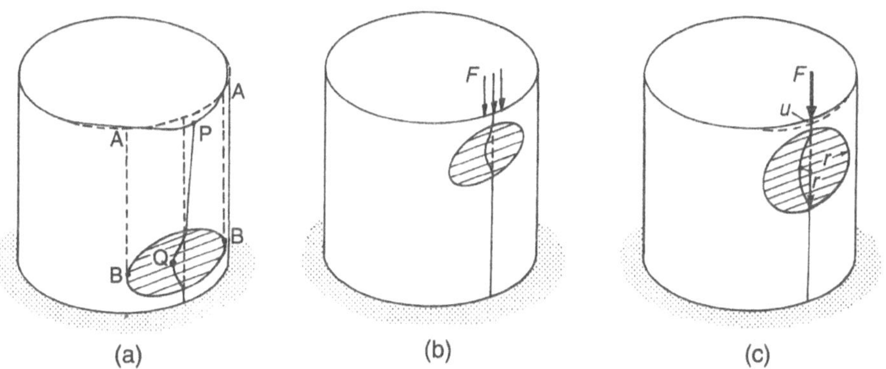


*Figure 4.* Schematic representations of the post-buckling modes of three thin-walled cylindrical shells, showing a common dimple *motif.* All shells are built-in at the base. (a) Open-topped shell loaded by gravity, as described by Mandal and Calladine (2000). (In section 10 it is postulated that the weight of portion ABBA of the shell provides the force that holds the dimple in place.) (b) Shell with its top closed by a diaphragm, loaded vertically by a localised force $F$ at the edge. This situation has been investigated by Guggenberger *et al.* (1999) in the context of localised support systems for silo structures. (c) As (b), but with an idealised small *circular* dimple of radius $r$, whose surface has been inverted to the same radius of curvature as that of the parent shell. Inextensional deformation within the dimple allows $F$ to move a small distance $u$.

thin cylindrical elastic shells, under the action of localised axial forces applied at an edge. The motivation for Guggenberger's study was an important problem in the design of cylindrical storage silos that are supported on a number of discrete columns at the base: it is obviously important for the engineer to understand the conditions under which local buckles may form.

Figure 6 shows a typical force-displacement curve for a shell with its end closed by a diaphragm, which is loaded as shown in Fig. 4(b). Buckling occurs when the peak load has been reached, whereupon a dimple forms. Guggenberger found that the load remains roughly constant as the dimple grows: there is a sort of first-order "plateau" in the force/deflection characteristic. He also found that the response depended primarily on the total force *F*, and only to a small extent upon the width of the shell's circumference over which the load is spread, up to a width of order $\sqrt{Rt}$.

The "plateau" values of Guggenberger's curves are consistent with the formula

$$F_h \approx 1.0 E t^{2.5} / R^{0.5}. \tag{3}$$

His shells were held circular at the loaded edge, but were free to rotate about the local tangent; that is, his edge was "hinged" (hence subscript *h*).. Zhu (see Zhu, Mandal and

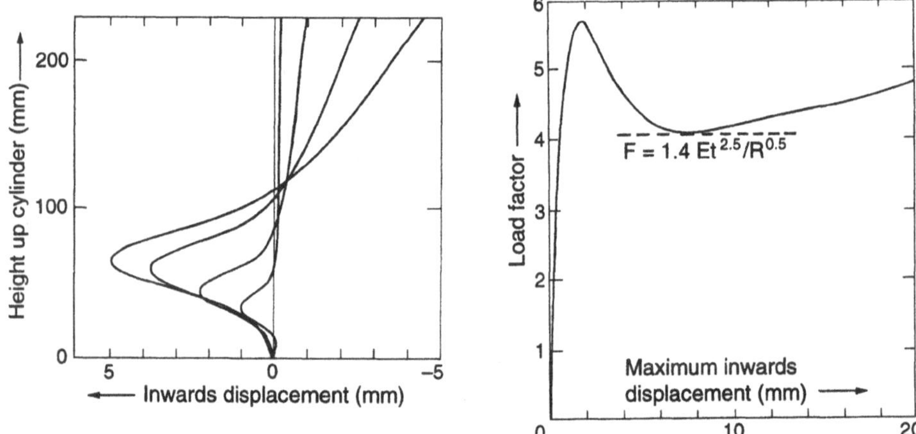


*Figure 5* (on the left). ABAQUS results for deflected profiles of the generator PQ in Fig. 4(a), at various stages of the development of the post-buckling mode. The shell had radius 120 mm, thickness 0.58 mm and height 230 mm. The post-buckling profiles were practically unchanged when the thickness of the shell was altered : see section 10.

*Figure 6* (on the right). ABAQUS results for a shell loaded as in Fig. 4(b) : load factor *versus* maximum inwards deflection in the dimple. The load is factored from an arbitrary "reference load", and a quantitative formula (4) for the minimum load is indicated. The behaviour is similar when the thickness of the shell is altered : see Zhu *et al.* (2000).

Calladine, 2000) made similar calculations, but with the edge of the shell restrained from rotation about the local tangent; and he also found that the force remained constant (to first order) as the dimple grew in size, in accordance with

$$F \approx 1.4Et^{2.5}/R^{0.5} \tag{4}$$

The form of equations (3) and (4) is the same; and the difference in the numerical constants doubtless reflects the difference in boundary-conditions.

These results suggest the beginnings of a simple explanation for the post-buckling mode shown in Figs 4(a) and 5. But it will be useful first to introduce another striking result which helps with the quantification of these physical effects.

## 8. Inversion of a thin spherical shell

Several authors have studied the mechanics of the inversion of a thin-walled elastic spherical shell under the action of a localised radial load, acting inwards : see Mescall (1965), Bushnell (1967), Ranjan and Steele (1977), Pogorelov (1988), Nowinka and Lukasiewicz (1991) and Holst and Calladine (1994). The general picture is the same in all these studies; and it agrees with our experience of pushing a finger into a toy rubber ball which has been pierced in order to eliminate any internal pressure.

The situation is most readily described with respect to Fig. 7, which shows schematically a diametral cross-section of the shell. The central portion of the shell has been inverted, and it has the same geometrical configuration as if an axisymmetric cap had been cut out, turned over, and re-united with the parent shell around its edge. (Here we are concerned only with axi-symmetric inversion, i.e. the stage before the edge of the indentation begins to go polygonal.) The actual inverted portion has some bending stress, of course; but the corresponding elastic strain energy is small. The key point is that the inverted surface is isometric with the original sphere : the deformation is *inextensional*. The above statements would be strictly true if the boundary of the inverted portion was a cut, and the junction between the two portions involved a sharp *crease*, as shown on the right in Fig. 7. But the deformed meridian actually makes a smooth transition between the original and the inverted portions, and the studies cited above have shown that the two portions of shell are separated by a *boundary layer* or "knuckle", whose width in the meridional direction is of order $\sqrt{at}$, where $a$ and $t$ are the radius and thickness of the spherical shell, respectively.

Practically all of the elastic strain energy of the distorted shell resides in this boundary layer. Now it is not difficult to see that if the width of the boundary layer were smaller, the strain energy of meridional *bending* would be larger, while the strain energy of circumferential *stretching* would be smaller; and *vice-versa* if the width were larger. (If the width became very small, as on the right in Fig. 7, the bending energy in the sharp crease would become very large, while the stretching energy – of the practically inextensional distortion – would be very small). The total elastic strain energy is minimum when the overall width of the boundary layer is equal to around $4\sqrt{at}$, with the precise value of the constant depending on the way in which the width is defined; but with the bending and stretching energies being equal in any case.

Holst's numerical studies (Holst and Calladine, 1994) show that the relationship between force $P$ and inwards deflection $w_o$ is given by

$$P \approx 1.7Et^{2.5} w_o^{0.5}/a \tag{5}$$

for moderate values of $w_o/a$. By evaluating $\int Pdw_o$ we can obtain an expression for the total elastic strain energy stored in the boundary layer. Hence, dividing this by the circumference of the boundary layer $2\pi r$,. and using the small-angle approximations

$$\psi \approx 2r/a \quad , \quad w_o a \approx r^2 \tag{6}$$

(where r is the radius of the boundary layer and $\psi$ is the "kink" angle), we obtain the following formula for the *elastic strain energy per unit length of boundary layer, Ω:*

$$\Omega \approx 0.03Et^{2.5}\psi^2(2/a)^{0.5}. \tag{7}$$

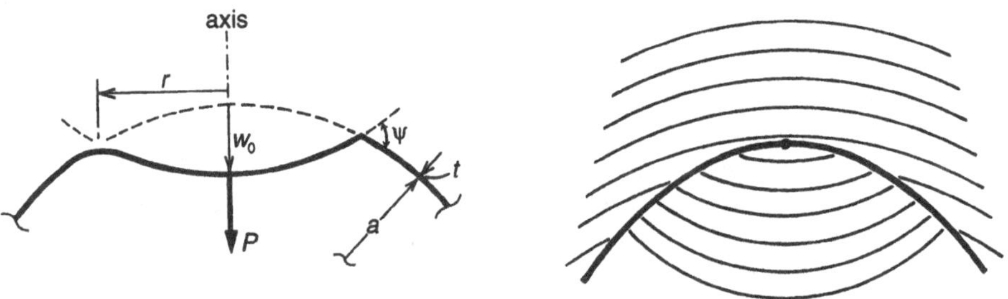


*Figure 7* (on the left). Cross-section of an elastic thin-walled spherical shell that is being inverted by a central force. To the right is shown a truly inextensional mode of deformation, with a sharp crease of angle $\psi$; while on the left the actual smooth *boundary-layer* or "knuckle" is shown.

*Figure 8* (on the right). Perspective general schematic sketch of a curved boundary-layer crease in a surface which has been inverted. The surface may originally have been plane, or cylindrical, or spherical. The bold line represents the crease, and families of light lines indicate the curvature of the outer and inner portions of the surface. At the point marked on the crease, $|\Delta\kappa|$ is defined as the jump in curvature (of lines parallel to the crease) as one crosses the crease. When the picture represents a portion of the boundary-layer ring in Fig. 7, $|\Delta\kappa| = 2/a$.

Of particular interest is the fact that $t$ appears in this expression, just as it does in Guggenberger's formula (3), with an exponent of 2.5. If, instead, the width of the boundary-layer had been constant, the strain energy of bending would have been proportional to $Et^3\psi^2$, because the curvature would then be proportional to $\psi$ and the bending stiffness of a shell element is proportional to $Et^3$ (e.g. Timoshenko and Gere, 1961). However, as we have seen already, the width of the boundary-layer is proportional to $\sqrt{at}$; and the upshot is that the exponent of $t$ in (7) is 2.5 instead. As we shall see below, the exponent 2.5 in (7) will lead to the sought-after exponent 1.5 in (2).

## 9. A simplified analysis of Guggenberger's problem

The establishment of the strain-energy formula (7) enables us to make a simple and illuminating re-derivation of Guggenberger's result that the force $F$ required to hold a dimple in place in a cylindrical shell of radius $R$ (Fig. 4(b)) is independent of the size of the dimple, to first order, and is proportional to $Et^{2.5}/R^{0.5}$.

For the sake of simplicity consider a dimple bounded by a circle of radius $r$, as shown in Fig. 4(c), and with the inverted portion in the form of a cylinder of radius $R$. Just as for the inversion of the spherical shell, we shall assume that the elastic strain energy of distortion resides only in the boundary layer.

In order to evaluate the total elastic strain energy we need to know values of $\psi$ around the circle; and for small values of $r/R$ ($r/R<1/3$, say) we find that $\psi$ is uniform ;

$$\psi \approx r/R \ . \tag{8}$$

At this stage it is not obvious how to proceed, because formula (7) includes $a$, the radius of the *spherical* shell for which it was derived. But further analysis shows that the same formula applies to a boundary-layer formed in an initially plane or cylindrical sheet, provided the term $(2/a)$ in (7) is replaced by $|\Delta\kappa|$, where $\Delta\kappa$ is the *jump* in curvature of the surface, measured *parallel* to the boundary layer, as we cross over the crease in the idealised form of the boundary layer. Figure 8 shows explicitly how $\Delta\kappa$ is defined. For the case of the spherical shell, there is a jump from $+1/a$ to $-1/a$ as we cross the crease; and so (7) is recovered. But for the dimple in the cylindrical shell there is a jump from $|1/R|$ to 0 at points on axial or transverse diameters of the dimple; and indeed

$$|\Delta\kappa| \approx 1/R \tag{9}$$

uniformly around the boundary layer.

For a dimple of radius $r$ we may therefore evaluate $U$, the total elastic strain energy in the boundary-layer:

$$U \approx 0.03Et^{2.5}(r/R)^2(1/R)^{0.5} \times 2\pi r \approx 0.2E(t/R)^{2.5}r^3. \qquad (10)$$

Next, let us evaluate the axial displacement $u$ of the top of the dimple, by using the kinematic condition that the axial generator in the dimple is inextensional. This gives

$$u = 2(r - R\sin(r/R)) \approx r^3/3R^2 \qquad (11)$$

to first order, by use of a Taylor expansion.

Now the dimple is held in position by an axial force $F$. Since there is negligible elastic strain energy stored in the inextensionally-deformed portions of the shell, $U$ is a good approximation to the total elastic energy stored, and so

$$F = dU/du \approx 0.6Et^{2.5}/R^{0.5}. \qquad (12)$$

Note, in particular, that since both $U$ and $u$ are proportional to $r^3$, the final expression for $F$ is independent of $r$.

Apart from the value of the constant, (12) is of precisely the same form as Guggenberger's approximate empirical expression (3), and (4). The difference in values of the constant is doubtless attributable to the over-simplified circular profile assumed for the boundary-layer that forms the outer rim of the dimple; and it seems likely that a more elaborate and realistic and kinematic exact version of the dimple shape would give a better value for the constant. But for present purposes that is a secondary matter : the important result is the way in which $F$ comes out as proportional to $Et^{2.5}/R^{0.5}$, and independent of the dimple-size $r$.

## 10. A model for self-weight buckling

We are now in a position to make a simple first-order theoretical analysis of the post-buckling mode for the open-topped shell. The basic idea (see Fig. 4(a)) is to equate the self-weight of the portion of the shell ABBA above (say) the centre-line of the dimple to the constant force $F$ necessary to hold the dimple in place. Since the weight of the portion ABBA is proportional to the area multiplied by the thickness $t$, it is immediately clear that the post-buckling behaviour will involve thickness as $t^{1.5}$, as in the experimental result (2).

Putting

$$F = CEt^{2.5}/R^{0.5} \tag{13}$$

and substituting for $F$ the weight above the centre of the dimple in the curves shown in Figs 3 and 5, we find that the value of $C$ rises from ~1.5 to ~2 as the inwards deflection at the centre of the dimple increases from $0.1R$ to $0.03R$. Since the distribution of membrane stress within the buckled shell is actually rather complicated (see Mandal and Calladine, 2000) the agreement between the two cases is encouraging. (Here, in computing the area ABBA, the width BB has been calculated from the maximum deflection $w_o$ by the conventional formula $\mathrm{BB} \approx (8w_oR)^{0.5}$, based on the theorem of intersecting chords of a circle. If, instead, the width is taken between points B of maximum outwards deflection, then $C$ turns out to be practically constant, but with a somewhat higher value.)

Although there is room for argument about the numerical constants in the various expressions above, the way in which the leading variables enter the formulae is well-defined: the physical relationship between the dimples in the different situations shown in Fig. 4 is clear.

We have built here a quantitative analysis of the post-buckling mode under self-weight of the open-topped shell onto the foundation of the simpler problem of the axisymmetric inversion of a thin, spherical shell. But there is also a strong *qualitative* connection between the spherical shell and the post-buckling mode. Thus, in all the modes of deformation shown in Figs 4, 5 and 7, the shells are behaving *inextensionally* throughout, *except* in the narrow boundary-layer which separates the inextensional moieties. The width of that boundary region – which is uniform in all our present examples – depends on a local interaction between bending and stretching effects in the shell. In this sense the spherical shell provides a good model for both the post-buckling mode in the self-weight buckling problem and also for Guggenberger's situation. The only way in which the mode changes as the thickness of the shell changes is that the width of the boundary layer changes. Thus, whereas in many shell problems (e.g. Koiter, 1945; Calladine, 1983) it is useful to work in terms of dimensionless variables made by normalising both deflections and characteristic modal wavelengths with respect to the thickness of the shell (raised to a suitable exponent), it is important in all of our modes to plot *absolute* values of these quantities (or, equivalently, to normalise them only with respect to the (constant) radius of the shell), because the overall patterns of deformation (i.e. apart from the details of the boundary-layer) do not depend upon the shell's thickness at all – just as in the case of the spherical shell

## 11. Experimental "scatter" of buckling loads for "ordinary" shells.

Having now explained the provenance of the mean experimental data (2), we are left with the problem of explaining the significant "scatter" in buckling loads that is found in the course of testing "ordinary" shells with conventional end-fittings, as seen in Figs 1 and 2(a).

Now it is clear that the post-buckling mode shown in Fig. 4(a) would not be able to occur if the top of the shell were to be held circular: the inextensional "storage-tank-like" mode of the upper part of the shell obviously requires the top edge to be free. The free upper edge also makes the shell statically determinate as a membrane (e.g. Calladine, 1983, §6.5.1). Thus our crude estimation of $F$ in Fig. 4(a) as the weight of the portion of the shell above the dimple would not be at all reasonable if the top were held circular. Hence it seems likely that the "scatter" observed in the testing of "ordinary" shells is somehow directly related to the statical indeterminacy of those structures.

This line of argument is supported by some unexpected results obtained by Lancaster, Palmer and Calladine (2000) in the testing of an "ordinary" cylindrical shell made from Melinex sheet ( $R/t \approx 1800$) to which the end discs were secured frictionally by circumferential belt-like clamps. The results of many tests on this shell (see the caption of Fig. 1) came unexpectedly high in comparison with otherwise comparable experiments, whether or not imperfections had been introduced deliberately; and it is most likely that the unusually good performance of the shell is attributable to the effective near static-determinacy of a shell having these rather unusual frictional boundary conditions.

If the absence of experimental scatter is indeed attributable to boundary conditions that make a shell statically determinate, then practical benefits may well accrue to studies of the detailed design of end-conditions, in order specifically to achieve situations close to static determinacy.

## 12. Summary of the chain of argument

The analysis which I have put forward may be summarised as follows.

1. The experimentally determined stress level at the base of an open-topped elastic shell that can just stand up without buckling under its own weight agrees very well with the mean buckling stress recorded in many experiments on "ordinary" cylindrical shells under uniform axial loading. In either set of experiments

$$\sigma_{\text{mean}}/E \approx 5(t/R)^{1.5} \qquad (2, bis)$$

2. This suggests a working hypothesis that the two problems are closely related.
3. Therefore, if we can explain empirical formula (2) in the context of self-weight buckling of open-topped shells, we shall also have explained the mean buckling strength of "ordinary" cylindrical shells.

4. Finite-element analysis of the self-weight buckling of shell specimens suggests that there is a post-buckling mode for which the load remains near a "plateau" value. The mode involves the growth of a dimple near the base, which allows the upper part of the shell to deform inextensionally, falling outwards over the dimple region.
5. Guggenberger's analysis of dimples near the edge of locally-loaded shells shows that the dimple supports a more-or-less constant force proportional to $Et^{2.5}/R^{0.5}$, irrespective of the size of the dimple. And essentially the same result may be obtained by a simple argument based on the energy of inversion of a thin spherical shell.
6. The experimental observations on self-weight post-buckling are consistent with the force holding the dimple in place being provided by the weight of the shell vertically above the dimple, so that the critical height depends on $t^{1.5}$, other things being equal.
7. Finally, the absence of "scatter" in experimental self-weight buckling assays on open-topped shells may be attributed to the absence of statical indeterminacy in such shells.

## 13. Closing remarks

In this paper I have attempted to resolve an ancient paradox in shell buckling theory. It is for the reader to judge whether or not I have succeeded in my aim. The style of my work is, of course, very different from that of many papers in the field of the theory of shell structures, which develop a rigorous mathematical trail from axiom to conclusion. Here, instead, I have assembled an argument in the form of discrete links in a chain. Whether or not I have succeeded depends, therefore, on each of the links holding good; for a chain is only as strong as its weakest link.

My key tool has been the idea that the post-buckling mode for an open-topped shell is *inextensional* everywhere except in a narrow *boundary-layer*, whose behaviour may be quantified approximately by analogy with the inversion of a spherical shell. In this connection we might recall the boundary-layer in small-deflection shell theory which likewise provided the key to the resolution of the famous controversy between A.E.H. Love and Lord Rayleigh in the 1880s (Calladine, 1988).

I hope that my arguments will not be assailed on the grounds that they lack absolute precision; for my "plateaux" are only flat "to first order", and my numerical constants are given only to about one significant figure. (That is why the usual term containing the Poisson ratio does not appear in (1)). I take comfort from Francis Crick's remark to the effect that some problems are so difficult that they can only be solved by *a process of over-simplification*. And indeed, my work is an example of what Robert May has called "the lie that tells the truth".

## References


Brush, D.O. and Almroth, B.O. (1975) *Buckling of bars, plates and shells.* McGraw-Hill Book Company, New York.

Bushnell, D. (1967). Bifurcation phenomena in spherical shells under concentrated and ring loads. *AIAA Journal* **5**, 2034-2040.

Calladine, C.R. (1983) *Theory of shell structures.* Cambridge University Press.

Calladine, CR. (1988) The theory of shell structures, 1888-1988. *Proceedings of the Institution of Mechanical Engineers* **202**, number 42.

Calladine, C.R. and Barber, J.N. (1970) Simple experiments on self-weight buckling of open cylindrical shells. *Journal of Applied Mechanics, Trans. ASME* **37**, 1150-1151.

Guggenberger, W., Greiner, R. and Rotter, J.M.(1999) The behaviour of locally-supported cylindrical shells : unstiffened shells. *Journal of Constructional Steel Research* (in press).

Holst, J.M.F.G. and Calladine, C.R. (1994) Inversion problems in elastic thin shells. *European Journal of Mechanics, A/Solids* **13**, no. 4, supplement pp 3-18.

Hibbit, Karlsson and Sorensen, Inc., (1995) *ABAQUS User's Manual* version 5.5.

Kamyab, H. and Palmer, S.C. (1989) Analysis of displacements and stresses in oil storage tanks caused by differential settlement. *Proceedings, Institution of Mechanical Engineers* **303**, Part C, 61-70.

Koiter, W.T. (1945) *On the stability of elastic equilibrium* (in Dutch with English summary). PhD thesis, Delft, H.J. Paris, Amsterdam. English translation, Air Force Dynamics Laboratory Technical Report, AFFDL-TR-70-25, Ohio, February 1970.

Lancaster, E.R., Palmer, S.C. and Calladine, C.R. (2000) Paradoxical buckling behaviour of a thin cylindrical shell under axial compression. *International Journal of Mechanical Sciences* **42**, 843-865.

Mandal, P. and Calladine, C.R. (2000) Buckling of thin cylindrical shells under axial compression. *International Journal of Solids and Structures* (in press).

Mescall, J.F. (1965) Large deflections of spherical shells under concentrated loads. *Journal of Applied Mechanics, Trans. ASME* **32**, 936-938.

Nowinka, J. and Lukasiewicz, S. (1991). Symmetric elements in geometrical analysis of large deformations of spherical shells. *International Journal of Non-Linear Mechanics* **26**, 151-168.

Pogorelov, A.. (1988) *Bending of surfaces and stability of shells.* Tr. From Russian by J.R. Schulenberger, ed. B. Silver, Providence : American Mathematical Society.

Ranjan, G.V. and Steele, CR. (1977) Large deflection of deep spherical shells under concentrated load. *Proceedings of 18th Structures, Structural Dynamics and Materials Conference*, 269-278, San Diego.

Timoshenko, S.P. and Gere, J.M. (1961) *Theory of elastic stability* (2nd edition). New York : McGraw-Hill Book Company.

von Kármán, T., Dunn, L.G. and Tsien, H. (1940) The influence of curvature on the buckling characteristics of structures. *Journal of the Aeronautical Sciences* **7**, 276-289.

von Kármán, T. and Tsien, H. (1941) The buckling of thin cylindrical shells under axial compression. *Journal of the Aeronautical Sciences* **8**, 303-312.

Zhu, E., Mandal, P. and Calladine, C.R. (2000) Buckling of thin cylindrical shells : an attempt to resolve a paradox. In J-G. Teng and J.M. Rotter (eds), *Buckling of thin metal shells*, E. and F.N. Spon, London (in press).

# BUCKLING ANALYSIS OF COMPOSITE PLATES

R. GILAT AND J. ABOUDI
*Faculty of Engineering, Academic College JAS, Ariel 44837, Israel*
*Faculty of Engineering, Tel Aviv University, Ramat Aviv 69978, Israel*

## 1. Introduction

The present paper summarizes a series of recent investigations that were conducted by the authors which address the analysis of bifurcation buckling, parametric stability, dynamic buckling and thermally induced dynamic buckling of composite plates and shells. Various types of material behavior are assumed including linearly elastic, nonlinearly elastic and thermo-inelastic.

## 2. Bifurcation buckling by global-local theory

In this section we consider the formulation of the bifurcation buckling of composite plates in the framework of the recently developed global-local plate theory of Williams (1999). In this theory the plate is divided into several sublaminae. The displacement at a point within the plate is given in terms of global coordinates that represent the location of this point, and local coordinates that refer to the location of that point with respect to the sublamina. The advantage of this theory is in its flexibility. It offers the possibility of using various combinations of local and global displacement components such that the accuracy and the computational expense could be optimized. Furthermore the theory does not rely on any specific constitutive law.

### 2.1. BASIC FORMULATION

Consider a rectangular laminated plate uniformly supported along the edges $0 \leq x \leq L_x$, $0 \leq y \leq L_y$, subjected to inplane loads. The $z$ coordinate is



*D. Durban et al. (eds.), Advances in the Mechanics of Plates and Shells,* 135–150.

perpendicular to the plane of the plate with its origin located in the mid-plane, and the thickness of the plate is $h$. The plate consists of $N$ layers such that the $k$'th layer occupies the region $z_{k-1} \leq z \leq z_k$.

According to Williams (1999), the displacement field can be considered to be composed of global and local contributions such that at a point which lies within the $k$'th layer

$$u_i(x,y,z) = u_i^{global}(x,y,z) + u_i^{(k)local}(x,y,z), \quad z_{k-1} \leq z \leq z_k \tag{1}$$

The global displacement field $u_i^{global}(x,y,z)$ is continuous over the plate thickness $-h/2 \leq z \leq h/2$, while $u_i^{(k)local}(x,y,z)$, which represents the local variation of the displacement within the $k$'th layer, is zero outside the layer namely for $z < z_{k-1}$ and $z > z_k$. In the framework of the global-local plate theory the global and local displacement fields are assumed to have the following general form

$$u_i^{global} = \sum_{r=J_{min}}^{J_{max}} U_i^r(x,y)P^r(z), \quad u_i^{(k)local} = \sum_{s=j_{min}}^{j_{max}} \mu_i^{(k)s}(x,y)g^{(k)s}(z) \tag{2}$$

where the functions $P^r(z)$, $g^{(k)s}(z)$ are arbitrary but yet independent of each other.

Using the principle of stationary potential energy, in conjunction with the von Karman nonlinear strain-displacement relations, the assumed displacement field, eqns. (1), (2), and in the absence of body forces, the following equilibrium equations are obtained (Williams, 1999)

$$\begin{aligned}
&\overline{N}^r_{x\alpha,\alpha} - \overline{R}^r_{xz} + \overline{\tau}^r_x = 0, \quad \overline{N}^r_{y\alpha,\alpha} - \overline{R}^r_{yz} + \overline{\tau}^r_y = 0 \\
&\overline{N}^r_{z\alpha,\alpha} - \overline{R}^r_{zz} + \overline{\tau}^r_z + \left[\sum_{p=J_{min}}^{J_{max}} \overline{M}^{pr}_{\alpha\beta} U^p_{z,\beta} + \sum_{k=1}^{N}\sum_{q=j_{min}}^{j_{max}} \hat{M}^{(k)qr}_{\alpha\beta}\mu^{(k)q}_{z,\beta}\right]_{,\alpha} = 0 \\
&\hat{N}^{(k)s}_{x\alpha,\alpha} - \hat{R}^{(k)s}_{xz} + \hat{\tau}^{(k)s}_x = 0, \quad \hat{N}^{(k)s}_{y\alpha,\alpha} - \hat{R}^{(k)s}_{yz} + \hat{\tau}^{(k)s}_y = 0 \\
&\hat{N}^{(k)s}_{z\alpha,\alpha} - \hat{R}^{(k)s}_{zz} + \hat{\tau}^{(k)s}_z + \left[\sum_{p=J_{min}}^{J_{max}} \hat{M}^{(k)sp}_{\alpha\beta} U^p_{z,\beta} + \sum_{q=j_{min}}^{j_{max}} \hat{m}^{(k)sq}_{\alpha\beta}\mu^{(k)q}_{z,\beta}\right]_{,\alpha} = 0
\end{aligned} \tag{3}$$

with $r = J_{min}, ..., J_{max}$, $s = j_{min}, ..., j_{max}$ $\alpha = x, y$ and the summation law is valid for repeating Greek subscript indices. In eqns. (3) the following definitions have been used

$$[\overline{N}^r_{ij}, \overline{R}^r_{ij}] = \int_{z_0}^{z_N} \sigma_{ij}[P^r, P^r_{,z}]dz, \quad [\hat{N}^{(k)s}_{ij}, \hat{R}^{(k)s}_{ij}] = \int_{z_{k-1}}^{z_k} \sigma_{ij}[g^{(k)s}, g^{(k)s}_{,z}]dz$$

$$\overline{M}^{rp}_{ij} = \int_{z_0}^{z_N} \sigma_{ij}P^r P^p dz, \quad \hat{M}^{(k)sr}_{ij} = \int_{z_{k-1}}^{z_k} \sigma_{ij}g^{(k)s}P^r dz \tag{4}$$

$$\hat{m}^{(k)qs}_{ij} = \int_{z_{k-1}}^{z_k} \sigma_{ij}g^{(k)q}g^{(k)s}dz \,, \ \overline{\tau}^r_i = \sigma_{iz}P^r|^{z_N}_{z_0} \,, \ \hat{\tau}^{(k)s}_i = \sigma^{(k)}_{iz}g^{(k)s}|^{z_k}_{z_{k-1}} \tag{5}$$

and $\sigma_{iz}^{(k)}|_{z_k}(x,y)$ represents the interfacial transverse stresses over the layer boundary $z = z_k$.

By adopting the classical linear stability theory, a critical load is sought under which a perturbed equilibrium state adjacent to the prebuckling equilibrium state position can exist. The equations that govern the change in the displacement field with respect to the prebuckling equilibrium state are eqns. (3a,b,d,e) together with the following linearized form of eqns. (3c,f) (Gilat et. al., 2000a)

$$\overline{N}^r_{z\alpha,\alpha} - \overline{R}^r_{zz} + \overline{\tau}^r_z + \sum_{p=J_{min}}^{J_{max}} (\overline{M}^{pr}_{\alpha\beta})^0 U^p_{z,\beta\alpha} + \sum_{k=1}^{N} \sum_{q=j_{min}}^{j_{max}} (\hat{M}^{(k)qr}_{\alpha\beta})^0 \mu^{(k)q}_{z,\beta\alpha} = 0$$

$$\hat{N}^{(k)s}_{z\alpha,\alpha} - \hat{R}^{(k)s}_{zz} + \hat{\tau}^{(k)s}_z + \sum_{p=J_{min}}^{J_{max}} (\hat{M}^{(k)sp}_{\alpha\beta})^0 U^p_{z,\beta\alpha} + \sum_{q=j_{min}}^{j_{max}} (\hat{m}^{(k)sq}_{\alpha\beta})^0 \mu^{(k)q}_{z,\beta\alpha} = 0$$

$$(\overline{M}^{rp}_{\alpha\beta})^0 = \int_{z_0}^{z_N} \sigma^0_{\alpha\beta} P^r P^p dz, \quad (\hat{M}^{(k)sr}_{\alpha\beta})^0 = \int_{z_{k-1}}^{z_k} \sigma^0_{\alpha\beta} g^{(k)s} P^r dz,$$

$$(\hat{m}^{(k)qs}_{\alpha\beta})^0 = \int_{z_{k-1}}^{z_k} \sigma^0_{\alpha\beta} g^{(k)q} g^{(k)s} dz \tag{6}$$

and the superscript 0 denotes the prebuckling state while terms free of it represent the change from the equilibrium state just prior to buckling.

The equations of equilibrium are accompanied by boundary conditions specifying either the essential conditions or the natural conditions on the boundaries $x = 0, L_x$ and $y = 0, L_y$. On the boundary surfaces $z = \pm\frac{h}{2}$, which are unloaded, the tractions vanish.

Since the assumed displacement field (2) is an arbitrary one, the continuity of tractions and the continuity/discontinuity of displacements still have to be imposed. The interfacial traction continuity conditions are

$$\sigma_{iz}^{(k)}|_{z_k} = \sigma_{iz}^{(k+1)}|_{z_k} = \sigma_{iz}^{(k)I}, \qquad k = 1, ..., N-1, \quad i = x, y, z \tag{7}$$

As the possibility of debonding between the layers is considered, let the jump in the displacement at the interface between two adjacent layers be denoted by

$$u_i^{(k+1)}|_{z_k} - u_i^{(k)}|_{z_k} = \sum_{s=\hat{J}_{min}}^{\hat{J}_{max}} (\mu_i^{(k+1)s} g^{(k+1)s}|_{z_k} - \mu_i^{(k)s} g^{(k)s}|_{z_k}) = [u_i]^{(k)} \tag{8}$$

In general, the displacement jump $[u_i]^{(k)}$ depends on the constitutive relations for delamination initiation and growth. An interfacial constitutive model that expresses the displacement jump $[u_i]^{(k)}$ in terms of the interfacial traction $\sigma_{iz}^{(k)I}$ can be represented the following general form

$$[u_i]^{(k)} = f(\sigma_{iz}^{(k)I}) \tag{9}$$

Equations (3a,b,d,e), (6) together with the constraints, (7)-(8), the linearized von-Karman strain-displacement relations and appropriate constitutive laws for the material behavior within the layers, and the interfacial behavior (9), establish the linear buckling problem in terms of the unknown field variables $U_i^r(x,y)$, $\mu_i^{(k)s}(x,y)$, and $\sigma_{iz}^{(k)I}$.

## 2.2. BUCKLING OF LINEARLY ELASTIC SPECIALLY ORTHOTROPIC PLATES

Consider a simply supported, specially orthotropic laminated plate. The plate edges $x = 0, L_x$ are subjected to uniform normal strains such that

$$\varepsilon_{xx}^0 = \lambda\varepsilon^0, \quad \varepsilon_{yy}^0 = \varepsilon_{xy}^0 = 0$$

where $\lambda$ is the critical buckling parameter to be determined, and $\sigma_{iz}^0 = 0$, $i = x, y, z$. The components of the displacement field, which represent the change from the prebuckling state are described in terms of expansions in the inplane coordinates $x$, $y$ and the thickness coordinate $z$. The expansion in the thickness direction is chosen to be

$$P^r = z^{r-1}\ , r = J_{min}, ..., J_{max}, \quad g^{(k)s} = \sum_{m=0}^{s} a_m^{(k)s} z^m\ , \quad s = j_{min}, ..., j_{max} \tag{10}$$

where $a_m^{(k)s}$ are coefficients. All but one of those coefficients are determined by an orthogonalization procedure which ensures that terms of the same power in the global and local polynomial expansion are independent.

The plane functional forms of the displacements which satisfy the simply supported boundary conditions, and the interlaminar transverse stresses are assumed to be of the following form

$$\begin{aligned}
[U_x^r, \mu_x^{(k)s}, \sigma_{xz}^{(k)}|_{z_k}] &= \sum_m \sum_n [\tilde{U}_{xmn}^r, \tilde{\mu}_{xmn}^{(k)s}, \tilde{\tau}_{xzmn}^{(k)}]\ cos(\frac{m\pi x}{L_x})\ sin(\frac{n\pi y}{L_y}) \\
\left[U_y^r, \mu_y^{(k)s}, \sigma_{yz}^{(k)}|_{z_k}\right] &= \sum_m \sum_n [\tilde{U}_{ymn}^r, \tilde{\mu}_{ymn}^{(k)s}, \tilde{\tau}_{yzmn}^{(k)}]\ sin(\frac{m\pi x}{L_x})\ cos(\frac{n\pi y}{L_y}) \\
\left[U_z^r, \mu_z^{(k)s}, \sigma_{zz}^{(k)}|_{z_k}\right] &= \sum_m \sum_n [\tilde{U}_{zmn}^r, \tilde{\mu}_{zmn}^{(k)s}, \tilde{\tau}_{zzmn}^{(k)}]\ sin(\frac{m\pi x}{L_x})\ sin(\frac{n\pi y}{L_y})
\end{aligned} \tag{11}$$

Furthermore, a linear interfacial constitutive model is adopted such that (Aboudi, 1991)

$$[u_x]^{(k)} = \frac{1}{R_t^{(k)}}\sigma_{xz}^{(k)I}, \quad [u_y]^{(k)} = \frac{1}{R_t^{(k)}}\sigma_{yz}^{(k)I}, \quad [u_z]^{(k)} = \frac{1}{R_n^{(k)}}\sigma_{zz}^{(k)I} \tag{12}$$

where for perfectly bonded layers $[u_i]^{(k)} = 0$, i.e. $R_t^{(k)} = R_n^{(k)} \to \infty$. By substituting eqns. (10)-(12) into eqns. (3a,b,d,e), (6) and eq. (8), in conjunction with eqn. (7), a set of $3(\bar{J} + N\hat{J} + N - 1)$ linear homogeneous algebraic equations is derived for each wave shape combination ($mn$). (Here $\overline{J} = J_{max} - J_{min} + 1$ and $\hat{J} = j_{max} - j_{min} + 1$ are the number of terms in the global and local expansions, respectively). This system of equations can be put in the following general form of an eigenvalue problem (Gilat et. al., 2000a)

$$(A + \lambda B)v = 0 \tag{13}$$

where $A$ and $B$ are matrices of coefficients, and $v$ is the vector of unknowns. The smallest eigenvalue $\lambda$ which corresponds to an eigenvector of a flexural mode, namely a displacement field with non zero out of plane components, is the critical buckling parameter.

This formulation has been implemented by Gilat et. al. (2000a) for the determination of the buckling loads of various types of laminated plates whose laminae are either perfectly bonded or debonded, and comparisons with available results have been shown.

## 3. Parametric stability of nonlinearly elastic composite plates

In this section the dynamic stability of nonlinearly elastic composite plates subjected to periodic in-plane loading is investigated. This stability analysis is performed by evaluating the largest Lyapunov exponent, the sign of which indicates whether the system is stable or not.

### 3.1. BASIC FORMULATION - CYLINDRICAL BENDING

Consider a rectangular nonlinearly elastic anisotropic plate of an infinite width in the $y$ direction. The plate is uniformly supported along the edges $x = 0, L$ which are subjected to a uniform normal in-plane periodic load. Neglecting the in-plane inertia, the response of the plate is governed by the following classical plate theory equations (Whitney, 1987)

$$N_{xx,x} = 0, \quad N_{xy,x} = 0, \quad M_{xx,xx} + N_{xx}u_{z,xx} = I_1\ddot{u}_z \tag{14}$$

Here $u_z$ is the displacement in the transverse direction $z$, $I_1 = \int_{-\frac{h}{2}}^{\frac{h}{2}} \rho dz$ where $\rho$ is the material effective density and $h$ is the plate thickness, and dot denote differentiation with respect to time $t$. The stress and moment resultants $N_{xx}$, $N_{xy}$, $M_{xx}$ are defined in the usual manner. For a simply supported plate the boundary conditions at $x = 0, L$ are

$$N_{xx} = N(t), \quad N_{xy} = 0, \quad u_z = 0, \quad M_{xx} = 0 \tag{15}$$

The strain-displacement relations for the present cylindrical bending situation are

$$\varepsilon_{xx} = \varepsilon^0_{xx} - zu_{z,xx}, \quad \varepsilon_{yy} = 0, \quad \varepsilon_{xy} = \varepsilon^0_{xy} \tag{16}$$

where $\varepsilon_{xx}^0$, $\varepsilon_{xy}^0$ are the strains of the midplane of the plate.

Let the plate layers consist of unidirectional linearly elastic anisotropic fibers reinforcing nonlinearly elastic resin matrix. The linearly anisotropic elastic fiber material behavior is governed by the generalized Hooke's law. The non-linearly elastic behavior of the elastic isotropic matrix is modeled by the generalized Ramberg-Osgood representation, which leads to strain-stress relations of the form

$$\varepsilon_{ij} = \frac{1+\nu}{E}\sigma_{ij} - \frac{\nu}{E}\sigma_{kk}\delta_{ij} + \frac{3}{2E}\hat{\sigma}_{ij}\left(\frac{\check{\sigma}}{\sigma_0}\right)^{n-1} \tag{17}$$

where $E$ is Young's modulus, $\nu$ is Poisson's ratio, $\sigma_0$ and $n$ are parameters characterizing the material nonlinearity, $\hat{\sigma}_{ij} = \sigma_{ij} - \frac{1}{3}\sigma_{kk}\delta_{ij}$ is deviatoric stress ($\delta_{ij}$ is the Kronecker delta) and $\check{\sigma} = \sqrt{\frac{3}{2}\hat{\sigma}_{ij}\hat{\sigma}_{ij}}$.

The overall behavior of the two-phase composite is obtained by the micromechanical method of cells (Aboudi, 1991) which provides the nonlinear anisotropic effective constitutive relations for the composite, relying on the material behavior of its constituents. By adopting an incremental formulation in conjunction with the micromechanical method of cells analysis, and through application of the standard transformation from the material to the plate coordinates, the instantaneous effective stiffness tensor $\boldsymbol{C}^I$ can be established such that the instantaneous response of the nonlinear composite is given by (Gilat and Aboudi, 2000b)

$$\Delta\sigma = \boldsymbol{C}^I \Delta\varepsilon \tag{18}$$

where $\Delta\varepsilon$ and $\Delta\sigma$ are the increments of strains and stresses, respectively, and all tensors are referred to the plate coordinates ($x$, $y$, $z$).

As a result of the above formulation of the instantaneous behavior of the composite, the incremental constitutive relations of the nonlinearly elastic plate under cylindrical bending can be expressed as follows

$$\begin{aligned}
\Delta N_{xx} &= A_{11}^I\,\Delta\varepsilon_{xx}^0 + A_{16}^I\,\Delta\varepsilon_{xy}^0 - B_{11}^I\,\Delta u_{z,xx} \\
\Delta N_{xy} &= A_{61}^I\,\Delta\varepsilon_{xx}^0 + A_{66}^I\,\Delta\varepsilon_{xy}^0 - B_{61}^I\,\Delta u_{z,xx} \\
\Delta M_{xx} &= B_{11}^I\,\Delta\varepsilon_{xx}^0 - B_{16}^I\,\Delta\varepsilon_{xy}^0 - D_{11}^I\,\Delta u_{z,xx}
\end{aligned} \tag{19}$$

where $A_{ij}^I$, $B_{ij}^I$, $D_{ij}^I$ are the instantaneous extensional coupling and bending stiffnesses of the plate.

The two first eqns. (14) imply that the inplane stress resultants are independent of $x$ such that taking into account the boundary conditions (15) their increments are given by

$$\Delta N_{xx} = \Delta N(t), \quad \Delta N_{xy} = 0 \tag{20}$$

A comparison between (20) and (19), yields the following expression for the increment of the moment, $\Delta M_{xx}$, in terms of the external load and the transverse displacement increments

$$\Delta M_{xx} = K_1 \Delta N + K_2 \Delta u_{z,xx} \tag{21}$$

where $K_1$ and $K_2$ are coefficients that can be expressed in terms of the instantaneous plate stiffnesses.

In order to investigate the parametric stability of the plate, we consider a time-dependent in-plane load which is the result of the following strain imposed to the edges $x = 0, L$

$$\varepsilon_{xx}^0 = \varepsilon_s + \varepsilon_d \; cos(\theta t) \tag{22}$$

where $\varepsilon_s$ and $\varepsilon_d$ are constants and $\theta$ is the load frequency.

Having established the incremental constitutive equations (19) for the nonlinearly elastic plate, the third eqn(14) can be transformed into an incremental form. Consequently, the variation of the displacement $\Delta u_z$ within a time increment $\Delta t$, $t^{(l)} \leq t \leq t^{(l+1)}$ is governed by

$$R + \Delta M_{xx,xx} + N^{(l)} \Delta u_{z,xx} + \Delta N (u_{z,xx}^{(l)} + \Delta u_{z,xx}) = I_1 \; \Delta \ddot{u}_z \tag{23}$$

where $R = M_{xx,xx}^{(l)} + N^{(l)} u_{z,xx}^{(l)}$, $\; u_z^{(l)} = u_z(x, t^{(l)})$ etc.

Using the separation of variables, the deflection is assumed to have the following form which satisfies the simply supported boundary conditions

$$\left[ u_z^{(l)}, \Delta u_z \right] = \sum_m \left[ W_m^{(l)} \; , \; \Delta W_m(t) \right] sin(m\pi x / L) \tag{24}$$

It should be noted that the boundary condition on $M_{xx}$ is satisfied as long as the initial configuration of the plate is symmetric with respect to $z = 0$. In such cases, due to the fact that the loading is constant throughout the plate thickness, $B_{ij}^I$ at the edges remain zero.

By employing the Galerkin method in conjunction with eqns. (21) and (24), eqn. (23) is reduced to the following set of ordinary nonlinear differential equations

$$R_k - \Delta N \alpha_k (W^{(l)} + H_k) - (N^{(l)} + \Delta N) \alpha_k \Delta W_k + \alpha_k \sum_m \frac{m^2 \pi^2}{L^2} J_{km} \Delta W_m = \Delta \ddot{W}_k \tag{25}$$

$$R_k = \frac{2}{L I_1} \int_0^L R \; sin \frac{k\pi x}{L} dx, \quad H_k = \frac{2}{L} \int_0^L K_1 \; sin \frac{k\pi x}{L} dx$$

$$J_{km} = \frac{2}{L} \int_0^L K_2 \; sin \frac{k\pi x}{L} \; sin \frac{m\pi x}{L} dx, \quad \alpha_k = \frac{k^2 \pi^2}{L^2 I_1} \qquad m, k = 1, ...M$$

## 3.2. PARAMETRIC STABILITY ANALYSIS

In order to investigate the stability of nonlinearly elastic plates under periodic in-plane loads, the concept of Lyapunov exponent is employed. Lyapunov stability analysis of a dynamical system consists of the evaluation of a corresponding set of characteristic numbers (e.g. Hahn, 1967). The negative values of these characteristic numbers are known as Lyapunov exponents. According to Lyapunov, the motion is asymptotically stable if all the exponents are negative. A positive Lyapunov exponent indicates an exponential separation between two initially close trajectories, namely instability of the system (Chetaev, 1961). The system is stable if the largest Lyapunov exponent is not greater then zero. Consequently, it is sufficient to evaluate the largest Lyapunov exponent in order to characterize the behavior of a dynamical system.

According to Goldhirsch et. al.(1987) the Lyapunov exponents can be determined by the following procedure. Consider the system of ordinary nonlinear differential equations

$$\dot{\boldsymbol{v}} = \boldsymbol{F}(\boldsymbol{v}) \tag{26}$$

The stability equation is defined to be

$$\dot{\boldsymbol{y}} = \boldsymbol{G}\boldsymbol{y}, \quad G_{ij} = \frac{\partial F_i}{\partial v_j}\Big|_{\boldsymbol{v}=\boldsymbol{v}(t)} \tag{27}$$

and $\boldsymbol{y}$ can be regarded as a small perturbation $\delta\boldsymbol{v}$. Within the time increment $0 < t < t^{(1)}$, the system (27) with $\boldsymbol{G}(t=0)$ is solved numerically for the normalized initial conditions $||\boldsymbol{y}(0)|| = 1$ where $||.||$ is the Euclidean norm. This yields $\boldsymbol{y}(t^{(1)})$.

Eqns. (27) with $\boldsymbol{G}(t = t^{(1)})$ and with the following initial conditions

$$\boldsymbol{y}(t^{(1)}) = \boldsymbol{z}(t^{(1)}), \quad \boldsymbol{z}(t^{(1)}) = \frac{\boldsymbol{y}(t^{(1)})}{||\boldsymbol{y}(t^{(1)})||}$$

are then solved within the second time interval $t^{(1)} < t < t^{(2)}$ yielding $\boldsymbol{v}(t^{(2)})$. The process is repeated for $n$ time intervals while correspondingly, the system (26) is solved to provide the values of $\boldsymbol{v}(t^{(l)})$ needed for the evaluation of $\boldsymbol{G}$. Namely, the incremental procedure is simultaneously used to get the nonlinear response and the Lyapunov exponents.

For the $n'$th time interval, let us define the value of the parameter $\mu_n$ as follows

$$\mu_n = \frac{\sum_{l=1}^{n} ln\left|\left|\boldsymbol{y}(t^{(l)})\right|\right|}{t^{(n)}} \tag{28}$$

It has been shown (Goldhirsch et. al.,1987) that for finite large time, the value of $\mu_n$ approaches to the value of the Lyapunov exponent.

Employing the above procedure to investigate the dynamic stability of a nonlinearly elastic infinitely wide plate, eqn. (26) is obtained by reducing eqn. (25) to a set of first order differential equations. The matrix $\boldsymbol{G}$ in the stability equations (27) can thus be written as follows

$$\boldsymbol{G} = \begin{bmatrix} [0] & [I] \\ [\alpha_k^2 I_1^2 J_{km} - N\alpha_k\delta_{km}] & [0] \end{bmatrix} \tag{29}$$

Note that the matrix $\boldsymbol{G}$ is a function of the instantaneous stiffnesses which depend on the current state of stress. The latter is the solution of eqn. (25) which (when reduced to a set of first order equations) is the relevant special case of eqn. (26). Hence eqns. (25), have to be solved simultaneously with the progressing of the stability analysis. Within a time interval $t^{(l)} < t < t^{(l+1)}$, the numerical integration of eqns. (25), with the initial conditions

$$\Delta W_k = 0, \quad \Delta \dot{W}_k = \dot{W}_k, \qquad at \quad t = t^{(l)} \tag{30}$$

yields the displacement field at $t^{(l+1)}$. On the basis of this, the stress field and the elements of the instantaneous stiffness tensor at every point of the plate can be re-evaluated. The coefficients of the stability equations $G_{ij}$ are updated accordingly, and the stability analysis is carried on.

Application of this methodology has been recently carried out by Gilat and Aboudi (2000b) where the parametric stability of nonlinearly elastic cross-ply plate has been investigate. It was shown that nonlinear effects due to the material behavior increase the stability of the the nonlinear plate as compared to the corresponding linearly elastic one.

## 4. Dynamic buckling of nonlinearly elastic composite cylindrical shell

In the present section the effect of the material nonlinearity on the dynamic buckling of cylindrical shells due to the application of a non-periodic time-dependent axial loading is investigated. To this end, the concept of dynamic buckling of Budiansky (1967) is adopted.

### 4.1. BASIC FORMULATION - AXISYMMETRIC RESPONSE

Consider a cylindrical shell of length $L$ radius $R$, wall thickness $h$, with $x$ and $y$ being respectively the axial and circumferential coordinates, and the $z$ coordinate is directed inward along the radial direction and its origin is located in the midsurface of the shell. For thin shallow cylindrical shells the theory of Donnell is utilized (Vinson, 1989). Due to the material nonlinearity, the previously described incremental formulation is adopted for the establishments of the material behavior by means of instantaneous stiffnesses as well as for the structural analysis. For a state of axisymmetric response of the shell, the incremental form of the von Karman strains have the following form (Gilat and Aboudi, 1995a)

$$\Delta\varepsilon_{xx}(x,z,t) = \Delta\varepsilon^0_{xx} + z\Delta\varepsilon^1_{xx}, \quad \Delta\varepsilon_{yy}(x,z,t) = \Delta\varepsilon^0_{yy}, \quad \Delta\varepsilon_{xy}(x,z,t) = \Delta\varepsilon^0_{xy}$$

$$\Delta\varepsilon^0_{xx}(x,t) = \Delta u^0_{x,x} + \frac{1}{2}(\Delta u^0_{z,x})^2 + \Delta u^0_{z,x}u^0_{z,x}, \quad \Delta\varepsilon^0_{yy}(x,t) = -\frac{\Delta u^0_z}{R}$$
$$\Delta\varepsilon^0_{xy}(x,t) = \frac{1}{2}\Delta u^0_{y,x}, \qquad \Delta\varepsilon^1_{xx}(x,t) = -\Delta u^0_{z,xx} \tag{31}$$

where $u^0_x(x,t)$, $u^0_y(x,t)$ and $u^0_z(x,t)$ denote the displacements of the mid-surface in the $x, y$ and $z$ directions respectively.

The equations based on Donnell's shell theory, which govern the axisymmetric motion of an anisotropic cylindrical shell subjected to time dependent axial load (within the time increment $t^{(l)} \leq t \leq t^{(l+1)}$), can be presented in the following incremental form

$$R_1 + \Delta N_{xx,x} = I_1\,\Delta\ddot{u}^0_x, \qquad R_2 + \Delta N_{xy,x} = I_1\,\Delta\ddot{u}^0_y$$
$$R_3 + \left[N^{(l)}_{xx}\,\Delta u^0_{z,x} + \Delta N_{xx}\left(u^{0(l)}_z + \Delta u^0_z + u_{z0}\right)_{,x}\right]_{,x}$$
$$+ \Delta M_{xx,xx} + \frac{1}{R}\Delta N_{yy} = I_1\,\Delta\ddot{u}^0_z \tag{32}$$
$$R_1 = N^{(l)}_{xx,x}, \quad R_2 = N^{(l)}_{xy,x}, \quad R_3 = M^{(l)}_{xx,xx} + \frac{1}{R}N^{(l)}_{yy} + \left(N^{(l)}_{xx}u^{0(l)}_{z,x}\right)_{,x}$$

These governing equations are accompanied by initial and boundary conditions. The substitution of the stress and moment resultants definition in conjunction with the strain-displacement relations eqn. (31), and the constitutive law (18) into the governing equations (32) yields a system of nonlinear partial differential equations in terms of the incremental displacement variables, $\Delta u^0_i$, $i = x, y, z$. These equations are solved successively for each of the small time increments $\Delta t = t^{(l+1)} - t^{(l)}$, $l = 0, 1, \ldots l_{max}$ by employing a spatial finite difference and temporal Runge-Kutta integration. The coefficients of the differential equations, which due to the material behavior are not constants, are assumed to remain unaltered within each time increment but are updated at its end.

### 4.2. DYNAMIC BUCKLING

The definition of the limit of stability for structures which are subjected to nonperiodic time-dependent loads requires the examination of the ability of the system to preserve a certain property under perturbation of a specific type (Bellman, 1953). In order to do this, the approach of Budiansky (1967) is adopted according to which the dynamic buckling is associated with the state at which small changes in the magnitude of loading lead to large changes in the structure response. To this end, the response of the structure to loads of various magnitudes is studied. Since the response depends both on time and space, it is necessary to characterize this response by a specific value. Buckling curves can then be constructed which exhibit the variation of the response characteristic with the loading magnitude. This criteria for instability has been widely used. Yet, it should be noted that this criterion for dynamic buckling is not necessarily unequivocal. As long as the slope of the buckling curve exhibits an abrupt change, from almost zero to almost infinity, the dynamic buckling load can be easily defined. But as the slope changes more gradually, the application of the Budiansky criterion becomes more ambiguous. It must be supported by an arbitrary definition of the critical slope, or according to Simitses (1990), by a definition of an allowed displacement.

Dynamic pulse buckling loads of nonlinearly elastic composite shells have been presented by Gilat and Aboudi (1995a). The influence of various parameters that control the applied load behavior and the geometry and materials of the composite shell have been investigated.

## 5. Thermally induced dynamic buckling of metal matrix composite plates

In this section the dynamic buckling of metal matrix composite plates induced by a rapid heating is analyzed. Here the temperature field is fully decoupled from the mechanical field and is solely governed by the transient heat equation. Due to the existence of the metallic phase, inelastic behavior occurs that must be incorporated into the analysis.

### 5.1. BASIC FORMULATION - CYLINDRICAL BENDING

Consider a laminated metal matrix composite rectangular plate of an infinite width in the $y$ direction, uniformly supported along the edges $x = 0, L$, and exposed to a rapid surface heating. The thickness of the plate is $h$ and the coordinate $z$ is perpendicular to the plane of the plate with its zero placed in the mid-plane.

If the effect of the mechanical field on the temperature filed is neglected, the temperature $T$ is governed by the heat conduction equation which can be solved independently of the mechanical problem.

In the framework of the classical plate theory, the von-Karman kinematic relations for the cylindrical bending state are

$$\varepsilon_{ij} = \varepsilon^0_{ij} + z\varepsilon^1_{ij} \tag{33}$$

$$\varepsilon^0_{xx} = u^0_{x,x} + \frac{1}{2}(u^0_{z,x})^2 + u^0_{z,x}\, u_{z0,x}, \quad \varepsilon^1_{xx} = -u^0_{z,xx}$$

$$\varepsilon^0_{yy} = \varepsilon^1_{yy} = 0, \quad \varepsilon^0_{xy} = \frac{1}{2}u^0_{y,x}, \quad \varepsilon^1_{xy} = 0$$

and $u_{z0}$ is the initial geometrical imperfection which is associated with the initial stress-free state.

The plate is made of layers consisting of unidirectional elastic anisotropic fibers embedded in an inelastic metallic matrix. The thermo-elastic fiber behavior is governed by the generalized Hooke's law, while the behavior of the thermo-elastic-viscoplastic matrix material is governed by the Bodner-Partom (1975) unified theory. The effective mechanical and thermal properties of the composite as well as its overall inelastic strains are determined by employing the method of cells.

Neglecting the inplane inertia, the equations which govern the motion of the plate are

$$N_{xx,x} = 0, \quad N_{xy,x} = 0, \quad M_{xx,xx} + N_{xx}(u_{z,xx} + u_{0z,xx}) = I\ddot{u}_z \tag{34}$$

These equations are associated with initial conditions which specify the initial displacements and velocities, and boundary conditions defining the displacements and moments at the plate's edges.

Integration of the first two of equations (34) in conjunction with the the stress-strain relations, the stress resultants definition for a state of plane stress, and the strain-displacement relation (33), results in

$$N_{xx} = A_{11}\,\varepsilon^0_{xx} + A_{16}\,2\varepsilon^0_{xy} - B_{11}\,u^0_{z,xx} - N^I_{xx} - N^T_{xx} = c_0$$

$$N_{xy} = A_{16}\,\varepsilon^0_{xx} + A_{66}\,2\varepsilon^0_{xy} - B_{16}\,u^0_{z,xx} - N^I_{xy} - N^T_{xy} = c_1 \tag{35}$$

where $c_0(t)$, $c_1(t)$ are integration functions and $N^I_{ij}$, $N^T_{ij}$ are inelastic and thermal stress resultants, respectively.

Let the temperature field be symmetric about $x = L/2$. Furthermore, it is assumed that the variation of the stiffnesses with respect $x$, due to the temperature dependence of the material properties, is negligible. The solution of (35) for $\varepsilon^0_{xx}$ and $\varepsilon^0_{xy}$ leads to

$$\varepsilon^0_{xx} = c_2 + c_3\,u^0_{z,xx} \qquad 2\varepsilon^0_{xy} = c_4 + c_5\,u^0_{z,xx} \tag{36}$$

$c_2(x,t)$, $c_3(t)$, $c_4(x,t)$ and $c_5(t)$ depend on the plate stiffnesses, the inelastic and thermal stress resultants, and $c_0$, $c_1$. Eqns. (36) in conjunction with strain-displacement relations (33) yield the inplane displacements of the midplane

$$\begin{aligned} u^0_x &= \int c_2\,dx + c_3\,u^0_{z,x} - \frac{1}{2}\int u^0_{z,x}\,(u^0_{z,x} + 2u_{0z,x})\,dx + c_7 \\ u^0_y &= \int c_4\,dx + c_5\,u^0_{z,x} + c_6 \end{aligned} \tag{37}$$

By imposing the boundary conditions of immovable edges, the following expressions for $c_0$ and $c_1$ are obtained

$$\begin{aligned} c_0 &= \frac{A_{16}c_5 + A_{11}c_3}{L}\left[u^0_{z,x}(0) - u^0_{z,x}(L)\right] + \frac{A_{11}}{2L}\int_0^L u^0_{z,x}(u^0_{z,x} + 2u_{0z,x})dx \\ &- \frac{1}{L}\int_0^L (N^I_{xx} + N^T_{xx})\Delta x \ , \quad c_1 = \frac{A_{66}c_5 + A_{16}c_3}{L}\left[u^0_{z,x}(0) - u^0_{z,x}(L)\right] \\ &+ \frac{A_{16}}{2L}\int_0^L u^0_{z,x}(u^0_{z,x} + 2u_{0z,x})dx - \frac{1}{L}\int_0^L (N^I_{xy} + N^T_{xy})dx \end{aligned}$$

Thus, the stress and moment resultants can be expressed in terms of the plate stiffnesses, the inelastic and thermal stress resultants and the unknown transverse displacement, such that the third of the equations of motion (34) becomes a partial differential equation in term of the transverse displacement, $u_z^0$, only.

Let the initial imperfection and the transverse displacement, which is expected to be symmetric with respect to $x = L/2$, be expressed by the following series

$$\begin{aligned} u_{0z} &= \sum_{j=1,3}^{J} W_{0j}\ sin(j\pi x/L) \\ u_z^0 &= \sum_{j=1,3}^{J} W_j(t)\ sin(j\pi x/L) + \tilde{W}(t)\ cos(2\pi x/L - 1) \end{aligned} \tag{38}$$

where $\tilde{W}(t) = B_{11}c_2(0,t) + B_{16}c_4(0,t) - M_{xx}^I - M_{xx}^T$ is defined such that the boundary condition $M_{xx} = 0$ at $x = 0, L$ is satisfied. When symmetry (with respect to $z$) of the layup and the temperature field exists, $\tilde{W}(t) = 0$.

Application the Galerkin method with respect to the spatial coordinate $x$ reduces the third equation (34) to the following set of nonlinear ordinary differential equations

$$\sum_j W_j^2\ W_k\ b_{1_{k_j}} + \sum_j W_j^2\ b_{2_{k_j}} + \sum_j W_j\ W_k\ b_{3_{k_j}} \tag{39}$$

$$+ \sum_j W_j\ b_{4_{k_j}} + W_k\ b_{5_k} + b_{6_k} = I_1 \ddot{W}_k + b_7 \ddot{\tilde{W}} \qquad k = 1, 3 ... J$$

where $b_{p_{k_j}}$, $b_{q_k}$, $b_r$ are functions of the plate stiffnesses, the inelastic and thermal resultants, the amplitudes of the initial deflection, and $\tilde{W}$.

The solution of eqns. (39), namely $W_k(t)$, $k = 1, ... J$ is obtained by using the Runge-Kutta integration scheme. This is done in conjunction with incrementally following both the development of plasticity and the change of the material properties due to the variation of the temperature field. At the end of each time increment, at all points of the structure, the micromechanical analysis is employed to obtain the plastic strains, plastic strain rates and the current stiffness matrix. The coefficients of eqn. (39) and their time derivatives are then evaluated. Assuming that the derivatives remain constant within the following time increment, eqns. (39) are integrated to obtain the amplitude of the transverse displacement $W_k$ at the end of the current time increment. It should be noted that $\dot{\tilde{W}}$ is also assumed to remain constant during each time increment such that the term including $\ddot{\tilde{W}}$ on the right hand side of eqns. (39), vanishes.

Results that exhibit the thermally induced dynamic buckling behavior of metal matrix composite plates have been presented by Gilat and Aboudi (1995b).

## 6. Fully coupled thermomechanical dynamic buckling of metal matrix composite plates

In this section we consider the dynamic buckling of metal matrix composite plates under circumstances of elevated temperature and high rate of loading. Here the mutual influence between the induced mechanical and thermal fields are significant. In such cases, the full thermomechanical coupling exists and must be taken into account.

### 6.1. BASIC FORMULATION - CYLINDRICAL BENDING

Consider an infinitely wide metal matrix composite plate. In the framework of the high order shear deformation theory of Reddy (1984), the von-Karman strain-displacement relations for the cylindrical bending state in an infinitely wide plate (described in the previous section) have the following form

$$\varepsilon\,(x,z,t) = \varepsilon^0 + z\varepsilon^1 + z^2\varepsilon^2 + z^3\varepsilon^3 \tag{40}$$

$$\begin{gathered}
\varepsilon^0_{xx} = u^0_{x,x} + \frac{1}{2} + (u^0_{z,x})^2 + u^0_{z,x}\,u_{z0,x}, \quad \varepsilon^1_{xx} = u^1_{x,x}, \quad \varepsilon^2_{xx} = 0 \\
\varepsilon^3_{xx} = -\frac{4}{3h^2}\left(u^0_{z,xx} + u^1_{x,x}\right), \quad \varepsilon^0_{yy} = \varepsilon^1_{yy} = \varepsilon^2_{yy} = \varepsilon^3_{yy} = 0 \\
\varepsilon^0_{xy} = \frac{1}{2}u^0_{y,x}, \quad \varepsilon^1_{xy} = \frac{1}{2}u^1_{y,x}, \quad \varepsilon^2_{xy} = 0, \quad \varepsilon^3_{xy} = -\frac{2}{3h^2}u^1_{y,x} \\
\varepsilon^0_{xz} = \frac{1}{2}\left(u^0_{z,x} + u^1_{x}\right), \quad \varepsilon^1_{xz} = 0, \quad \varepsilon^2_{xz} = -\frac{4}{h^2}\varepsilon^0_{xz}, \quad \varepsilon^3_{xz} = 0 \\
\varepsilon^0_{yz} = \frac{1}{2}u^1_{y}, \quad \varepsilon^1_{yz} = 0, \quad \varepsilon^2_{yz} = -\frac{4}{h^2}\varepsilon^0_{xy}, \quad \varepsilon^3_{yz} = 0
\end{gathered}$$

Here $u^0_i$, $i = x, y, z$ denote the displacements of a point on the mid-plane, $u^1_i$, $i = x, y$ are the rotations of normals to the mid-plane and $u_{z0}$ is the initial geometrical imperfection associated with the initial stress-free state. Using the definitions of stress and moment resultants in conjunction with the effective constitutive relations for the thermo-visco-plastic composite with $\sigma_{zz} = 0$, the following plate constitutive relations are established (Gilat and Aboudi, 1996)

$$\begin{bmatrix} N \\ M \\ R \\ P \end{bmatrix} = \begin{bmatrix} A & B & D & E \\ B & D & E & F \\ D & E & F & G \\ E & F & G & H \end{bmatrix} \begin{bmatrix} \varepsilon^0 \\ \varepsilon^1 \\ \varepsilon^2 \\ \varepsilon^3 \end{bmatrix} - \begin{bmatrix} N^I \\ M^I \\ R^I \\ P^I \end{bmatrix} - \begin{bmatrix} N^T \\ M^T \\ R^T \\ P^T \end{bmatrix} \tag{41}$$

where $N^I$, $M^I$, $R^I$, $P^I$ and $N^T$, $M^T$, $R^T$, $P^T$ are the corresponding inelastic and thermal resultants.

The plate equations of motion which govern the behavior of the plate under a state of cylindrical bending, are

$$N_{xx,x} = I_1 \ddot{u}_x^0, \qquad N_{xy,x} = I_1 \ddot{u}_y^0$$

$$\left[N_{xx}\left(u_z^0 + u_{z0}\right)_{,x}\right]_{,x} + N_{xz,x} + \beta k_1 P_{xx,xx} - \beta k_2 R_{xz,x} = I_1 \ddot{u}_z^0 + \beta \overline{I}_5 \ddot{u}_{x,x}^1 - \beta k_1^2 I_7 \ddot{u}_{z,xx}^0$$

$$M_{xx,x} - \beta k_1 P_{xx,x} - N_{xz} + \beta k_2 R_{xz} = \overline{I}_3 \ddot{u}_x^1 - \beta \overline{I}_5 \ddot{u}_{z,x}^0$$

$$M_{xy,x} - \beta k_1 P_{xy,x} - N_{yz} + \beta k_2 R_{yz} = \overline{I}_3 \ddot{u}_y^1 \tag{42}$$

$$I_i = \int_{-h/2}^{h/2} \rho z^{i-1} dz \qquad i = 1,3,5,7$$

$$\overline{I}_3 = I_3 - \beta(2k_1 I_5 - k_1^2 I_7), \quad \overline{I}_5 = k_1 I_5 - k_1^2 I_7, \quad k_1 = \frac{4}{3h^2}$$

The coupled energy equation for anisotropic thermo-inelastic media is (Allen, 1991)

$$-(k_{ij}T_{,j})_{,i} + \rho c_v \dot{T} + C_{ijkl}\alpha_{kl}T(\dot{\varepsilon}_{ij} - \dot{\varepsilon}_{ij}^I) - \eta \sigma_{ij}\dot{\varepsilon}_{ij} = 0 \tag{43}$$

where $\eta$ is a positive scalar not greater than 1 (usually around 0.9, see Hunter (1983) for example) representing the part of inelastic energy loss which is transformed into heat.

Following McQuillen and Brull (1970), the Galerkin method is used in order to derive the plate energy equations from the 3-D energy equation (43). For a state of cylindrical bending, the deviation from the initial temperature $T_R$ is assumed to have the following $z$ dependent form

$$\Delta T = T_0(1 + \tau_2 z^2) + T_1(z + \tau_4 z^3) + \tau_1 z^2 + \tau_3 z^3 \tag{44}$$

where the unknown functions $T_0(x,t)$ and $T_1(x,t)$ are the temperature at the midsurface and the constant part of the temperature gradient in the $z$ direction. The functions $\tau_i(x,t)$, $i = 1,...4$ are defined such the boundary conditions, specifying either the temperature or the heat flux over the upper and lower surfaces of the plate, are satisfied.

The expression (44) is substitute into equation (43) and the error is required to be orthogonal to the weighting functions $(1 + \tau_2 z^2)$ and $(z + \tau_4 z^3)$ within the interval $-\frac{h}{2} \le z \le \frac{h}{2}$. These weighting functions, unlike those used by McQuillen and Brull, ensure the derivation of variationally consistent plate energy equations, having the following form

$$\dot{T}_0 G_{11} + \dot{T}_1 G_{12} + S_1 = 0, \quad \dot{T}_0 G_{21} + \dot{T}_1 G_{22} + S_2 = 0 \tag{45}$$

Here $G_{ij}$ and $S_i$, $i,j = 1,2$ are defined in terms of the the thermal and mechanical fields, the thermal and mechanical material properties and the inelastic strains.

The mechanical equations of motion (42) and the two energy equations (45) which govern the displacements $u_i^0$, $i = x, y, z$, $u_i^1$, $i = x, y$ and the unknown temperature variables $T_i$, $i = 0, 1$, are associated with initial and boundary conditions for both the thermal and the mechanical fields.

Dynamic buckling behavior of metal matrix composite plates under combined thermomechanical loading has been presented by Gilat and Aboudi (1996). The effects of coupling due to mechanical energy generation with both its reversible and irreversible parts has been studied.

## References


Aboudi, J. (1991) *Mechanics of composite materials - a unified micromechanical approach.* Elsevier, Amsterdam.

Allen, D. H. (1991) Thermomechanical coupling in inelastic solids. *Appl. Mech. Rev.* **44**, 361-373.

Bellman, R. (1953) *Stability theory of differential equations.* McGraw-Hill, New York.

Bodner, S. R. and Partom, Y. (1975) Constitutive equations for elastic-viscoplastic strain-hardening materials. *J. Appl. Mech.* **42**, 385-89.

Budiansky, B. (1967) Dynamic buckling of elastic structures: criteria and estimates, *Dynamic stability of structures*, edited by G. Herrmann, Pergamon Press, Oxford, 83.

Chetaev, N. G. (1961) *Stability of motion.* Oxford: Pergamon Press.

Gilat, R. and Aboudi, J. (1995a) Dynamic buckling of resin matrix composite structures. *Composite Struct.* **32**, 81-88.

Gilat, R. and Aboudi, J. (1995b) Dynamic inelastic response and buckling of metal matrix composite infinitely wide plates due to thermal shocks. *Mech. Composite Mat. Struct.* **2**, 257-271.

Gilat, R. and Aboudi, J. (1996) Thermomechanical coupling effects on the dynamic inelastic response and buckling of metal matrix composite infinitely wide plates. *Composite Struct.* **35**, 49-63.

Gilat, R., Williams, T.O., and Aboudi, J. (2000a) Buckling of composite plates by global-local plate theory. (to appear).

Gilat, R. and Aboudi, J. (2000b) Parametric stability of nonlinearly elastic composite plates by Lyapunov exponents. (to appear).

Goldhirsch, I., Sulem, P.L. and Orszag, S.A. (1987) Stability and Lyapunov stability of dynamical systems: a differential approach and a numerical method. *Physica* **27D**, 311-337.

Hahn, W. (1967) *Stability of motion.* Berlin: Springer-Verlag.

Hunter, S. C. (1983) *Mechanics of continuous media.* 2nd ed., Ellis Horwood Ltd, Chichester.

McQuillen, E. J., and Brull, M. A. (1970) Dynamic thermoelastic response of cylindrical shells. *J. Appl. Mech.* **38**, 661-670.

Reddy, J. N. (1984) A refined nonlinear theory of plates with transverse shear deformation. *Int. J. Solids Struct.* **20**, 881-896.

Simitses, G. J. (1990) *Dynamic Stability of Suddenly Loaded Structures.* Springer-Verlag, Berlin.

Vinson, J. R. (1989) *The behavior of thin walled structures: beams, plates, and shells.* Kluwer Academic Publishers, Dordrecht.

Whitney, J. M. (1987) *Structural analysis of laminated anisotropic plates.* Technomic Publishing Co. Inc., Lancaster.

Williams, T.O. (1999) A generalized multilength scale nonlinear composite plate theory with delamination. *Int. J. Solids Struct.* **36**, 3015-3050.

# STATIC OPTIMAL CONTROL OF THE LARGE DEFORMATION OF A HYPERELASTIC PLATE

Dan Givoli
*Department of Aerospace Engineering*
*and Asher Center for Space Research*
*Technion — Israel Institute of Technology*
*Haifa 32000, Israel**
givolid@aerodyne.technion.ac.il

Igor Patlashenko
*Altair Engineering*
*1757 Maplelawn Drive*
*Troy, MI 48084-4603 USA*
igorpt@better.net

**Abstract** The reduction of the large in-plane static deformation of a thin hyperelastic plate using control loads is considered. This problem has important applications in the control of flexible space structures. A mathematical model leads to an elliptic optimal control problem in nonlinear elasticity. A numerical optimal control method, based on Finite Element (FE) discretization and Sequential Quadratic Programming (SQP), is employed to minimize the deformation of the plate. Results are presented for a specific example.

*Partial funding provided by the Adler Fund for Space Research managed by the Israel Academy of Sciences, and by the Fund for the Promotion of Research at the Technion.



*D. Durban et al. (eds.), Advances in the Mechanics of Plates and Shells,* 151–166.

## Introduction

The optimal control of flexible structures is an active area of research. The main body of work in this area is concerned with the control of time-dependent displacements and stresses, and assumes linear elastic conditions, namely linear elastic material behavior and small deformation. See, e.g., [1]–[3], the collections of papers [4, 5], and references therein.

On the other hand, in the present paper we consider the *static* optimal control of a structure made of a *nonlinear elastic* material and undergoing *large deformation.* An important application is the suppression of static or quasi-static elastic deformation in flexible space structures such as parts of satellites by the use of control loads [6]. Solar radiation and radiation from other sources induce a temperature field in the structure, which in turn generates an elastic displacement field. The displacements must usually satisfy certain limitations dictated by the allowed working conditions of various orientation-sensitive instruments and antennas in the space vehicle. For example, a parabolic reflector may cease to be effective when undergoing large deflection. The elastic deformation can be reduced by use of control loads, which may be implemented via mechanically-based actuators or more modern piezoelectric devices. When the structure under consideration is made of a rubber-like material and is undergoing large deformation, nonlinear material and geometric effects must be taken into account in the analysis.

Finite Element (FE) methods for elliptic optimal control of structures in a variational setting have been considered in [7] for plastic deformation problems in metal forming, in [8] for the von Karman elastic plate equations, in [9] for viscoelasticity, and in [10] for a cracked linear elastic structure. Most of the numerical methods employed in these optimal control formulations involve the "adjoint state," which appears when the Pontryagin maximum principle is employed [11]. The problem's variables include, in addition to the primary state variables, the adjoint variables. This leads to a mixed FE formulation involving a linear algebraic system of dimension $2N^u$, where $N^u$ is the number of primary degrees of freedom.

Recently, a new general framework has been developed by Givoli for the FE solution of optimal control problems governed by nonlinear elliptic partial differential equations [12]. In contrast to the FE schemes mentioned above, the approach in [12] is a direct one, which does not involve adjoint variables. Computationally, this has the effect of leading to a simpler formulation and reducing the number of variables by a factor of two. The solution of the final discrete minimization problem is performed via Sequential Quadratic Programming (SQP). This formulation does not employ the standard tools of classical control theory, but fits naturally into the framework of computational continuum mechanics.

In this paper we apply the direct optimal control methodology to the problem of reducing the large in-plane static deformation of a thin hyperelastic plate using control loads. First we develop the general FE-SQP formulation associated with the problem, including the constrained and unconstrained cases, and then we present some results for a specific example.

# 1. THE OPTIMAL-CONTROL HYPERELASTIC PROBLEM

## 1.1 CONSTITUTIVE MODEL

We consider a thin deformable plate in a plane-stress state. We fix a reference cartesian coordinate system $\boldsymbol{X} = (X_1, X_2, X_3)$, where $X_3$ is in the plate's thickness direction. Let $\boldsymbol{x} = (x_1, x_2, x_3)$ be the position vector in the current configuration. Since the deformation of the plate is assumed to be planar, only $x_1$ and $x_2$ are of interest. Let $\boldsymbol{F}$ be the three-dimensional deformation tensor, i.e., $\boldsymbol{F} = \partial \boldsymbol{x}/\partial \boldsymbol{X}$, and let

$$J = \det \boldsymbol{F} = \lambda_1 \lambda_2 \lambda_3 \ . \tag{1}$$

Here $J$ is the Jacobian associated with the three-dimensional deformation (which is also the volume ratio in the current and initial configurations, $J = dv/dV$), and the $\lambda_i$ are the principal stretches. The latter are the eigenvalues of the tensor $\boldsymbol{F}$, or more precisely,

$$\boldsymbol{F} = \sum_{i=1}^{3} \lambda_i \ \boldsymbol{n}_i \otimes \boldsymbol{N}_i \ . \tag{2}$$

Here, the $\boldsymbol{N}_i$ and $\boldsymbol{n}_i$ are, respectively, the eigenvectors of the right Cauchy-Green tensor (in the reference configuration) and the eigenvectors of the left Cauchy-Green tensor (in the current configuration).

Now, we consider a stretch-based hyperelastic material whose stored elastic potential is (see, e.g., [13]):

$$\psi(\lambda_1, \lambda_2, \lambda_3) = \frac{\lambda}{2}(\ln J)^2 + \mu \sum_{i=1}^{3} (\ln \lambda_i)^2 \ . \tag{3}$$

Here $\lambda$ and $\mu$ are given Lamé coefficients. This elastic potential leads to the Cauchy stress tensor $\boldsymbol{\sigma}$, which can be expressed in principal directions $\boldsymbol{m}_i$ as

$$\boldsymbol{\sigma} = \sum_{i=1}^{3} \sigma_i \, \boldsymbol{m}_i \otimes \boldsymbol{m}_i \ . \tag{4}$$

Here the $\sigma_i$ and the $\boldsymbol{m}_i$ are the eigenvalues and eigenvectors of the tensor $\boldsymbol{\sigma}$. The principal Cauchy stresses are obtained from the elastic potential $\psi$ via

$$\sigma_i = \frac{\lambda_i}{J} \frac{\partial \psi}{\partial \lambda_i} \ . \tag{5}$$

Here and elsewhere we *do not* enforce the summation rule on repeated indices. In particular, we obtain from (5) and (3):

$$\sigma_3 = \frac{2\mu}{J} \ln \lambda_3 + \frac{\lambda \ln J}{J} \ . \tag{6}$$

Eq. (5) and thus (6) hold in the general case. In the plane-stress case we have $\sigma_3 = 0$. We also define the Jacobian $j$ associated with the *two-dimensional* deformation in the $(X_1, X_2)$ plane (which is also the area ratio in the current and initial configurations, $j = da/dA$), i.e.,

$$j = \det{}_{2\times 2} \boldsymbol{F} = \lambda_1 \lambda_2 \qquad , \qquad J = \lambda_3 j \ . \tag{7}$$

By using these facts in (6) we get

$$\ln \lambda_3 = -\frac{\lambda}{\lambda + 2\mu} \ln j \ . \tag{8}$$

We substitute this into (3) and, after some algebra, obtain a plane-stress form for $\psi$ analogous to (3), i.e.,

$$\psi(\lambda_1, \lambda_2) = \frac{\bar{\lambda}}{2} (\ln j)^2 + \mu \sum_{\alpha=1}^{2} (\ln \lambda_\alpha)^2 \ . \tag{9}$$

Here, $\bar{\lambda}$ is the effective plane-stress Lamé coefficient defined by

$$\bar{\lambda} = \gamma \lambda \qquad , \qquad \gamma = \frac{2\mu}{\lambda + 2\mu} \ . \tag{10}$$

Also, from (7), (8) and (10) it is easy to show that $\ln J = \gamma \ln j$, namely,

$$J = j^{\gamma} \ . \tag{11}$$

To obtain an expression for the principal stress, we first differentiate $\psi$ given by (9) with respect to the stretches, and obtain (for $\alpha = 1, 2$)

$$\frac{\partial \psi}{\partial \lambda_\alpha} = \frac{\bar{\lambda} \ln j}{\lambda_\alpha} + \frac{2\mu \ln \lambda_\alpha}{\lambda_\alpha} \ . \tag{12}$$

Then from (5) and (12) we finally get

$$\sigma_\alpha = j^{-\gamma}(\bar{\lambda} \ln j + 2\mu \ln \lambda_\alpha) \ . \tag{13}$$

This is the stress-stretch relation in the principal directions. To obtain Cauchy stresses in other directions one may use the stress transformation formula

$$\sigma_{kl} = \sum_{\alpha=1}^{2} \sigma_{\alpha\alpha} T_{\alpha k} T_{\alpha l} \quad , \qquad T_{\alpha k} = \boldsymbol{n}_\alpha \cdot \boldsymbol{e}_k \ . \tag{14}$$

Here the $\boldsymbol{n}_\alpha$ are the principal stress directions in the current configuration, and $\boldsymbol{e}_k$ is the unit vector in the $x_k$ direction.

## 1.2 STATEMENT OF THE OPTIMAL-CONTROL PROBLEM

The statement of the problem consists of four ingredients: (1) governing equations, (2) boundary conditions, (3) objective functional, and (4) constraints on control.

**1.2.1 Governing Equations.** Let $\Omega$ be the finite spatial domain representing the plate, and let $\Gamma$ be its boundary. Let $\boldsymbol{U}(\boldsymbol{X})$ be the unknown displacement field with respect to the reference configuration. The governing equations in $\Omega$ are

$$\nabla \cdot \boldsymbol{\sigma} + \boldsymbol{f} = 0 \ , \tag{15}$$

$$\sigma_\alpha = j^{-\gamma}(\bar{\lambda} \ln j + 2\mu \ln \lambda_\alpha) \ , \tag{16}$$

$$\boldsymbol{\sigma} = \sum_{\alpha=1}^{2} \sigma_\alpha \, \boldsymbol{m}_\alpha \otimes \boldsymbol{m}_\alpha \quad , \qquad \boldsymbol{F} = \sum_{\alpha=1}^{2} \lambda_\alpha \, \boldsymbol{n}_\alpha \otimes \boldsymbol{N}_\alpha \ , \tag{17}$$

$$\boldsymbol{F} = \boldsymbol{I} + \frac{\partial \boldsymbol{U}}{\partial \boldsymbol{X}} \ . \tag{18}$$

Here $\boldsymbol{f}$ is the (two-dimensional) body-force vector, and $\boldsymbol{I}$ is the unit second-order tensor. All the other variables have been defined in the previous section. Eqs. (15)–(18) are, respectively, the equation of equilibrium, the hyperelastic constitutive equation (see (13)), the definitions of principal Cauchy stresses and stretches, and the kinematic (deformation-displacement) relation.

If the stresses and stretches are eliminated from these equations, one is led to a system of nonlinear equilibrium equations in terms of the displacements $U_\alpha$, which may be written abstractly in the form

$$NU + f = 0 \quad \text{in} \quad \Omega \, . \tag{19}$$

Here $N$ is a nonlinear elliptic differential operator.

**1.2.2 Boundary Conditions.** To fix ideas, we suppose that the plate is clamped on part of its boundary, $\Gamma_1$, and is initially free on the rest of its boundary, $\Gamma_2$. Our goal is to reduce the size of the displacement $U(X)$ of the plate generated by the body force $f$. To this end, we apply control load $s$ along the part of the boundary $\Gamma_c \subset \Gamma_2$. The rest of $\Gamma_2$, denoted $\Gamma_f \equiv \Gamma_2 - \Gamma_c$, remains traction-free. Thus, the boundary conditions are

$$U = 0 \quad \text{on} \quad \Gamma_1 \, , \tag{20}$$

$$\sigma \cdot n = s \quad \text{on} \quad \Gamma_c \, , \tag{21}$$

$$\sigma \cdot n = 0 \quad \text{on} \quad \Gamma_f \, . \tag{22}$$

The subdivision $\Gamma = \Gamma_1 + \Gamma_c + \Gamma_f$ is assumed to be given a-priori. Note that the control load function $s(x)$, for $x \in \Gamma_c$, is a primal unknown in the problem, accompanying the other unknown function $U(X)$, for $X \in \Omega$.

**1.2.3 Objective Functional.** The requirement for minimal displacement $U$ is enforced in the least-squares sense. Thus, we define the quadratic functional:

$$C[U] = \int_\Omega |U(X)|^2 \, dX \, . \tag{23}$$

Other objective functions may be defined too [12], but here we concentrate on the $L_2$-norm (23).

**1.2.4 Constraints on Control.** We define bounds which limit the *size* of the components of the control load $s$, i.e., $s_\alpha \leq z_\alpha$, or more generally,

$$ts \leq z \, , \tag{24}$$

where $t$ is a transformation matrix, $z$ is a given constant bound vector, and the vector inequality is to be interpreted entry-wise. We remark that restricting the control load to have a prescribed *direction* can be viewed as a special case of the constraint (24).

One may be interested also in the *unconstrained* case, which simplifies the formulation. We shall relate to this case later when considering the numerical optimal control scheme.

**1.2.5 Statement of the Problem.** The problem to be solved is: *Find $U(X)$, $X \in \Omega$ and $s(x)$, $x \in \Gamma_c$, which satisfy the nonlinear equation (19), the boundary conditions (20)–(22), and the constraint (24), such that $C[U]$ given by (23) is minimized.*

# 2. COMPUTATIONAL SCHEME

## 2.1 FINITE ELEMENT DISCRETIZATION

The Galerkin Finite Element (FE) method is applied to the problem under consideration. Both $U$ and $s$ are approximated via FE shape function expansions, i.e.,

$$U(X) \simeq U^h(X) = \sum_{I \in E_u} d_I \psi_I(X) , \tag{25}$$

$$s(x) \simeq s^h(x) = \sum_{A \in E_s} S_A \phi_A(x) . \tag{26}$$

Here $E_u$ and $E_s$ are the set of displacement nodes and control nodes, $\psi_I$ and $\phi_A$ are the displacement and control shape functions, and $d_I$ and $S_A$ are the displacement and control nodal values, respectively. The global vectors whose entries are all the nodal displacements and all control loads are denoted $d$ and $S$, respectively. We denote the total number of displacement degrees of freedom by $N^u$, and the total number of control degrees of freedom by $N^s$.

Applying the approximations in (25) and (26) to the variational (or weak) form of (19)–(22) results in a system of nonlinear algebraic equations, of the form (see [12])

$$G_0(d) = F + QS . \tag{27}$$

Here $G_0$ is the vector of internal forces (which is a *nonlinear* function of the displacements $d$), and $F$ is the background load vector, both standard in nonlinear FE analysis. The term $QS$ is the control load contribution to the equilibrium equations. Note that both $d$ and $S$ are unknown. The matrix $Q$ in (27) has the form

$$Q = [Q_{IB}] = [Q_{(P,\alpha)(Y,\beta)}] \qquad (N^u \times N^s) ,$$

$$Q_{(P,\alpha)(Y,\beta)} = \delta_{\alpha\beta} \int_{\Gamma_c} \psi_P \phi_Y \, dx . \tag{28}$$

Here, the $\delta_{\alpha\beta}$ is the Kronecker delta. Also, in (28) and elsewhere, we have represented the degree of freedom $I$ by the pair $(P,\alpha)$, where $P$ is the nodal point and $\alpha$ is the direction corresponding to $I$. Likewise, the control degree of freedom $B$ is represented by the pair $(Y,\beta)$.

The approximation of (23) using (25) leads to the discrete objective function,

$$C^h[\boldsymbol{d}] = \boldsymbol{d}^T \boldsymbol{M} \boldsymbol{d} \tag{29}$$

where

$$\boldsymbol{M} = [M_{IJ}] = [M_{(P,\alpha)(Q,\beta)}] \qquad (N^u \times N^u) ,$$
$$M_{(P,\alpha)(Q,\beta)} = \delta_{\alpha\beta} \int_\Omega \psi_P \psi_Q \, d\boldsymbol{X} . \tag{30}$$

In (29), the superscript $^T$ denotes transposition.

The discrete counterpart of the constraint (24) is

$$\boldsymbol{T}\boldsymbol{S} \leq \boldsymbol{Z} , \tag{31}$$

where $\boldsymbol{T}$ is a given constant transformation matrix, and $\boldsymbol{Z}$ is a given constant bound vector.

Now the discrete optimal control problem can be posed as follows: *Find* $\boldsymbol{d}$ *and* $\boldsymbol{S}$ *which satisfy the nonlinear system (27) and the constraint (31), such that* $C^h$ *given by (29) is minimized.*

## 2.2 SEQUENTIAL QUADRATIC PROGRAMMING

The Newton iteration procedure is now applied to the nonlinear system (27). We denote the vector $\boldsymbol{d}$ at iteration $i$ by $\boldsymbol{d}^{(i)}$. At iteration $i+1$ the solution vector is updated via

$$\boldsymbol{d}^{(i+1)} = \boldsymbol{d}^{(i)} + \Delta\boldsymbol{d}^{(i)} \tag{32}$$

The increment $\Delta\boldsymbol{d}^{(i)}$ is found by solving the linear system of equations,

$$\boldsymbol{K}^{(i)} \Delta\boldsymbol{d}^{(i)} = \boldsymbol{R}^{(i)} \tag{33}$$

Here $\boldsymbol{K}^{(i)}$ is the tangent stiffness matrix,

$$\boldsymbol{K}^{(i)} \equiv \frac{\partial \boldsymbol{G}_0(\boldsymbol{d})}{\partial \boldsymbol{d}} \Big|_{\boldsymbol{d}=\boldsymbol{d}^{(i)}} \tag{34}$$

and $\boldsymbol{R}^{(i)}$ is the residual vector obtained from (27),

$$\boldsymbol{R}^{(i)} = \boldsymbol{F} - \boldsymbol{G}_0(\boldsymbol{d}^{(i)}) + \boldsymbol{Q}\boldsymbol{S} . \tag{35}$$

Eqs. (32), (33) and (35) can be written as

$$\boldsymbol{d}^{(i+1)}(\boldsymbol{S}) = \hat{\boldsymbol{d}}^{(i)} + \left(\boldsymbol{K}^{(i)}\right)^{-1} \boldsymbol{Q}\boldsymbol{S} \tag{36}$$

where

$$\hat{\boldsymbol{d}}^{(i)} = \boldsymbol{d}^{(i)} + \left(\boldsymbol{K}^{(i)}\right)^{-1} \left(\boldsymbol{F} - \boldsymbol{G}_0(\boldsymbol{d}^{(i)})\right) . \tag{37}$$

The vector $\hat{\boldsymbol{d}}^{(i)}$ as defined by (37) is the current solution *with no control.*

Now we substitute (36) into (29), and after some algebra obtain,

$$\bar{C}^h[\boldsymbol{S}] = \boldsymbol{S}^T \boldsymbol{P}^{(i)} \boldsymbol{S} - 2\boldsymbol{S}^T \boldsymbol{B}^{(i)} + \text{ const.} \tag{38}$$

where

$$\boldsymbol{P}^{(i)} = \boldsymbol{A}^{(i)T} \boldsymbol{M} \boldsymbol{A}^{(i)} \qquad (N^s \times N^s) \tag{39}$$

$$\boldsymbol{A}^{(i)} = \left(\boldsymbol{K}^{(i)}\right)^{-1} \boldsymbol{Q} \qquad (N^u \times N^s) \tag{40}$$

$$\boldsymbol{B}^{(i)} = -\boldsymbol{A}^{(i)T} \boldsymbol{M} \hat{\boldsymbol{d}}^{(i)} \qquad (N^s \times 1) \tag{41}$$

From (38) and (31) we then obtain the Quadratic Programming (QP) problem:

$$\textit{Given} \quad \boldsymbol{d}^{(i)}, \quad \textit{find} \quad \min_{\boldsymbol{TS} \leq \boldsymbol{Z}} [\boldsymbol{S}^T \boldsymbol{P}^{(i)} \boldsymbol{S} - 2\boldsymbol{S}^T \boldsymbol{B}^{(i)}] . \tag{42}$$

The QP problem (42) is solved using a standard QP algorithm. In the numerical of the next section, we shall use the Goldfarb-Idnani QP algorithm [14] for this purpose.

To summarize, the proposed method reduces the original optimal control problem into a sequence of QP problems, one in each Newton iteration. These problems are in turn solved by applying a standard QP algorithm. See [12, 15] for a detailed discussion on the computational aspects of this scheme.

## 2.3 COMPUTATIONAL ASPECTS

We now make a few remarks regarding the computational aspects of this formulation.

**2.3.1 Remark 1.** Equations (37) and (40) involve the inverse of the tangent stiffness matrix $\boldsymbol{K}^{(i)}$. In practice, the inverse is never actually computed, but $\boldsymbol{K}^{(i)}$ is factorized, and then back substitution is performed to obtain $\boldsymbol{K}^{-1}(\boldsymbol{F} - \boldsymbol{G}_0)$ in (37) and $\boldsymbol{K}^{-1}\boldsymbol{Q}$ in (40). The latter involves back substitution for each column of the "right hand side vector" $\boldsymbol{Q}$.

**2.3.2 Remark 2.** The matrix $\boldsymbol{P}^{(i)}$ appearing in the quadratic form in (42) is *symmetric and positive semidefinite.* Symmetry follows from (39) and from the symmetry of $\boldsymbol{M}$ defined in (30). Positivity is obtained from the simple calculation

$$\boldsymbol{v}^T \boldsymbol{P} \boldsymbol{v} = \boldsymbol{v}^T \boldsymbol{A}^T \boldsymbol{M} \boldsymbol{A} \boldsymbol{v} = (\boldsymbol{A}\boldsymbol{v})^T \boldsymbol{M} (\boldsymbol{A}\boldsymbol{v}) \geq 0 \ . \tag{43}$$

The first equality in (43) follows from (39), and the last inequality follows from the positivity of $\boldsymbol{M}$, which can easily be shown. Strict positive-definiteness of $\boldsymbol{P}^{(i)}$ is not obtained in general. QP algorithms for the problem (42) with a symmetric positive semidefinite matrix $\boldsymbol{P}$ are widely known [16, 17].

**2.3.3 Remark 3.** The matrix $\boldsymbol{P}$ and the operations in (39)–(41) are global in nature. This may have an undesirable effect on the computational effort needed in forming $\boldsymbol{P}$ and in the actual solution of the problem (42). However, this becomes a difficulty only when $N^s$ (i.e., the dimension of the discrete control space) is large. Typically, $N^s$ is much smaller than $N^u$. In other words, the total number of nodal control variables is much smaller than the total number of $u$ degrees of freedom. Thus, the computational effort associated with the matrix $\boldsymbol{P}$ in (39) is not necessarily large, even when the discrete problem at hand is large.

**2.3.4 Remark 4.** It is important to note that operations with the $N^u$-dimensional arrays are local in nature, and can be performed on the element level. The matrices and vectors $\boldsymbol{Q}$, $\boldsymbol{M}$, $\boldsymbol{G}_0^{(i)}$, $\boldsymbol{F}$ and $\boldsymbol{K}^{(i)}$, calculated in the proposed scheme, are formed in practice by the assembly of analogous element-level matrices and vectors, as usual in finite element analysis. The calculation of $\boldsymbol{M}\hat{\boldsymbol{d}}$ in (41) is also performed on the element level.

**2.3.5 Remark 5.** The QP problem (42) is solved *in each* Newton iteration by using a QP algorithm. All QP algorithms are iterative, and include some stopping criteria [16, 17]. Since only the QP step in the *last* Newton iteration yields the final optimal control, it is reasonable to modify the QP stopping criterion tolerance during the Newton process, so that it becomes tighter towards the end of this process. This would guarantee that the computational effort associated with the QP step is not too large when the solution $\boldsymbol{d}^{(i)}$ is not sufficiently close to the converged solution.

**2.3.6 Remark 6.** The shape functions $\psi_I$ are standard $C^0$ finite element functions, e.g., linear on triangular elements or bilinear on quadrilateral elements. On the other hand, the control shape functions $\phi_A$ need not be so regular. In the formulation above they appear only in the definition of the matrix $\boldsymbol{Q}$ (see (28)), and thus they are allowed to be *piecewise-continuous*. For example, one may use piecewise-constant $\phi_A$'s, where $A$ indicates the midpoint of element $A$. This enables one to represent, for example, a spatial "bang-bang control" as the approximate solution. In fact, the $\phi_A$ may even be Dirac delta functions, since the integral in (28) exists in this case. Then $u^h = \sum_A U_A \phi_A$ represents "concentrated forces" with intensities $U_A$ acting at the control nodal points.

**2.3.7 Remark 7.** If the governing equations are those of *linear* elasticity, i.e., the material behaves linearly and the deformation is small, then there is no need in Newton iterations, and the formulation above reduces to a *single QP problem* of the form (42). All the expressions given previously are valid, except that the superscript $(i)$ is omitted everywhere, $\boldsymbol{K}$ is a constant stiffness matrix, and $\hat{\boldsymbol{d}}$ in (37) is simply $\hat{\boldsymbol{d}} = \boldsymbol{K}^{-1}\boldsymbol{F}$, i.e., the solution with no control.

## 2.4 THE UNCONSTRAINED CASE

Now we consider the case where there are *no constraints* except those related to the required regularity of the control functions. (Regarding the latter requirement, see [12].) In this case, the controls may be "constrained" through penalty terms in the objective functional. Thus, we replace the objective functional $C[\boldsymbol{U}]$ in (23) by

$$C[\boldsymbol{U}] = \int_\Omega |\boldsymbol{U}(\boldsymbol{X})|^2 \, d\boldsymbol{X} + \int_{\Gamma_c} W(\boldsymbol{x}) |\boldsymbol{s}(\boldsymbol{x})|^2 \, d\boldsymbol{x} \; . \tag{44}$$

Here $W(\boldsymbol{x}) \geq 0$ is a given weight function.

The continuous-level optimal control problem is now:
*Find $\boldsymbol{U}(\boldsymbol{X})$, $\boldsymbol{X} \in \Omega$ and $\boldsymbol{s}(\boldsymbol{x})$, $\boldsymbol{x} \in \Gamma_c$, which satisfy the nonlinear equation (19), and the boundary conditions (20)–(22), such that $C[\boldsymbol{U}]$ given by (44) is minimized.*

We introduce the FE approximations (25) and (26), and obtain, after some algebra, the following algebraic minimization problem analogous to (42):

$$\textit{Given} \quad \boldsymbol{d}^{(i)}, \quad \textit{find} \quad \min\, [\boldsymbol{S}^T \boldsymbol{P}^{(i)} \boldsymbol{S} - 2\boldsymbol{S}^T \boldsymbol{B}^{(i)} + \boldsymbol{S}^T \boldsymbol{N} \boldsymbol{S}] \; , \tag{45}$$

where

$$\boldsymbol{N} = [N_{AB}] \qquad (N^s \times N^s) \qquad , \qquad N_{AB} = \int_{\Gamma_c} \phi_A W \phi_B \, d\boldsymbol{x} \; . \tag{46}$$

A necessary condition for a minimum is that $\partial\bar{C}^h/\partial S_A = 0$, for $A = 1, ..., N^s$, where $\bar{C}^h$ is the expression in brackets in (45). Hence (45) yields the $N^s$-dimensional linear system of equations,

$$(\boldsymbol{P}^{(i)} + \boldsymbol{N})\,\boldsymbol{S} = \boldsymbol{B}^{(i)}\;. \tag{47}$$

This linear system has to be solved anew in each Newton iteration.

## 3. NUMERICAL RESULTS

We consider a square hyperelastic plate occupying the domain $\Omega = [0, 10] \times [0, 10]$ and with thickness 0.1. The right side of the plate ($x_1 = 10,\ 0 \leq x_2 \leq 10$) is free, and its other three sides are clamped. The material parameters (see (3)) are $\lambda = 100$ and $\mu = 100$. The plate is loaded with the background body force $\boldsymbol{f} = (1, 0)^T$.

Control loads are applied in order to minimize the the in-plane deformation of the plate, in accordance with the problem stated previously. We consider four control-load cases:

(a) A single concentrated control load $\boldsymbol{s}_0 = (-s_0, 0)^T\delta(\boldsymbol{x}-\boldsymbol{x}_0)$ is applied at $\boldsymbol{x}_0 = (10, 5)^T$, with no constraints;

(b) Two concentrated control loads $\boldsymbol{s}_1 = (-s_1, 0)^T\delta(\boldsymbol{x} - \boldsymbol{x}_1)$ and $\boldsymbol{s}_2 = (-s_2, 0)^T\delta(\boldsymbol{x} - \boldsymbol{x}_2)$ are applied at $\boldsymbol{x}_1 = (10, 3)^T$ and $\boldsymbol{x}_2 = (10, 7)^T$, respectively, with no constraints;

(c) Three concentrated control loads $\boldsymbol{s}_0 = (-s_0, 0)^T\delta(\boldsymbol{x} - \boldsymbol{x}_0)$, $\boldsymbol{s}_1 = (-s_1, 0)^T\delta(\boldsymbol{x} - \boldsymbol{x}_1)$ and $\boldsymbol{s}_2 = (-s_2, 0)^T\delta(\boldsymbol{x} - \boldsymbol{x}_2)$ are applied at $\boldsymbol{x}_0 = (10, 5)^T$, $\boldsymbol{x}_1 = (10, 3)^T$ and $\boldsymbol{x}_2 = (10, 7)^T$, respectively, with no constraints;

(d) Like (c), but with the constraint $|s_i| \leq s_{\max} = 13$.

Thus there are between one and three control degrees of freedom, as the case may be. The $\delta(\boldsymbol{x} - \boldsymbol{x}_0)$ appearing above is the Dirac delta.

We employ our modified version of the hyperelasticity finite element code FLAGSHYP [13] to solve this optimal control problem. A finite element mesh of $10 \times 10 = 100$ square bilinear elements is used. In the Newton iteration process we use a convergence tolerance of $10^{-4}$, and incremental loading with 10 loading steps. For the solution of the QP problem in each Newton iteration we use the Goldfarb-Idnani algorithm [14].

Without control, the background load causes the deformation illustrated in Fig. 1.1.

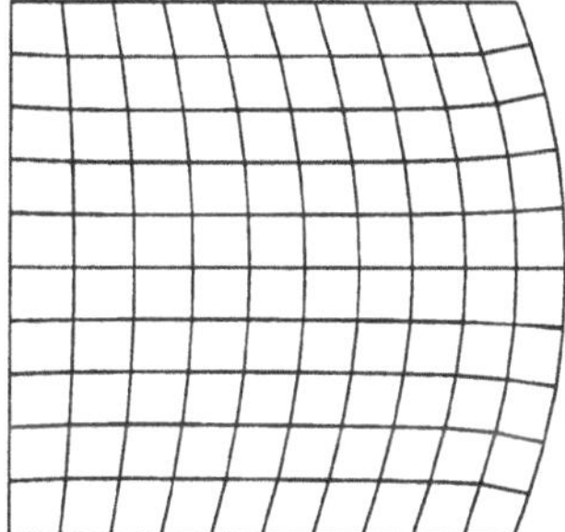

*Figure 1.1* Deformation generated by the background load without control.

Now we add control loads, to reduce this deformation. Figs. 1.2(a)–(d) show the deformed meshes in cases (a)–(d), respectively. It is clear that the control loads cause a large local deformation (indeed not well resolved by the coarse mesh used), but significantly reduces the deformation away from the right boundary.

Table 1.1 gives the values of the optimal control loads $s_0$, $s_1$ and $s_2$, and the corresponding values of the objective function $C^h$ in cases (a)–(d) as well as in the uncontrolled case. The arrangement of the optimal control loading in cases (b)–(d) is symmetric as expected. The uncontrolled value of $C^h$ decreases significantly when control is introduced in case (a), with only a slight further reduction in cases (b) and (c). In case (c), the upper and lower loads are seemingly more effective than the middle one in reducing the global deformation. Also, it is apparent that the constraint $s_i \leq s_{\max}$ in case (d) hardly affects the optimal value of $C^h$ and the deformation (cf. Figs. 1.2(c)–(d)), although it limits the values of $s_1$ and $s_3$ to the allowed maximum.

*Table 1.1* Results obtained for the optimal control loads and the objective function value.

| Case | $s_0$ | $s_1$ | $s_2$ | $C^h$ |
|---|---|---|---|---|
| Uncontrolled | — | — | — | 35.7 |
| (a) | 29.77 | — | — | 5.97 |
| (b) | — | 20.43 | 20.43 | 4.11 |
| (c) | 9.87 | 14.19 | 14.19 | 3.53 |
| (d) | 11.45 | 13.00 | 13.00 | 3.55 |

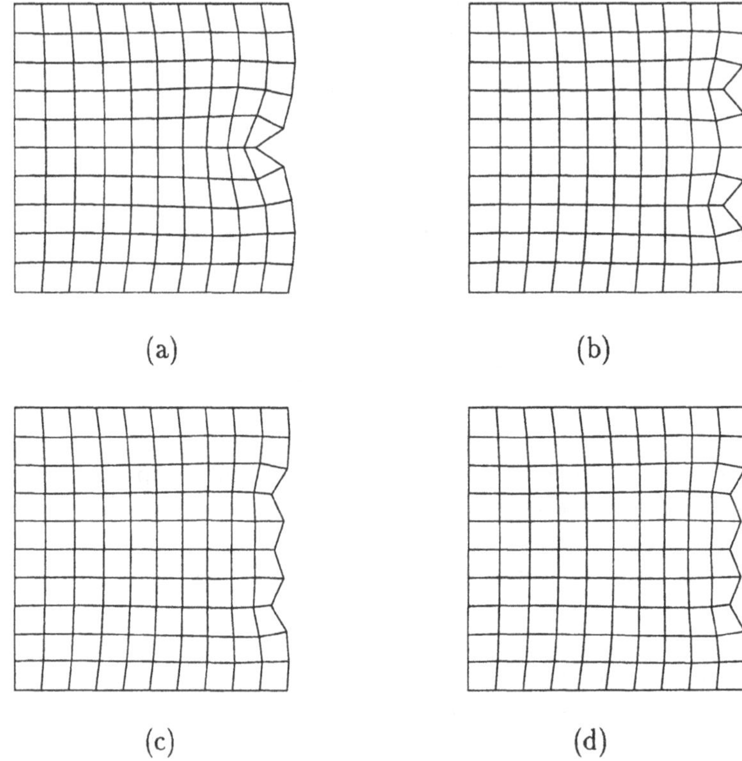


*Figure 1.2* Deformation generated by the combined action of the background and optimal control loads in cases (a)–(d).

## 4. SUMMARY

We have presented a numerical scheme for the static optimal control of the in-plane deformation of a flexible plate. The governing equations are nonlinear due to the nonlinear behavior of the plate material and due to the large deformation involved. A stretch-based hyperelastic material model has been employed.

The numerical method used has been developed in [12] in a theoretical setting. It has been demonstrated that the Sequential Quadratic Programming (SQP) approach for static (elliptic) optimal control problems of flexible structures is indeed effective. The method has the advantage that it does not make use of adjoint variables and thus leads to a simpler formulation with a reduced number of unknown variables. Other details, examples, extensions and computational issues are discussed in [12, 15].

## References

[1] S.M. Joshi, *Control of Large Flexible Space Structures*, Springer-Verlag, Berlin, 1989.

[2] A.K. Chatterjee, "Optimal Orbit Transfer Suitable for Large Flexible Structures," *J. Astro. Sci.*, **37**, 261–280, 1989.

[3] M.G. Safonov, R.Y. Chiang and H. Flashner, "H-infinity Robust Control Synthesis for a Large Space Structure," *J. Guidance, Control and Dynamics*, **14**, 513–520, 1991.

[4] M.P. Kamat, ed., *Optimization Issues in the Design and Control of Large Space Structures*, ASCE publication, New York, 1985.

[5] S.N. Atluri and A.K. Amos, eds., *Large Space Structures: Dynamics and Control*, Springer-Verlag, Berlin, 1988.

[6] D. Givoli and O. Rand, "Minimization of the Thermoelastic Deformation in Space Structures Undergoing Periodic Motion," *AIAA J. of Spacecraft and Rockets*, **32**, 662-669, 1995.

[7] R.V. Grandhi, A. Kumar and A. Chaudhary, "State-space Representation and Optimal Control of Non-linear Material Deformation using the Finite Element Method," *Int. J. Num. Meth. Engng*, **36**, 1967–1986, 1993.

[8] L.S. Hou and J.C. Turner, "Finite Element Approximation of Optimal Control Problems for the Von Karman Equations," *Numer. Methods Partial Differ. Equations*, **11**, 111–125, 1995.

[9] M. Brokate and M. Sprekels, "Existence and Optimal Control of Mechanical Processes with Hysteresis in Viscous Solids," *IMA J. Applied Math*, **43**, 219, 1989.

[10] G.E. Stavroulakis, "Optimal Prestress of Cracked Unilateral Structures: Finite Element Analysis of an Optimal Control Problem for Variational Inequalities," *Comp. Meth. Applied Mech. Engng*, **123**, 231–246, 1995.

[11] G. Knowles, *An Introduction to Applied Optimal Control*, Academic Press, New York, 1981.

[12] D. Givoli, "A Direct Approach to the Finite Element Solution of Elliptic Optimal Control Problems," *Numer. Methods Partial Differ. Equations*, **15**, 371–388, 1999.

[13] J. Bonet and R.D. Wood, *Nonlinear Continuum Mechanics for Finite Element Analysis*, Cambridge University Press, 1997.

[14] D. Goldfarb and A. Idnani, "A Numerically Dual Method for Solving Strictly Convex Quadratic Programs," *Mathematical Programming*, **27**, 1–33, 1983.

[15] D. Givoli and I. Patlashenko, "Solution of Static Optimal Control Problems in Nonlinear Elasticity via Quadratic Programming," *Commun. Numer. Meth. Engng.*, submitted.

[16] P.E. Gill, W. Murray and M.H. Wright, *Practical Optimization*, Academic Press, New York, 1981.

[17] D.G. Luenberger, *Linear and Nonlinear Programming*, Addison-Wesley, Reading, MA, 1984.

# COMPUTER SIMULATION OF NONISOTHERMAL ELASTOPLASTIC SHELL RESPONSES

W. B. KRÄTZIG, U. MONTAG
*Institute for Statics and Dynamics, Ruhr University Bochum,*
*Universitätsstraße 150, 44 780 Bochum, Germany*

J. SORIĆ, Z. TONKOVIĆ
*Faculty of Mechanical Engineering and Naval Architecture,*
*University of Zagreb,*
*I. Lučića 5, 10000 Zagreb, Croatia*

**Abstract.** Shell structures are extremly efficient, thin walled load-carrying components, in the elastic as well as in the inelastic regime. Realistic and efficient computational strategies lately are in rapid development. Such computational strategy for modelling of nonisothermal, highly nonlinear hardening responses in elastoplastic shell analysis has been proposed in this article. Therein, the closest point projection algorithm employing the Reissner-Mindlin type kinematic model, completely formulated in tensor notation, is applied. A consistent elastoplastic tangent modulus ensures high convergence rates in the global iteration approach. The integration algorithm has been implemented into a layered assumed strain isoparametric finite shell element, which is capable of geometrical nonlinearities including finite rotations. Under the assumption of an adiabatic process, the increase of the temperature is analysed during elastoplastic deformation. Finally, numerical examples illustrate robustness and efficiency of the proposed algorithms.

## 1. Introduction

Thin-walled shell structures and structural components play an important role in modern technical applications. Recently, inelastic phenomena and their numerical simulations have gained an increasing attention in shell research, e.g. in industrial crash-worthiness problems. For such applications the temperature change during plastic flux plays a significant role for mapping reality in sufficient quality.

Especially for shell-like metal structures, numerically efficient computational strategies for elastoplastic deformation processes have received much attention during recent years. Accurate modelling of the nonlinear hardening responses represents the key for realistic material modelling. Employing thermomechanical coupling in the model can significantly contribute to the accuracy of the numerical simulation. The influence of the temperature on the material behaviour is encompassed in the evolution



*D. Durban et al. (eds.), Advances in the Mechanics of Plates and Shells,* 167–180.

laws for internal variables describing combined kinematic and isotropic hardening. Hardening rules formulated by differential equations will be as usually applied [1], [5], [10], [13]. The temperature effect may be embedded in the corresponding model parameters. Numerical simulations up to now have been mainly performed for two dimensional plane stress and plane strain problems [6-8]. An algorithm for modelling of nonlinear hardening responses describing cyclic plasticity of shell structures is proposed in [19].

The present paper is concerned with numerical modelling of nonisothermal hardening responses in elastoplastic analyses of shell structures employing the Reissner-Mindlin type kinematic model. The elastoplastic material model is taken from [10], in which all coefficients of the hardening evolution laws are temperature dependent. Small strain and an associative flow rule are assumed, and an adiabatic process is considered. Temperature path history dependence is implicitly included in the formulation. The employed material functions were obtained experimentally for mild steel in [20]. Further, the computational algorithm, entirely formulated in tensor notation, is based completely on the multi-scale-simulation-strategy as proposed in [9]. Additionally, a closest point projection scheme [18] is applied, and the consistent elastoplastic tangent operator is used with great benefit.

The computational algorithm has been implemented into a four-noded isoparametric, assumed strain layered finite shell element [3], which allows for geometrically nonlinear analyses considering finite displacements and rotations. Efficiency of the proposed algorithm will be demonstrated by several numerical examples, which also display changes of temperature during elastoplastic deformation. All computations have been performed within the finite element system FEMAS [4] developed at the Institute for Statics and Dynamics, Ruhr-University Bochum.

## 2. Constitutive Law on Material Point Level

The presently applied elastoplastic material model employs an associative flow rule and the evolution laws for hardening variables as proposed by Lehmann [10] with material functions determined by experimental studies in [20]. The associative expression for the plastic strain rate is written as

$$\dot{\gamma}^{p}_{ij} = \lambda \frac{\partial F\left(\sigma^{ij}, \rho^{ij}, a, T\right)}{\partial \sigma^{ij}}, \tag{1}$$

where $F\left(\sigma^{ij}, \rho^{ij}, a\right)$ represents the von Mises-type yield function. As usually, $\lambda$ is the plastic multiplier and $\sigma^{ij}$ denotes the stress tensor components. $\rho^{ij}$ and $a$ are internal variables describing kinematic and isotropic hardening, respectively. $T$ denotes the process temperature. The kinematic hardening is expressed by the following nonlinear evolution equation [10]

$$\dot{\rho}'^{ij} = \varsigma \dot{\gamma}^{p\,ij} - \chi \rho'^{ij} \sqrt{\dot{\gamma}^{pk}_{l} \dot{\gamma}^{pl}_{k}} \tag{2}$$

with $\varsigma$ and $\chi$ as given material functions. $\rho'^{ij}$ are the deviatoric parts of the back stress tensor components $\rho^{ij}$. The material function $\chi$ is expressed in dependence on the process temperature $T$ by a polynomial in [20], while the function $\varsigma$ may be obtained by the relation

$$\varsigma = c_1 + c_2 e^{-c_3 A} . \tag{3}$$

Herein $A$ denotes the second invariant of the back stress deviator, and $c_1$, $c_2$, $c_3$ are temperature dependent coefficients. The internal variable describing isotropic hardening is assumed in the form

$$\dot{a} = \left(S'^{IJ} - \rho'^{IJ}\right)\dot{\gamma}^{P}_{IJ} , \tag{4}$$

where $S'^{IJ}$ stands for the deviatoric components of the stress tensor. According to [20], the following isotropic hardening model is adopted

$$k^2(a,T) = b_1 + b_2 a + b_3\left(1 - e^{b_4 a}\right) , \tag{5}$$

in which $b_1$, $b_2$, $b_3$ and $b_4$ are again temperature-dependent coefficients. The von Mises-type yield condition has the form

$$F\left(\sigma^{IJ}, \rho^{IJ}, a, T\right) = \left(S'^{IJ} - \rho'^{IJ}\right)\left(S'_{IJ} - \rho'_{IJ}\right) - k^2(a,T) = 0 , \tag{6}$$

and the following consistency condition generally has to be fulfilled

$$\dot{F} = \frac{\partial F}{\partial \sigma^{IJ}}\dot{\sigma}^{IJ} + \frac{\partial F}{\partial \rho^{IJ}}\dot{\rho}^{IJ} + \frac{\partial F}{\partial a}\dot{a} + \frac{\partial F}{\partial T}\dot{T} = 0 . \tag{7}$$

The temperature changes during the elastoplastic deformation process is expressed by the relation

$$\dot{T} = \frac{\xi}{c_p \rho}\left(S'^{IJ} - \rho'^{IJ}\right)\dot{\gamma}^{P}_{IJ} , \tag{8}$$

in which $\xi$ denotes the dissipation function, while $c_p$ and $\rho$ abbreviate the specific heat capacity and the mass density, respectively. The values $c_p$ and $\rho$ depend on the process temperature $T$, while $\xi$ is assumed to be constant [20]. Introducing the relative stress deviator

$$\eta'^{IJ} = S'^{IJ} - \rho'^{IJ} \tag{9}$$

and further its second invariant

$$J_2 = \frac{1}{2}\eta'^{IJ}\eta'_{IJ} , \tag{10}$$

the yield criterion may be rewritten in the form

$$F = 2J_2 - k^2(a,T) = 0 . \tag{11}$$

The components of the relative stress deviator are expressed in terms of relative stress components by the relation

$$\eta'^{IJ} = \mu'^{IJ}_{kl}\,\eta^{kl} , \tag{12}$$

where $\mu'^{IJ}_{kl}$ represents the transformation tensor

$$\mu'^{IJ}_{kl} = \delta^{I}_{k}\delta^{J}_{l} - \frac{1}{3}a_{kl}a^{IJ} . \tag{13}$$

Finally, the relative stress tensor components are defined as

$$\eta^{ij} = \sigma^{ij} - \rho^{ij} . \tag{14}$$

In equation (13) $\delta^i_k$ exhibits the Kronecker delta, while $a_{kl}$ and $a^{ij}$ are the covariant and contravariant components of the metric tensor [2,12]. Analogously to (12), the deviatoric components of the back stress tensor shall now be written as

$$\rho'^{ij} = \mu'^{ij}_{kl} \rho^{kl} . \tag{15}$$

According to (1) and after differentiation of the yield function, the plastic strain rate can be broken down as follows

$$\dot{\gamma}^p_{ij} = 2\lambda \mu_{ijkl} \eta^{kl} . \tag{16}$$

By means of equations (9)-(16), the isotropic hardening variable may be written in terms of the second invariant of the relative stress deviator

$$\dot{a} = 4\lambda J_2 , \tag{17}$$

and the expression for the back stress tensor components can be transformed into the following relation

$$\dot{\rho}^{ij} = 2\lambda \varsigma \eta^{ij} - 2\lambda \chi k \rho^{ij} . \tag{18}$$

After comparison with (4), the temperature rate may be reformulated in terms of the isotropic hardening variable as follows

$$\dot{T} = \frac{\xi}{c_p \rho} \dot{a} . \tag{19}$$

## 3. Re-Formulation for Reissner-Mindlin Shell Kinematics

### 3.1. INTEGRATION ALGORITHM

The constitutive relations for thermo-plasticity sketched out in the previous chapter hold in each material point $X^i$ of the shell continuum. They shall now be transformed into shell space $\theta^\alpha, \theta^3$. For Reissner-Mindlin type shell kinematics, the stress and strain measures are described by eight tensor components, $\sigma \in \mathrm{R}^8$ and $\gamma \in \mathrm{R}^8$. $\rho \in \mathrm{R}^8$ portrays the back stress tensor describing the kinematic hardening response, as well. In contrast to the standard matrix notation, all deviatoric components of the stress and back stress tensor, $\boldsymbol{S} \in \mathrm{R}^9$ and $\rho' \in \mathrm{R}^9$, are included in the present formulation. In the following, the rate of all measures is replaced by their incremental values.

According to (16), the plastic strain increment (upper index $...^p$) is expressed by the in-plane and shear components separately as follows

$$\overset{+}{\gamma}{}^p_{\alpha\beta} = 2\lambda \mu_{\alpha\beta\gamma\delta} \eta^{\gamma\delta} ,$$

$$\overset{+}{\gamma}{}^p_{\delta 3} = 2\lambda a_{\delta\varepsilon} \eta^{\varepsilon 3} . \tag{20}$$

The transverse normal plastic strain component $\overset{+}{\gamma}{}^{p}_{33}$ is computed from the incompressibility condition $\overset{+}{\gamma}{}^{pJ}_{J} = 0$. We remind our readers, that in contrast to the Latin indices which represent numbers 1, 2 and 3, the Greek indices take the values 1 and 2. In order to explain the algorithmic steps of the future computations, starting with equation (17), the isotropic hardening variable in the time interval $\left({}^{t-1}t, {}^{t}t\right)$ will be updated by the relation

$$ {}^{t}a = {}^{t-1}a + 4\,{}^{t}\lambda\,{}^{t}J_2 . \tag{21} $$

By means of (17) and (19), the increase in temperature is described by the following equation

$$ {}^{t}T = {}^{t-1}T + \frac{4\,{}^{t}\lambda\,\xi\,{}^{t}J_2}{{}^{t}\left(c_p \rho\right)} . \tag{22} $$

Applying (18), the back stress tensor components at time ${}^{t}t$ may be written as

$$ \begin{aligned} {}^{t}\rho^{\alpha\beta} &= {}^{t}R\,{}^{t-1}\rho^{\alpha\beta} + 2\,{}^{t}R\,{}^{t}\lambda\,{}^{t-1}\varsigma\,{}^{t}\eta^{\alpha\beta}, \\ {}^{t}\rho^{\delta 3} &= {}^{t}R\,{}^{t-1}\rho^{\delta 3} + 2\,{}^{t}R\,{}^{t}\lambda\,{}^{t-1}\varsigma\,{}^{t}\eta^{\delta 3}, \end{aligned} \tag{23} $$

where ${}^{t}R$ is the auxiliary variable given by

$$ {}^{t}R = \frac{1}{1 + 2\chi\,{}^{t}\lambda\,{}^{t}k} \tag{24} $$

with ${}^{t}k$ as the isotropic hardening function ${}^{t}k = k\left({}^{t}a, {}^{t}T\right)$. To avoid computational difficulties, one should note that the kinematic hardening function $\varsigma$ has already been computed at the end of the previous time step at ${}^{t-1}t$.

For integration of the constitutive relations, the closest point projection scheme [18] is adopted in such a way, that the predictor phase is expressed as

$$ \begin{aligned} {}^{t}\eta^{\alpha\beta}_{trial} &= {}^{t}\sigma^{\alpha\beta}_{trial} - {}^{t}R\,{}^{t-1}\rho^{\alpha\beta}, \\ {}^{t}\eta^{\delta 3}_{trial} &= {}^{t}\sigma^{\delta 3}_{trial} - {}^{t}R\,{}^{t-1}\rho^{\delta 3} \end{aligned} \tag{25} $$

with the following relations for the trial stress components:

$$ \begin{aligned} {}^{t}\sigma^{\alpha\beta}_{trial} &= {}^{t-1}\sigma^{\alpha\beta} + C^{\alpha\beta\delta\varepsilon}\,{}^{t}\overset{+}{\gamma}_{\delta\varepsilon}, \\ {}^{t}\sigma^{\delta 3}_{trial} &= {}^{t-1}\sigma^{\delta 3} + 2G a^{\delta\varepsilon}\,{}^{t}\overset{+}{\gamma}_{\varepsilon 3}. \end{aligned} \tag{26} $$

In (26) the total strain increments $\overset{+}{\gamma}_{\delta\varepsilon}$ and $\overset{+}{\gamma}_{\varepsilon 3}$ consist of elastic, plastic and thermal parts. $C^{\alpha\beta\delta\varepsilon}$ is the elastic material tensor [11,12], while $G$ denotes the shear modulus. The total strain components are then decomposed in the form

$$ \begin{aligned} \overset{+}{\gamma}_{\alpha\beta} &= \overset{+}{\gamma}{}^{e}_{\alpha\beta} + \overset{+}{\gamma}{}^{p}_{\alpha\beta} + \overset{+}{\gamma}{}^{T}_{\alpha\beta}, \\ \overset{+}{\gamma}_{\varepsilon 3} &= \overset{+}{\gamma}{}^{e}_{\varepsilon 3} + \overset{+}{\gamma}{}^{p}_{\varepsilon 3}. \end{aligned} \tag{27} $$

Herein the upper indices $e$ ($p$) denote elastic (plastic) strain parts, and $\overset{+}{\gamma}{}^{T}_{\alpha\beta}$ represents the thermal strain components

$$\overset{+}{\gamma}{}^{T}_{\alpha\beta} = \alpha_T \overset{+}{T} \delta_{\alpha\beta} \tag{28}$$

with $\alpha_T$ as coefficient of thermal expansion, depending on temperature [20]. Employing the usual additive decomposition (27) of the strain tensor, the stress components at the end of the time step are computed correspondingly as

$$ {}^{t}\sigma^{\alpha\beta} = {}^{t}\sigma^{\alpha\beta}_{trial} - C^{\alpha\beta\gamma\delta}\left( {}^{t}\overset{+}{\gamma}{}^{p}_{\gamma\delta} + {}^{t}\overset{+}{\gamma}{}^{T}_{\gamma\delta} \right),$$

$$ {}^{t}\sigma^{\delta 3} = {}^{t}\sigma^{\delta 3}_{trial} - 2Ga^{\delta\varepsilon}\, {}^{t}\overset{+}{\gamma}{}^{p}_{\varepsilon 3} \,. \tag{29}$$

By means of (14) and of equations (20), (23), (25) and (29), the following expressions for the relative stress tensor components are obtained

$$ {}^{t}\eta^{\alpha\beta} = {}^{t}A^{\alpha\beta}_{\gamma\delta}\left( {}^{t}\eta^{\gamma\delta}_{trial} - \frac{2\xi\, {}^{t}\alpha_T\, {}^{t}\lambda\, {}^{t}k^2}{{}^{t}(c_p\rho)} C^{\gamma\delta\varepsilon\varepsilon} \right),$$

$$ {}^{t}\eta^{\delta 3} = \frac{1}{1+2\,{}^{t}\lambda\left({}^{t}R^{\,t-1}\varsigma + 2G\right)}\, {}^{t}\eta^{\delta 3}_{trial} \,, \tag{30}$$

where the transformation tensor ${}^{t}A^{\alpha\beta}_{\gamma\delta}$ can be broken down in the form

$$ {}^{t}A^{\alpha\beta}_{\gamma\delta} = \frac{1}{1+2\,{}^{t}\lambda\left({}^{t}R^{\,t-1}\varsigma+2G\right)}\left[ \delta^{\alpha}_{\gamma}\delta^{\beta}_{\delta} + \frac{4\,{}^{t}\lambda G(1-2\nu)}{4\,{}^{t}\lambda G(1+\nu)+3(1-\nu)\left(1+2\,{}^{t}\lambda\,{}^{t}R^{\,t-1}\varsigma\right)} a_{\gamma\delta}a^{\alpha\beta} \right]. \tag{31}$$

After inserting (30) into the yield criterion (11), the nonlinear scalar equation is obtained

$$ {}^{t}F = 2\,{}^{t}J_2\left({}^{t-1}\varsigma, {}^{t}R, {}^{t}\lambda, {}^{t}\eta^{\alpha\beta}_{trial}, {}^{t}\eta^{\delta 3}_{trial}\right) - k^2\left({}^{t}a,\, {}^{t}T\right) = 0\,, \tag{32}$$

which has to be solved for ${}^{t}\lambda$. For this task, the Newton iteration method has been applied. During the iteration process, the unknown auxiliary variable ${}^{t}R$ must be computed, which is performed also numerically by means of the following nonlinear equation obtained from equations (5), (21) and (11)

$$ {}^{t-1}a - a + 2\lambda\left[ b_1 + b_2 a + b_3\left(1-e^{b_4 a}\right)\right] = 0. \tag{33}$$

For a given value of $\lambda$ this equation is to be solved for the isotropic hardening variable $a$ in each iteration step by applying a local iteration scheme. After determining $a$, the values of the variable ${}^{t}R$ and the temperature ${}^{t}T$ can be evaluated.

After determination of the plastic multiplier, the updated values of the stresses as well as all internal variables can be calculated. To avoid spurious unloadings, all state variables are updated with respect to the previous equilibrium state. During the computational process, the loading/unloading criterion is expressed by the Kuhn-Tucker condition [17]. In order to preserve numerical efficiency of the global iteration strategy, the elastoplastic tangent modulus consistent with the integration algorithm has to be derived and applied.

## 3.2. CONSISTENT ELASTOPLASTIC TANGENT MODULUS

By differentiation of the updated relations presented in the previous section using the consistency condition and after some tedious and suitable formulae manipulations, the following relations between the differential stress and strain components are obtained

$$d\sigma^{\alpha\beta} = C_{ep}^{\alpha\beta\gamma\delta} d\gamma_{\gamma\delta} + 2C_{ep}^{\alpha\beta\varepsilon 3} d\gamma_{\varepsilon 3},$$

$$d\sigma^{\alpha 3} = C_{ep}^{\alpha 3\gamma\delta} d\gamma_{\gamma\delta} + 2C_{ep}^{\alpha 3\varepsilon 3} d\gamma_{\varepsilon 3}, \tag{34}$$

which deliver the desired tensor components of the consistent elastoplastic tangent modulus:

$$C_{ep}^{\alpha\beta\gamma\delta} = B_{\varepsilon\varsigma}^{\alpha\beta}\left(C^{\varepsilon\varsigma\gamma\delta} - \frac{K_2 \mu_{\vartheta\kappa\lambda\nu} D^{\varepsilon\varsigma} B_{\xi\pi}^{\vartheta\kappa} C^{\xi\pi\gamma\delta} \eta^{\lambda\nu}}{K_4 + H_1 + 2H_2}\right),$$

$$C_{ep}^{\alpha\beta\varepsilon 3} = -\frac{K_1 K_2 B_{\gamma\delta}^{\alpha\beta} D^{\gamma\delta} \eta^{\varepsilon 3}}{K_4 + H_1 + 2H_2},$$

$$C_{ep}^{\alpha 3\gamma\delta} = -K_1 K_2 \frac{\mu_{\vartheta\kappa\lambda\nu} L^{\alpha 3} B_{\xi\pi}^{\vartheta\kappa} C^{\xi\pi\gamma\delta} \eta^{\lambda\nu}}{K_4 + H_1 + 2H_2},$$

$$C_{ep}^{\alpha 3\varepsilon 3} = \frac{K_1}{2}\left(a^{\alpha\varepsilon} - \frac{2K_1 K_2 L^{\alpha 3} \eta^{\varepsilon 3}}{K_4 + H_1 + 2H_2}\right). \tag{35}$$

Herein the following abbreviations have been introduced

$$B_{\gamma\delta}^{\alpha\beta} = \frac{1}{1 + 4GK_2\lambda}\left[\delta_\gamma^\alpha \delta_\delta^\beta - \frac{4GK_2\lambda(2\nu - 1)}{4GK_2\lambda(\nu + 1) - 3(\nu - 1)} a_{\gamma\delta} a^{\alpha\beta}\right],$$

$$D^{\alpha\beta} = C^{\alpha\beta\gamma\delta} \mu_{\gamma\delta\varepsilon\varsigma} L^{\varepsilon\varsigma} + \frac{4J_2 \xi \alpha_T C^{\alpha\beta\rho\rho}}{c_p \rho\left(1 - 2\nu \dfrac{\partial k^2}{\partial a}\right) - 2\nu\xi \dfrac{\partial k^2}{\partial T}},$$

$$H_1 = \mu_{\alpha\beta\gamma\delta}\left(K_2 B_{\varepsilon\varsigma}^{\alpha\beta} D^{\varepsilon\varsigma} + M^{\alpha\beta}\right)\eta^{\gamma\delta},$$

$$H_2 = a_{\alpha\beta}\left(K_1 K_2 L^{\alpha 3} + M^{\alpha 3}\right)\eta^{\beta 3},$$

$$K_1 = \frac{2G(2\chi k\lambda + 1)}{2\chi k\lambda + 1 - 4G\lambda\left(2K_2\lambda^{\,t-1}\zeta - 2\chi k\lambda - 1\right)},$$

$$K_2 = \frac{1 + 2\chi\lambda k}{1 + 2\lambda\left({}^{t-1}\zeta + \chi k\right)},$$

$$K_3 = 2k \frac{1 - \lambda\left(\dfrac{\partial k^2}{\partial a} + \dfrac{\xi}{c_p \rho}\dfrac{\partial k^2}{\partial T}\right)}{1 - 2\lambda\left(\dfrac{\partial k^2}{\partial a} + \dfrac{\xi}{c_p \rho}\dfrac{\partial k^2}{\partial T}\right)},$$

$$K_4 = \frac{2J_2\left(\frac{\partial k^2}{\partial a} + \frac{\xi}{c_p\rho}\frac{\partial k^2}{\partial T}\right)}{1-2\lambda\left(\frac{\partial k^2}{\partial a} + \frac{\xi}{c_p\rho}\frac{\partial k^2}{\partial T}\right)}. \tag{36}$$

The appearing tensor components $L^{\alpha\beta}$, $L^{\alpha 3}$, $M^{\alpha\beta}$ and $M^{\alpha 3}$ are further expressed by the relations

$$L^{ij} = 2K_2\left(\eta^{ij} + \frac{K_3\lambda\chi}{1+2\lambda\chi k}\rho^{ij}\right),$$

$$M^{ij} = \frac{K_2}{1+2\lambda\chi k}\left(2^{i-1}\zeta\eta^{ij} - K_3\chi\rho^{ij}\right). \tag{37}$$

In the above given equations the left upper index $i$ referring to all state variables at time ${}^{i}t$ is omitted due to notational simplicity. It is not difficult to verify that the presented tensor components $C_{ep}^{ijkl}$ are unsymmetrical with respect to the couple of indices ($i$ $j$) and ($k$ $l$). This fact has its origin in the assumed nonlinear kinematic hardening response. For the evaluation of the stiffness matrix, the tangent operator has been symmetrized by use of arithmetic mean values, as successfully applied in [19].

## 4. Computed Examples

The integration algorithm presented based on the consistent tangent modulus has been implemented into the formulation of one of the assumed-strain layered finite elements [3] within the finite element code FEMAS [4]. By use of this highly modular software code, a series of numerical simulations, demonstrating the excellent performances of the proposed algorithm, has been executed. Material nonlinearity therein is combined with the modelling of geometrically nonlinear responses. For the tracing of the deformation paths, Newton-Raphson and Riks-Wempner-Wessels iteration schemes, both enhanced by a special line search procedure [14,15], are applied. The termination criterion of the iterations is expressed in terms of the energy norm [18]. For the material model applied, all material parameters, obtained experimentally for German mild steel St37.12, are taken from [20]. The Young's modulus of elasticity $E$ and the Poisson's ratio $\nu$ at the initial temperature of $25\,^{\circ}\mathrm{C}$ (298.15 K) have the values of $E$=212 GPa and $\nu$=0.285. The initial yield stress is $\sigma_Y$=240 MPa. All material parameters depend on the temperature and they are changed during the deformation process. All computations have been performed at the initial temperature, and the temperature changes have been considered by assumption of adiabatic deformation processes.

### 4.1. CIRCULAR TUBE UNDER CYCLIC TORSION

A clamped circular tubular shell subjected to torsion is analysed as a first example. Its geometry with the finite element mesh is presented in Figure 1. The complete cylinder has been discretized by 20x20 elements. The strain controlled cyclic loading is modelled by the twisting moment $M_T$ at the right end of the tube producing a strain amplitude of $\pm 0.03$. The simulated stress-strain diagram presenting four hysteresis loops for the free end of the cylinder is shown in Figure 2. The temperature increases at the end of each hysteresis are displayed in Table 1. As obvious, the total increase of the temperature at the end of the computed cyclic loading process is 26.5 K. To asses the convergence rate of the algorithm during the global iteration procedure, the numerical values of the residual energy norm of the Newton-Raphson approach at load level at the end of the first hysteresis loop are presented in Table 2. As may clearly be observed, quadratic convergence is exhibited.

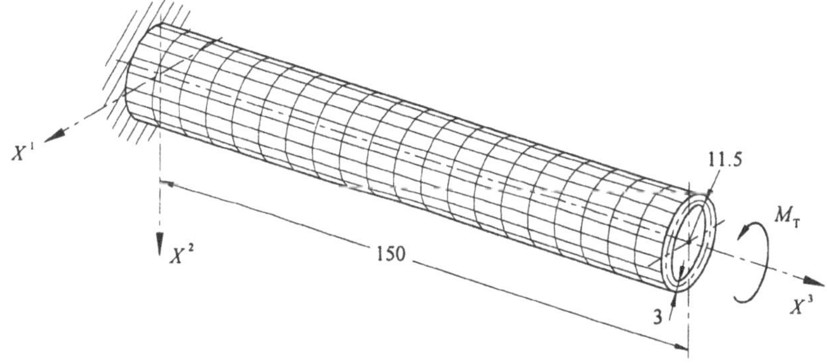


*Figure 1.* Geometry and finite element mesh for circular tube (dimensions in [mm])

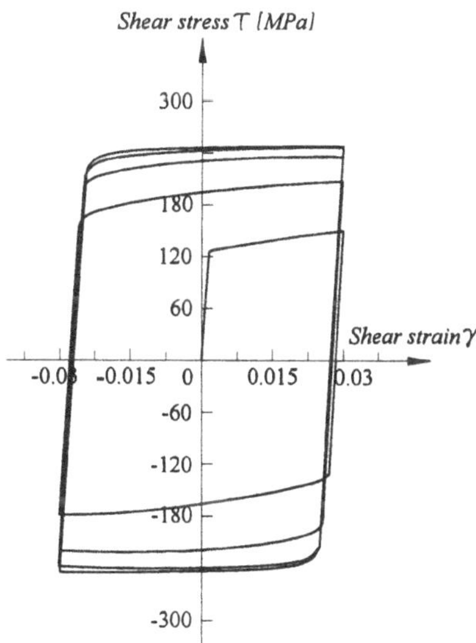


*Figure 2.* Cyclic response of circular tube

TABLE 1. Temperature increase during deformation process

| Hysteresis loop | 1 | 2 | 3 | 4 |
|---|---|---|---|---|
| Increase of temperature [$K$] | 8.8 | 7.3 | 6.2 | 4.2 |

TABLE 2. Convergence of global energy error norms for circular tube under cyclic torsion

| Iteration | Energy error norms |
|---|---|
| 1 | 2.2383D-03 |
| 2 | 1.2215D-06 |
| 3 | 1.0737D-09 |
| 4 | 5.3673D-12 |
| 5 | 9.5771D-15 |

## 4.2. AXIAL COMPRESSION OF A THIN CYLINDRICAL SHELL

As second example, a thin vertical cylinder is considered. This axially compressed structural element may dissipate vast amounts of energy during its plastic deformation, and it is thus a very competitive energy absorbing element for crash situations [16]. Around the top end, the cylinder is compressed by a uniform reference line load $q_0$ corresponding to total equivalent force of 625 N. The shell is clamped along the bottom end and only axial displacements are allowed on the upper shell boundary. Employing symmetry, one quarter of the shell is discretized by 20x80 finite elements. The geometry and the finite element mesh are shown in Figure 3, all dimensions therein are given in [mm].

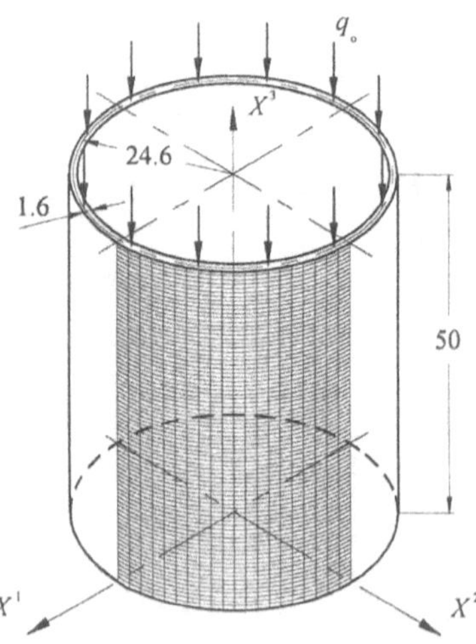


*Figure 3.* Geometry and finite element mesh for axially compressed cylindrical shell

The load factor scaling the reference load versus the axial displacement on the upper boundary is presented in Figure 4. The convergence of the global energy error norms of the Riks-Wempner-Wessels iteration strategy for the load level at point B is displayed in Table 3. As may be seen, a buckling problem occurs. The cylinder collapses at limit point A where the shell buckling is initiated, as depicted in Figure 5 showing also the plastic zones throughout the deformed shell thickness. After exceeding the limit point, unloading is exhibited and the shortening of the shell occurs by plastic folding in an axisymmetric buckling mode. The further deformed configuration at load factor of 38.924, noted by point B in Figure 4 and the spread of plastic zones are presented in Figure 6. Evidently, a redistribution of plastic zones has been appeared. The initial plastic regions are converted into elastic unloading zones.

By the elastoplastic buckling process, the temperature has been increased considerably. The changes of the temperature along the outer shell generatrix for the limit point and for the load level at point B are presented in Figure 7. The temperature distributions are plotted on the undeformed shell configuration. The largest increase in temperature is produced in the plastic folding regions undergoing large plastic deformation.

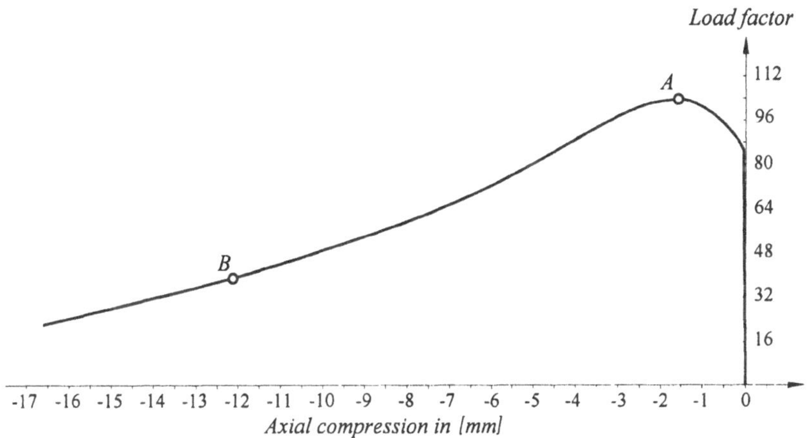


*Figure 4.* Load- axial displacement curve for the top end of cylindrical shell

TABLE 3. Convergence of global energy error norms for the load level at point B

| Iteration | Energy error norms |
|---|---|
| 1 | 1.6052D-06 |
| 2 | 2.8537D-09 |
| 3 | 8.2349D-12 |
| 4 | 6.2238D-14 |

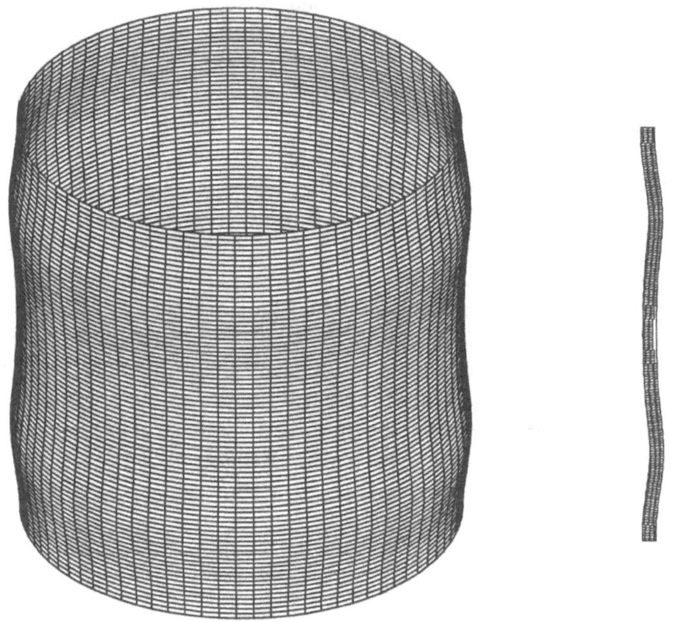

*Figure 5.* Deformed configuration and spread of plastic zones throughout shell thickness at limit point

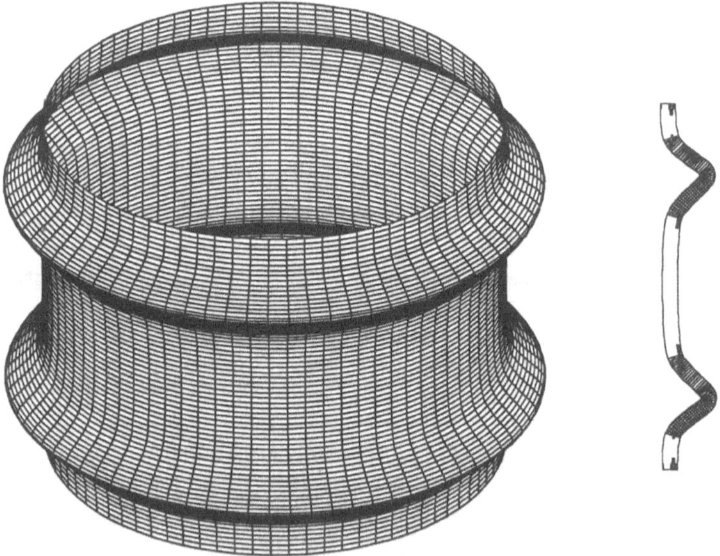

*Figure 6.* Deformed configuration and spread of plastic zones throughout shell thickness for the load level at point B

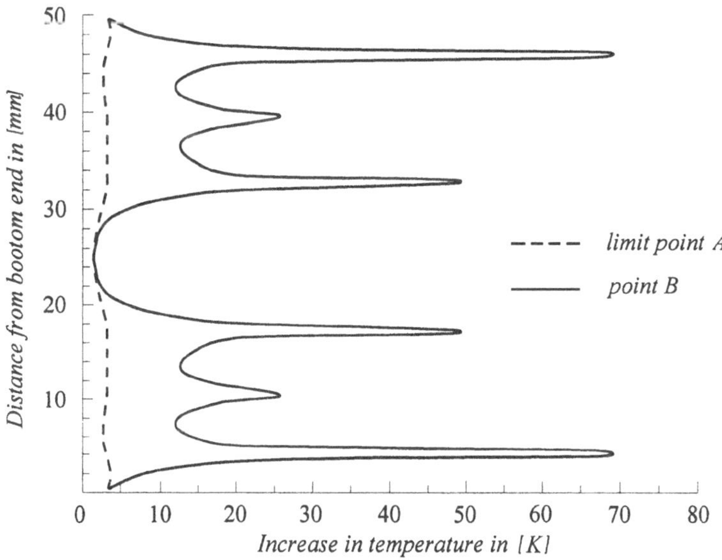


*Figure 7.* Temperature increase along outer shell generatrix at limit point and for the load level at point B

## 5. Conclusions

An efficient numerical simulation technique for nonisothermal elastoplastic responses of shell structures has been presented, employing a realistic material model for German mild steel St37.12. Additionally, a highly nonlinear isotropic and kinematic hardening model depending on temperature is incorporated. The yield condition is expressed in the space of stress and temperature as well. The closest point projection algorithm for a Reissner-Mindlin type kinematic model has been successfully employed in order to have a highly efficient algorithm at disposal. The tensor formulation applied therein allows all nine stress deviator components to explicitly include in the formulation, which turns out to be an advantage over the classical matrix notation. Using consistent linearization of the projection algorithm, the derived elastoplastic tangent modulus ensures quadratic convergence in global solution procedures. Robustness and numerical stability of the proposed algorithms are demonstrated by numerical examples. Assuming adiabatic deformation processes, the increase in temperature is evaluated and monitored. As expected, the largest temperature increase occurs in the regions undergoing gross plastic deformation. An accurate and efficient modelling of nonisothermal hardening responses of ductile metals can significantly contribute to the realistic description of the crash behaviour of shell structural components and to the prediction of energy absorption during collapse processes.

**Acknowledgement.** The authors feel obliged to express their deep gratitude to the Volkswagen–Stiftung, German Science Foundation for financial support.

## 6. References


1. Armstrong, P.J. and Frederick, C.O.: *A mathematical representation of the multiaxial Bauschinger effect,* GEGB Report No. RD/B/N731, Berceley Nuclear Laboratories, 1966.
2. Basar, Y. and Krätzig, W.B.: *Mechanik der Flächentragwerke,* Vieweg, Braunschweig, 1985.
3. Basar, Y., Montag, U. and Ding Y.: On an isoparametric finite element for composite laminates with finite rotations, *Comp. Mech.* 12 (1993), 329-348.
4. Beem, H., Könke, C., Montag, U. and Zahlten, W.: *FEMAS 2000-Finite Element Moduls of Arbitrary Structures, Users Manual,* Institute for Statics and Dynamics, Ruhr-University Bochum, 1996.
5. Chaboche, J.L.: Time-independent constitutive theories for cyclic plasticity, *Int. J. Plast.* 2 (1986), 149-188.
6. Doghri, I.: Fully implicit integration and consistent tangent modulus in elasto-plasticity, *Int. J. Numer. Meth. Engng.* 36 (1993), 3915-3932.
7. Hartmann, S. and Haupt, P.: Stress computation and consistent tangent operator using non-linear kinematic hardening models, *Int. J. Numer. Meth. Engng.* 36 (1993), 3801-3814.
8. Hopperstad, O.S. and Remseth S.: A return mapping algorithm for a class of cyclic plasticity models, *Int. J. Numer. Meth. Engng.* 38 (1995), 549-564.
9. Krätzig, W.B.: Multi-level modeling techniques for elasto-plastic structural responses, in: Owen, D.R.J. et al. (eds.), *Computational Plasticity, Part I,* CIMNE, Barcelona (1997), 457-468.
10. Lehmann, Th.: On a generalized constitutive law for finite deformations in thermo-plasticity and thermo-viscoplasticity, in: Desai, C.S. et al. (eds.), *Constitutive Laws for Engineering Materials, Theory and Applications,* Elsevier, New York (1987), 173-184.
11. Libai, A. and Simmonds, J.G.: Nonlinear elastic shell theory, *Adv. Appl. Mechanics* 23 (1983), 271-371.
12. Libai, A. and Simmonds, J.G.: *The Nonlinear Theory of Elastic Shells,* 2nd ed., Cambridge University Press, Cambridge, 1998.
13. McDowell, D.L.: A nonlinear kinematic hardening theory for cyclic thermoplasticity and thermoviscoplasticity, *Int. J. Plast.* 8 (1992), 695-728.
14. Montag, U., Krätzig, W.B. and Sorić, J.: On stable numerical simulation strategies for elastoplastic deformation processes of shell structures, in: Topping, B.H.V. (ed.), *Advances in Computational Methods for Simulation,* Civil-Comp Press, Edinburgh (1996), 61-71.
15. Montag, U., Krätzig, W.B. and Sorić, J.: Increasing solution stability for finite-element modeling of elasto-plastic shell response, *Adv. Engg. Software* 30 (1999), 607-619.
16. Reid, S.R.: Plastic deformation mechanisms in axially compressed metal tubes used as impact energy absorbers, *Int. J. Mech. Sci.* 35 (1993), 1035-1052.
17. Simo, J.C. and Hughes, T.J.R.: *Computational Inelasticity,* Springer, New York, 1998.
18. Sorić, J., Montag, U. and Krätzig, W.B.: An efficient formulation of integration algorithms for elastoplastic shell analyses based on layered finite element approach, *Comp. Meth. Appl. Mech. Eng.* 148 (1997), 315-328.
19. Sorić, J., Tonković, Z. and Krätzig, W.B.: On numerical simulation of cyclic elastoplastic deformation processes of shell structures, in: Topping, B.H.V. (ed.), *Advances in Finite Element Procedures and Techniques,* Civil-Comp Press, Edinburgh (1998), 221-228.
20. Szepan, F.: *Ein elastisch-viskoplastisches Stoffgesetz zur Beschreibung großer Formänderungen unter Berücksichtigung der thermomechanischen Kopplung (An elasto-plastic constitutive law for mild steel under large deformations and thermo-mechanical coupling),* IfM 70, Institute for Applied Mechanics, Ruhr-University Bochum, 1989.

# ON APPLICATION OF THE EXACT THEORY OF ELASTIC BEAMS

P. LADEVÈZE AND P. SANCHEZ
*LMT-Cachan (ENS Cachan, CNRS, Paris 6 Univ.) (France)*
AND
J.G. SIMMONDS
*Department of Civil Engineering, Univ. of Virginia at Charlottesville (USA)*

**Abstract.** A new approach is developed for the analysis and calculation of straight prismatic beams of piecewise constant cross-section under arbitrary loads. The material can be anisotropic and composite; it is only supposed that the beam is $x$-homogeneous, $x$ being the abscissa. This theory can be called "exact" because it determines exact static and kinematic generalized quantities. Contrary to classical theories, it is not limited to high aspect ratio (*i.e.* relatively slender beams). The paper is focused on how to use the exact theory of elastic beam for computing 3D stresses. It is shown in particular how to compute the basic operators which depend on the cross-section geometry, the material and the loading which are the basic building blocks of the theory. An example is of an elastic tube with a small thickness submitted to nearly concentrated extremity loads.

## 1. Introduction

Even within the confines of linearized elasticity, there are numerous works from the last half-century that develop beam theories either by asymptotic techniques (Ciarlet P.G., 1990), (Rigolot A., 1972) or by the introduction of a *priori* hypotheses (Ladevèze J., Ladevèze P., Mantion M., Pecastaings F., Pelle J.P., 1979). The principle of Saint-Venant, in one way or another, plays a central role. Since the work of Toupin (and those who have followed, see (Horgan C.O., 1989) and (Horgan C.O., Simmonds J.G., 1994)), this principle has taken the status of a theorem that specifies the conditions on the data assuring that the displacement and stresses are localized in neighborhoods of the extremities.

The key question, it seems to us, is to divide the solution into:

- a long wavelength part;
- and a short wavelength, localized part.



*D. Durban et al. (eds.), Advances in the Mechanics of Plates and Shells,* 181–196.

These effects must be separated during calculation. However, current theories of beams are not based on such an underlying partition, but on approximations valid for large aspect ratios.

This point of view has already been introduced and an answer given in the case of Saint-Venant's problem, that is to say, in the special situation where the lateral surface of the beam is free and body forces are absent (Ladevèze P., 1983), (Ladevèze P., 1985). It has been extended in (Ladevèze P., Simmonds J.G., 1996), (Ladevèze P., Simmonds J.G., 1998) to arbitrary loads and to piecewise constant cross-sections, that is, to most problems encountered in practice. The final result is a general method for beam calculation named an "exact" theory of beams which, contrary to the classical theories, is not limited to relatively slender beams. Its domain of interest contains also composite beams, beams with rapidly varying loads, connection between beam and 3D media.

The fundamental result is expressed in terms of $s(x)$, the displacement-normal stress pair defined on a cross-section of abscissa $x$ by:

$$\forall x \in [0,\, L] \qquad s(x) = s_{sv}(x) + \int_0^L s^{\pm}(x-t,t)\,dt$$

where $s$ is the solution to the 3D reference elasticity problem and $s_{sv}$ the long wavelength part of the solution that we always call the Saint-Venant solution.

$s^{\pm}$ is a sort of Green function; $t \to s^{\pm}(x-t,t)$ is a "local effect" density which includes a regular part and Dirac distributions centered at the abscissa of different discontinuities and, in particular, at the extremities.

A major property of a localized effect is that the associated generalized displacement-stress pair vanishes. The generalized stress is formed from the resultant and moment of the normal stress acting on the cross-section. We introduce the notion of a generalized displacement starting from invariants along the abscissa; it is constructed from a mean displacement and a mean rotation in the cross-section.

An important consequence is that the generalized displacement-stress pair of the Saint-Venant solution and that of the solution to the reference 3D problem are equal. The generalized displacement-stress pair can be calculated directly; it comes from a solution to an "exact" theory of beams where the constitutive relation is obtained after solving a series of 2D problems in the cross-section(s). This approach represents a departure from current approaches in which the accuracy depends on the aspect ratio of the beam. The localized effects are computed in a second pass which requires the solution to some 3D problems defined on short portions of the beam.

The paper is focused on how to use the exact theory of elastic beam. A first step is concerned with the technique that we propose for the calculation of the operators depending of the cross-section geometry, the material and the loading which are the basic building blocks of the theory. In fact, we do not solve a series of 2D problems on the cross-section but a 3D problem which can be handled very easily with a standard finite element code. A second step is related to the calculation of extremity effect corrections which are not necessarily truly localized at the extremities, for example in the case of cylinders with small thickness. The paper ends with an example: the 3D analysis of an elastic tube with a small thickness submitted to nearly concentrated extremity loads.

## 2. The 3D elasticity problem to be solved

Under the hypotheses (linear elasticity, small displacements, statics), one studies the equilibrium of a straight prismatic beam of piecewise constant cross-section under arbitrary loads (see *Fig.* 1).

The reference line is oriented by $\underline{N} = \underline{N}_1$ ; $S_x$ denotes the cross-section of the $x$-abscissa. The domain occupied by the beam is then:

$$\Omega = \{ \underline{M} = x\underline{N} + \underline{X} \mid \underline{X} \in S_x \ , \ x \in ]0, L[ \}$$

where $L$ is the beam's length.

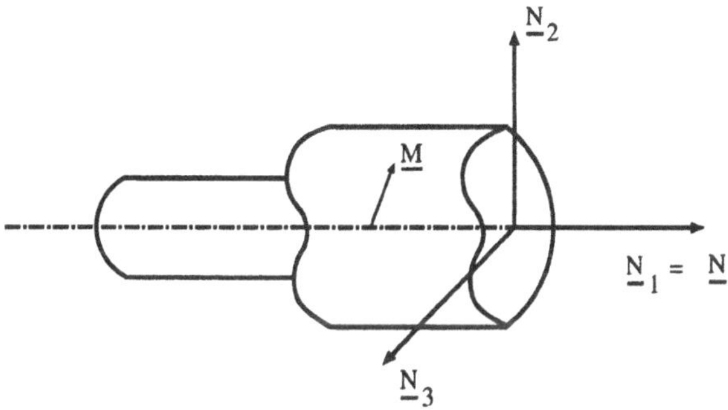


*Figure 1.* Beam geometry

The beam environment is defined by a body force density $\underline{f}_d$ on $\Omega$ and a surface force density $\underline{F}_d$ on the lateral surface $\Sigma_l$. Moreover, let us prescribe a displacement $\underline{U}_d$ on the cross-section $S_0$ and a surface force density $\underline{H}_d$ on $S_L$. Other usual extremity conditions can be introduced without any additional difficulty. Here, we work within the usual mathematical framework (Duvaut G., Lions J.L., 1976). Let $\mathcal{U}$ and $\mathcal{S}$ be the finite energy spaces containing, respectively, the displacement field $\underline{U}$ and the stress field $\sigma$. The 3D elasticity problem to be solved can then be written:

Find $\underline{U}(\underline{M}), \sigma(\underline{M}), \underline{M} \in \Omega$ such that the following conditions are statisfied:

- kinematic constraints:

$$\underline{U} \in \mathcal{U}, \qquad \underline{U}_{|x=0} = \underline{U}_d \tag{1}$$

- equilibrium equations:

$$\text{div}(\sigma) + \underline{f}_d = 0 \text{ on } \Omega, \quad \sigma\underline{n} = \underline{F}_d \text{ on } \Sigma_l, \quad \sigma \in \mathcal{S} \tag{2}$$

$$\sigma\underline{n} = \underline{H}_d \text{ on } S_L \tag{3}$$

– constitutive relation:

$$\sigma = \mathbf{K}\,\varepsilon\,(\underline{U}) \tag{4}$$

$\underline{n}$ is the unit normal to the boundary of the domain $\Omega$. $\mathbf{K}$ is the Hooke's tensor and $\varepsilon$ the strain operator. Regarding $\mathbf{K}$, we suppose that the beam can be split into subbeams such that the Hooke's tensor is $x$-constant on each subbeam.

## 3. The Exact Beam Theory (EBT)

The main results concerning the EBT are recalled here without giving the proofs which can be found in (Ladevèze P., Simmonds J.G., 1996).

### 3.1. SPECIFIC BEAM NOTATIONS AND PRELIMINARY RESULTS

In the study of beams, it is helpful to introduce some specific notations, such as the quantity $s(x)$ associated with the cross-section $S_x$ defined by:

$$s(x) = (\underline{U}\,,\, \sigma\,\underline{N}\,)_{|S_x}$$

The corresponding space is denoted $\mathbf{S}$ ($\mathbf{S} = [\mathbf{H}^{1/2}\,(S)]^3 \times [\mathbf{H}^{-1/2}\,(S)]^3$); the space of the associated fields defined on $[0,\, L]$ is $\mathbf{S}_{[0,\,L]}$.

3.1.1. *Saint-Venant solution for the particular case of $x$-constant data*
The main operators $\mathbb{A}$, $\mathbb{B}$, $\mathbb{A}^\circ$, $\mathbb{B}^\circ$, $\wedge$ which are the basic building blocks of the EBT are defined through the Saint-Venant solution for the particular case of $x$-constant data. It is the exact 3D $x$-polynomial solution.

**Theorem 1** *The Saint-Venant solution for $x$-constant data (cross-section, loads, $\mathbf{K}$) can be written:*

$$\begin{aligned} \underline{U}_{sv} &= \underline{\tilde{u}} + \underline{\tilde{\omega}} \wedge \underline{X} + \mathbb{A}\underline{\tilde{T}} + \mathbb{B}\underline{\tilde{M}} + \underline{W}_d \\ \sigma_{sv}\,\underline{N} &= \mathbb{A}^\circ\,\underline{\tilde{T}} + \mathbb{B}^\circ\,\underline{\tilde{M}} + \underline{C}_d \end{aligned} \tag{5}$$

*where:*

- $\mathbb{A}$, $\mathbb{B}$, $\mathbb{A}^\circ$, $\mathbb{B}^\circ$ *are $x$-constant linear operators depending on the material and the cross-section* $(\mathbb{A}(\underline{X}\,), \mathbb{B}(\underline{X}\,), \mathbb{A}^\circ(\underline{X}\,), \mathbb{B}^\circ(\underline{X}\,))$.
- $\underline{W}_d$, $\underline{C}_d$ *are $x$-constant vectors depending on the material, the cross-section and the load densities* $\underline{f}_d$, $\underline{F}_d$, $(\underline{W}_d\,(\underline{X}), \underline{C}_d\,(\underline{X}))$.
- $(\underline{\tilde{T}}, \underline{\tilde{M}})$ *and* $(\underline{\tilde{u}}, \underline{\tilde{\omega}})$ *are respectively the generalized stress and the generalized displacement. They depend only on $x$ and will be specified later* $(\underline{\tilde{T}}(x), \underline{\tilde{M}}(x))$ *and* $\underline{\tilde{u}}(x), \underline{\tilde{\omega}}(x))$
- *The generalized displacement and stress related to* $(\underline{W}_d\,, \underline{C}_d)$ *are zero.*

3.1.2. *Generalized displacement-stress*
Here, we consider again the general case.

**Definition 1** *The generalized displacement-stress associated with* $s(x) = (\underline{U}, \sigma\underline{N})_{|x}$ *is:*

$$\tilde{u}(x) = \int_{S_x} (\mathbb{A}^{\circ t}\underline{U} - \mathbb{A}^t\sigma\underline{N})\,dS \qquad \underline{\tilde{\omega}}(x) = \int_{S_x} (\mathbb{B}^{\circ t}\underline{U} + \mathbb{B}^t\sigma\underline{N})\,dS$$

$$\underline{\tilde{T}}(x) = \int_{S_x} \sigma\underline{N}\,dS \qquad \underline{\tilde{M}}(x) = \int_{S_x} \underline{X}\wedge(\sigma\underline{N})\,dS$$

Moreover here, $(\underline{\tilde{T}}, \underline{\tilde{M}})$ will systematically denote an equilibrated generalized stress, *i.e.* a solution to the beam equilibrium equations:

$$\forall x \in ]0, L[ \qquad \begin{cases} \underline{\tilde{T}}_{,x} + \int_{S_x}\underline{f}_d\,dS + \int_{\partial S_x}\underline{F}_d\,d\tau = 0 \\ \underline{\tilde{M}}_{,x} + \underline{N}\wedge\underline{\tilde{T}} + \int_{S_x}\underline{X}\wedge\underline{f}_d\,dS + \int_{\partial S_x}\underline{X}\wedge\underline{F}_d\,d\tau = 0 \end{cases} \tag{6}$$

3.1.3. *The basic localized effect (or solution) and the Saint-Venant principle*
For the free infinite beam, one prescribes at $x = 0$ a discontinuity of both the displacement and the normal stress vector which is denoted by $s_0 = (\underline{W}_0, \underline{C}_0)$ ($s_0 \in \mathbf{S}$). The solution $x \to s^{\pm}(x)$ equilibrating the given perturbation $(\underline{W}_0, \underline{C}_0)$, formally defined by the following problem, extends what is called the "extremity effect": (see (Ladevèze P., 1983) (Ladevèze P., 1985))

Find $s^{\pm}$ equal to $s^+$ on $S\times]0, +\infty[$ and to $s^-$ on $S\times]-\infty, 0[$ such that:

$$(\underline{U}^+ - \underline{U}^-)_{|x=0} = \underline{W}_0 \qquad (\sigma^+\underline{N} - \sigma^-\underline{N})_{|x=0} = \underline{C}_0 \tag{7}$$

$$\begin{cases} \mathrm{div}\sigma^+ = 0 \text{ on } S\times]0, +\infty[, & \sigma^+\underline{n} = 0 \text{ on } \partial S\times]0, +\infty[ \\ \mathrm{div}\sigma^- = 0 \text{ on } S\times]-\infty, 0[, & \sigma^-\underline{n} = 0 \text{ on } \partial S\times]-\infty, 0[ \\ \sigma^+ = \mathbf{K}\varepsilon(\underline{U}^+) \text{ on } S\times]0, +\infty[, & \sigma^- = \mathbf{K}\varepsilon(\underline{U}^-) \text{ on } S\times]-\infty, 0[ \end{cases} \tag{8}$$

This problem in fact describes the equilibrium of two connected semi-infinite free beams, with the connection being defined by the relations in (7). The total elastic energy is the sum of the energies of the two semi-infinite parts.

We are interested herein with finite energy solutions. Among the solutions to problem (7) (8), we distinguish those which correspond to the following definition:

**Definition 2** *The solution* $s^{\pm}$ *is localized in the neighborhood of* $x = 0$ *if:*

- *its energy is finite,*
- $\lim_{x\to+\infty} ||\underline{U}^+||_{L^2(S)} = \lim_{x\to-\infty} ||\underline{U}^-||_{L^2(S)} = 0$

**Theorem 2** *The solution* $s^{\pm}$ *produced by* $s_0 = (\underline{W}_0, \underline{C}_0) \in \mathbf{S}$ *is localized in the neighborhood of* $x = 0$ *if and only if the generalized displacement-stress associated* *with $s_0$ is zero i.e. if and only if:*

$$\int_S \underline{C}_0 \, dS = 0 \qquad \int_S \underline{X} \wedge \underline{C}_0 \, dS = 0$$

$$\int_S (\mathbb{A}^{ot}\, \underline{W}_0 - \mathbb{A}^t\, \underline{C}_0 )\, dS = 0 \qquad \int_S (\mathbb{B}^{ot}\, \underline{W}_0 - \mathbb{B}^t\, \underline{C}_0 )\, dS = 0 \tag{9}$$

**Property 1** *If the solution $s^{\pm}$ is localized in the neighborhood of $x = 0$, one has:*

$$\forall x \in ]-\infty, +\infty[ \quad \forall s^*_{sv} \qquad [s^{\pm},\, s^*_{sv}]_{|x} = 0$$

*where $s^*_{sv}$ denotes a Saint-Venant solution for no external loads and $[s,\, s^*_{sv}]_{|x}$ is the following square brackets:*

$$[s,\, s^*_{sv}]_{|x} = \int_{S_x} [\sigma^*_{sv} \underline{N} \cdot \underline{U} - \underline{U}^*_{sv} \cdot \sigma \underline{N}\,]\, dS$$

The localization property is more precisely defined by:

**Property 2** *The elastic energy of a solution $s^{\pm}$ which is localized in the neighborhood of $x = 0$ satisfies:*

$$E_D(x) \leq exp\left(\frac{-x}{l}\right) E_D(0)$$

*where:*

$$E_D(x) = \frac{1}{2}\int_x^{+\infty} dx \int_S \mathrm{Tr}[\mathbf{K}\,\varepsilon\,(U^+)\varepsilon\,(U^+)]dS + \frac{1}{2}\int_{-\infty}^{-x} dx \int_S \mathrm{Tr}[\mathbf{K}\,\varepsilon\,(U^-)\varepsilon\,(U^-)]dS$$

*$l$ is a constant length characterizing the cross-section geometry and the material.*

**Definition 3** *The linear operator $\mathbf{L}$ characterizing a localized effect is defined by:*

$$s^{\pm}(x, 0) = -\mathbf{L}(x - 0)s_0$$

*$\mathbf{L}$, which depends on the cross-section geometry and the material, is a sort of Green's function.*

<u>Remark</u>
The Saint-Venant principle is in fact an orthogonality property of the family of Saint-Venant solutions (Ladevèze P., 1983), (Ladevèze P., 1985).

3.1.4. *The exact 1D-theory of elastic beams*
The generalized displacement ($\underline{\tilde{u}}(x)$, $\underline{\tilde{\omega}}(x)$) and the generalized stress are the solution of the following 1D-problem defined on $[0,\, L]$:

Find ($\underline{\tilde{u}}$, $\underline{\tilde{\omega}}$) and ($\underline{\tilde{T}}$, $\underline{\tilde{M}}$), $x \in [0,\, L]$ such the following are satisfied:

- the beam equilibrium equations (6)
- the beam constitutive relation

$$\begin{bmatrix} \underline{\tilde{\gamma}} \\ \underline{\tilde{\chi}} \end{bmatrix} = \mathbb{\Lambda} \begin{bmatrix} \underline{\tilde{\mathrm{T}}} \\ \underline{\tilde{\mathrm{M}}} \end{bmatrix} + \begin{bmatrix} \underline{\tilde{\gamma}}_d \\ \underline{\tilde{\chi}}_d \end{bmatrix}$$

with (10)

$$\underline{\tilde{\gamma}}(x) = [\underline{\tilde{u}}_{,x} + \underline{\mathrm{N}} \wedge \underline{\tilde{\omega}}]_{|x}$$

$$\underline{\tilde{\chi}}(x) = [\underline{\tilde{\omega}}_{,x}]_{|x}$$

- some 1D extremity conditions at $x = 0$ and $x = L$

The operator $\mathbb{\Lambda}$ and $(\underline{\tilde{\gamma}}_d, \underline{\tilde{\chi}}_d)$ are determined in terms of the basic buildings blocks $\mathbb{A}$, $\mathbb{B}$, $\mathbb{A}^\circ$, $\mathbb{B}^\circ$, $\underline{\mathrm{W}}_d$, $\underline{\mathrm{C}}_d$; they will be specified later. The 1D extremity conditions are derived from the 3D ones (see (Ladevèze P., Simmonds J.G., 1996)).

3.1.5. *Fundamental result*

**Theorem 3** *The solution to the reference 3D elastic problem can be written in an unique way* $(s = (\underline{\mathrm{U}}, \sigma\underline{\mathrm{N}}))$*:*

$$\forall x \in [0, L] \qquad s(x) = s_{sv}(x) + \int_0^L s^{\pm}(x-t, t)\, dt$$

*where*

- $s_{sv}$ *is the Saint-Venant solution (5) associated with the* $x$*-values of the data and of the generalized displacement and stress, solution of (10).*

$$\underline{\mathrm{U}}_{sv} = \underline{\tilde{u}}(x) + \underline{\tilde{\omega}}(x) \wedge \underline{\mathrm{X}} + \mathbb{A}\underline{\tilde{\mathrm{T}}}(x) + \mathbb{B}\underline{\tilde{\mathrm{M}}}(x) + \underline{\mathrm{W}}_d(x)$$

$$\sigma_{sv}\underline{\mathrm{N}} = \mathbb{A}^\circ\, \underline{\tilde{\mathrm{T}}}(x) + \mathbb{B}^\circ\, \underline{\tilde{\mathrm{M}}}(x) + \underline{\mathrm{C}}_d(x)$$

- $x \to s^{\pm}(x-t,\, t)$ *is an effect which is localized in the neighborhood of* $x = t$. $t \to s^{\pm}(x-t,\, t)$ *consists of a regular part and Dirac distributions at the extremities and at the load discontinuities.*

Moreover, one has:

**Property 3** *The integral of the localized effects is equal to:*

$$\int_0^L s^{\pm}(x-t, t)\, dt = \int_0^L \mathbf{L}(x-t)[h_d]_{|t}\, dt + s^{+}(x) + s^{-}(L-x)$$

*where :*

- $s^{+}(x)$ *and* $s^{-}(L-x)$ *are effects localized in the neighborhood of extremities;*
- $\mathbf{L}$ *is a linear operator that depends on the cross-section and the material,*
- $h_d(x) = \frac{d}{dx}(\underline{\mathrm{W}}_d, \underline{\mathrm{C}}_d)$

*Remark*

The Saint-Venant solution does not exactly satisfy the various equations of the 3D elasticity problem. Two residuals can be distinguished apart from the extremity conditions:

- a residual associated with the non-satisfaction of the equilibrium equations: $\underline{C}_{d,x}$
- a residual associated with the non-satisfaction of the compatibility equations: $\underline{W}_{d,x}$.

Another remarkable point is:

**Property 4** *The values of the generalized displacement and the generalized stress for the Saint-Venant solution and the reference 3D elasticity solution are equal.*

## 4. Calculation of the operators associated with the exact beam theory

A general computational technique is proposed to calculate the operators and vectors:

- $\mathbb{A}$, $\mathbb{B}$, $\mathbb{A}^o$, $\mathbb{B}^o$, $\mathbb{\Lambda}$
- $\underline{W}_d$, $\underline{C}_d$, $\tilde{\underline{\gamma}}_d$, $\tilde{\underline{\chi}}_d$

These quantities depend on the cross-section geometry, the material and the load densities ($\underline{f}_d$, $\underline{F}_d$). The basic problem from which they are defined is the determination of the Saint-Venant solution for the particular case of $x$-constant data; it is an $x$-polynomial of degree 4.

### 4.1. CALCULATION OF $\mathbb{A}$, $\mathbb{B}$, $\mathbb{A}^o$, $\mathbb{B}^o$, $\underline{C}_d$, $\underline{W}_d$

One considers a piece of beam limited by two cross-sections $x = 0$ and $x = L$. It is described with 3D finite elements of degree 4 in the $x$-direction, their length being $L$. Therefore, the displacement is written:

$$\underline{U}_h = \sum_{i=1}^{n} \underline{\varphi}_i u_i = [\varphi]\underline{u}^h$$

where $\underline{\varphi_i}$ are the shape functions and $u_i$ the nodal displacement components. The corresponding displacement subspace is denoted by $\mathcal{U}_h$ (classical additional constraints on the displacement are added to get for one strain field only one displacement field). The first equation is:

$$\int_\Omega \mathrm{Tr}(\mathbf{K}\,\varepsilon(\underline{U}_h)\varepsilon(\underline{U}_h^*))\,d\Omega = \int_{S_L} \mathbf{K}\,\varepsilon(\underline{U}_h)\underline{N}\cdot\underline{U}_h^*\,dS - \int_{S_0} \mathbf{K}\,\varepsilon(\underline{U}_h)\underline{N}\cdot\underline{U}_h^*\,dS + \int_0^L dx[\int_S \underline{f}_d\cdot\underline{U}_h^*\,dS + \int_{\partial S}\underline{F}_d\cdot\underline{U}_h^*\,d\tau] \quad (11)$$

$$\forall \underline{U}_h^* \in \mathcal{U}_h$$

One gets:

$$\mathbb{K}\underline{u}^h = \Omega\,\underline{u}^h + \underline{R}_d \tag{12}$$

One has also:

$$\begin{aligned}\underline{\tilde{\mathrm{T}}} = \underline{\tilde{\mathrm{T}}}(0) &= \int_{So} \mathbf{K}\,\varepsilon\,(\underline{\mathrm{U}}_h)\underline{\mathrm{N}}\;dS = \mathbb{T}\,\underline{u}^h \\ \underline{\tilde{\mathrm{M}}} = \underline{\tilde{\mathrm{M}}}(0) &= \int_{So} \underline{\mathrm{X}} \wedge (\mathbf{K}\,\varepsilon\,(\underline{\mathrm{U}}_h)\underline{\mathrm{N}})\,dS = \mathbb{M}\underline{u}^h\end{aligned} \tag{13}$$

$\mathbb{K}$, $\Omega$, $\mathbb{T}$, $\mathbb{M}$ are matrices depending on the cross-section geometry and the material. $\underline{R}_d$ depends also on the load densities on $S$. The rigidity matrix $\mathbb{K}$ is positive definite and symmetric.

There are several ways for solving the problem defined by the equations (12) and (13). Our method consists in introducing as unknowns Lagrange multipliers $\underline{\lambda}$ and $\underline{\mu}$ related to the constraints (13), the quantities $\underline{\tilde{\mathrm{T}}}$ and $\underline{\tilde{\mathrm{M}}}$ being given. Precisely, we solve numerically:

$$\begin{bmatrix} \mathbb{K}-\Omega & \mathbb{T}^t & \mathbb{M}^t \\ \mathbb{T} & 0 & 0 \\ \mathbb{M} & 0 & 0 \end{bmatrix} \begin{bmatrix} \underline{u}^h \\ \underline{\lambda} \\ \underline{\mu} \end{bmatrix} = \begin{bmatrix} \underline{R}_d \\ \underline{\tilde{\mathrm{T}}} \\ \underline{\tilde{\mathrm{M}}} \end{bmatrix} \tag{14}$$

Let us note that the right side is given. Even if the matrix $(\mathbb{K}-\Omega)$ is not positive, the resolution of (14) does not involve any numerical difficulty. The computed values of $\underline{\lambda}$ and $\underline{\mu}$ are very small; they can be used as error indicators.

The solution can be expressed as:

$$\begin{aligned}\underline{\mathrm{U}}_h &= \underline{\hat{u}} + \underline{\hat{\omega}} \wedge \underline{\mathrm{X}} + \hat{\mathbb{A}}\,\underline{\tilde{\mathrm{T}}} + \hat{\mathbb{B}}\,\underline{\tilde{\mathrm{M}}} + \underline{\hat{\mathrm{W}}}_d \\ \sigma^h\,\underline{\mathrm{N}} &= \mathbb{K}\varepsilon\,(\underline{\mathrm{U}}_h)\underline{\mathrm{N}} \simeq \hat{\mathbb{A}}^o\,\underline{\tilde{\mathrm{T}}} + \hat{\mathbb{B}}^o\,\underline{\tilde{\mathrm{M}}} + \underline{\hat{\mathrm{C}}}_d\end{aligned}$$

It is the Saint-Venant solution modulo a rigid body displacement. Consequently, noting (5), we get:

$$\begin{aligned}\mathbb{A}^o &= \hat{\mathbb{A}}^o \\ \mathbb{B}^o &= \hat{\mathbb{B}}^o \\ \underline{\mathrm{C}}_d &= \underline{\hat{\mathrm{C}}}_d\end{aligned}$$

To go further, let us recall a characteristic property of the generalized displacement proved in (Ladevèze P., Simmonds J.G., 1996).

**Property 5** *The operators $\mathbb{A}$, $\mathbb{B}$ and the vector $\underline{\mathrm{W}}_d$ can be chosen such that $s_d = (\underline{\mathrm{W}}_d, \underline{\mathrm{C}}_d)$ satisfies the localization condition i.e.:*

$$\forall s^*_{sv} \qquad \int_S [\sigma^*_{sv}\,\underline{\mathrm{N}} \cdot \underline{\mathrm{W}}_d - \underline{\mathrm{U}}^*_{sv} \cdot \underline{\mathrm{C}}_d\,]\,dS = 0 \tag{15}$$

*and*

$$\forall s^*_{sv} \qquad \forall s_{sv} \qquad \int_S \sigma^*_{sv}\, \underline{\mathrm{N}} \cdot < \underline{\mathrm{U}}_{sv} > \; dS = 0 \tag{16}$$

*with*

$$< \underline{\mathrm{U}}_{sv} > = \mathbb{A}\tilde{\underline{\mathrm{T}}} + \mathbb{B}\tilde{\underline{\mathrm{M}}}$$

Let us start with an additional rigid body displacement to the Saint-Venant solution for zero-value load densities:

$$\underline{u}' + \underline{\omega}' \wedge \underline{\mathrm{X}}$$

which depends on $\tilde{\underline{\mathrm{T}}}$, $\tilde{\underline{\mathrm{M}}}$:

$$\begin{bmatrix} \underline{u}' \\ \underline{\omega}' \end{bmatrix} = \mathbb{Z} \begin{bmatrix} \tilde{\underline{\mathrm{T}}} \\ \tilde{\underline{\mathrm{M}}} \end{bmatrix}$$

$\mathbb{Z}$ is a constant operator to be determined from (16). One obtains:

$$\mathbb{Z} + \int_S \begin{bmatrix} \mathbb{A}^{ot}\hat{\mathbb{A}} & \mathbb{A}^{ot}\hat{\mathbb{B}} \\ \mathbb{B}^{ot}\hat{\mathbb{A}} & \mathbb{B}^{ot}\hat{\mathbb{B}} \end{bmatrix} dS = 0 \tag{17}$$

It follows:

$$\begin{aligned} \mathbb{A} &= \hat{\mathbb{A}} + \mathbb{Z}_{11} - \underline{\mathrm{X}} \wedge \mathbb{Z}_{21} \\ \mathbb{B} &= \hat{\mathbb{B}} + \mathbb{Z}_{12} - \underline{\mathrm{X}} \wedge \mathbb{Z}_{22} \end{aligned} \tag{18}$$

To determine $\underline{\mathrm{W}}_d$, let us introduce one more additional rigid body displacement to the Saint-Venant solution which depends on the load densities:

$$\underline{u}_d + \underline{\omega}_d \wedge \underline{\mathrm{X}}$$

The localization condition (15) leads to:

$$\begin{aligned} \underline{u}_d &= - \int_S (\mathbb{A}^{ot}\, \hat{\underline{\mathrm{W}}}_d - \mathbb{A}^{t}\, \underline{\mathrm{C}}_d)\, dS \\ \underline{\omega}_d &= - \int_S (\mathbb{B}^{ot}\, \hat{\underline{\mathrm{W}}}_d - \mathbb{B}^{t}\, \underline{\mathrm{C}}_d)\, dS \end{aligned} \tag{19}$$

It follows:

$$\underline{\mathrm{W}}_d = \hat{\underline{\mathrm{W}}}_d + \underline{u}_d + \underline{\omega}_d \wedge \underline{\mathrm{X}} \tag{20}$$

## 4.2. CALCULATION OF THE OPERATOR $\mathbb{A}$

First, from the computed displacement, one determines the generalized displacement on $]0, L[$:

$$\tilde{\underline{u}} = \int_S (\mathbb{A}^{ot}\underline{\mathrm{U}}_h - \mathbb{A}^{t}\sigma^h\, \underline{\mathrm{N}})\, dS \qquad \tilde{\underline{\omega}} = \int_S (\mathbb{B}^{ot}\underline{\mathrm{U}}_h - \mathbb{B}^{t}\sigma^h\, \underline{\mathrm{N}})\, dS \tag{21}$$

and one determines:

$$\underline{\tilde{\gamma}}_{x=0} = (\underline{\tilde{u}}_{,x} + \underline{\mathrm{N}} \wedge \underline{\tilde{\omega}})_{|x=0} \qquad \underline{\tilde{\chi}}_{x=0} = (\underline{\tilde{\omega}}_{,x})_{|x=0} \tag{22}$$

These quantities depend linearly on $\underline{\tilde{\mathrm{T}}}, \underline{\tilde{\mathrm{M}}}$ and the load densities. Therefore, they can be written as:

$$\begin{bmatrix} \underline{\tilde{\gamma}} \\ \underline{\tilde{\chi}} \end{bmatrix}_{x=0} = \hat{\mathbb{A}} \begin{bmatrix} \underline{\tilde{\mathrm{T}}} \\ \underline{\tilde{\mathrm{M}}} \end{bmatrix}_{x=0} + \begin{bmatrix} \underline{\hat{\gamma}}_d \\ \underline{\hat{\chi}}_d \end{bmatrix} \tag{23}$$

From the exact beam constitutive relation:

$$\begin{bmatrix} \underline{\tilde{\gamma}} \\ \underline{\tilde{\chi}} \end{bmatrix} = \mathbb{A} \begin{bmatrix} \underline{\tilde{\mathrm{T}}} \\ \underline{\tilde{\mathrm{M}}} \end{bmatrix} + \begin{bmatrix} \underline{\tilde{\gamma}}_d \\ \underline{\tilde{\chi}}_d \end{bmatrix} \tag{24}$$

where $\mathbb{A}$ depends only on the cross-section geometry and the material, one gets:

$$\begin{gathered} \mathbb{A} = \hat{\mathbb{A}} \\ \underline{\tilde{\gamma}}_d = \underline{\hat{\gamma}}_d \qquad \underline{\tilde{\chi}}_d = \underline{\hat{\chi}}_d \end{gathered} \tag{25}$$

## 4.3. EXAMPLE

The numerical technique for computing the basic operators associated with the exact beam theory has been implemented in the f.e. code CASTEM 2000 for arbitrary cross-section and composite material. For cylindrical tubes (radii $a$, $b$) made with an isotropic material, the basic operators can be defined explicitely. One gets:

$$\sigma_{sv}\, \underline{\mathrm{N}} = \mathbb{A}^{\circ}(\underline{\mathrm{X}})\, \underline{\tilde{\mathrm{T}}}(x) + \mathbb{B}^{\circ}(\underline{\mathrm{X}})\, \underline{\tilde{\mathrm{M}}}(x)$$

with:

$$\mathbb{A}^{\circ} = \begin{bmatrix} \frac{1}{S} & 0 & 0 \\ 0 & f_1(y,z) & f_2(z,y) \\ 0 & f_2(y,z) & f_1(z,y) \end{bmatrix} \qquad \mathbb{B}^{\circ} = \begin{bmatrix} 0 & \frac{z}{I} & -\frac{y}{I} \\ -\frac{z}{2I} & 0 & 0 \\ \frac{y}{2I} & 0 & 0 \end{bmatrix}$$

where $f_1(y,z)$ and $f_2(y,z)$ are:

$$f_1(y,z) = \frac{1}{8I(1+\nu)} \left[ (3+2\nu)(a^2+b^2-y^2-a^2b^2\frac{(y^2-z^2)}{(y^2+z^2)^2}) - (1-2\nu)z^2 \right]$$

$$f_2(y,z) = \frac{-yz}{4I(1+\nu)} \left[ 1+2\nu+\frac{(3+2\nu)a^2b^2}{(y^2+z^2)^2} \right]$$

For the constitutive relation, we get:

$$\begin{bmatrix} \underline{\tilde{\gamma}}_1 \\ \underline{\tilde{\gamma}}_2 \\ \underline{\tilde{\gamma}}_3 \\ \underline{\tilde{\chi}}_1 \\ \underline{\tilde{\chi}}_2 \\ \underline{\tilde{\chi}}_3 \end{bmatrix} = \begin{bmatrix} \frac{1}{ES} & 0 & 0 & 0 & 0 & 0 \\ 0 & \frac{1}{kGS} & 0 & 0 & 0 & 0 \\ 0 & 0 & \frac{1}{kGS} & 0 & 0 & 0 \\ 0 & 0 & 0 & \frac{1}{GJ} & 0 & 0 \\ 0 & 0 & 0 & 0 & \frac{1}{EI} & 0 \\ 0 & 0 & 0 & 0 & 0 & \frac{1}{EI} \end{bmatrix} \begin{bmatrix} \tilde{\mathrm{T}}_1 \\ \tilde{\mathrm{T}}_2 \\ \tilde{\mathrm{T}}_3 \\ \tilde{\mathrm{M}}_1 \\ \tilde{\mathrm{M}}_2 \\ \tilde{\mathrm{M}}_3 \end{bmatrix}$$

where:

$$S = \pi(b^2 - a^2) \qquad\qquad I = \frac{\pi}{4}(b^4 - a^4)$$

$$G = \frac{E}{2(1+\nu)} \qquad\qquad J = 2I$$

and one gets for $k$:

$$k = \frac{6(a^2+b^2)^2(1+\nu)^2}{(a^4+b^4)(7+14\nu+8\nu^2)+2a^2b^2(17+34\nu+16\nu^2)}$$

For the displacement:

$$\underline{U}_{sv} = \underline{\tilde{u}}(x) + \underline{\tilde{\omega}}(x) \wedge \underline{X} + \mathbb{A}(\underline{X})\,\underline{\tilde{T}}(x) + \mathbb{B}(\underline{X})\,\underline{\tilde{M}}(x)$$

the operator $\mathbb{A}$ is:

$$\mathbb{A} = \begin{bmatrix} 0 & g_1(y,z) - C_2 y & g_1(z,y) - C_2 z \\ -\frac{\nu}{ES} y & 0 & 0 \\ -\frac{\nu}{ES} z & 0 & 0 \end{bmatrix}$$

and the operator $\mathbb{B}$ is:

$$\mathbb{B} = \begin{bmatrix} 0 & 0 & 0 \\ 0 & -\frac{\nu}{EI} yz & \frac{\nu}{2EI}(y^2 - z^2) + C_1 \\ 0 & \frac{\nu}{2EI}(y^2 - z^2) - C_1 & \frac{\nu}{EI} yz \end{bmatrix}$$

where:

$$g_1(y,z) = \frac{y}{4EI}\left[(3+2\nu)(a^2+b^2+\frac{a^2b^2}{y^2+z^2}) - y^2 - z^2\right]$$

and:

$$C_1 = \frac{\nu((a^4+b^4)(1+2\nu)+2a^2b^2(5+4\nu))}{3(b^2-a^2)(a^2+b^2)^2\pi E(1+\nu)}$$

$$C_2 = \frac{(a^4+b^4)(7+6\nu)+2a^2b^2(17+12\nu)}{3(b^2-a^2)(a^2+b^2)^2\pi E}$$

The exact 1D constitutive relation coincides with the one proposed recently in (Renton J.D., 1997) but is different from (Cowper G.R., 1966).

Using the classical shell theory, approximations have been derived in (Ladevèze P., Sanchez P., Simmonds J.G., 2000b) for thin elastic tubes of arbitrary cross-section.

## 4.4. CALCULATION OF EXTREMITY EFFECT CORRECTIONS

### 4.4.1. *Principle*

One has to solve the problem giving the basic localized effect (or solution) which is defined by equations (7) and (8). A finite element code can be used.

However a difficulty can arise with cross-section of very thin tubes: the wavelength of the localized effect can be very large compared to the cross-section diameter and then such an effect contains a part which is not practically "localized".

Let us consider a beam like elastic tube of arbitrary cross-section (thickness: $2l$; diameter: $2R$). It is well known (McDevitt T.J., Simmonds J. G., 1999) that the localized effect can be split in two parts:

– a part with a decay length, $O(R\sqrt{\frac{h}{R}})$

– a part with a very long decay length, $O(R\sqrt{\frac{R}{h}})$

To get the first part, one has to solve a 3D problem for which one can used a finite element code. The splitting is done here by introducing an extension of the Saint-Venant concept.

4.4.2. *Calculation of extremity effect corrections with a very long decay length*
A efficient way for computing such an effect is to use the set of functions introduced in (Ladevèze P., 1983) which extends the Papkovitch one for the semi-infinite strip. For the sake of simplicity, let us consider that the cross-section plane is a plane of material symmetry. For an $x$-homogeneous beam $[0,\ L]$, it has been proved in (Ladevèze P., 1983) that the extremity effects can be written:

$$\sum_{i=1}^{+\infty} a_i^+ \underline{\psi}_i^+(\underline{X}) \exp(\lambda_i^+ x) + \sum_{j=1}^{+\infty} a_j^- \underline{\psi}_j^-(\underline{X}) \exp(\lambda_j^-(L-x))$$

where $a_i^+$, $a_j^-$ are coefficients (complex numbers). $\underline{\psi}_i^+$ and $\underline{\psi}_j^-$ are two biorthogonal sets of functions which satisfy:

– $\lambda_i^+ = \lambda_j^-$ for $i = j$; $Re(\lambda_i^+) > 0$

– $\underline{\psi}_i^- = \mathbb{Z}\,\underline{\psi}_i^+$ where $\mathbb{Z}$ is the cross-section plane symmetry operator.

– $(\underline{\psi}_i^+,\ \underline{\psi}_j^-) = -\int_S \sigma_i^+ \underline{N} \cdot \underline{\psi}_j^-\, dS = \delta_{ij}$

Introducing classical shell theory, one gets very easily the set $\{\underline{\psi}_i : i \in 1, 2, ...\}$ by solving a single 1D eigenvalue problem (Ladevèze P., Sanchez P., Simmonds J.G., 2000b) (Ladevèze P., Sanchez P., Simmonds J.G., 2000a). Practically, one need keep only the functions for which the decay length is larger than $R$; let be $m$ the corresponding index. One notes:

$$\begin{aligned} \underline{U}^m &= \sum_{i=1}^{m} a_i^+ \underline{\psi}_i^+ \exp(\lambda_i^+ x) + \sum_{j=1}^{m} a_j^- \underline{\psi}_j^- \exp(\lambda_j^-(L-x)) \\ s^m_{|x} &= (\underline{U}^m,\ \mathbf{K}\,\varepsilon\,(\underline{U}^m)\underline{N})_{|x} \end{aligned} \tag{26}$$

The corresponding displacement space is $\mathcal{U}^m$. For equilibrating the residues at the extremities, two separate 3D problems located over the domains $S \times [0,\ R]$ and $S \times [L-R,\ L]$ have to be solved, the displacement being prescribed as 0 at the sections $S_R$ and $S_{L-R}$. In other words, we suppose that the corrections associated with the two extremity residues have practically no interactions. The residue is in fact:

$$s_r = (\underline{U}_r, \mathbf{K}\varepsilon(\underline{U}_r)\underline{N}) \qquad \text{with} \qquad \underline{U}_r = \underline{U}_{ex} - \underline{U}_{sv} - \underline{U}^m \tag{27}$$

For defining the parameter of the Saint-Venant solution, *i.e.* the generalized displacement and stress, one writes the Saint-Venant principle which is an orthogonality condition (Ladevèze P., 1985) (Ladevèze P., Simmonds J.G., 1996) at $\bar{x} \in [0, L]$:

$$[s_{ex} - s_{sv}, s^*_{sv}]_{|\bar{x}} = 0 \qquad \forall s^*_{sv} \tag{28}$$

where

$$[s, s^*]_{|\bar{x}} = \int_{S_{\bar{x}}} [\sigma^*\underline{N} \cdot \underline{U} - \underline{U}^* \cdot \sigma\underline{N}]\, dS$$

If it is satisfied for $\bar{x}$, the orthogonality condition is also satisfied for any abscissa belonging to $[0, L]$. In our case, for defining both $s_{sv}$ and $s^m$, we introduce the following orthogonality condition which can be interpreted as an extension of the Saint-Venant principle:

$$[s_r, s^*_{sv} + s^{m*}]_{|\bar{x}} = 0 \qquad \forall s^*_{sv} \quad \forall s^{m*} \tag{29}$$

If the condition holds for $\bar{x}$, it also holds for any $\bar{x}$ belonging to $[0, L]$. For introducing the data at the extremities and for eliminating $s_{ex}$, we follow the technique that we described in (Ladevèze P., Simmonds J.G., 1996), technique for which some additional but separate 3D finite element calculation have to be done at the extremities. It follows a little equation system which gives:

- the generalized quantities
- $a_i^+$, $a_j^-$ with $i \in \{1, 2, \dots m\}$ and $j \in \{1, 2, \dots m\}$

## 4.5. AN EXAMPLE

Let us consider an isotropic circular tube submitted to two extremity loads which are equilibrated by an uniform axial load density. Using classical notations, the data are $E = 100000MPa, \nu = 0.3$ and the figure (2) gives the loading.

Figures (3) and (4) give the Mises stress computed with the exact beam theory following the proposed numerical method.

## 4.6. CONCLUSION

The exact beam theory is a very straight forward way for calculating 3D displacements and stresses even if there is a strong interaction between the "localized" effects which occur in the neighborhood of the extremities and of the cross-section discontinuities. Further works will extend it to non piecewise constant cross-section, elastic constants and to curved beams.

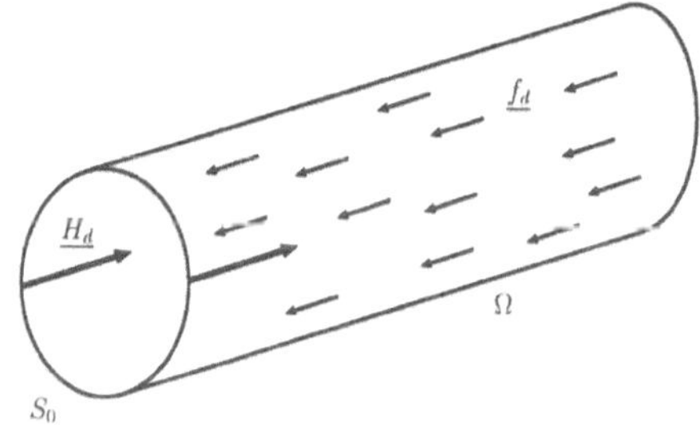


*Figure 2.* Loading

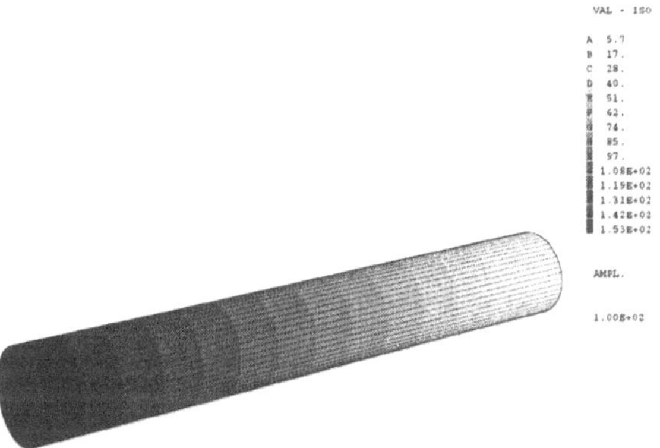


*Figure 3.* Mises stress distribution of the Saint-Venant solution.

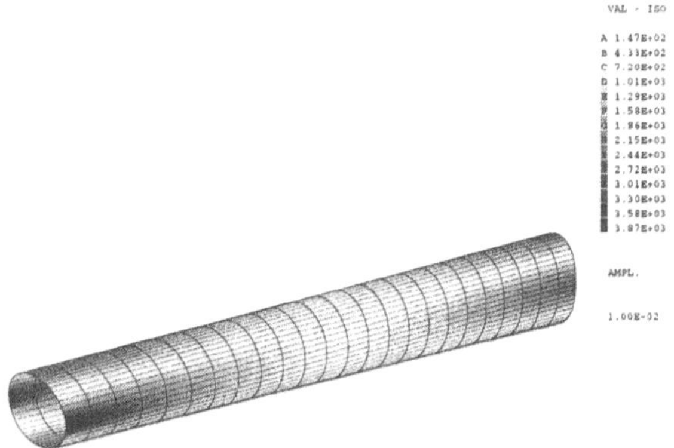


*Figure 4.* Mises stress distribution of "localized" effects defined on the deformed beam.

## References

Ciarlet P.G. (1990). *Plates and junctions in elastic multi-structures.* MASSON.

Cowper G.R. (1966). The shear coefficient in Timoshenko's beam theory. *J. of Applied Mechanics*, 33:335–440.

Duvaut G., Lions J.L. (1976). *Inequalities in Mechanics and Physics*. Springer.

Horgan C.O. (1989). Recent developments concerning Saint-Venant's principle: An update. *Appl. Mech. Rev.*, 42:295–303.

Horgan C.O., Simmonds J.G. (1994). Saint-Venant end effects in composite structures. *Composites Engineering*, 4:279–286.

Ladevèze J., Ladevèze P., Mantion M., Pecastaings F., Pelle J.P. (1979). Sur les fondements de la théorie linéaire des poutres élastiques I. Méthode. Construction du prolongement admissible optimal. *J. de Mécanique*, 18(1):131–173.

Ladevèze P. (1983). Sur le principe de Saint-Venant en élasticité. *J. de Mécanique Théorique et Appliquées*, 1(2):161–184.

Ladevèze P. (1985). On Saint-Venant principle in elasticity. In Ladevèze P., editor, *Local Effects in the Analysis of Structures*, pages 3–34. Elvesier.

Ladevèze P., Sanchez P., Simmonds J.G. (2000a). Anisotropic elastic tubes of arbitrary cross section under arbitrary end loads: separation of beamlike and decaying solutions. *(to appear)*.

Ladevèze P., Sanchez P., Simmonds J.G. (2000b). Beamlike (Saint-Venant) solutions for fully anisotropic elastic tubes of arbitrary cross section. *(to appear)*.

Ladevèze P., Simmonds J.G. (1996). De nouveaux concepts en théorie des poutres pour des charges et des géométries quelconques. *C.R. Acad. Sci. Paris, t. 322, séries II b*, pages 455–462.

Ladevèze P., Simmonds J.G. (1998). New concepts for linear beam theory with arbitrary geometry and loading. *Eur. J. Mech., A/Solids*, 17(3):377–402.

McDevitt T.J., Simmonds J. G. (1999). Reduction of the Sanders-Koiter equations for fully anisotropic circular cylindrical shells to two coupled equations for a stress and a curvature function. *J. of Applied Mechanics*, 66:593–597.

Renton J.D. (1997). A note on the form of the shear coefficient. *Int. J. Solids Structures*, 34(14):1681-1685.

Rigolot A. (1972). Sur une théorie asymptotique des poutres. *J. de Mécanique*, 11(4):673–703.

# EFFECTS OF ECCENTRIC STIFFENING ON STATIC AND DYNAMIC NONLINEAR RESPONSE OF RECTANGULAR PANELS EXPOSED TO THERMOMECHANICAL LOADING

LIVIU LIBRESCU[a] and MARCO A. SOUZA[b]

[a] *Department of Engineering Science and Mechanics*
*Virginia Polytechnic Institute and State University*
*Blacksburg, VA 24061, USA*

[b] *Brown & Root Energy Services*
*10200 Bellaire Boulevard*
*Houston, TX 77072-5299, USA*

## 1. INTRODUCTION

Advanced supersonic and hypersonic space vehicles will experience very high temperatures and pressure gradients during their flight missions.

Moreover, these vehicles will typically experience these loadings in a dynamic environment. A problem of crucial importance toward the rational design of their structural subcomponents consists of the possibility to accurately determine their load carrying capacity. The exhaustive use of the load carrying capacity of such structures can dramatically contribute to the increase, without weight penalties, of the performance of such vehicles. Due to thermomechanical load interaction, changes in the vibration characteristics of structures and implicity of their dynamic response and flutter are likely to occur.

As a result, a better understanding of the effects of thermomechanical loads on the behavior of vibrating flat panel constitutes a fundamental step in determining and understanding the overall structural behavior of structural subcomponents.

A great part of flight vehicle structures consists of plates and shells reinforced by stiffeners. Such stiffened structures contribute, among others, to achieve low weight and high stiffness, reduced deflection and an enhanced load carrying capacity.

In spite of the great importance upon the analysis and design of advanced flight vehicle structures, results on static and vibrational non-linear response of stiffened panels under thermomechanical loading, appear to be extremely scarce.

The surveys in Refs. 1 through 5, were comprehensive accounts on the effort carried out in this area are provided, underline in full the absence of results in this area. To the best of authors' knowledge, the only results obtained via the finite element method are contained in Ref. 6.



*D. Durban et al. (eds.), Advances in the Mechanics of Plates and Shells,* 197–212.

## 2. Analysis Description. Basic Equations

Consider the case of a rectangular isotropic thin plate of edge lengths $L_1$ and $L_2$, and thickness of $h$, eccentrically reinforced by orthogonal stiffeners. One assumes that the stiffener spacing $l_s$ and $l_r$ in the direction $\xi_1$ and $\xi_2$, respectively are constant and small enough so that the smeared out procedure may be applied. One assumes that the panel is exposed to a uniform through the thickness temperature

$$T(\xi_1, \xi_2, \xi_3) \equiv \overset{o}{T} (\xi_1, \xi_2) \tag{1}$$

measured from a stress-free temperature $T_r$. This temperature field will be referred in the sequel to as the membrane temperature. Such a temperature distribution can be experienced during the steady-flight of a high-speed flight vehicle. In Eq. (1), $\xi_1$ and $\xi_2$ are the in-plane Cartesian orthogonal coordinates of the mid-plane of the plate, while $\xi_3$ is the transversal coordinate, positive in the inward direction.

We also will assume the existence of an initial geometric imperfection $\overset{0}{V}_3 (\xi_1, \xi_2, \xi_3) = \overset{o}{v}_3 (\xi_1, \xi_2)$ that refers to the transverse displacement in the unstressed configuration of the panel.

In the context of the 3-D geometrically non-linear elasticity theory, the strain-displacement relationships in Lagrangian description specialized to the case of the von-Kármán's approximation is:

$$2e_{ij} = V_{i,j} + V_{j,i} + V_{3,i}V_{3,j} + \overset{o}{V}_{3,i} V_{3,j} + V_{3,i} \overset{o}{V}_{3,j}, \quad (i, j = 1, 2, 3) \tag{2}$$

where $V_i \equiv V_i(\xi_1, \xi_2, \xi_3, t)$ denote the 3-D displacement components. By convention, the transverse deflection $V_3(\xi_1, \xi_2, \xi_3, t) \equiv v_3(\xi_1, \xi_2, t)$ is measured from the imperfect surface, in the positive inward direction. Consistent with the Kirchhoff's hypothesis, the expression of the strain tensor components is

$$e_{\alpha\beta} = \epsilon_{\alpha\beta} + \xi_3 \kappa_{\alpha\beta}, \tag{3}$$

where

$$2\epsilon_{\alpha\beta} = v_{\alpha,\beta} + v_{\beta,\alpha} + v_{3,\alpha}v_{3,\beta} + v_{3,\alpha} \overset{o}{v}_{3,\beta} + \overset{o}{v}_{3,\alpha} v_{3,\beta}, \tag{4a}$$

and

$$\kappa_{\alpha\beta} = -v_{3,\alpha\beta}, \tag{4b}$$

define the membrane and bending strains, respectively. Herein $v_\alpha$ and $v_3$ denote the in-plane and transverse displacement components, respectively, while $(\cdot)_{,\alpha} \equiv \partial(\cdot)/\partial\xi_\alpha$. In addition, the Greek and Latin indices range from 1 to 2, and from 1 to 3, respectively.

The compatibility equation that will be useful in the formulation of the governing equations, obtained by eliminating $v_\alpha$ in Eq. (4a) is given by:

$$\epsilon_{11,22} + \epsilon_{22,11} - 2\epsilon_{12,12} + v_{3,11}v_{3,22} - v_{3,12}v_{3,12} + \overset{o}{v}_{3,11} v_{3,22} + \overset{o}{v}_{3,22} v_{3,11} - 2 \overset{o}{v}_{3,12} v_{3,12} = 0. \tag{5}$$

Upon retaining the transverse inertia term only, the equations of motion, are expressed as:

$$L_{\alpha\beta,\beta} = 0, \tag{6a}$$

$$L_{\alpha\beta}(v_{3,\beta} + \overset{o}{v}_{3,\beta})_{,\alpha} + M_{\alpha\beta,\alpha\beta} + q_3 = m_0 \ddot{v}_3. \tag{6b}$$

Herein $L_{\alpha\beta}$ and $M_{\alpha\beta}$ denote the stress-resultant and stress-couple components, respectively, $q_3 (\equiv q_3(\xi_1, \xi_2, t))$ the lateral pressure, $m_0$ the reduced mass term, while the superposed dots denote time derivatives.

Within the *eccentrically* and *smeared* stiffener concept which is adopted here (see e.g. Refs. 7 and 8), due to the asymmetry of the resulting panel, the constitutive equations feature bending-stretching coupling.

The relevant constitutive equations, where the above mentioned coupling terms are underscored by a discontinuous line, are displayed next:

$$\begin{aligned}
L_{11} &= A(\epsilon_{11} + \nu\epsilon_{22}) + \frac{E_s A_s}{\ell_s}(\epsilon_{11} + \underline{z_s \kappa_{11}}) + (h + \frac{A_s}{\ell_s})\tilde{\lambda}_s \overset{o}{T}, \\
L_{22} &= A(\epsilon_{22} + \nu\epsilon_{11}) + \frac{E_r A_r}{\ell_r}(\epsilon_{22} + \underline{z_r \kappa_{22}}) + (h + \frac{A_r}{\ell_r})\tilde{\lambda}_r \overset{o}{T}, \\
L_{12} &= 2A\epsilon_{12} \\
M_{11} &= -D(v_{3,11} + \nu v_{3,22}) + \frac{E_s}{\ell_s}(A_s \underline{z_s \epsilon_{11}} - I_s v_{3,11}) + \underline{\frac{F_s}{\ell_r}\tilde{\lambda}_s \overset{o}{T}}, \\
M_{22} &= -D(v_{3,11} + \nu v_{3,11}) + \frac{E_r}{\ell_r}(A_r \underline{z_r \epsilon_{ss}} - I_r v_{3,22}) + \underline{\frac{F_r}{\ell_r}\tilde{\lambda}_r \overset{o}{T}}, \\
M_{12} &= -D(1-\nu)v_{3,12}.
\end{aligned} \tag{7a-f}$$

In these equations $A$ and $D$ denote the stretching and bending stiffness quantities of the panel skin; $\nu$ denotes Poisson's ratio of the panel skin material; $z_s$, $z_r$ are referred to as the stiffener eccentricities and are defined as the distances from the stiffener centroidal axis to the plate mid-plane (positive for the inside stiffeners); $E_s$, $E_r$ are the moduli of elasticity of stiffeners $\ell_s$, $\ell_r$ denote the stiffener spacings measured between their axial lines; $A_s$, $A_r$ denote the stiffener cross-sectional areas $((A_r, A_s) = \int_{\xi_3} (\beta_r, \beta_s) d\xi_3))$; $I_s$, $I_r$ are the moments of inertia of stiffeners about the plate skin mid-plane; $(F_s, F_r) \equiv \int_{\xi_3} (\beta_s, \beta_r)\xi_3 d\xi_3)$ are the first moments of stiffeners about the mid-plane plate skin, where $\beta_s$ and $\beta_r$ denote the stiffener widths; $\tilde{\lambda} = -E\alpha/(1-\nu)$, where $E$ and $\alpha$ denote the Young's modulus and thermal expansion coefficient, which can be associated with the materials of plate skin and stiffeners, and $s$ and $r$ denote the quantities relevant to stiffeners running in the $\xi_1$ and $\xi_2$ directions, respectively. The main geometrical characteristics of stiffened panels are depicted in Fig. 1.

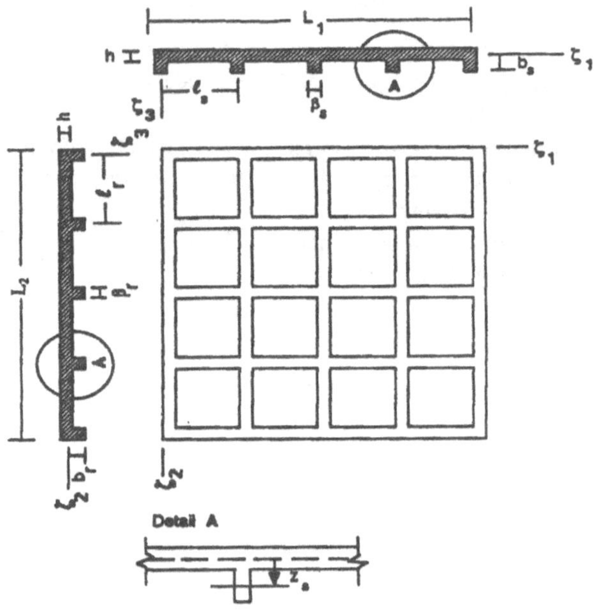

Figure 1: Geometry of the stiffened plate and coordinate system.

The coupling appearing in the constitutive equations will further be reflected in the governing equations, implying, among others, that a temperature field, uniformly distributed throughout the thickness of the panel and stiffeners will induce bending from the onset of heating. This reverts to the conclusion that in contrast to the case of *geometrically perfect panels*, (being at the same time non-stiffened, or symmetrically stiffened), in the present case, the panel will not exhibit the thermal buckling bifurcation in the classical Saint-Venant sense. It should be mentioned that consistent with the *smeared stiffener concept*, the stiffeners in both directions should be relatively closely spaced implying that the spacings $\ell_r$ and $\ell_s$ are assumed to be small enough.

## 2. Governing System

In the present study, the nonlinear dynamic equations governing the thermomechanical response of stiffened flat panels are represented in a form which generalizes the classical von Kármán-Marguerre nonlinear plate theory, in the sense that the relevant equations include the effects of initial geometric imperfections and the presence of uni/biaxial stiffeners. By using a similar procedure to that developed e.g. in Ref. 10, the governing equations are reduced to two partial differential equations in terms of the Airy stress function $F(\equiv F(\xi_1, \xi_2, t))$ and the transverse deflection $v_3(\xi_1, \xi_2, t)$ as:

$$\begin{aligned}
&\tilde{\mathcal{A}}_1 F_{,2222} + \tilde{\mathcal{A}}_2 F_{,1111} + 2\left(\tilde{\mathcal{B}} + \tilde{\mathcal{F}}\right) F_{,1212} \\
&- \left(\tilde{\mathcal{C}}_1 + \tilde{\mathcal{C}}_2\right) v_{3,1122} - \tilde{\mathcal{D}}_1 v_{3,2222} - \tilde{\mathcal{D}}_2 v_{3,1111} + v_{3,11} v_{3,22} \\
&- v_{3,12} v_{3,12} + \overset{\circ}{v}_{3,11} v_{3,22} + \overset{\circ}{v}_{3,22} v_{3,11} - 2\overset{\circ}{v}_{3,12} v_{3,12} \\
&+ \tilde{\lambda}\left(\Omega_1 \overset{\circ}{T}_{,22} + \Omega_2 \overset{\circ}{T}_{,11}\right) = 0,
\end{aligned} \tag{8a}$$

$$\begin{aligned}
&D\Delta\Delta v_3 - F_{,22}\left(v_{3,11} + \overset{\circ}{v}_{3,11}\right) - F_{,11}\left(v_{3,22} + \overset{\circ}{v}_{3,22}\right) \\
&+ 2F_{,12}\left(v_{3,12} + \overset{\circ}{v}_{3,12}\right) - \mathcal{O}_1 F_{,1122} - \mathcal{O}_2 F_{,1111} \\
&- \mathcal{O}_3 F_{,2222} + \mathcal{O}_4 v_{3,1122} + \mathcal{O}_5 v_{3,1111} + \mathcal{O}_6 v_{3,2222} \\
&- \tilde{\lambda}(\mathcal{O}_7 \overset{\circ}{T}_{,11} + \mathcal{O}_8 \overset{\circ}{T}_{,22}) - q_3 + m_0 \ddot{v}_3 = 0.
\end{aligned} \tag{8b}$$

These equations can be viewed as the extended counterpart for stiffened plates of the classical von-Kármán-Marguerre equations. Herein $\Delta\,(\cdot)$ denotes the 2-D Laplace operator, $m_0$ is the reduced mass of the reinforced plate, whereas the Airy's function, $F$ is defined as $L_{\alpha\beta} = c_{\alpha\omega} c_{\beta\delta} F_{,\omega\delta}$ where $c_{\alpha\beta}$ is the permutation symbol and the Einstein summation convention over repeated indices is implied. The coefficients appearing in Eqs. (8) can be found in Ref. 9.

A simple inspection of the linearized version of Eqs. (8) reveals that $\tilde{\mathcal{C}}_1, \tilde{\mathcal{C}}_2, \tilde{\mathcal{D}}_1, \tilde{\mathcal{D}}_2$ play the role of coupling stiffness quantities, in the sense that these are associated with the bending occuring in the stretching governing equation, Eq. (8a). On the other hand, the coefficients $\mathcal{O}_1$ through $\mathcal{O}_3$ are associated with the stretching occuring in the bending governing equation Eq. (8b); the coefficients $\mathcal{O}_4$ through $\mathcal{O}_6$ are associated with the newly induced bending terms occuring in the bending governing equations, whereas the coefficients $\mathcal{O}_7$ and $\mathcal{O}_8$ reflect the stretching influence induced by the membrane temperature in conjunction with the non-symmetric reinforcements, occuring in the bending governing equation. In view of their expression it is readily seen that in the absence of any reinforcement, these coefficients vanish, and the *linearized* counterpart of Eqs. (8a) and (8b) become decoupled.

In the present study the edges are considered to be simply supported. It is supposed also that the tangential motion of the panel in the normal direction to the edge is unconstrained, i.e. the edges are freely movable. For these boundary conditions, the transverse displacement at each edge, the tangential stress resultant and the bending stress-couple are zero- valued quantities. Denoting by $n$ and $t$ the in-plane directions normal and tangential to the panel edge, the pertinent boundary conditions are expressed as:

$$v_3 = 0; \; M_{nn} = 0; \; L_{nt} = L_{nn} = 0. \tag{9}$$

where, when $n = 1, t = 2$ and vice-versa.

## 3. Solution of the Nonlinear Equations

The nonlinear boundary-value problem in the present study is solved using Galerkin's method. First, the transverse deflection $v_3$ is expressed in terms of functions that satisfy the simply supported boundary conditions

$$v_3(\xi_1, \xi_2, t) = f_{mn}(t) \sin \lambda_m \xi_1 \sin \mu_n \xi_2, \tag{10}$$

where $\lambda_m = m\pi/L_1$, $\mu_n = n\pi/L_2$ and $f_{mn}$ (t) are the modal amplitudes, whereas the initial geometric imperfection $\overset{o}{v}_3$ is expressed as

$$\overset{o}{v}_3\ (\xi_1, \xi_2) = \overset{o}{f}_{mn} \sin \lambda_m \xi_1 \sin \mu_n \xi_2, \tag{11}$$

where $\overset{o}{f}_{mn}$ are the modal amplitudes of the initial geometric imperfection shape. Similarly, the applied temperature and lateral pressure are most generally represented by Navier-type double Fourier sine series. In the present study, the temperature and lateral pressure are represented as:

$$\overset{o}{T}\ (\xi_1, \xi_2) = \overset{o}{T}_{mn} \sin \lambda_m \xi_2 \sin \mu_n \xi_2, \tag{12a}$$

$$q_3(\xi_1, \xi_2) = q_{mn} \sin \lambda_m \xi_1 \sin \mu_n \xi_2. \tag{12b}$$

where $\overset{0}{T}_{mn}$ and $q_{mn}$ are the modal amplitudes of $\overset{0}{T}$ and $q_3$, respectively.

The displacement and temperature expansions are substituted into the Eq. (8a), and the Airy's stress function is obtained by solving the resulting linear non-homogeneous partial differential equation (see in this sense Ref. 10). Its solution is

$$F(\xi_1, \xi_2, t) = S_1 \cos 2\lambda_m \xi_1 + S_2 \cos 2\mu \xi_2 + S_3 \sin\lambda_m \xi_1, \sin\mu_n \xi_2$$
$$+ \sum_{p,q} D_{pq} \sin \lambda_p \xi_1 \sin\mu_q \xi_2, \tag{13}$$

where the expressions of the coefficients are not displayed here.

The remaining nonlinear partial differential equation, Eq. (8b), is converted into a set of nonlinear ordinary algebraic equations using Galerkin's method. To this end, substitution of $v_3, \overset{o}{v}_3, F, \overset{o}{T}$ and $q_3$ as expressed by Eqs. (10), (11), (13) and (12), respectively, in (8b) followed by its multiplication by $\sin \lambda_p \xi_1 \sin \mu_r \xi_2$ and integration of the obtained equation over the panel area yields the following set of $M \times N$ nonlinear ordinary differential equations for each set of wave forms determined by the index pair $(m, n)$

$$A_{mn}\ddot{f}_{mn} + R_{mn} f_{mn} + q_{mn} B_{mn} + P_1[f_{mn}, \overset{o}{f}_{mn}] + \hat{P}_1[f_{mn}, \overset{o}{f}_{mn}, \overset{o}{T}_{mn}]$$
$$+ P_2[f^2_{mn}, \overset{o}{f}_{mn}] + P_3[f^3_{mn}, \overset{o}{f}_{mn}] = 0, \tag{14}$$

where the indices $m$ and $n$ are not summed and have the values $m = 1, 2, ..., M$ and $n = 1, 2, ..., N$.

In equation (14) $P_1$ and $\hat{P}_1$, are linear, while $P_2$ and $P_3$ are quadratic and cubic polynomials in the unknown modal amplitudes $f_{mn}(t)$, respectively. The coefficients $A_{mn}$, $B_{mn}$ and $R_{mn}$ are constants that depend on the material and geometric properties of the plate and stiffeners. The explicit form of Eq. (14) is not displayed in this paper.

## 4. Vibrational Behavior About a Mean Static Equilibrium Configuration

Following the procedure used in a number of previoius papers, (see Refs. 11 and 12), the unknown modal amplitudes are expressed as

$$f_{mn}(t) = \overline{f}_{mn} + \hat{f}_{mn}(t) \tag{15}$$

where $\hat{f}_{mn}(t)$ represents small vibrations about a mean static equilibrium configuration described by $\overline{f}_{mn}$.

In this equation, the time-dependent part $\hat{f}_{mn}$ is considered small as compared to $\overline{f}_{mn}$ and $\mathring{f}_{mn}$, in the sense of

$$\hat{f}^2_{mn} << (\overline{f}_{mn}, \mathring{f}_{mn}) \tag{16}$$

The equations for the static prebuckling and postbuckling equilibrium states are obtained by discarding the inertia terms given by $A_{mn}\ddot{f}_{mn}$ in equation (14), and recognizing that the solution to the resulting equation is $\overline{f}_{mn}$. The equations for small vibrations about the static equilibrium state are then obtained by substituting equation (15) into equation (14) and enforcing the smallness condition given by equation (16). The resulting equations of motion are

$$A_{mn}\ddot{\hat{f}}_{mn}(t) + G_{mn}\hat{f}_{mn}(t) = 0 \tag{17}$$

where

$$G_{rs} = G_{mn}(\overline{f}_{mn}, \overline{f}^2_{mn}, \overline{f}^3_{mn}, \mathring{f}_{mn}, q_{mn}, \mathring{T}_{mn}) \tag{18}$$

for values of $m = 1, 2, ..., M$ and $n = 1, 2..., N$. The constant coefficients $A_{mn}$ are functions of the material and the geometric properties of the panel.

Equations (17) govern the small vibrations about a given static equilibrium state and are solved for synchronous motion by expressing

$$\hat{f}_{mn}(t) = \tilde{f}_{mn} \exp(i\omega_{mn}t), \qquad (i = \sqrt{-1}) \tag{19}$$

Substitution of equation (19) into (17) yields an algebraic eigenvalue problem given by

$$\tilde{f}_{mn}(G_{mn} - \omega^2_{mn}A_{mn}) = 0 \tag{20}$$

for values of $m = 1, 2, ..., M$ and $n = 1, 2, ..., N$. The frequencies $\omega_{mn}$ in equation (20) are the unknown quantities to be determined, and the corresponding amplitudes $\tilde{f}_{mn}$ are indeterminate.

The static equilibrium configuration for a given flat panel is obtained by solving the static counterpart of the nonlinear algebraic equations given by Eqs. (14) via Newton's method. Every solution to Eqs. (14) represents a possible stable or unstable equilibrium configuration. The stability of each equilibrium configuration is determined by evaluating the second variation of the total potential energy of the panel.

**5. Possibilities to Use the Derived Governing System and Solution Methodology as to Investigate the Supersonic Flutter and Postflutter of Stiffened Panels**

The postflutter behavior is a result of the coupling between bending and stretching induced by the large deflection of the panel. The balance of the destabilizing dynamic pressure and stabilizing membrance stresses results in a sustained limit-cycle motion. However, depending upon the specific geometric and elastic characteristics of the panel, and/or the level of the supersonic flight measured in terms of the flight Mach number, a violent transition from the undisturbed equilibrium state to finite motion may occur even at the pre-flutter flight velocities. The flutter boundary is correspondingly referred to as benign (soft) or dangerous (hard), depending on whether the transgression of it is accompanied by a monotonous increase of the oscillation amplitude, or by an explosive failure of the structure, respectively. It clearly appears that identification of the circumstances under which the latter type of flutter occurs constitutes a vital problem in the aeroelastic design of supersonic/hypersonic flight vehicles.

The increased flexibility of next generation of high speed space vehicles compounded with the high temperatures induced by aerodynamic heating can result not only in the diminishion of the flutter speed, but in a conversion of the flutter boundary from benign to catastrophic.

In addition to the previously mentioned circumstances, the *nonlinear unsteady aerodynamic* loading occuring at high supersonic Mach numbers can further contribute to the conversion of the flutter boundary to a catastrophic one. In order to address the problem of the postflutter behavior, the geometrically nonlinear equations of the panel, considered together with the nonlinear unsteady aerodynamic loads at high Mach numbers should be considered.

For the approach of the supersonic/hypersonic flutter of stiffened reinforced panels we assume that the flow takes place over the upper face of the reinforced panel, in the direction parallel to the $\xi_1$-coordinate. In these conditions the unsteady aerodynamic load obtained within the third-order piston aerodynamic theory can be expressed as $q_3(\xi_1, \xi_2, t) = \mathcal{A}(v_3 + \overset{\circ}{v}_3)$, where $\mathcal{A}(\cdot)$ is the aerodynamic operator defined as

$$\mathcal{A}(\cdot) = -\kappa p_\alpha \left[ \frac{1}{a_\infty} \left( \frac{\partial(\cdot)}{\partial t} + U_\alpha \frac{\partial(\cdot)}{\partial \xi_1} \right) + \frac{\kappa+1}{4} M^2 \left( \frac{\partial(\cdot)}{\partial \xi_1} \right)^2 + \frac{\kappa+1}{12} M^3 \left( \frac{\partial(\cdot)}{\partial \xi_1} \right)^3 \right], \quad (21)$$

$U_\infty$ being the velocity of the undesturbed flow, $M (\equiv U_\infty / a_\infty)$ is the flight Mach number, $a_\infty$ and $p_\infty$ the undisturbed velocity of sound and pressure, while $\kappa$ is the polytropic gas coefficient.

The method previously presented can be used to address also the linear and nonlinear flutter of stiffened panels. However, in contrast to the representations (10) in this case, as it was documented in the specialized literature (see e.g Refs. 10 and 13), for simply supported boundary conditions, the transversal deflection should be expressed as:

$$v_3(\xi_1, \xi_2, t) = \sin \mu_n \xi_2 \sum_m f_{mn}(t) \sin \lambda_m \xi_1. \quad (22)$$

Further, the procedure to determine the Airy's function and the nonlinear, system of equations, similar to that provided in the absence of aerodynamic flow should be followed.

Moreover, representation (15) will be used, where $\bar{f}_{mn}$ and $\hat{f}_{mn}(t)$ provide in this case the thermoaeroelastic static equilibrium and the self-excited aeroelastic vibration, respectively. There is no question that in this case the counterpart of Eq. (14) will contain the flight speed (in terms of the Mach number).

The aeroelatic counterpart of Eq. (20) can supply, when is linearized, the flutter boundary, and in the nonlinear form the postflutter behavior as influenced by the stiffeners, the temperature field and the external load.

## 6. Numerical Simulations and Discussion

Using the nonlinear governing equations, an assessment of the effects played by the uniform through the thickness temperature field and a lateral pressure on the frequency-temperature interaction of stiffened, geometrically imperfect flat panels will be accomplished. Throughout the numerical applications, the case of a simply supported panel of a square planform ($L_1 = L_2 \equiv L$) is considered. One also assumes that both the panel and stiffeners are of aluminum i.e. that $E_r \equiv E_s \equiv E = 10.4 \times 10^6$psi ($\equiv 73GPa$), $\nu = 0.32$ and $\alpha = 13.15 \times 10^{-6}$in/in/$^0F$($\equiv 23 \times 10^{-6}$ mm/mm/$^0C$). It is also assumed, unless otherwise specified, that $L_1/h = 40$. For an uniaxially stiffened panel, say in the $\xi_1$ - direction, one should consider in the $\xi_2$ - direction that $b_r = \beta_r = 0$ and$\ell_r \Longrightarrow \infty$.

In the displayed plot $(\delta_{11}, \overset{\circ}{\delta}_{11}) = (f_{11}, \overset{\circ}{f}_{11})/h$, denote the dimensionless amplitude in the mode (1,1) of the transversal deflection and initial geometric imperfection, evaluated at the center, $\xi_1 = \xi_2 = L/2$ of the panel, while $\tilde{q}_{11} = q_{11}L_1^4/(Dh)$ denotes the dimensionless amplitude of mode (1,1) of the lateral pre-load.

The response of reinforced panels to a temperature rise measured in $^\circ F$ , as well as the effects of the amplitude of the membrane temperature $\overset{0}{T}_{11}$ evaluated at the center of the panel, $\xi_1 = \xi_2 = L/2$ and of the ratio $b_s/h$ of uniaxial stiffeners on the dimensionless fundamental vibration frequency (squared) $\overline{\omega}^2(\equiv \omega^2 m_0 L_1^4/\pi^4 D))$ of geometrically perfect unloaded panel are displayed in Fig. 2. In this figure a number of three scenarios, related with the values of $b_s/h$ in the sequence $b_s/h = 3, 4$ and 5 are considered.

An important trend emerging from this plot lies on the fact that, even in the presence of a uniform through thickness temperature field and in the absence of the pre-load, the geometrically perfect *stiffened panel* does not experience buckling bifurcation. In other words, under such conditions and in contrast to the case of an unstiffened panel, a stiffened one will deflect from the onset of the temperature rise.

Another results is concerned with the implication of the stiffener height $\hat{b}_s(\equiv b_s/h)$ on the non-linear response. As it becomes evident, the increase of $\hat{b}_s$, results in a noticeable increase of the thermal loading capacity of the panel.

The results displayed in the companion graph, Fig. 3, obtained for the same geometric and physical conditions reveal that the fundamental frequency squared decreases linearly with increasing $\overset{0}{T}_{11}$, reaches a minimum and afterwards increases monotonically with the further temperature rise. This trend is due to the stiffening caused by the increased participation of the membrane stiffness as the deflection becomes larger. From the same graph it becomes evident that: i) the eigenfrequencies (corresponding to $\overset{0}{T}_{11} = 0$)

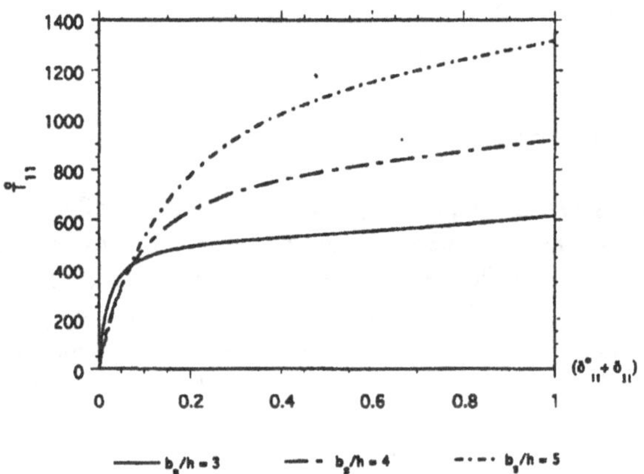


Figure 2: Effect of the uniaxial stiffener hight $\hat{b}_s(\equiv b_s/h)$ on the nonlinear response of a geometrically prefect plate under a temperature rise

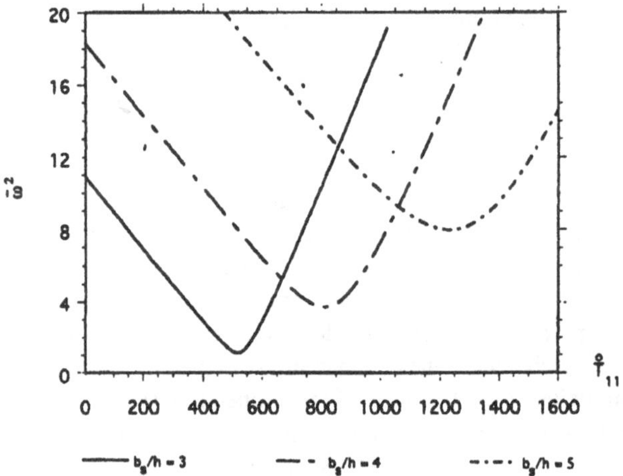


Figure 3: Frequency-temperature interaction $(\bar{\omega}^2 - \overset{\circ}{T}_{11})$ for the panel descibed in Fig. 2.

increase with the increase of $\hat{b}_s$, ii) with the increase of $\hat{b}_s$, a shift of the minimum of $\overline{\omega}^2$ toward larger temperature amplitudes, accompanied by an increase of the frequency squared, is experienced.

In Fig. 4 the case of a square uniaxially reinforced plate was considered, featuring the characteristics $\hat{b}_s = 2.5, \hat{\beta}_s(\equiv \beta_s/h) = 1$. In the plot, various scenarios indicated by the characteristics provided in the brackets $(L_1/h, \tilde{q}_{11})$ in the same sequences have been provided. Here $\tilde{q}_{11}$ is the amplitude of the transversal load applied to the plate in the conditions of $T_{11} = 0$. The results emerging from this plot reveal that the thermal loading capacity increases significantly with the decrease of the ratio $L_1/h$. In addition, as is readily seen, the transversal load plays a similar role as an initial geometric imperfection. It is also seen that even in the absence of the transversal load, due to the inherent bending-stretching coupling induced by the asymmetry of the structure, the panel does not exhibit the buckling bifurcation in the Saint-Venant sense.

The dynamic counterpart of Fig. 4 is displayed in Fig. 5. Due to the absence of buckling bifurcation, with the increase of the temperature the frequency does not become a zero-valued quantity.

Moreover, this plot reveals that for smaller $L_1/h$, the minimum frequncy occurs a larger temperatures as compared to the case of larger $L_1/h$. Moreover, also in this context the lateral pre-load plays a similar role to an initial geometric imperfection.

In Figs. 6 and 7 the effects of the parameters contained in the brackets $(\beta_s/h,\ b_s/h)$ on the thermal loading carrying capacity and the freqency-temperature interaction are revealed. One considers the uniaxially stiffened square panel without pre-load featuring $L_1/\ell_s = 5$. The values of the parameters in the brackets characterize the various curves in Figs. 6 and 7.

The results reveal that the increase of the relative hight of the stiffener $\hat{b}_s$ yields a larger increase of the load carrying capacity than that induced by the increase of $\hat{\beta}_s$.

The same trend occurs in the case of the frequency-temperature interaction, in the sense that the increase of $\hat{b}_s$ has a more powerful effect toward the increase of the fundamental frequencies than the increase of the parameters $\hat{\beta}_s$.

In Figs. 8 and 9 one considers the case of a square panel featuring $L_1/h = 35, b_s/h = 2.5, \beta_s/h = 1$, that is subjected to a pre-load of amplitude $\tilde{q}_{11} = 10$. Both case of uniaxially and biaxially reinforced panels are included in the analysis. In the latter case, $b_s = b_r, \beta_s = \beta_r$. The following scenorios identified by the abbreviations (R1, $\overset{\circ}{\delta}_{11}$) and (R2, $\overset{\circ}{\delta}_{11}$) are displayed, where R1 and R2 concern the uniaxially and biaxially reinforced panels, respectively.

Figure 8 reveals that a biaxially reinforced panel can carry a larger temperature panel than a uniaxially reinforced one.

The results from Fig. 9 show that the initial geometic imperfection plays on the frequency-temperature interaction, a similar role to that of the lateral pre-load. It is interesting also to see than an initial geometric imperfection of fixed amplitude plays a strong role toward increasing the minimum fundamental frequency in bio-axially reinforced panels than in their uniaxially reinforced panel counterparts.

Finally, is worthwhile to recall a results obtained in Ref. 14 according to which the smeared out theory of stiffened panels can not be replaced by an equivalent orthotropic

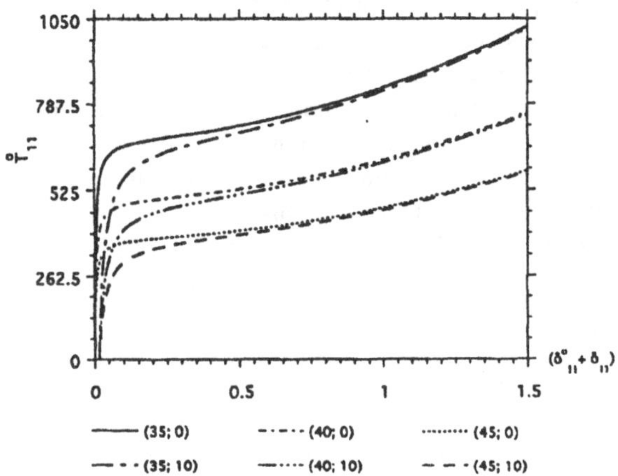


Figure 4: Effect of the parameters $(L_1/h, \tilde{q}_{11})$ on the nonlinear response of a geometrically perfect uniaxially stiffened panel ($\hat{b}_s = 2.5$; $\hat{\beta}_s = 1$).

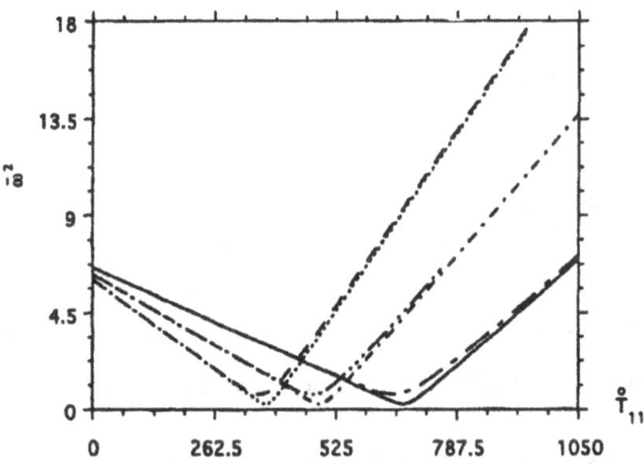


Figure 5: Frequency-temperature interaction for the panel described in Fig. 4.

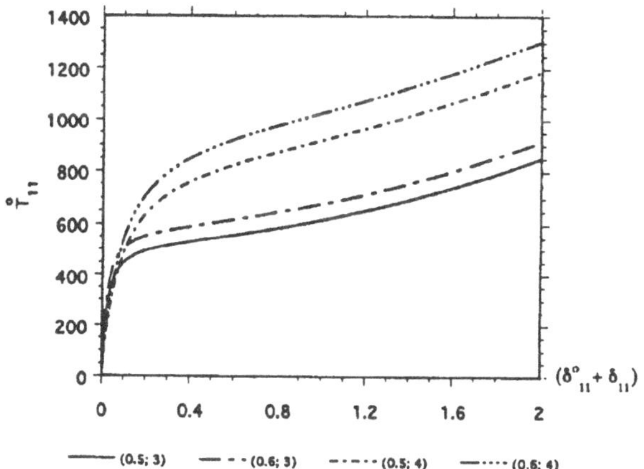


Figure 6: Influence of parameters included in the bracket, $(\hat{\beta}_s,\ \hat{b}_s)$ on the nonlinear response of a uniaxially reinforced panel. Geometrically perfect panel, $L_1/\ell_s = 5$.

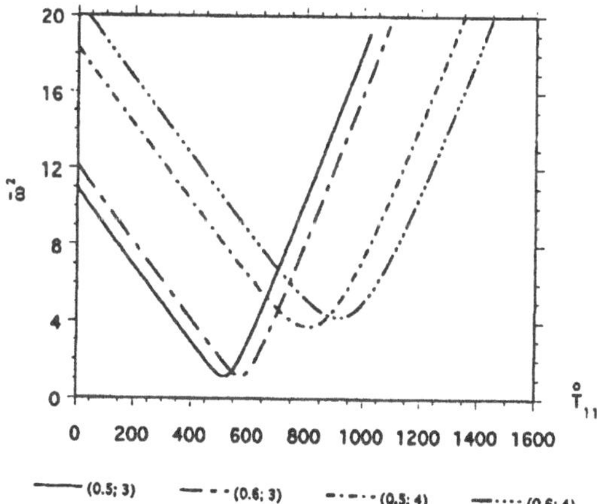


Figure 7: Frequency-temperature interaction for the panel described in Fig. 6. (The curves are identified by the values of the parameters included in the bracket $(\hat{\beta}_s,\ \hat{b}_s)$ and indicated in previous figure.

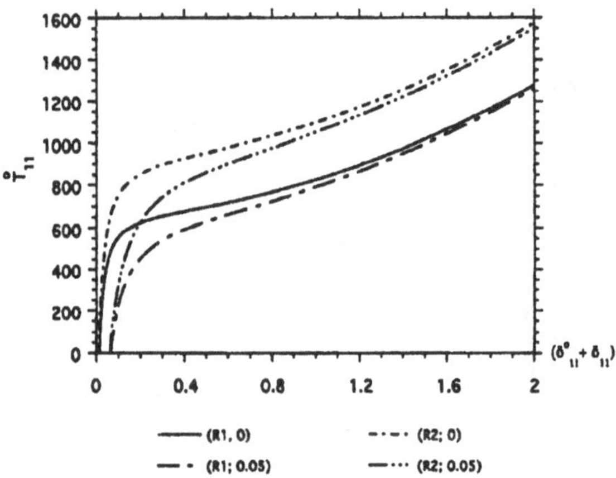


Figure 8: The effects of the uni (R1) and biaxially (R2) stiffeners, and of the initial geometric imperfection (R1, $\overset{\circ}{\delta}_{11}$), (R2, $\overset{\circ}{\delta}_{11}$) on the thermal load carrying capacity of the panel ($L_1/h = 35$).

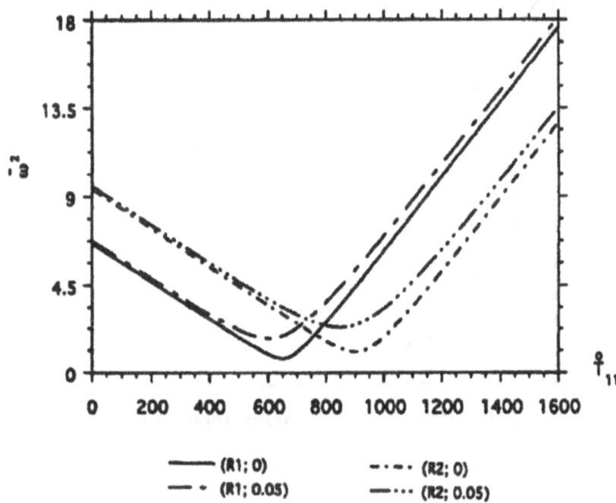


Figure 9: Frequency-temperature interaction for the panel described in Fig. 8. The curves are identified by the values of parameters included in the brackets (R1, $\overset{\circ}{\delta}_{11}$), (R2, $\overset{\circ}{\delta}_{11}$) indicated in previous figure.

theory. In this sense, the results concerning a comparison of predictions obtained by the application of the two theories have revealed significant differences, even in linear problems.

## 7. Conclusions

A parametric study of the non-linear static response and vibration behavior of eccentrically reinforced flat panels featuring initial geometric imperfections and subjected to thermal and mechanical loads has been presented. The loads considered in this study consists of a lateral pressure and a non-uniform membrane temperature field. Throughout this study simply-supported panels are considered. The results show that the reinforcements, initial geometric imperfections, and tranverse lateral pressure are all significant factors that should be considered in the dynamic design of panels subjected to a thermal field. Moreover, the beneficial contribution played by the reinforcements on the thermomechanical load carrying capacity and the vibrational behavior of flat panels was also put into evidence.

Apart from this, several elements enabling one to address various problem of the supersonic/hypersonic flutter instability of reinforced flat panels subjected to a membrane temperature and a mechanical pre-load have also been presented.

It is hoped that the results of this study will be useful toward a better understanding of the effects induced by the reinforcements on the thermal load carrying capacity and dynamic behavior of flat panels, and provide a basis for the approach of the supersonic/hypersonic linear flutter and postflutter of stiffened panels under a temperature and transversal load.

## 8. References

Noor, A.K. and Burton, W.S., "Computational Models for High-Temperature Multilayered Composite Plates and Shells," *Applied Mechanics Reviews*, 45 (10), 414-446, (1992).

Tauchert, T. R., (1991), "Thermally Induced Flexure, Buckling and Vibration of Plates," *Applied Mechanics Reviews*, 44(8), 347-360.

Librescu, L. and Lin. W., "Thermomechanical Postbuckling of Plates and Shells Incorporating Non-Classical Effects," *Thermal Stresses IV*, Editor, R.B. Hetnarski, Elsevire, Amsterdam, Lausanne, New York, Oxford, Shanon, Tokyo, 379-452, (1996).

Argyris, J. and Tenek, L., "Recent Advances in Computational Thermostructural Analysis of Composite Plates and Shells with Strong Nonlinearities," *Applied Mechanics Reviews*, Vol. 50, 5, 285-306 (1997).

Bismark-Nasr, M. N., "Structural Dynamics in Aeronautical Engineering," AIAA, Inc., Reston, VA, 1999.

Lee, D. M. and Lee, I.,"Vibration Analysis of Stiffened Laminated Plates Including Thermally Postbuckling Deflection Effect," *J. Reinforced Plastics and Composites*, **16**, 12, pp. 1138-1154, (1997).

Bogdanovich, A.E., "Nonlinear Problems of the Dynamic Buckling of Reinforced Laminar Cylindrical Shells," *Prikladnaya Mechanika*, Vol 22, 8, 57-66, 1986, Plenum Press, (1987).

Birman, V. and Bert, C. W., , "Dynamic Stability of Reinforced Composite Cylindrical Shells in Thermal Fields," *Journal of Sound and Vibration*, **142**, 2, 183-190, (1990).

Librescu, L. and Souza, M. A., "Non-Linear Response of Geometrically Stiffened Flat Panels Under Thermomechanical Loadings," *Journal of Thermal Stresses*, Vol. 21, Jan., No. 1, 3-19, (1998).

Librescu, L., *Elasto-Statics and Kinetics of Anisotropic and Heterogeneous Shell-Type Structures*, Noordhoff International Publ., The Netherlands, Leyden, (1975).

Librescu, L., Lin, W., Nemeth, M.P. and Starnes, Jr., J.H., (1996a), "Frequency-Load Interaction of Geometrically Imperfect Curved Panels Subjected to Heating," *AIAA Journal*, Vol. 34, No. 1, January, pp. 166-177.

Librescu, L., Lin, W., Nemeth, M.P. and Starnes, Jr., J.H., (1996b), "Vibration of Geometrically Imperfect Panels Subjected to Thermal and Mechanical Loads," *Journal of Spacecraft and Rockets*. Vol. 33, No. 2, March-April, pp. 285-291.

Librescu, L., "Aeroelastic Stability of Orthotropic Heterogeneous Thin Panels in the Vicinity of the Flutter Critical Boundary," (I) *Journal de Mécanique,* 4, 1, pp. 51-76, Paris, (1965).

McElman, J. A., Mikulas, M. M. and Stein, M. "Static and Dynamic Effects of Eccentric Stiffening of Plates and Cylindrical Shells," *AIAA Journal*, Vol. 4, No. 5, pp. 887-894, (1966).

# Bending and Twisting Effects in the Three-Dimensional Finite Deformations of an Inextensible Network

CHENG LUO and DAVID J. STEIGMANN
*Department of Mechanical Engineering, University of California*
*Berkeley, CA 94720-1740, USA*

**Abstract.** A theory is presented for bending and twisting effects in three-dimensional deformations of an inextensible network. The networks are modeled as material surfaces endowed with kinematical variables representing bending and non-standard fiber twisting effects. By using the minimum-energy principle, the Euler-Lagrange equations and boundary conditions are derived. Also, the compatibility conditions are obtained. Finally, the Euler-Lagrange equations are simplified and then specialized to obtain the equilibrium equations of Wang and Pipkin (1986a) and those for an inextensible rod.

## 1. Introduction

Recently, Simmonds (1985, §2) considered elastic surfaces with resistance to strain and flexure. Wang and Pipkin (1986a, b) considered inextensible nets with bending stiffness. In a series of papers by Hilgers and Pipkin (1992a,b, 1993, 1996), the theory of elastic sheets was developed independently by introducing the second derivatives of the deformation as well as the first derivatives into the strain-energy density. Hilgers (1997) also examined dynamic effects.

In Luo and Steigmann (2000), a theory is established by considering the effects of bending and twisting in an extensible network for three-dimensional deformations. In the present work we elucidate the structure of the simplest purely mechanical theory of networks that models both bending and twisting effects for finite deformations in 3-space. The model may be viewed as a generalized plate/shell theory. The network is composed of two families of inextensible elastic fibers. We assume that the cross sections of each fiber remain plane and suffer no deformation. We assign a triad of embedded orthonormal vectors to every material point of the fiber line to characterize the properties and orientations of the cross section at that point. By using these vectors we give a clear physical interpretation and mathematical representation of the curvature and twisting angle associated with each fiber. In Section 2, we establish the notation and discuss the kinematical and constitutive hypotheses. In Section 3, we postulate a particular form of the potential energy for conservatively-loaded networks and determine its variational form. We adopt variational techniques used in Steigmann (1996). In Section 4, by applying the minimum-energy principle to the potential energy, we derive the Euler-Lagrange equations and boundary conditions. We also derive the compatibility relations. In Section 5, the Euler-Lagrange equations are simplified and then reduced to obtain those of Wang and Pipkin (1986a) and those of an inextensible rod.



*D. Durban et al. (eds.), Advances in the Mechanics of Plates and Shells,* 213–228.

## 2. Kinematical and Constitutive Hypotheses

We consider a sheet of fabric that initially occupies a region $D$ with the boundary curve $C$ in the x-y plane. The sheet is composed of two families of inextensible fibers, which initially lie parallel to the x and y axes; thus every line x=constant or y=constant in $D$ is regarded as a fiber. The two families of fibers are orthogonal in the reference configuration. They are assumed to be continuously distributed and fastened together at their points of intersection to prevent slipping of one fiber family relative to the other. The sheet is treated as a continuum.

We use $\mathbf{R}$ and $\mathbf{r}$ to denote the respective position vectors in the reference and current configurations. Let $\mathbf{A}$ and $\mathbf{B}$ be the unit vectors tangential to the fibers in the reference configuration, and $\mathbf{a}$ and $\mathbf{b}$ be the unit vectors tangential to the fibers on the deformed surface. Then

$$\mathbf{A} = \frac{\partial \mathbf{R}}{\partial x}, \quad \mathbf{B} = \frac{\partial \mathbf{R}}{\partial y}, \quad \mathbf{a} = \frac{\partial \mathbf{r}}{\partial x}, \quad \mathbf{b} = \frac{\partial \mathbf{r}}{\partial y}. \tag{2.1}$$

The vector $\mathbf{A}$ is tangential to the curve occupied by a fiber y=constant in the reference configuration and $\mathbf{B}$ is tangential to a fiber x=constant. We call the fibers in the reference configuration $\mathbf{A}$- and $\mathbf{B}$-lines, respectively. Likewise, the vectors $\mathbf{a}$ and $\mathbf{b}$ are, respectively, tangential to the deformed $\mathbf{A}$- and $\mathbf{B}$-lines. We call the deformed $\mathbf{A}$- and $\mathbf{B}$-lines, respectively, $\mathbf{a}$- and $\mathbf{b}$-lines. In this paper, Greek indices, except $\alpha$ and $\beta$, range from 1 to 2, Latin indices, except a and b, range from 1 to 3, and the summation convention for repeated indices is employed. The Greek indices $\alpha$ and $\beta$ range from 2 to 3. The Latin index a or b will be used as a superscript to indicate that the parameters are associated with $\mathbf{a}$- or $\mathbf{b}$-lines. They are not the summation indices when they are repeated in an expression.

After deformation, the point at $\mathbf{R}(x,y)$ displaces to $\mathbf{r}(x,y)$ and each fiber may suffer twisting. We associate two orthonormal bases $\{\mathbf{A}_i(x,y)\}$ and $\{\mathbf{B}_i(x,y)\}$ with $\mathbf{A}$- and $\mathbf{B}$-lines, respectively. They satisfy $\mathbf{A}_1 = \dfrac{\partial \mathbf{R}}{\partial x}$, $\mathbf{A}_1 \bullet \mathbf{A}_2 \times \mathbf{A}_3 = 1$, $\mathbf{B}_1 = \dfrac{\partial \mathbf{R}}{\partial y}$, and $\mathbf{B}_1 \bullet \mathbf{B}_2 \times \mathbf{B}_3 = 1$. Then we invoke the Bernoulli-Euler hypotheses for each fiber: First we suppose that cross-sections of each fiber remain plane, suffer no strain, and are normal to the fiber in every configuration. We further assume that deformations from $\{\mathbf{R}, \mathbf{A}_i, \mathbf{B}_i\}$ to $\{\mathbf{r}, \mathbf{a}_i, \mathbf{b}_i\}$ are inextensional and orientation preserving. These hypotheses are equivalent to

$$\mathbf{a}_i \bullet \mathbf{a}_j = \delta_{ij}, \quad \mathbf{a}_i \bullet \mathbf{a}_j \times \mathbf{a}_k = e_{ijk}, \quad \mathbf{b}_i \bullet \mathbf{b}_j = \delta_{ij}, \quad \mathbf{b}_i \bullet \mathbf{b}_j \times \mathbf{b}_k = e_{ijk}, \tag{2.2}$$

where $\delta_{ij}$ is the Kronecker delta and $e_{ijk}$ is the permutation symbol, together with the nonholonomic constraints

$$\frac{\partial \mathbf{r}}{\partial x} = \mathbf{a}_1, \qquad \frac{\partial \mathbf{r}}{\partial y} = \mathbf{b}_1. \tag{2.3}$$

The two equations in (2.3) imply that x and y, respectively, measure arc lengths of $\mathbf{a}$ - and $\mathbf{b}$ - lines and $\mathbf{a}_1$ and $\mathbf{b}_1$, respectively, coincide with the unit tangents $\mathbf{a}$ and $\mathbf{b}$ in the current configuration. The two families of fibers are initially orthogonal. However after deformation they are not necessarily orthogonal. The fiber shear angle $\gamma$ is defined by

$$\sin\gamma = \mathbf{a}_1 \bullet \mathbf{b}_1. \tag{2.4}$$

The unit normal to the deformed sheet is

$$\mathbf{n} = \frac{\mathbf{a}_1 \times \mathbf{b}_1}{|\mathbf{a}_1 \times \mathbf{b}_1|}. \tag{2.5}$$

Next, let

$$\kappa_i^a = \frac{1}{2} e_{ijk} \frac{\partial \mathbf{a}_k}{\partial x} \bullet \mathbf{a}_j, \qquad \kappa_i^b = \frac{1}{2} e_{ijk} \frac{\partial \mathbf{b}_k}{\partial y} \bullet \mathbf{b}_j. \tag{2.6}$$

We will show that the twists and curvatures of the $\mathbf{a}$ - and $\mathbf{b}$ -lines can be expressed in terms of $\kappa_i^a$ and $\kappa_i^b$, respectively. Let us introduce the Frenet triad of each fiber line, and let $\{\mathbf{a}_1, \mathbf{n}^a, \mathbf{b}^a\}$ and $\{\mathbf{b}_1, \mathbf{n}^b, \mathbf{b}^b\}$ be the Frenet triads of the $\mathbf{a}$ - and $\mathbf{b}$ -lines, respectively. Then, we have

$$\frac{\partial \mathbf{a}_1}{\partial x} = \eta^a \mathbf{n}^a, \quad \frac{\partial \mathbf{n}^a}{\partial x} = -\eta^a \mathbf{a}_1 + \tau^a \mathbf{b}^a, \quad \frac{\partial \mathbf{b}^a}{\partial x} = -\tau^a \mathbf{n}^a, \tag{2.7}$$

$$\frac{\partial \mathbf{b}_1}{\partial y} = \eta^b \mathbf{n}^b, \quad \frac{\partial \mathbf{n}^b}{\partial y} = -\eta^b \mathbf{b}_1 + \tau^b \mathbf{b}^b, \quad \frac{\partial \mathbf{b}^b}{\partial y} = -\tau^b \mathbf{n}^b. \tag{2.8}$$

The symbols $\eta^a$ and $\eta^b$ in (2.7) and (2.8) denote the principal curvatures of the $\mathbf{a}$ - and $\mathbf{b}$ - lines, respectively. The symbols $\tau^a$ and $\tau^b$ in (2.7) and (2.8) represent the torsions of the $\mathbf{a}$ - and $\mathbf{b}$ -lines, respectively. It can be readily shown from (2.7) and (2.8) that

$$\eta^a = \frac{\partial \mathbf{a}_1}{\partial x} \bullet \mathbf{n}^a, \ \eta^b = \frac{\partial \mathbf{b}_1}{\partial y} \bullet \mathbf{n}^b. \tag{2.9}$$

Note that since the fibers are straight in the reference configuration, their curvatures there are identically zero. We now define an angle $\theta^a(x, y)$ of the $\mathbf{a}$ -line and an angle $\theta^b(x, y)$ of the $\mathbf{b}$ -line such that

$$\mathbf{a}_2 = \cos\theta^a \mathbf{n}^a + \sin\theta^a \mathbf{b}^a, \ \mathbf{a}_3 = -\sin\theta^a \mathbf{n}^a + \cos\theta^a \mathbf{b}^a, \tag{2.10}$$

$$\mathbf{b}_2 = \cos\theta^b \mathbf{n}^b + \sin\theta^b \mathbf{b}^b, \ \mathbf{b}_3 = -\sin\theta^b \mathbf{n}^b + \cos\theta^b \mathbf{b}^b, \tag{2.11}$$

where y in $\theta^a(x,y)$ and x in $\theta^b(x,y)$ assume constant values along the respective **a** - and **b**-lines. Substituting (2.10) and (2.11), respectively, into (2.6), and making use of (2.7), we find

$$\kappa_1^a = \tau^a + \frac{\partial \theta^a}{\partial x}, \quad \kappa_2^a = \sin\theta^a \eta^a, \ \kappa_3^a = \cos\theta^a \eta^a, \tag{2.12}$$

$$\kappa_1^b = \tau^b + \frac{\partial \theta^b}{\partial y}, \quad \kappa_2^b = \sin\theta^b \eta^b, \ \kappa_3^b = \cos\theta^b \eta^b. \tag{2.13}$$

Equations $(2.12)_2$, $(2.12)_3$, $(2.13)_2$ and $(2.13)_3$ lead to

$$\eta^a = \sqrt{(\kappa_2^a)^2 + (\kappa_3^a)^2}, \ \eta^b = \sqrt{(\kappa_2^b)^2 + (\kappa_3^b)^2}. \tag{2.14}$$

We use $\beta^a$ and $\beta^b$ to denote the respective twists of the **a** - and **b**-lines (see Art. 253 of Love (1927)). Then, we have

$$\beta^a = \kappa_1^a, \ \beta^b = \kappa_1^b. \tag{2.15}$$

Equations (2.14) and (2.15) give the explicit expressions for the twists and curvatures of the **a** - and **b** -lines in terms of $\kappa_1^a$ and $\kappa_1^b$, respectively. In the reference configuration, every fiber line is a plane curve and each fiber suffers no twisting. Therefore, in the reference configuration, the twists of every fiber are identically zero.

Next, we introduce another triad for each fiber line. In Section 4, where we derive the compatibility conditions, we will make use of this triad. Define auxiliary unit vectors

$$\mathbf{p}^a = \mathbf{n} \times \mathbf{a}_1, \ \mathbf{p}^b = \mathbf{n} \times \mathbf{b}_1 \tag{2.16}$$

associated with the current configuration. Then $\{\mathbf{a}_1, \mathbf{p}^a, \mathbf{n}\}$ and $\{\mathbf{b}_1, \mathbf{p}^b, \mathbf{n}\}$ are right-handed orthonormal bases for 3-space. In addition we define an angle $\vartheta^a(x,y)$ of the **a** -line and another angle $\vartheta^b(x,y)$ of the **b** -line such that

$$\mathbf{n}^a = \cos\vartheta^a \mathbf{p}^a + \sin\vartheta^a \mathbf{n}, \ \mathbf{b}^a = -\sin\vartheta^a \mathbf{p}^a + \cos\vartheta^a \mathbf{n}, \tag{2.17}$$

$$\mathbf{n}^b = \cos\vartheta^b \mathbf{p}^b + \sin\vartheta^b \mathbf{n}, \ \mathbf{b}^b = -\sin\vartheta^b \mathbf{p}^b + \cos\vartheta^b \mathbf{n}, \tag{2.18}$$

where y in $\vartheta^a(x,y)$ and x in $\vartheta^b(x,y)$ assume constant values along the respective **a** - and **b** -lines. Then, the partial derivatives of $\mathbf{a}_1$ and $\mathbf{b}_1$ can be expressed as

$$\frac{\partial \mathbf{a}_1}{\partial x} = \eta^a \cos\vartheta^a \mathbf{p}^a + \eta^a \sin\vartheta^a \mathbf{n}, \ \frac{\partial \mathbf{b}_1}{\partial y} = \eta^b \cos\vartheta^b \mathbf{p}^b + \eta^b \sin\vartheta^b \mathbf{n}, \tag{2.19}$$

$$\frac{\partial \mathbf{a}_1}{\partial y} = \phi^a \mathbf{p}^a + \tau \mathbf{n}, \quad \frac{\partial \mathbf{b}_1}{\partial x} = \phi^b \mathbf{p}^b + \tau \mathbf{n}. \tag{2.20}$$

The coefficients $\phi^a$ and $\phi^b$ in (2.20) are the so-called Tchebychev curvatures of $\mathbf{a}$- and $\mathbf{b}$-lines, respectively, according to the terminology of Kuznetsov (1982). The coefficient $\phi^a$ measures the tangential part of the rate of change of $\mathbf{a}_1$ with respect to arc length along a $\mathbf{b}$-line. The coefficient $\phi^b$ has a similar interpretation. The coefficient $\tau$ in (2.20) is the torsion of the deformed network and is defined by

$$\tau = \frac{\partial^2 \mathbf{r}}{\partial x \partial y} \bullet \mathbf{n}. \tag{2.21}$$

### 3. Potential Energy and its Variation

In this section, we will determine the potential energy of the sheet of fabric and its variation. Let $W = w(\sin\gamma, \eta^a, \eta^b, \beta^a, \beta^b)$ denote the strain energy per unit initial area, where we suppress the values of $\eta^a$, $\eta^b$, $\beta^a$, and $\beta^b$ at the reference configuration and the dependence of the function $w$ on position. We remark that the list of arguments of the function $w(\cdot)$ reflects that notion that each fiber is transversely isotropic with respect to its tangent (see (2.14)). Note that in this paper, the variables $w$, $\eta^a$, $\eta^b$, $\beta^a$, and $\beta^b$ at the reference configuration are assumed to be zero. The potential energy functional $E(\mathbf{r}, \mathbf{a}_i, \mathbf{b}_i)$ is of the form

$$\begin{aligned} E(\mathbf{r}, \mathbf{a}_i, \mathbf{b}_i) = & \iint_D w(\sin\gamma, \eta^a, \eta^b, \beta^a, \beta^b)\mathrm{dA} - \iint_D \mathbf{f} \bullet \mathbf{r}\mathrm{dA} \\ & - \oint_C (\mathbf{t} \bullet \mathbf{r} + \mathbf{m} \bullet \frac{\partial \mathbf{r}}{\partial n} + \mathbf{n}_1^a \bullet \mathbf{a}_2 + \mathbf{n}_2^a \bullet \mathbf{a}_3 + \mathbf{n}_1^b \bullet \mathbf{b}_2 + \mathbf{n}_2^b \bullet \mathbf{b}_3)ds, \end{aligned} \tag{3.1}$$

where $\mathrm{dA} = \mathrm{dxdy}$, the symbol s represents arc length along the boundary curve $C$, and $\frac{\partial \mathbf{r}}{\partial n}$ denotes the partial derivative of $\mathbf{r}$ with respect to arc length in the outward normal direction in the x-y plane. The prescribed functions $\mathbf{f}(x, y)$ and $\mathbf{t}(s)$ are used to represent dead loads with $\mathbf{f}(x, y)$ per unit initial area and $\mathbf{t}(s)$ per unit initial length on the boundary. The symbols $\mathbf{n}_1^a(s)$, $\mathbf{n}_2^a(s)$, $\mathbf{n}_1^b(s)$, and $\mathbf{n}_2^b(s)$ denote the dead loads per unit initial length on the boundary and are related to the boundary conditions for the twisting couples, which can be seen in Section 4. Using the same methods as those used in Section 3 of Hilgers and Pipkin [5], we can show that $\mathbf{f}(x, y)$ and $\mathbf{t}(s)$ must satisfy the equilibrium condition

$$\iint_D \mathbf{f}\mathrm{dA} + \oint_C \mathbf{t}\mathrm{ds} = \mathbf{0} \tag{3.2}$$

and $\mathbf{m}$ satisfies the equilibrium condition

$$\iint_D \mathbf{r} \times \mathbf{f} dA + \oint_C (\mathbf{r} \times \mathbf{t} + \frac{\partial \mathbf{r}}{\partial n} \times \mathbf{m}) ds = \mathbf{0}, \tag{3.3}$$

where (3.2) and (3.3), respectively, correspond to equations (3.8) and (3.9) in Hilgers and Pipkin (1992a). If we interpret $\frac{\partial \mathbf{r}}{\partial n} \times \mathbf{m}$ as a couple per unit length applied to the edge of the sheet, then $\mathbf{m}$ will give us the information about the couple $\frac{\partial \mathbf{r}}{\partial n} \times \mathbf{m}$ after the deformation has been determined.

Let $\varepsilon \in (-\varepsilon_0, \varepsilon_0)$ for some positive number $\varepsilon_0$, and consider a smooth one-parameter family of kinematically admissible configurations $\{\mathbf{r}^*(x, y; \varepsilon),\ \mathbf{a}_1{}^*(x, y; \varepsilon), \mathbf{b}_1{}^*(x, y; \varepsilon)\}$ with $\{\mathbf{r}^*(x, y;0),\ \mathbf{a}_1{}^*(x, y;0), \mathbf{b}_1{}^*(x, y;0)\} = \{\mathbf{r}(x, y),\ \mathbf{a}_1(x, y), \mathbf{b}_1(x, y)\}$. Here kinematic admissibility means that, for each fixed $\varepsilon$, $\mathbf{r}^*(\cdot;\varepsilon)$, $\mathbf{a}_1{}^*(\cdot;\varepsilon)$ and $\mathbf{b}_1{}^*(\cdot;\varepsilon)$ are at least piecewise $C^2$ and satisfy

$$\frac{\partial \mathbf{r}^*}{\partial x} = \mathbf{a}_1^*, \quad \frac{\partial \mathbf{r}^*}{\partial y} = \mathbf{b}_1^*. \tag{3.4}$$

If $\{\mathbf{r}, \mathbf{a}_1, \mathbf{b}_1\}$ is a minimizer of the energy, then we require that

$$E(\mathbf{r}^*, \mathbf{a}_1{}^*, \mathbf{b}_1{}^*) \geq E(\mathbf{r}, \mathbf{a}_1, \mathbf{b}_1). \tag{3.5}$$

Let superimposed dots denote derivatives of functions with respect to $\varepsilon$, evaluated at $\varepsilon = 0$. As $\varepsilon \to 0$, we have

$$\mathbf{r}^* = \mathbf{r} + \varepsilon \mathbf{u} + o(\varepsilon), \quad \mathbf{a}_1{}^* = \mathbf{a}_1 + \varepsilon \dot{\mathbf{a}}_1 + o(\varepsilon), \quad \mathbf{b}_1{}^* = \mathbf{b}_1 + \varepsilon \dot{\mathbf{b}}_1 + o(\varepsilon), \tag{3.6}$$

where

$$\mathbf{u} = \dot{\mathbf{r}}. \tag{3.7}$$

Let $\mathbf{Q}^{a^*}$ be the rotation that maps $\{\mathbf{A}_1\}$ onto $\{\mathbf{a}_1^*\}$:

$$\mathbf{Q}^{a^*} = \mathbf{a}_1^* \otimes \mathbf{A}_1. \tag{3.8}$$

Then

$$\frac{d}{d\varepsilon}\mathbf{a}_1^* = \alpha^{a^*}\mathbf{a}_1^* = \mathbf{c}^{a^*} \times \mathbf{a}_1^*, \tag{3.9}$$

where

$$\alpha^{a^*} = (\frac{d}{d\varepsilon}\mathbf{Q}^{a^*})(\mathbf{Q}^{a^*})^T \tag{3.10}$$

is a skew tensor and $\mathbf{c}^{a^*}$ is its axial vector. Consequently,

$$\dot{\mathbf{a}}_i = \mathbf{c}^a \times \mathbf{a}_i, \tag{3.11}$$

where $\mathbf{c}^a$ is the value of $\mathbf{c}^{a^*}$ at $\varepsilon = 0$.

Let $\mathbf{Q}^{b^*}$ be the rotation that maps $\{\mathbf{B}_i\}$ onto $\{\mathbf{b}_i^*\}$:

$$\mathbf{Q}^{b^*} = \mathbf{b}_i^* \otimes \mathbf{B}_i. \tag{3.12}$$

Then

$$\frac{d}{d\varepsilon}\mathbf{b}_i^* = \alpha^{b^*}\mathbf{b}_i^* = \mathbf{a}^{b^*} \times \mathbf{b}_i^*, \tag{3.13}$$

where

$$\alpha^{b^*} = (\frac{d}{d\varepsilon}\mathbf{Q}^{b^*})(\mathbf{Q}^{b^*})^T \tag{3.14}$$

is a skew tensor and $\mathbf{c}^{b^*}$ is its axial vector. Likewise, we obtain

$$\dot{\mathbf{b}}_i = \mathbf{c}^b \times \mathbf{b}_i, \tag{3.15}$$

where $\mathbf{c}^b$ is the value of $\mathbf{c}^{b^*}$ at $\varepsilon = 0$.

Moreover, with the aid of (3.7), (3.11) and (3.15), we establish the variational version of the constraints (2.3):

$$\frac{\partial \mathbf{u}}{\partial x} = \mathbf{c}^a \times \mathbf{a}_1, \quad \frac{\partial \mathbf{u}}{\partial y} = \mathbf{c}^b \times \mathbf{b}_1. \tag{3.16}$$

With the aid of (3.11) and (3.15), from (2.6) we obtain the variation of $\kappa_i$:

$$\dot{\kappa}_1^a = \mathbf{a}_1 \bullet \frac{\partial \mathbf{c}^a}{\partial x}, \quad \dot{\kappa}_2^a = \mathbf{a}_2 \bullet \frac{\partial \mathbf{c}^a}{\partial x}, \quad \dot{\kappa}_3^a = \mathbf{a}_3 \bullet \frac{\partial \mathbf{c}^a}{\partial x}, \tag{3.17}$$

and differentiation of $(2.12)_2$ and $(2.12)_3$ leads to

$$\dot{\eta}^a \sin\theta^a + \cos\theta^a \,\dot{\theta}^a \eta^a = \dot{\kappa}_2^a, \quad \dot{\eta}^a \cos\theta^a - \sin\theta^a \,\dot{\theta}^a \eta^a = \dot{\kappa}_3^a. \tag{3.18}$$

From (3.17), (3.18) and $(2.15)_1$, we obtain the variations of $\eta^a$ and $\beta^a$

$$\dot{\eta}^a = \mathbf{b}^a \bullet \frac{\partial \mathbf{c}^a}{\partial x}, \quad \dot{\beta}^a = \mathbf{a}_1 \bullet \frac{\partial \mathbf{c}^a}{\partial x}. \tag{3.19}$$

Likewise, we have

$$\dot{\eta}^b = \mathbf{b}^b \bullet \frac{\partial \mathbf{c}^b}{\partial y}, \quad \dot{\beta}^b = \mathbf{a}_1 \bullet \frac{\partial \mathbf{c}^b}{\partial y}. \tag{3.20}$$

Using (3.11), (3.15), and (3.16), from (2.4) we also obtain the variation of $\sin\gamma$

$$\dot{\sin\gamma} = \mathbf{b}_1 \bullet \frac{\partial \mathbf{u}}{\partial x} + \mathbf{a}_1 \bullet \frac{\partial \mathbf{u}}{\partial y}. \tag{3.21}$$

We further define

$$\mathbf{M}^a = \frac{\partial w}{\partial \eta^a}\mathbf{b}^a, \ \mathbf{M}^b = \frac{\partial w}{\partial \eta^b}\mathbf{b}^b, \ \mathbf{N}^a = \frac{\partial w}{\partial \beta^a}\mathbf{a}_1, \ \mathbf{N}^b = \frac{\partial w}{\partial \beta^b}\mathbf{b}_1. \tag{3.22}$$

Since $\mathbf{M}^a$ and $\mathbf{N}^a$ are directly related to $\eta^a$ and $\beta^a$ respectively, they can be interpreted as the bending and torsional moments on the cross sections of $\mathbf{a}$-lines. There is a similar interpretation for $\mathbf{M}^b$ and $\mathbf{N}^b$.

According to the multiplier rule of the calculus of variations (see Bliss (1946) and Elsgolts (1977)), an admissible configuration that renders $E(\cdot,\cdot,\cdot)$ stationary is also a stationary configuration for the functional

$$\hat{E}(\mathbf{r}, \mathbf{a}_1, \mathbf{b}_1) = E(\mathbf{r}, \mathbf{a}_1, \mathbf{b}_1) + \iint_D \mathbf{f}^a \bullet (\frac{\partial \mathbf{r}}{\partial x} - \mathbf{a}_1) + \mathbf{f}^b \bullet (\frac{\partial \mathbf{r}}{\partial y} - \mathbf{b}_1) dA, \tag{3.23}$$

where $\mathbf{f}^a$ and $\mathbf{f}^b$ are vectors of Lagrange multipliers.

Set

$$\mathbf{F}^a = \mathbf{f}^a + \frac{\partial w}{\partial \sin\gamma}\mathbf{b}_1, \ \mathbf{F}^b = \mathbf{f}^b + \frac{\partial w}{\partial \sin\gamma}\mathbf{a}_1. \tag{3.24}$$

Let $F(\varepsilon) = \hat{E}(\mathbf{r}^*, \mathbf{a}_1^*, \mathbf{b}_1^*)$, where $\mathbf{r}^*, \mathbf{a}_1{}^*$ and $\mathbf{b}_1{}^*$ are defined by (3.6). According to the stationary-energy principle, we require that

$$\dot{F} = 0. \tag{3.25}$$

With the help of (3.16), (3.19), (3.20), (3.22) and (3.24), from (3.1) we reduce the variation of $F$ to

$$
\begin{aligned}
\dot{F} &= \iint_D \frac{\partial w}{\partial \eta^a}\dot{\eta}^a + \frac{\partial w}{\partial \eta^b}\dot{\eta}^b + \frac{\partial w}{\partial \beta^a}\dot{\beta}^a + \frac{\partial w}{\partial \beta^b}\dot{\beta}^b + \frac{\partial w}{\partial \sin\gamma}\dot{\sin}\gamma \, dA \\
&+ \iint_D \mathbf{f}^a \bullet (\frac{\partial \mathbf{u}}{\partial x} - \mathbf{c}^a \times \mathbf{a}_1) + \mathbf{f}^b \bullet (\frac{\partial \mathbf{u}}{\partial y} - \mathbf{c}^b \times \mathbf{b}_1) dA - \iint_D \mathbf{f} \bullet \mathbf{u} dA \\
&- \oint_C \mathbf{t} \bullet \mathbf{u} ds - \oint_C \mathbf{m} \bullet \frac{\partial \mathbf{u}}{\partial n} ds - \oint_C (\mathbf{n}_1^a \bullet \dot{\mathbf{a}}_2 + \mathbf{n}_2^a \bullet \dot{\mathbf{a}}_3 + \mathbf{n}_1^b \bullet \dot{\mathbf{b}}_2 + \mathbf{n}_2^b \bullet \dot{\mathbf{b}}_3) ds \\
&= -\iint_D \mathbf{c}^a \bullet (\frac{\partial \mathbf{M}^a}{\partial x} + \frac{\partial \mathbf{N}^a}{\partial x} + \frac{\partial w}{\partial \sin\gamma} \mathbf{b}_1 \times \mathbf{a}_1 - \mathbf{F}^a \times \mathbf{a}_1) dA \\
&- \iint_D \mathbf{c}^b \bullet (\frac{\partial \mathbf{M}^b}{\partial y} + \frac{\partial \mathbf{N}^b}{\partial y} \frac{\partial w}{\partial \sin\gamma} \mathbf{a}_1 \times \mathbf{b}_1 - \mathbf{F}^b \times \mathbf{b}_1) dA \\
&- \iint_D \mathbf{u} \bullet [\frac{\partial}{\partial x}(\mathbf{F}^a + \frac{\partial w}{\partial \sin\gamma} \mathbf{b}_1) + \frac{\partial}{\partial y}(\mathbf{F}^b + \frac{\partial w}{\partial \sin\gamma} \mathbf{a}_1) + \mathbf{f}] dA \\
&+ \oint_C \frac{dy}{ds}(\mathbf{M}^a + \mathbf{N}^a) \bullet \mathbf{c}^a ds - \oint_C \frac{dx}{ds}(\mathbf{M}^b + \mathbf{N}^b) \bullet \mathbf{c}^b ds - \oint_C (\mathbf{t} - \frac{dy}{ds}\mathbf{F}^a + \frac{dx}{ds}\mathbf{F}^b) \bullet \mathbf{u} ds \\
&- \oint_C \mathbf{m} \bullet \frac{\partial \mathbf{u}}{\partial n} ds - \oint_C (\mathbf{n}_1^a \bullet \mathbf{c}^a \times \mathbf{a}_2 + \mathbf{n}_2^a \bullet \mathbf{c}^a \times \mathbf{a}_3 + \mathbf{n}_1^b \bullet \mathbf{c}^b \times \mathbf{b}_2 + \mathbf{n}_2^b \bullet \mathbf{c}^b \times \mathbf{b}_3) ds.
\end{aligned}
\tag{3.26}
$$

## 4. Equilibrium Equations, Boundary Conditions and Compatibility Conditions

In this section, we will first derive the equilibrium equations and boundary conditions from (3.25) and then derive the compatibility conditions.

First, we will simplify the line integrals in (3.26). Let us confine attention to the line integral $\oint_C (\frac{dy}{ds}\mathbf{M}^a \bullet \mathbf{c}^a - \frac{dx}{ds}\mathbf{M}^b \bullet \mathbf{c}^b) ds$.

Multiplying both sides of $(3.16)_1$ by $\mathbf{n}^a$ and noting that $\{\mathbf{a}_1, \mathbf{n}^a, \mathbf{b}^a\}$ is orthonormal, we find

$$
\mathbf{c}^a \bullet \mathbf{b}^a = \frac{\partial \mathbf{u}}{\partial x} \bullet \mathbf{n}^a. \tag{4.1}
$$

The derivative of $\mathbf{u}$ with respect to arc length along the boundary curve $C$ is $\frac{\partial \mathbf{u}}{\partial s}$ and its derivative with respect to arc length in the outward normal direction is $\frac{\partial \mathbf{u}}{\partial n}$. Then

$$
\frac{\partial \mathbf{u}}{\partial s} = \frac{\partial \mathbf{u}}{\partial x}\frac{dx}{ds} + \frac{\partial \mathbf{u}}{\partial y}\frac{dy}{ds}, \quad \frac{\partial \mathbf{u}}{\partial n} = \frac{\partial \mathbf{u}}{\partial x}\frac{dy}{ds} - \frac{\partial \mathbf{u}}{\partial y}\frac{dx}{ds} \tag{4.2}
$$

and thus

$$
\frac{\partial \mathbf{u}}{\partial x} = \frac{\partial \mathbf{u}}{\partial s}\frac{dx}{ds} + \frac{\partial \mathbf{u}}{\partial n}\frac{dy}{ds}, \quad \frac{\partial \mathbf{u}}{\partial y} = \frac{\partial \mathbf{u}}{\partial s}\frac{dy}{ds} - \frac{\partial \mathbf{u}}{\partial n}\frac{dx}{ds}. \tag{4.3}
$$

Substituting $(4.3)_1$ into (4.1) yields

$$\mathbf{c}^a \bullet \mathbf{b}^a = \frac{dx}{ds}\frac{\partial \mathbf{u}}{\partial s}\bullet \mathbf{n}^a + \frac{dy}{ds}\frac{\partial \mathbf{u}}{\partial n}\bullet \mathbf{n}^a . \tag{4.4}$$

Similarly, we have

$$\mathbf{c}^b \bullet \mathbf{b}^b = \frac{dy}{ds}\frac{\partial \mathbf{u}}{\partial s}\bullet \mathbf{n}^b - \frac{dx}{ds}\frac{\partial \mathbf{u}}{\partial n}\bullet \mathbf{n}^b . \tag{4.5}$$

With the aid of (4.4), (4.5), $(3.22)_1$ and $(3.22)_2$, we obtain

$$\begin{aligned}\frac{dy}{ds}\mathbf{M}^a \bullet \mathbf{c}^a - \frac{dx}{ds}\mathbf{M}^b \bullet \mathbf{c}^b &= (\frac{dx}{ds}\frac{dy}{ds}\frac{\partial w}{\partial \eta^a}\mathbf{n}^a - \frac{dx}{ds}\frac{dy}{ds}\frac{\partial w}{\partial \eta^b}\mathbf{n}^b)\bullet \frac{\partial \mathbf{u}}{\partial s} \\ &+ [(\frac{dy}{ds})^2\frac{\partial w}{\partial \eta^a}\mathbf{n}^a + (\frac{dx}{ds})^2\frac{\partial w}{\partial \eta^b}\mathbf{n}^b]\bullet \frac{\partial \mathbf{u}}{\partial n}.\end{aligned} \tag{4.6}$$

Then

$$\begin{aligned}\oint_C (\frac{dy}{ds}\mathbf{M}^a \bullet \mathbf{c}^a - \frac{dx}{ds}\mathbf{M}^b \bullet \mathbf{c}^b)ds = \oint_C \{[(\frac{dy}{ds})^2\frac{\partial w}{\partial \eta^a}\mathbf{n}^a + (\frac{dx}{ds})^2\frac{\partial w}{\partial \eta^b}\mathbf{n}^b)]\bullet \frac{\partial \mathbf{u}}{\partial n} \\ \frac{\partial}{\partial s}[(\frac{dx}{ds}\frac{dy}{ds}\frac{\partial w}{\partial \eta^a}\mathbf{n}^a - \frac{dx}{ds}\frac{dy}{ds}\frac{\partial w}{\partial \eta^b}\mathbf{n}^b)\bullet \mathbf{u}] \\ -[\frac{\partial}{\partial s}(\frac{dx}{ds}\frac{dy}{ds}\frac{\partial w}{\partial \eta^a}\mathbf{n}^a - \frac{dx}{ds}\frac{dy}{ds}\frac{\partial w}{\partial \eta^b}\mathbf{n}^b)]\bullet \mathbf{u}\}ds.\end{aligned} \tag{4.7}$$

Suppose the boundary curve $C$ is piecewise smooth with n corners. We use $C_i$ to denote the ith corner. Let $\mathbf{u}_i$ denote the first variation of **r** evaluated at $C_i$. We obtain from (4.7)

$$\begin{aligned}\oint_C (\frac{dy}{ds}\mathbf{M}^a \bullet \mathbf{c}^a - \frac{dx}{ds}\mathbf{M}^b \bullet \mathbf{c}^b)ds = \oint_C \{[(\frac{dy}{ds})^2\frac{\partial w}{\partial \eta^a}\mathbf{n}^a + (\frac{dx}{ds})^2\frac{\partial w}{\partial \eta^b}\mathbf{n}^b)]\bullet \frac{\partial \mathbf{u}}{\partial n} \\ -[\frac{\partial}{\partial s}(\frac{dx}{ds}\frac{dy}{ds}\frac{\partial w}{\partial \eta^a}\mathbf{n}^a - \frac{dx}{ds}\frac{dy}{ds}\frac{\partial w}{\partial \eta^b}\mathbf{n}^b)]\bullet \mathbf{u}\}ds + \sum_{i=1}^{n}\mathbf{F}_i \bullet \mathbf{u}_i,\end{aligned} \tag{4.8}$$

where

$$\mathbf{F}_i = [(\frac{dx}{ds}\frac{dy}{ds}\frac{\partial w}{\partial \eta^a}\mathbf{n}^a - \frac{dx}{ds}\frac{dy}{ds}\frac{\partial w}{\partial \eta^b}\mathbf{n}^b)_+ - (\frac{dx}{ds}\frac{dy}{ds}\frac{\partial w}{\partial \eta^a}\mathbf{n}^a - \frac{dx}{ds}\frac{dy}{ds}\frac{\partial w}{\partial \eta^b}\mathbf{n}^b)_-]|_{C_i} \tag{4.9}$$

is the force at $C_i$ required to support the deformation. Note that if $C$ is smooth, then $\mathbf{F}_i$ vanishes.

Making use of $(3.22)_3$ and $(3.22)_4$, we find

$$\frac{dy}{ds}\mathbf{N}^a \bullet \mathbf{c}^a = \frac{dy}{ds}\frac{\partial w}{\partial \beta^a}\mathbf{a}_1 \bullet \mathbf{c}^a\,,\ -\frac{dx}{ds}\mathbf{N}^b \bullet \mathbf{c}^b = -\frac{dx}{ds}\frac{\partial w}{\partial \beta^b}\mathbf{b}_1 \bullet \mathbf{c}^b\,. \tag{4.10}$$

Then,

$$\int_C \left(\frac{dy}{ds}\mathbf{N}^a \bullet \mathbf{c}^a - \frac{dx}{ds}\mathbf{N}^b \bullet \mathbf{c}^b\right)ds = \int_C \left(\frac{dy}{ds}\frac{\partial w}{\partial \beta^a}\mathbf{a}_1 \bullet \mathbf{c}^a - \frac{dx}{ds}\frac{\partial w}{\partial \beta^b}\mathbf{b}_1 \bullet \mathbf{c}^b\right)ds\,. \tag{4.11}$$

Next, we restrict attention to the last line integral in (3.26). Let $n^a_{\alpha i}$ denote the component of $\mathbf{n}^a_\alpha$ along $\mathbf{a}_i$. Define $n^b_{\alpha i}$ to be the component of $\mathbf{n}^b_\alpha$ along $\mathbf{b}_i$. Then, we have

$$\mathbf{n}^a_1 = n^a_{1i}\mathbf{a}_i\,,\ \mathbf{n}^a_2 = n^a_{2i}\mathbf{a}_i\,,\ \mathbf{n}^b_1 = n^b_{1i}\mathbf{b}_i\,,\ \mathbf{n}^b_2 = n^b_{2i}\mathbf{b}_i\,. \tag{4.12}$$

With the aid of (4.12), the last line integral in (3.26) can be expressed as

$$\begin{aligned}
&\oint_C \left(\frac{dy}{ds}\mathbf{n}^a_1 \bullet \mathbf{c}^a \times \mathbf{a}_2 + \frac{dy}{ds}\mathbf{n}^a_2 \bullet \mathbf{c}^a \times \mathbf{a}_3 + \frac{dx}{ds}\mathbf{n}^b_1 \bullet \mathbf{c}^b \times \mathbf{b}_2 + \frac{dx}{ds}\mathbf{n}^b_2 \bullet \mathbf{c}^b \times \mathbf{b}_3\right)ds = \\
&\oint_C \left[(-n^a_{13} + n^a_{22})\frac{dy}{ds}\mathbf{c}^a \bullet \mathbf{a}_1 + (-n^b_{13} + n^b_{22})\frac{dx}{ds}\mathbf{c}^a \bullet \mathbf{b}_1\right]ds + \oint_C \left(n^a_{11}\frac{dy}{ds}\mathbf{c}^a \bullet \mathbf{a}_3 - n^a_{21}\frac{dy}{ds}\mathbf{c}^a \bullet \mathbf{a}_2\right)ds \\
&+ \oint_C \left(n^b_{11}\frac{dx}{ds}\mathbf{c}^b \bullet \mathbf{b}_3 - n^b_{21}\frac{dx}{ds}\mathbf{c}^b \bullet \mathbf{b}_2\right)ds.
\end{aligned} \tag{4.13}$$

At each material point of the network, $\mathbf{c}^a$ and $\mathbf{c}^b$ are independent of $\mathbf{u}$ and respectively related to $\frac{\partial \mathbf{u}}{\partial x}$ and $\frac{\partial \mathbf{u}}{\partial y}$ through (3.16). If we observe $(3.16)_1$, we find that only $\mathbf{c}^a \bullet \mathbf{a}_1$ is independent of $\frac{\partial \mathbf{u}}{\partial x}$, and $\mathbf{c}^a \bullet \mathbf{a}_2$ and $\mathbf{c}^a \bullet \mathbf{a}_3$ are determined by $\frac{\partial \mathbf{u}}{\partial x}$. There is a similar interpretation for the relation of $\mathbf{c}^b \bullet \mathbf{b}_i$ to $\frac{\partial \mathbf{u}}{\partial y}$, which may be deduced from $(3.16)_2$. Therefore, with the aid of (4.3), the integrands in the second and third integrals of (4.13) may finally be expressed in the form of the terms on the right-hand side of (4.6). To avoid redundancy, we set

$$n^a_{11} = 0\,,\ n^a_{21} = 0\,,\ n^b_{11} = 0\,,\ n^b_{21} = 0\,,\ n^a = -n^a_{13} + n^a_{22}\,,\ n^b = -n^b_{13} + n^b_{22}\,. \tag{4.14}$$

Thus, we have

$$\begin{aligned}
&\oint_C \left(\frac{dy}{ds}\mathbf{n}^a_1 \bullet \mathbf{c}^a \times \mathbf{a}_2 + \frac{dy}{ds}\mathbf{n}^a_2 \bullet \mathbf{c}^a \times \mathbf{a}_3 + \frac{dx}{ds}\mathbf{n}^b_1 \bullet \mathbf{c}^b \times \mathbf{b}_2 + \frac{dx}{ds}\mathbf{n}^b_2 \bullet \mathbf{c}^b \times \mathbf{b}_3\right)ds = \\
&\oint_C \left(n^a\frac{dy}{ds}\mathbf{c}^a \bullet \mathbf{a}_1 + n^b\frac{dx}{ds}\mathbf{c}^a \bullet \mathbf{b}_1\right)ds.
\end{aligned} \tag{4.15}$$

Substituting (4.8), (4.11) and (4.15) into (3.26) yields

$$\dot{F} = -\iint_D \mathbf{c}^a \bullet [(\frac{\partial \mathbf{M}^a}{\partial x} + \frac{\partial \mathbf{N}^a}{\partial x} + \frac{\partial w}{\partial \sin\gamma}\mathbf{b}_1 \times \mathbf{a}_1) - \mathbf{F}^a \times \frac{\partial \mathbf{r}}{\partial x}]dA$$
$$- \iint_D \mathbf{c}^b \bullet [(\frac{\partial \mathbf{M}^b}{\partial y} + \frac{\partial \mathbf{N}^b}{\partial y} + \frac{\partial w}{\partial \sin\gamma}\mathbf{a}_1 \times \mathbf{b}_1) - \mathbf{F}^b \times \frac{\partial \mathbf{r}}{\partial y}]dA$$
$$- \iint_D \mathbf{u} \bullet (\frac{\partial \mathbf{F}^a}{\partial x} + \frac{\partial \mathbf{F}^b}{\partial y} + \mathbf{f})dA$$
$$+ \oint_C [(\frac{dy}{ds})^2 \frac{\partial w}{\partial \eta^a}\mathbf{n}^a + (\frac{dx}{ds})^2 \frac{\partial w}{\partial \eta^b}\mathbf{n}^b - \mathbf{m}] \bullet \frac{\partial \mathbf{u}}{\partial n} ds$$
$$+ \oint_C [(\frac{\partial w}{\partial \beta^a} - n^a)\frac{dy}{ds}\mathbf{a}_1 \bullet \mathbf{c}^a + (\frac{\partial w}{\partial \beta^b} - n^b)\frac{dx}{ds}\mathbf{b}_1 \bullet \mathbf{c}^b]ds$$
$$- \oint_C [\mathbf{t} - \frac{dy}{ds}\mathbf{F}^a + \frac{dx}{ds}\mathbf{F}^b + \frac{\partial}{\partial s}(\frac{dx}{ds}\frac{dy}{ds}\frac{\partial w}{\partial \eta^a}\mathbf{n}^a - \frac{dx}{ds}\frac{dy}{ds}\frac{\partial w}{\partial \eta^b}\mathbf{n}^b)] \bullet \mathbf{u} ds + \sum_{i=1}^{n} \mathbf{F}_i \bullet \mathbf{u}_i . \tag{4.16}$$

Next, with the aid of (4.16), we derive the equilibrium equations and boundary conditions by using the requirement $\dot{F} = 0$, given by (3.25). We first apply the condition to cases in which $\mathbf{u}_i = \mathbf{0}$, $\mathbf{u} = \mathbf{0}$, $\frac{\partial \mathbf{u}}{\partial n} = \mathbf{0}$, $\mathbf{a}_1 \bullet \mathbf{c}^a = 0$ and $\mathbf{b}_1 \bullet \mathbf{c}^b = 0$ on the boundary curve *C*. We then obtain

$$\iint_D \mathbf{c}^a \bullet [(\frac{\partial \mathbf{M}^a}{\partial x} + \frac{\partial \mathbf{N}^a}{\partial x} + \frac{\partial w}{\partial \sin\gamma}\mathbf{b}_1 \times \mathbf{a}_1) - \mathbf{F}^a \times \frac{\partial \mathbf{r}}{\partial x}]dA$$
$$+ \iint_D \mathbf{c}^b \bullet [(\frac{\partial \mathbf{M}^b}{\partial y} + \frac{\partial \mathbf{N}^b}{\partial y} + \frac{\partial w}{\partial \sin\gamma}\mathbf{a}_1 \times \mathbf{b}_1) - \mathbf{F}^b \times \frac{\partial \mathbf{r}}{\partial y}]dA \tag{4.17}$$
$$+ \iint_D \mathbf{u} \bullet (\frac{\partial \mathbf{F}^a}{\partial x} + \frac{\partial \mathbf{F}^b}{\partial y} + \mathbf{f})dA$$
$$= 0.$$

Let $a_i$ and $f_i^a$ respectively denote the components of $(\frac{\partial \mathbf{M}^a}{\partial x} + \frac{\partial \mathbf{N}^a}{\partial x} + \frac{\partial w}{\partial \sin\gamma}\mathbf{b}_1 \times \mathbf{a}_1) - \mathbf{F}^a \times \frac{\partial \mathbf{r}}{\partial x}$ and $\mathbf{f}^a$, which is a vector of Lagrange multipliers, along $\mathbf{a}_i$. Then, we may choose $f_\alpha^a$ $(\alpha = 2, 3)$ in $\mathbf{F}^a$, whose expression is given by $(3.24)_1$, such that

$$\mathrm{a}_{\alpha} = 0\,. \tag{4.18}$$

Let $\mathrm{b}_1$ and $\mathrm{f}_1^{\mathrm{b}}$ respectively denote the components of $(\frac{\partial \mathbf{M}^{\mathrm{b}}}{\partial \mathrm{y}} + \frac{\partial \mathbf{N}^{\mathrm{b}}}{\partial \mathrm{y}} + \frac{\partial w}{\partial \sin\gamma}\mathbf{a}_1 \times \mathbf{b}_1) - \mathbf{F}^{\mathrm{b}} \times \frac{\partial \mathbf{r}}{\partial \mathrm{y}}$ and $\mathbf{f}^{\mathrm{b}}$, which is also a vector of Lagrange multipliers, along $\mathbf{b}_1$. Similarly, we choose $\mathrm{f}_{\alpha}^{\mathrm{b}}$ $(\alpha = 2, 3)$ in $\mathbf{F}^{\mathrm{b}}$, whose expression is given by $(3.24)_2$, such that

$$\mathrm{b}_{\alpha} = 0\,. \tag{4.19}$$

The components $\mathrm{f}_1^{\mathrm{a}}$ and $\mathrm{f}_1^{\mathrm{b}}$ will be determined by the equilibrium equations. With the aid of (4.18) and (4.19), Equation (4.17) is reduced to

$$\iint_D (\mathrm{a}_1 \mathbf{c}^{\mathrm{a}} \bullet \mathbf{a}_1)\mathrm{dA} + \iint_D (\mathrm{b}_1 \mathbf{c}^{\mathrm{b}} \bullet \mathbf{b}_1)\mathrm{dA} + \iint_D \mathbf{u} \bullet (\frac{\partial \mathbf{F}^{\mathrm{a}}}{\partial \mathrm{x}} + \frac{\partial \mathbf{F}^{\mathrm{b}}}{\partial \mathrm{y}} + \mathbf{f})\mathrm{dA} = 0. \tag{4.20}$$

Since $\mathbf{c}^{\mathrm{a}} \bullet \mathbf{a}_1$, $\mathbf{c}^{\mathrm{b}} \bullet \mathbf{b}_1$ and $\mathbf{u}$ are arbitrary and independent from one another identically over $D$, together with the aid of (4.18) and (4.19), we obtain the Euler-Lagrange equations as follows:

$$\mathbf{F}^{\mathrm{a}} \times \frac{\partial \mathbf{r}}{\partial \mathrm{x}} = \frac{\partial \mathbf{M}^{\mathrm{a}}}{\partial \mathrm{x}} + \frac{\partial \mathbf{N}^{\mathrm{a}}}{\partial \mathrm{x}} + \frac{\partial w}{\partial \sin\gamma}\mathbf{b}_1 \times \mathbf{a}_1,\ \mathbf{F}^{\mathrm{b}} \times \frac{\partial \mathbf{r}}{\partial \mathrm{y}} = \frac{\partial \mathbf{M}^{\mathrm{b}}}{\partial \mathrm{y}} + \frac{\partial \mathbf{N}^{\mathrm{b}}}{\partial \mathrm{y}} + \frac{\partial w}{\partial \sin\gamma}\mathbf{a}_1 \times \mathbf{b}_1,$$
$$\frac{\partial \mathbf{F}^{\mathrm{a}}}{\partial \mathrm{x}} + \frac{\partial \mathbf{F}^{\mathrm{b}}}{\partial \mathrm{y}} + \mathbf{f} = \mathbf{0}. \tag{4.21}$$

Observing (4.21), we find that $\mathbf{F}^{\mathrm{a}}$ and $\mathbf{F}^{\mathrm{b}}$ may be interpreted as the respective forces on cross sections of $\mathbf{a}$- and $\mathbf{b}$-lines. Given (4.21), the integral over $D$ in (4.16) is absent.

We next consider variations such that $\mathbf{u}_{\mathrm{i}}$, $\mathbf{u}$, $\frac{\partial \mathbf{u}}{\partial \mathrm{n}}$, $\mathbf{a}_1 \bullet \mathbf{c}^{\mathrm{a}}$ and $\mathbf{b}_1 \bullet \mathbf{c}^{\mathrm{b}}$ are respectively arbitrary on the boundary curve $C$. We obtain from (4.16) the following boundary conditions:

$\mathbf{F}_{\mathrm{i}} = \mathbf{0}$ ( i=1,…,n), if $\mathbf{u}_{\mathrm{i}}$ is arbitrary; (4.22)

$\mathbf{t} - \frac{\mathrm{dy}}{\mathrm{ds}}\mathbf{F}^{\mathrm{a}} + \frac{\mathrm{dx}}{\mathrm{ds}}\mathbf{F}^{\mathrm{b}} + \frac{\mathrm{d}}{\mathrm{ds}}(\frac{\mathrm{dx}}{\mathrm{ds}}\frac{\mathrm{dy}}{\mathrm{ds}}\frac{\partial w}{\partial \eta^{\mathrm{a}}}\mathbf{n}^{\mathrm{a}} - \frac{\mathrm{dx}}{\mathrm{ds}}\frac{\mathrm{dy}}{\mathrm{ds}}\frac{\partial w}{\partial \eta^{\mathrm{b}}}\mathbf{n}^{\mathrm{b}}) = \mathbf{0}$, if $\mathbf{u}$ is arbitrary; (4.23)

$$(\frac{dy}{ds})^2\frac{\partial w}{\partial \eta^a}\mathbf{n}^a+(\frac{dx}{ds})^2\frac{\partial w}{\partial \eta^b}\mathbf{n}^b-\mathbf{m}=\mathbf{0}\text{, if } \frac{\partial \mathbf{u}}{\partial n} \text{ is arbitrary;} \tag{4.24}$$

$$(\frac{\partial w}{\partial \beta^a}-n^a)\frac{dy}{ds}=0\text{, if } \mathbf{a}_1\bullet\mathbf{c}^a \text{ is arbitrary;} \tag{4.25}$$

$$(\frac{\partial w}{\partial \beta^b}-n^b)\frac{dx}{ds}=0\text{, if } \mathbf{b}_1\bullet\mathbf{c}^b \text{ is arbitrary.} \tag{4.26}$$

Finally, compatibility conditions are derived from the relations:

$$\frac{\partial^2\mathbf{r}}{\partial x\partial y}=\frac{\partial^2\mathbf{r}}{\partial y\partial x},\ \frac{\partial^2\mathbf{a}_1}{\partial x\partial y}=\frac{\partial^2\mathbf{a}_1}{\partial y\partial x},\ \frac{\partial^2\mathbf{b}_1}{\partial x\partial y}=\frac{\partial^2\mathbf{b}_1}{\partial y\partial x}. \tag{4.27}$$

The first vector equation in (4.27) includes two scalar conditions:

$$\phi^b=\phi^a\sin\gamma,\ \phi^a\,|\cos\gamma|=0, \tag{4.28}$$

where use is made of (2.20), (2.16) and (2.4). Following the same procedure as that in Section 5(c) of Steigmann and Pipkin (1991), with the aid of (2.20), (2.16), (2.4) and (2.19), we find that the remaining two vector equations of (4.27) include the following three scalar conditions:

$$\frac{\eta^a\eta^b\sin\vartheta^a\sin\vartheta^b-\tau^2}{|\cos\gamma|}+\frac{\partial(\eta^b\cos\vartheta^b)}{\partial x}+\frac{\cos\gamma}{|\cos\gamma|}\frac{\partial^2\gamma}{\partial x\partial y}=0, \tag{4.29}$$

$$|\cos\gamma|\,[\frac{\partial\tau}{\partial x}-\frac{\partial(\eta^a\sin\vartheta^a)}{\partial y}]=\eta^a\cos\vartheta^a(\eta^b\sin\vartheta^b-\tau\sin\gamma)+\phi^a(\eta^a\sin\vartheta^a\sin\gamma-\tau), \tag{4.30}$$

$$|\cos\gamma|\,[\frac{\partial(\eta^b\sin\vartheta^b)}{\partial x}-\frac{\partial\tau}{\partial y}]=\eta^b\cos\vartheta^b(\eta^a\sin\vartheta^a-\tau\sin\gamma)+\phi^b(\eta^b\sin\vartheta^b\sin\gamma-\tau). \tag{4.31}$$

## 5. Simplification of Euler-Lagrange Equations and Two Special Cases

In this final section, we will simplify the Euler-Lagrange equations given in (4.21) and then reduce them to obtain the equilibrium equations of Wang and Pipkin (1986a), and those of an inextensible rod.

By $(4.21)_1$, we have

$$\mathbf{F}^a\times\frac{\partial\mathbf{r}}{\partial x}=[-\frac{\partial}{\partial x}(\frac{\partial w}{\partial \eta^a}\mathbf{n}^a)+\frac{\partial w}{\partial \beta^a}\eta^a\mathbf{b}^a+\frac{\partial w}{\partial \sin\gamma}\mathbf{b}_1]\times\mathbf{a}_1+\frac{\partial}{\partial x}(\frac{\partial w}{\partial \beta^a})\mathbf{a}_1. \tag{5.1}$$

Further, we have

$$[\mathbf{F}^a+\frac{\partial}{\partial x}(\frac{\partial w}{\partial \eta^a}\mathbf{n}^a)-\frac{\partial w}{\partial \beta^a}\eta^a\mathbf{b}^a-\frac{\partial w}{\partial \sin\gamma}\mathbf{b}_1]\times\mathbf{a}_1-\frac{\partial}{\partial x}(\frac{\partial w}{\partial \beta^a})\mathbf{a}_1=\mathbf{0}. \tag{5.2}$$

This implies that the sum of the terms in the brackets is parallel to $\mathbf{a}_1$, with a value $T^a\mathbf{a}_1$ (say). Consequently,

$$\mathbf{F}^a = -\frac{\partial}{\partial x}(\frac{\partial w}{\partial \eta^a}\mathbf{n}^a) + \frac{\partial w}{\partial \sin\gamma}\mathbf{b}_1 + \frac{\partial w}{\partial \beta^a}\eta^a\mathbf{b}^a + T^a\mathbf{a}_1, \quad \frac{\partial}{\partial x}(\frac{\partial w}{\partial \beta^a}) = 0. \quad (5.3)$$

Likewise, we have

$$\mathbf{F}^b = -\frac{\partial}{\partial y}(\frac{\partial w}{\partial \eta^b}\mathbf{n}^b) + \frac{\partial w}{\partial \sin\gamma}\mathbf{a}_1 + \frac{\partial w}{\partial \beta^b}\eta^b\mathbf{b}^b + T^b\mathbf{b}_1, \quad \frac{\partial}{\partial y}(\frac{\partial w}{\partial \beta^b}) = 0. \quad (5.4)$$

Thus, we have

$$\mathbf{F}^a = -\frac{\partial}{\partial x}(\frac{\partial w}{\partial \eta^a}\mathbf{n}^a) + \frac{\partial w}{\partial \sin\gamma}\mathbf{b}_1 + \frac{\partial w}{\partial \beta^a}\eta^a\mathbf{b}^a + T^a\mathbf{a}_1,$$

$$\mathbf{F}^b = -\frac{\partial}{\partial y}(\frac{\partial w}{\partial \eta^b}\mathbf{n}^b) + \frac{\partial w}{\partial \sin\gamma}\mathbf{a}_1 + \frac{\partial w}{\partial \beta^b}\eta^b\mathbf{b}^b + T^b\mathbf{b}_1,$$

$$\frac{\partial}{\partial x}(\frac{\partial w}{\partial \beta^a}) = 0, \quad \frac{\partial}{\partial y}(\frac{\partial w}{\partial \beta^b}) = 0, \quad \frac{\partial \mathbf{F}^a}{\partial x} + \frac{\partial \mathbf{F}^b}{\partial y} + \mathbf{f} = \mathbf{0}. \quad (5.5)$$

Wang and Pipkin (1986a) considered an inextensible network with shearing resistance and bending stiffness. They did not consider twisting effects. They assumed a special form of the strain-energy function given by

$$w = w_0(\mathbf{a}_1 \bullet \mathbf{b}_1) + \frac{1}{2}\Gamma(\frac{\partial \mathbf{a}_1}{\partial x} \bullet \frac{\partial \mathbf{a}_1}{\partial x} + \frac{\partial \mathbf{b}_1}{\partial y} \bullet \frac{\partial \mathbf{b}_1}{\partial y}), \quad (5.6)$$

where $w_0(\mathbf{a}_1 \bullet \mathbf{b}_1)$ is the energy due to the shearing stress and $\Gamma$, a positive constant, is the associated stiffness. By using (2.4) and (2.7), equation (5.6) may be written as

$$w = w_0(\sin\gamma) + \frac{1}{2}\Gamma[(\eta^a)^2 + (\eta^b)^2]. \quad (5.7)$$

Substituting (5.7) into (5.5), we obtain the following three conditions:

$$\mathbf{F}^a = T^a\mathbf{a}_1 - \Gamma\frac{\partial^2 \mathbf{a}_1}{\partial x^2} + \frac{dw_0}{d(\sin\gamma)}\mathbf{b}_1, \ \mathbf{F}^b = T^b\mathbf{b}_1 - \Gamma\frac{\partial^2 \mathbf{b}_1}{\partial y^2} + \frac{dw_0}{d(\sin\gamma)}\mathbf{a}_1, \ \frac{\partial \mathbf{F}^a}{\partial x} + \frac{\partial \mathbf{F}^b}{\partial y} + \mathbf{f} = \mathbf{0}. \quad (5.8)$$

The three equations in (5.8) coincide with their counterparts (4.1), (4.5) and (3.4) in Wang and Pipkin (1986a), respectively.

If we eliminate both the equations and terms pertaining to either the shear angle $\gamma$ or to y or **b**-lines in (5.5), we obtain the equilibrium equations for an inextensible rod

$$\mathbf{F}^a = -\frac{\partial}{\partial x}(\frac{\partial w}{\partial \eta^a}\mathbf{n}^a) + \frac{\partial w}{\partial \beta^a}\eta^a\mathbf{b}^a + T^a\mathbf{a}_1, \quad \frac{\partial}{\partial x}(\frac{\partial w}{\partial \beta^a}) = 0, \quad \frac{\partial \mathbf{F}^a}{\partial x} + \mathbf{f} = \mathbf{0}, \quad (5.9)$$

where x is the arc length, and $w$ is of the form $w(\eta^a, \beta^a)$ with $\eta^a$ and $\beta^a$, respectively, representing the curvature and torsion of the rod.

## References


Bliss G A 1946 *Lectures on the Calculus of Variations.* University of Chicago Press, Chicago

Elsgolts L 1977 *Differential Equations and the Calculus of Variations* MIR, Moscow

Green W A and Shi J , 1990 Plane deformations of membranes formed with elastic cords *Q Jl Mech Appl Math* **43**, 317-333

Hilgers, M G 1997 Plane infinitesimal waves in elastic sheets with bending stiffness. *Math. Mech Solids.* **2**, *75-89*

Hilgers, M G & Pipkin, A C 1992a Elastic sheets with bending stiffness *Q Jl Mech Appl Math* **45**, 57-75.

Hilgers, M G & Pipkin, A C 1992b Bending energy of highly elastic membranes *Q Appl Math* **50**, 389-400

Hilgers, M G & Pipkin, A C 1993 Energy minimizing deformations of elastic sheets with bending stiffness *J Elasticity* **31**, 125-139

Hilgers, M G & Pipkin, A C 1996 Bending energy of highly elastic emembranes II. *Q. Appl Math* **54**, 307-316

Kuznetsov, E N 1982 Axisymmetic static nets. *Int J Solids Structures* **18**, 1103.

Love A E H, 1927 *A treatise on the mathematical theory of elasticity* Cambridge University Press

Luo, C and Steigmann, D , 2000 Bending and twisting effects in an extensible network for three-dimensional deformations (In preparation)

Simmonds, J G 1985 The strain energy density of rubber-like shells. *Int J. Solids Struc.* **21**, *67-77*

Steigmann D J 1996 The variational structure of a nonlinear theory for spatial lattices, *Meccanica* **31**, 441-455

Steigmann D J and Faulkner M G 1993 Variational theory for spatial rods, *J Elast* , **33**, 1-26

Steigmann D J and Pipkin A C 1991 Equilibrium of elastic nets, *Phil Trans R Soc Lond* A **335**, 419-454

Wang, W -B and Pipkin, A. C 1986a Inextensible networks with bending stiffness, *Q Jl Mech. Appl Math* **39**, 343 - 359

Wang, W -B and Pipkin, A C 1986b Plane deformations of nets with bending stiffness. *Acta Mechanica.* **65**, 263-279

# PLASTIC BUCKLING OF RECTANGULAR PLATES WITH RANDOM MATERIAL PROPERTIES AND RANDOM LOADING: A DEMONSTRATION OF PROBABILISTIC STRUCTURAL ANALYSYS

G. MAYMON
*RAFAEL R&D*
*P.O.Box 2250*
*Haifa 31021, ISRAEL*

## 1. Introduction

An engineering design is a process of decision-making under constraints of uncertainty. The uncertainty in this process is due to the lack of deterministic knowledge of different physical parameters and the uncertainty concerning models with which the design is performed. This is true for all the engineering disciplines such as electronics, mechanics, aerodynamics and structural analysis involved in any design.

The uncertainty approach to the design of subsystems and complete systems was enhanced by the concept of reliability. Systems are analyzed for possible failure processes and criteria, probability of occurrence, reliability of components, redundancy, possible human errors in the production, and other uncertainties. Consequently, the required reliability of a given design is defined with proper reliability appropriation for subsystems. This required reliability certainly influences both the design and the product cost.

Nevertheless, in most cases structural analyst is still required to supply a design with absolute reliability, and most structural designs are performed using deterministic solutions. To compensate for many uncertainties, structural designers use a safety factor (lack of knowledge factor?), thus recognizing de facto the stochastic nature of many of the design parameters and models.

During the last decade, the need for application of probabilistic methods for non-deterministic structures has started to gain acceptance within the structural design community. Many designers have started to adopt the stochastic approach and the concepts of structural the stochastic approach and the concepts of structural reliability to present designs. It is likely that in the near future, this approach will start to dominate structural analysis procedures. Thus, the structural design will become increasingly integrated into the total system design, where reliability concepts have already been implemented.

The main sources of uncertainties in structural analysis and design are the model that is used, the loads that are applied to the structure and the uncertainties inherent in various structural parameters. In this paper only the last uncertainties will be discussed and demonstrated. The treatment of these uncertainties, which is part of the probabilistic



*D. Durban et al. (eds.), Advances in the Mechanics of Plates and Shells,* 229–244.

analysis of structures, has become extensive and practical only in the last 10-20 years, although pioneering studies were published earlier, in the late 1960s and the early 1970s (especially by scientific journals of the civil engineering community). Tremendous progress has been made in the formulation of mathematical models and the establishment of several algorithms for the determination of the behavior of stochastic structures submitted to excitation of stochastic loads (i.e. in [1], [2], [3] and [4]). This progress was followed by the adaptation by the industry of these relatively new models and algorithms in the structural design process of practical structures. It is most likely that in the near future increasingly more design codes and specifications will include the use of probabilistic analysis. The introduction of these methods into practical applications will be quicker than the introduction of finite element programs, due to the large infrastructure that already exists in computational analysis and the rapid advances in computational power.

For many years non-deterministic structures were analyzed using simulation methods (such as the Monte Carlo simulation). These simulations are time consuming, and sometimes their use becomes prohibitive for practical industrial applications due to the deal of time required for the simulation of a complex structural design, which interferes with the schedule of any project.

In the last two decades, numerical algorithms using non-simulative methods were developed. Also, more advanced simulation methods, which decrease tremendously the number of required computations, were developed (i.e. in [5], [6] and [7]). These developments led to the elaboratin of several computer programs that solve the probabilistic structural analysis problem within a reasonable and practical time frame, and are also suitable for industrial use (i.e. in [8], [9] and [10]).

It is the purpose of this paper to describe the basic concepts involved in a stochastic structural analysis, and to demonstrate these concepts by applying stochastic analysis to the problem of plastic buckling of rectangular plates whose material properties are not deterministic. The selection of random material properties does not imply that the same methods cannot be used for other non-deterministic properties such as geometry, dimensions and loads.

## 2. Failure Surface: Basic Concepts

A failure surface (or failure function) is a multidimensional function of all the random variables of a design problem, which separates between the failure space and the success space of a structural design (see [11]). On one side of this multidimensional surface the structure is safe, and on the other side it fails. The basic case includes two random variables. $R$ is the allowable quantity (stress, displacement, etc.) in the structure and is designated "the resistance" term. $S$ is the actual quantity in the structure and is designated "the load effect" term. Traditionally, a negative value of the failure function means failure and a positive value provide a safe region. The failure surface is then

$$\begin{aligned} &\mathbf{g(R,S) = R - S = 0} \\ &\textbf{so that}\quad \mathbf{g(R,S) \le 0 \equiv Failure} \end{aligned} \tag{1}$$

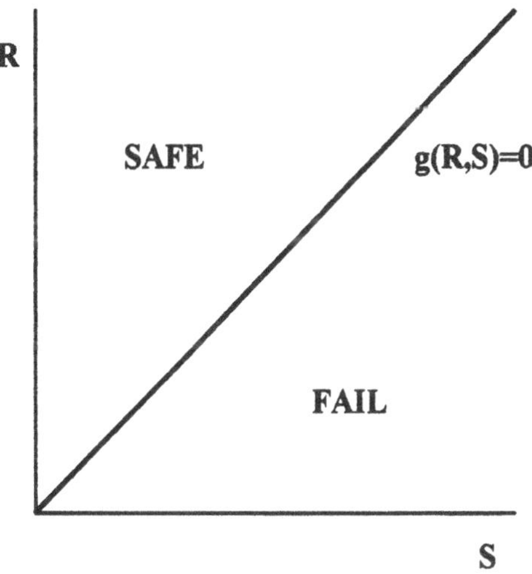


Figure 1: Failure Function of Two Random Variables

In Figure 1 the "Fail" and "Safe" regions for Eq.(1) are shown. When each of the two terms of Eq.(1) is a function of many other structural parameters the failure surface is not necessarily a linear line like that in Figure 1, but a multi-dimensional surface of all the random variables.

$$g(R,S) = g(X_1, X_2, ...X_k; Y_1, Y_2, ...Y_m) = 0 \qquad (2)$$

where $X_i$ are the variables influencing **R** and $Y_i$ are the variables influencing S. Sometimes a variable can influence both **R** and S. Therefore it is more convenient to express the failure function as

$$g(X_1, X_2, \cdots, X_n) = 0 \qquad (3)$$

where **n** is the total number of random variables.

It can be shown that the probability of failure is the result of the following integral

$$P_f = \int_{g \le 0} \phi_{X_1, X_2 ... X_n}(x_1, x_2, ..., x_n) dx_1 dx_2 ... dx_n \qquad (4)$$

where $\phi$ is the joint probability density function (JPDF) of the **n** random variables $X_n$ and the integral is performed over the hyper-space where g is negative. The area over which integration is done is shown in Figure 2 (for a two-variables problem with normal distribution).

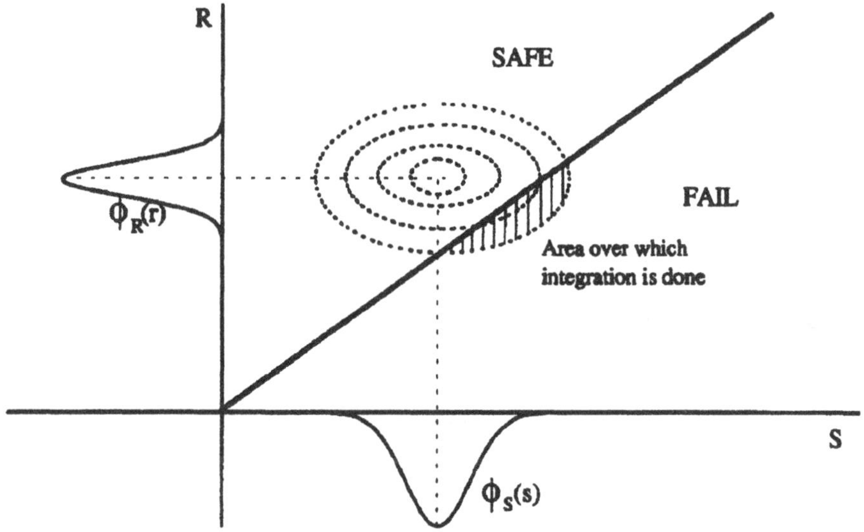


Figure 2: Basic Case, Two Normally Distributed Random Variables

The integral in Eq.(4) can only be solved explicitly in very few cases, and therefore other methods must be used.

A common practice in the analysis of stochastic structures is the transformation of the physical random variables vector $X$ into a standard normal vector $\mathbf{u}$. The transformation is (see [1]):

$$\mathbf{u}_i = \Phi^{-1}\left[\mathbf{F}_{X_i}(\mathbf{x}_i)\right] \tag{5}$$

where $\mathbf{\Phi}$ is the standard normal distribution function and $\mathbf{F}$ is the cumulative distribution of the physical random variable $\mathbf{X}_i$. When dependant random variables exist, the Rosenblatt transformation [7] is used to further transform the variables into independent standard normal variables. The probability density function (PDF) of a standard normal space decays exponentially with the square of the distance of this variable from the origin. In the case of a practical structure, very low probabilities of failure are allowed, and the integration boundary of the JPDF in Eq.(4) can be replaced by another, approximated, boundary. In first-order reliability methods (FORM), the failure function is replaced by a hyper-plane, tangent to the function at the point closest to the origin. In second-order reliability methods (SORM), the failure function is replaced by a second-order (quadratic) hyper-surface.

The point of the transformed limit state function closest to the origin of the **u** space is called the most probable point of failure (the MPP, sometimes called "the design point"), and is denoted $\mathbf{u}^*$. The distance from the origin to this point, denoted $\beta$, is called the reliability index. When $\beta$ is obtained, the first-order probability of failure $\mathbf{P_f}$ can be calculated by

$$\mathbf{P_f} = \Phi(-\beta) \tag{6}$$

Second-order probabilities of failure can be calculated by an extension of Eq.(6) creating an expression which includes the radii of curvatures of the hyper-space as cited in [12],[13]. An example of the transformation into a **u** space can be seen in figure 3.

Most of the non-simulative computational algorithms used today to find the point of most probable point of failure (the MPP) $\mathbf{u}^*$ utilize the gradient vector of the transformed random variable **u**.

## 3. Closed-Form Expressions for the Failure Function

In Eq.(1) the load effects term S represents the structural property which is assumed to be critical in the failure analysis. Assume the simple case of a beam loaded with a distributed force. Suppose that the failure criterion is defined as: "the beam fails when a stress higher than the yield stress exists in the beam". The load effects term **S** is the maximum stress in the beam, which depends on its length, cross-section dimensions and the magnitude of the applied forces. The resistance term **R** in this case is the yield stress. Assume that the failure criterion is defined as: "the beam fails when a mid-span displacement is higher than a given value $\delta_{max}$". Then the load effect term is the mid-span displacement, which depends on the beam length, cross-section dimensions, elastic modulus of the beam material and the applied force. The resistance term is then the allowed displacement $\delta_{max}$. In both these cases there is a closed-form expression that describes the load effect term as a function of the structural parameters (the mechanical transformation), thus the failure function can be expressed in a closed-form expression.

Combined failure criteria can also be defined. Such a criterion may be, for instance, "the beam fails when the stress is higher than the yield stress, and the displacement is higher than the allowed one". Another criterion may be "the beam fails when the stress is higher than the yield stress, or the displacement is higher than the allowed one". The definition of an appropriate failure criterion is one of the most important aspects of any design, and must be based on the designer skills and experience and the expected behavior of the designed system.

Structural analysis of practical structures designed in the industry can seldom be made by using closed-form expressions. Instead, a numerical algorithm such as a finite element code is used. When the finite element computation time is very short, Monte Carlo simulations can be used by running many thousands of finite element computations. This is a very impractical method when the finite element computation is long, which is usually the case for practical designs. For these cases, an approximated closed-form expression can be used for the failure function. This approximation can be

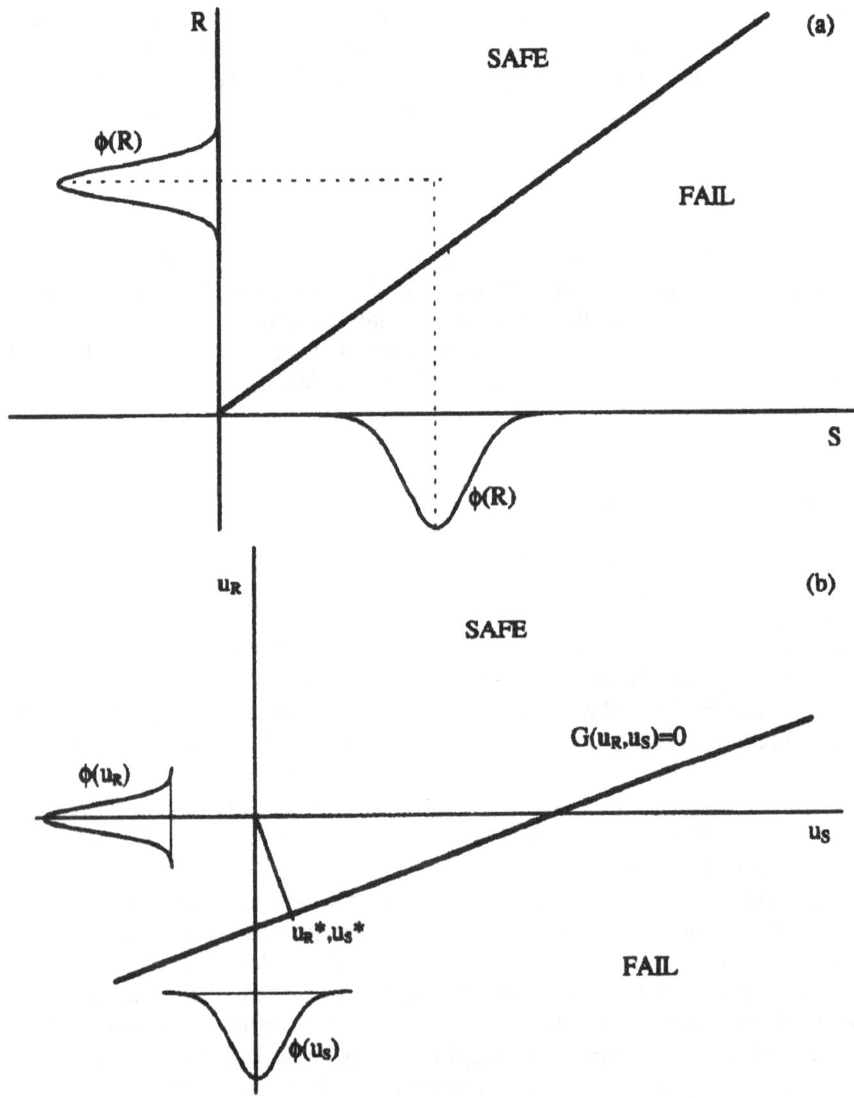


Figure 3: Transformaion from **X** space (a) into **u** space

obtained by solving a relatively small number of deterministic cases, changing slightly the values of those parameters which are assumed to be random variables. These solutions are then applied to form an approximate surface (i.e. in [14] and [15]), by using least mean square techniques. When a second-order surface is required, at least **2n+1** deterministic solutions are required, where **n** is the total number of the random variables. The approximated surface will be more accurate when more deterministic solutions are performed. When an approximate surface (sometimes called "response surface") is computed, it is used as an approximate closed-form expression for the probabilistic computation. Most of the commercially available probabilistic structural analysis computer programs include the computation of the approximated response surface as part of the whole computational process. Thus the user does not have to compute separately the approximate response surface, but only to list the deterministic solutions performed earlier by a finite element code.

Another method that was suggested (see [16], is the direct computation of the structural probabilistic problem using only an existing finite element code. Adopting this method (called the "modified joint probability density function") requires some manipulations of the optimization modules of the finite element program. It enables the determination of the MPP directly by the finite element code, thus enabling the computation of the first order probability of failure without any probabilistic analysis program.

## 4. Plastic Buckling of Rectangular Plate Under Biaxial Loading

The structural probabilistic analysis is demonstrated on the problem solved deterministically in [17].

The theory of elastic stability predicts an increase in the buckling load of a simply supported rectangular plate (of length **a**, width **b** and thickness **h**), loaded by a compression force **P** in the **y** direction, due to tension $\xi \cdot \mathbf{P}$ in the perpendicular direction (positive $\xi$ means tension). The model is described in Figure 4. The perpendicular tension has thus a strengthening effect. This effect becomes less effective when a plastic behavior is accounted for. Thus the plastic behavior of the rectangular plate under compression/tension is governed by two competing mechanisms, and an optimization problem can emerge here in one way or another.

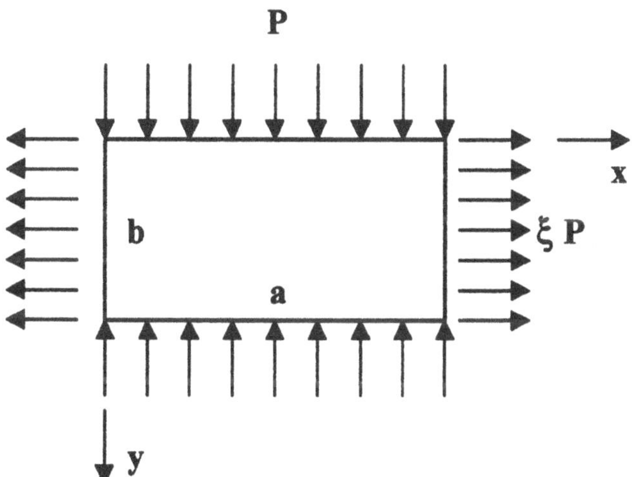


Figure 4: Simply Supported Rectangular Plate Under Biaxial Loading (negative $\xi$ means compression in the **x** Direction)

The problem was solve analytically in [17], and three elasto-plastic solutions were presented: (a) a general formulation; (b) a general formulation neglecting the elastic compressibility ($\nu$=0.5); (c) a closed form approximation for (b).

Without any loss of generality, the structural probabilistic analysis demonstrated here is based on the closed form solution. Two kinds of effects are shown: (a) the expected dispersion in the buckling load due to random material properties; (b) the expected probability of failure of this structure due to dispersion in both the material properties and the applied stress **P**.

The approximated closed-form expression for the buckling stress $\mathbf{P_b}$ is (see [17]):

$$\mathbf{P_b} = \mathbf{E} \cdot \left[ \frac{\frac{8}{3}(1+\xi) \cdot \left(1+\sqrt{1-C^2}\right)}{\left(1+\xi+\xi^2\right)^{\frac{N-1}{N}}} \right]^{\frac{1}{N}} \cdot \left(\frac{\mathbf{E}}{\mathbf{Y}}\right)^{\frac{1-N}{N}} \alpha^{\frac{1}{N}} \tag{7}$$

where **E** is the material elastic Young Modulus, **Y** is the material linear yield stress, **N** is a material parameter, and

$$C = \frac{\left(\frac{\sqrt{3}}{2}\right)\sqrt{\frac{N-1}{N}}}{\sqrt{1+\xi+\xi^2}} \quad ; \qquad \alpha = \frac{\pi^2 h^2}{12a^2} \tag{8}$$

Note that according to Eq.(7) the magnitude of the buckling load does not depend on the ratio **b/a**. This ratio influences the number of half waves **n** in the **y** direction, a parameter which is not included in the approximate closed form solution of [17] and in the following probabilistic analysis.

The parameter **N** describes the plastic behavior of the material stress-strain curve. This behavior is expressed as

$$\varepsilon = \frac{Y}{E}\left(\frac{\sigma}{Y}\right)^N \qquad \text{for } \sigma \geq Y \tag{9}$$

In Figure 5, stress-strain curves are shown for two values of **N** for a material with **E**=700000 kg/cm$^2$ and **Y**=4000 kg/cm$^2$.

In Figure 6, results calculated deterministically using Eq.(7) for a plate of **h**=0.5 cm are shown.

Some parametric computations were made to check the influence of the plate thickness and the parameter $N$ on the behavior of the buckling stress. Naturally, the thicker the plate, the higher the buckling stress. The maximum values of the buckling stress in these computations (not shown in this paper) was found to be in the vicinity of $\xi=0$. As **N** increase (for given values of **E** and **Y**), the drop in the buckling stress with positive $\xi$ values is enhanced. These results can be concluded intuitively by inspection of the stress-strain curves shown in Figure 5.

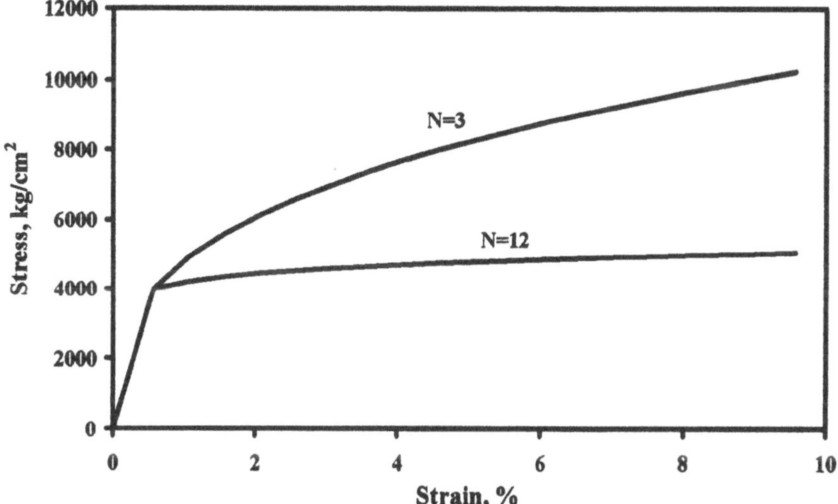


Figure 5: The Influence of N on the Stress-Strain Curve

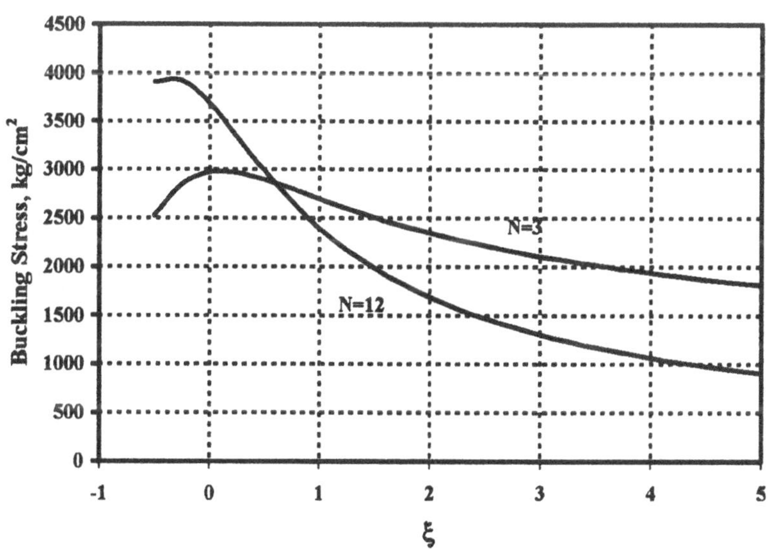


Figure 6: Buckling Stress vs. $\xi$ for **E**=700000 kg/cm$^2$, **Y**=4000 kg/cm$^2$

## 5. Stochastic Buckling Stress due to Random Material Properties

The statistical behavior of the buckling stress due to the randomness of the material properties **E**, **Y** and **N** were computed using (7). The computations were performed using the probabilistic structural analysis program FPI (Fast Probability Integration), developed under NASA contract in Southwest Research Institute, in cooperation with NASA Lewis Research Center and with several industries in the USA. One of the solution options in this program allows the use of a closed-form expression like (7), substituting random variables by their probability density functions (PDF), their mean values and their standard deviations. In Table 1, the distributions, mean values and standard deviations used for the demonstrated computations are listed.

**Table 1: Characteristics of Random Material Properties**

| | Distribution | Mean | Standard deviation | Range from $-3\sigma$ to $+3\sigma$ |
|---|---|---|---|---|
| N | normal | 3 | 0.03 | 2.91 – 3.09 |
| | | 12 | 0.12 | 11.64 – 12.36 |
| E | normal | 700000 | 16700 | 649900 - 750100 |
| | | 2100000 | 50100 | 1949700 – 2250300 |
| Y | normal | 4000 | 70 | 3790 – 4210 |
| | | 2000 | 35 | 1895 – 2105 |

It should be emphasized that the selection of normal distribution is made only to demonstrate the stochastic effects. Results can be obtained using many other types of statistical probability density functions, and may include correlation effects between any two random variables.

In the following figures, a dispersion range of the buckling stress (computed using Eq.(7) with the data of Table 1) are shown. The lower line represents a bound below which lie 0.135% of the buckling stresses. The upper line represents a bound under which lie 99.865% of the buckling stresses. In the range between these two limiting lines lie 99.73% of all the expected results. This is equivalent to the normal dispersion between $-3\sigma$ and $+3\sigma$ of the normal distribution. Other ranges can also be computed. In Figure 7, the ranges are shown for **E** with $\mathbf{E}_{\mathbf{mean}}$=700000 kg/cm$^2$, **Y** with $\mathbf{Y}_{\mathbf{mean}}$=4000 kg/cm$^2$ , for two mean values of normal random **N**, with standard deviations of all variables given in Table 1. In Figure 8, the ranges are shown for **N** with $\mathbf{N}_{\mathbf{mean}}$=3, **Y** with $\mathbf{Y}_{\mathbf{mean}}$=4000 kg/cm$^2$, for two mean values of normal random **E**, with standard deviations of all variables also given in Table 1. In Figure 9, the ranges are shown for **N** with $\mathbf{N}_{\mathbf{mean}}$=3, **E** with $\mathbf{E}_{\mathbf{mean}}$=700000 kg/cm$^2$, with the standard deviations of all variables also given in Table 1.

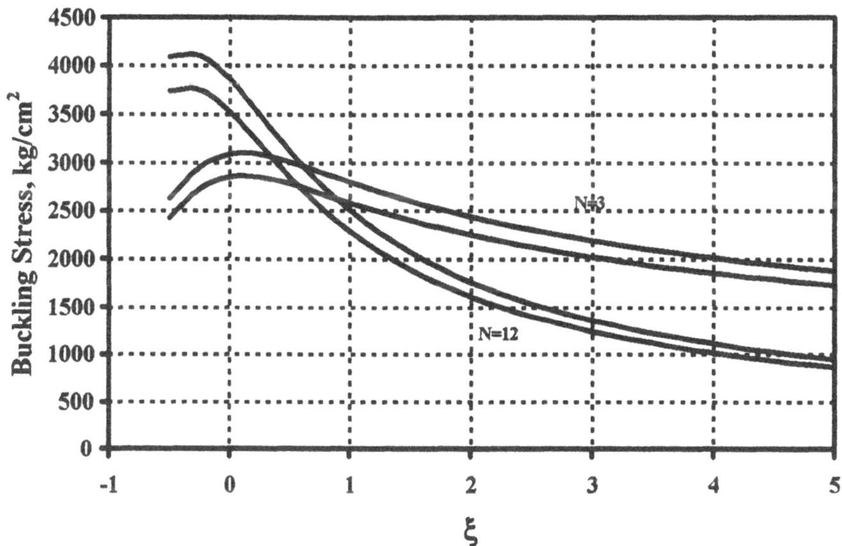


Figure 7: Buckling Load Dispersion Due to Dispersion in E and Y, for Two values of N

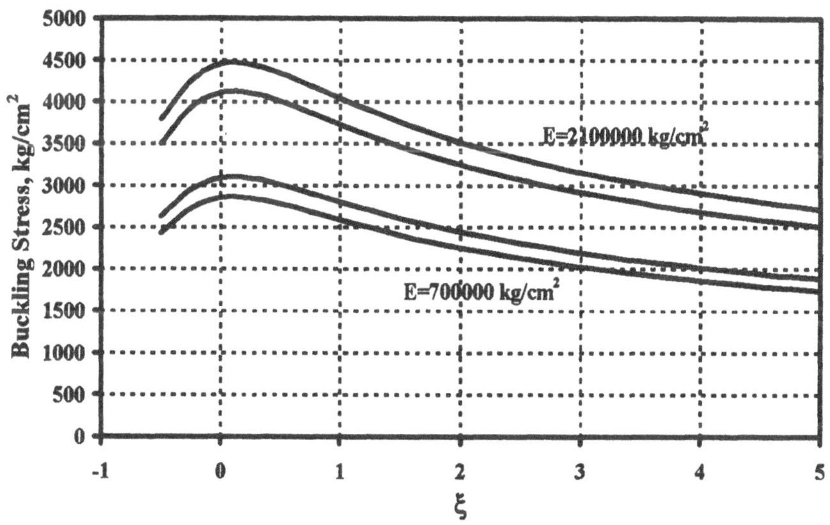


Figure 8: Buckling Load Dispersion Due to Dispersion in Y and N, for Two values of E

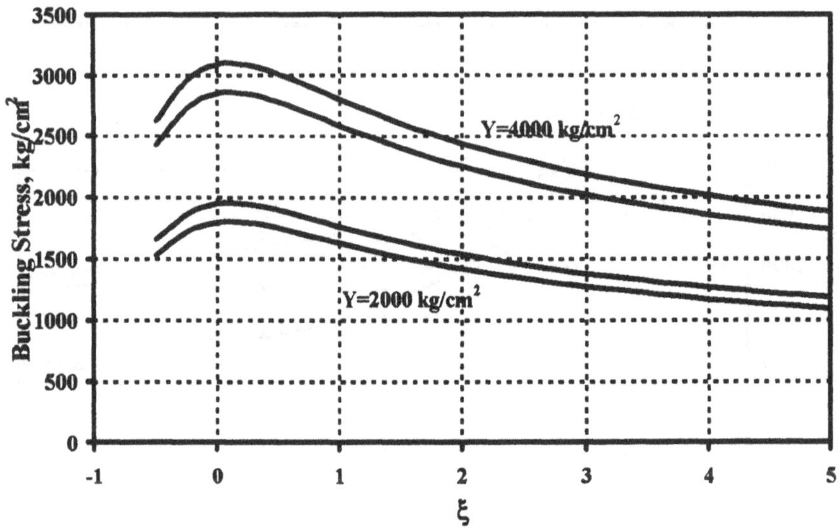


Figure 9: Buckling Load Dispersion Due to Dispersion in E and N for Two Values of Y

## 6. Probability of Failure

The probability of failure of a structure can be computed only when loading is also applied. Therefore a loading stress **P**, which is also considered a random variable has to be applied.

The mean value of the load $P$ was determined using deterministic engineering considerations. A safety factor of 1.1 was selected, so that the calculated deterministic buckling stress should be 10% higher than the applied load, leaving a deterministic margin of safety of 10%.

In Table 2, the mean values for the deterministic buckling stresses and the selected means for the applied stress are shown for two values of **N** and for two values of $\xi$.

Table 2: Deterministic Buckling Stresses and Means for the Applied Stress

| | $\xi=0$ | | $\xi=2$ | |
|---|---|---|---|---|
| N | $P_b$, kg/cm² | P, kg/cm² | $P_b$, kg/cm² | P, kg/cm² |
| 3 | 2970 | 2700 | 2350 | 2135 |
| 12 | 3685 | 3350 | 1685 | 1532 |

The resistance term of the generic failure function Eq.(1) is the allowed stress, which is equal to the buckling stress $\mathbf{P_b}$, Eq.(7). The load effect term is the load **P**, thus the failure function is

$$\mathbf{g(E,Y,N,P)=E\cdot\left[\frac{\frac{8}{3}(1+\xi)\cdot\left(1+\sqrt{1-C^2}\right)}{\left(1+\xi+\xi^2\right)^{\frac{N-1}{N}}}\right]^{\frac{1}{N}}\cdot\left(\frac{E}{Y}\right)^{\frac{1-N}{N}}\alpha^{\frac{1}{N}}\text{-}P=0} \tag{10}$$

and the probability of failure $\mathbf{P_f}$ is

$$\mathbf{P_f=Prob(g\leq 0)} \tag{11}$$

The external load **P** is assumed to be normally distributed, with a mean value given in Table 2. In order to examine the effect of the dispersion of the external load, different values of standard deviations were selected, and the influence of the coefficient of variation (COV = standard deviation divided by the mean value) on the probability of failure was examined.

In Figure 10, results for the 0.5 cm thick plate, made of material with mean elastic modulus **E**=700000 kg/cm$^2$, and mean yield stress **Y**=4000 kg/cm$^2$, were calculated for mean values of **N**=3 and 12. The standard deviations of these three random variables are those tabulated in Table 1. All the numerical computations were performed using the FPI program. Several analysis methods were used, such as FORM, SORM and Monte Carlo simulations. All methods yield similar results.

It can be seen that when the material properties are stochastic but the loading is deterministic (COV=0), the probability of failure is $1\cdot 10^{-8}\,\%$, or $1\cdot 10^{-10}$ for **N**=3. This probability is increased tremendously when the loading is also stochastic. For COV=2%, probability of failure is $5\cdot 10^{-3}\%$ or $5\cdot 10^{-5}$ for **N**=3. When the dispersion in the load is increased (i.e. COV=4%) the probability of failure is 1%, or 0.01, etc.

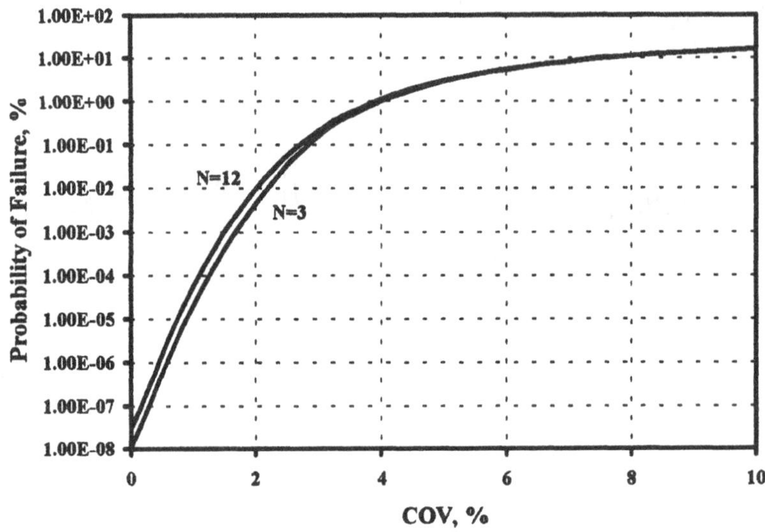


Figure 10: Effect of COV of the External Load on the Probability of Failure

Using the described computational scheme, the effects of variation in any other structural parameter (material properties, geometry, dimensions) can be calculated. The sensitivity of the success or failure of a design to dispersion in any structural variable can be computed. These sensitivities are of great importance to the designer when a robust design is required.

## 7. Concluding Remarks

The purpose of this paper is to demonstrate the concepts of probabilistic structural analysis. The probabilistic (stochastic) approach enables a designer to check the effects of uncertainties in the physical design parameters of a structural problem on the outcome stress (or other results such as displacements, accelerations etc.) for a given design. Also, probability of failure can be estimated using the described process, instead of the traditional "Factor-of-Safety" concept.

The stochastic approach was demonstrated on the problem of plastic buckling of rectangular plates under biaxial loading, whose deterministic solution is given in the literature. Reasonable statistical dispersions were assumed for Young's modulus **E**, the linear yield stress **Y** and the stress-starin curve parameter **N**. The dispersion in buckling stress was calculated. Also, the influence of the coefficient of variation of the external load on the probability of failure was demonstrated. It was shown that the probability of buckling (failure) of the rectangular plate can vary between a very low value ($10^{-8}$%) to very high value (10%), when the uncertainty in the external load is increased, and the material's properties have statistical dispersions.

It is believed that in the near future the use of probabilistic structural analysis will dominate engineering applications, and will enable more optimal, non-conservative and robust designs.

## 8. References


1. Madsen, H.O., Krenk, S., and Lind, N.C., *Methods of Structural Safety*, Prentice-Hall, Englewood Cliffs, NJ, 1986
2. Melcher, R.E., *Structural Reliability Analysis and Prediction*, Ellis Horwood, Chichester, England, UK, 1987.
3. Bjerager, P., "On Computational Methods for Structural Reliability Analysis", *New Directions in Structural Systems Reliability*, edited by D. Frangopol, Univ. of Colorado Press, Boulder, CO, 1988, pp.52-67.
4. Shinozuka, M., "Basic Analysis of Structural Safety," *Journal of Structural Engineering*, Vol. 109, No. 3, 1983, pp.721-740.
5. Hasofer, A.M., and Lind, N.C., "Exact and Invariant Second Moment Code for Materials," *Journal of Engineering Mechanics*, Vol. 100, No. 1, 1974, pp.111-121.
6. Rackwitz, R., and Fiessler, B., "Structural Reliability Under Combined Random Load Sequences," *Computers and Structures*, Vol. 9, No. 5, 1983, pp.269-276.
7. Hoenbichler, M, and Rackwitz, R., "Non Normal Dependent Vectors in Structural Safety," *Journal of Engineering Mechanics*, Vol. 107, No. 6, 1981, pp.1127-1238.
8. "PROBAN - The Probabilistic Analysis Program Manuals," Det Norske Veritas Research, Rept. Nos. 89-2022, 89-2024, 89-2025, 89-2026, 89-2027, Hovik, Norway,1989.
9. Liu, P.L, Lin, H.Z., and Der-Kiureghian, A., "CALREL User Manual", Report No. UCB/SEMM-89/18, University of California at Berkeley, 1989.
10. FPI Users Manual
11. Maymon, G., *Some Engineering Applications in Random Vibrations and Random Structures*, AIAA Publications (Progress in Astronautics and Aeronautics Vol. 178),Reston, VA, 1998.
12. Breitung, K., "Asymptotic Approximations for Multinormal Integrals," *Journal of Engineering Mechanics*, Vol. 110, No. 3, 1984, pp. 357-366.

13. Tvedt, L., "Distribution of Quadric Forms in Normal Space Applications to Structural Reliability," *Journal of Engineering Mechanics*, Vol. 116, No. 6, 1990, pp. 1183-1199.

14. Wu, Y. T., "Efficient Methods for Mechanical and Structural Reliability and Design," Ph.D. Dissertation, Dept. of Mechanical Engineering, Univ. of Arizona, Tucson, AZ, April 1984.

15. Maymon, G., "Probability of Failure of Structures Without a Closed-Form Failure Function," *Computers and Structures*, Vol. 49, No. 2, 1993, pp. 301-313.

16. Maymon, G., "Direct Computation of the Design Point of a Stochastic Structure Using a Finite Element Code,", Structural Safety, Vol. 14, 1994, pp. 185-202.

17. Durban, D., "Plastic Buckling of Rectangular Plates Under Biaxial Loading," *Buckling of Structures, Theory and Experiments*, edited by I. Elishakoff, J. Arbocz, C.D. Babcock, Jr., and A. Libai, Elsevier, Amsterdam, the Netherlands, 1988, pp. 183-194.

# ON USING ROTATIONS AS PRIMARY VARIABLES IN THE NON-LINEAR THEORY OF THIN IRREGULAR SHELLS

W. PIETRASZKIEWICZ
*Polish Academy of Sciences, Institute of Fluid-Flow Machinery*
*Ul. Gen. J. Fiszera 14, 80-952 Gdańsk, Poland*

## 1. Introduction

A non-linear theory of thin shell-like structures with irregularities of geometry, material properties and deformation along singular curves was developed in Makowski *et al.* [1,2]. In these papers an irregular structure was modelled by a reference network being a union of piecewise smooth surfaces and surface curves resisting only the stretching and bending. The resulting boundary value problem was expressed through displacement vector as the only independent field variable.

In the general approach to the non-linear theory of shells presented in the book of Libai and Simmonds [3] the finite rotation field appears naturally as one of primary variables of the boundary value problem. This statically and geometrically exact formulation of shell theory, which grew from early ideas of Reissner [4] and Simmonds [5], allowed one to develop effective computational procedures based on the finite element method for both the regular shells [6] and the irregular shell-like structures [7]. The classical thin shell theory was defined in [3] with the help of the Kirchhoff hypothesis regarded as a constitutive hypothesis and not as a kinematic one. It was confirmed in [3] that in the classical theory the rotations become expressible through displacements and are no longer independent field variables. Thus, in order to regard them again as primary variables some additional constraint conditions with Lagrange multipliers should be imposed.

The rotation angle as one of primary variables of thin shell theory was first introduced by Reissner [8] to describe a one-dimensional axisymmetric deformation state of a thin shell of revolution. Simmonds and Danielson [9,10] formulated two-dimensional thin shell relations in terms of the finite rotation and stress function vectors, and derived an appropriate variational principle. Several alternative forms of relations for thin shells expressed in terms of rotations were developed by Pietraszkiewicz [11-14], Shkutin [15], Valid [16,17], Atluri [18], and Libai and Simmonds [19]. In particular, within the geometrically non-linear theory of thin, regular, isotropic, elastic shells many such relations were summarised in Chapter 5 of [14], where references to earlier papers can be found.



*D. Durban et al. (eds.), Advances in the Mechanics of Plates and Shells,* 245–258.

The aim of this report is to extend the results presented in Chapter 5 of [14] in three directions:

a) In place of the reference surface we introduce the reference network defined in [1,2]. This allows one, also within the rotational formulation, to take into account various irregularities of shell geometry, deformation and mechanical properties along singular curves.

b) In all shell relations large surface strains are admitted. This allows one to discuss within the same formulation also large strain problems of irregular shells made, for example, of a rubber-like material with constitutive equations proposed in [3,20].

c) The boundary terms for each regular surface element are discussed in more detail, which allows one to derive an appropriate form of jump conditions at singular curves and points representing the irregularities.

The deformation of the reference network models entirely the deformation of a thin, irregular, shell-like structure. The network consists of a finite number of regular surface elements connected together along singular spatial curves. The equilibrium conditions of the entire structure are given in Chapter 3 by the postulated principle of virtual work (PVW) in which the internal surface stress and strain fields are associated only with stretching and bending of the reference network. Then appropriate constraints with Lagrange multipliers are introduced into the PVW in order to regard also the rotations as primary variables. Transforming the so modified PVW we obtain the known, [14], local forms of equilibrium equations and boundary conditions. We also derive the local forms of jump conditions at the singular curves (45) and at the singular points (43) and (46). The jump conditions seem to be new in the literature.

## 2. Geometry and deformation of a regular surface element

In this report we shall apply primarily the system of notation used in [14] and remind here only basic relations.

Let $\mathcal{M}^{(k)}$ be a connected, oriented and regular surface element of class $C^n, n \geq 2$, in the three-dimensional Euclidean point space $\mathcal{E}$ whose translation (three-dimensional vector) space is $E$. The position vector of a point $M \in \mathcal{M}^{(k)}$ is given by

$$\mathbf{r} = \overrightarrow{OM} = \mathbf{r}(\theta^\alpha), \tag{1}$$

where $O \in \mathcal{E}$ is a reference origin and $\theta^\alpha, \alpha = 1,2$, are surface co-ordinates. At $M \in \mathcal{M}^{(k)}$ we have the natural base vectors $\mathbf{a}_\alpha = \partial \mathbf{r} / \partial \theta^\alpha \equiv \mathbf{r}_{,\alpha}$, the dual base vectors $\mathbf{a}^\beta$ such that $\mathbf{a}^\beta \cdot \mathbf{a}_\alpha = \delta^\beta_\alpha$, where $\delta^\beta_\alpha$ is the Kronecker symbol, the components $a_{\alpha\beta} = \mathbf{a}_\alpha \cdot \mathbf{a}_\beta$ and $a^{\alpha\beta} = \mathbf{a}^\alpha \cdot \mathbf{a}^\beta$ of the surface metric tensor $\mathbf{a}$ with $a = \det(a_{\alpha\beta}) > 0$, the unit normal vector $\mathbf{n} = (1/\sqrt{a})\mathbf{a}_1 \times \mathbf{a}_2$ orienting $\mathcal{M}^{(k)}$, the components $b_{\alpha\beta} = -\mathbf{n}_{,\alpha} \cdot \mathbf{a}_\beta$

of the surface curvature tensor **b**, and the components $\varepsilon_{\alpha\beta}=(\mathbf{a}_\alpha\times\mathbf{a}_\beta)\bullet\mathbf{n}$ of the surface permutation tensor such that $\varepsilon_{12}=-\varepsilon_{21}=\sqrt{a}$, $\varepsilon_{11}=\varepsilon_{22}=0$.

The boundary $\partial\mathcal{M}^{(k)}$ of $\mathcal{M}^{(k)}$ consists of a finite number of closed, piecewise smooth curves that do not meet in cusps, each described parametrically by $\mathbf{r}(s)=\mathbf{r}[\theta^\alpha(s)]$, where $s$ is the arc length along any regular part of $\partial\mathcal{M}^{(k)}$. At each regular point $M\in\partial\mathcal{M}^{(k)}$ we have the unit tangent vector $\boldsymbol{\tau}=d\mathbf{r}/ds\equiv\mathbf{r}'=\tau^\alpha\mathbf{a}_\alpha$ and the outward unit normal vector $\boldsymbol{\nu}=\mathbf{r},_\nu=\boldsymbol{\tau}\times\mathbf{n}=\nu^\alpha\mathbf{a}_\alpha$, where $(\cdot),_\nu$ is the external surface derivative normal to $\partial\mathcal{M}^{(k)}$.

The deformed, regular surface element $\overline{\mathcal{M}}^{(k)}$ with the boundary $\partial\overline{\mathcal{M}}^{(k)}$ is described relative to the same origin $O\in\mathcal{E}$ by the relations

$$\bar{\mathbf{r}}(\theta^\alpha)=\chi[\mathbf{r}(\theta^\alpha)]=\mathbf{r}(\theta^\alpha)+\mathbf{u}(\theta^\alpha),\quad \bar{\mathbf{r}}(s)=\chi[\mathbf{r}(s)]=\mathbf{r}(s)+\mathbf{u}(s), \tag{2}$$

where $\theta^\alpha$ and $s$ are convected surface co-ordinates, $\chi:\mathcal{M}^{(k)}\to\overline{\mathcal{M}}^{(k)}$ is the deformation function, and $\mathbf{u}\in E$ is the displacement vector.

In the convected surface co-ordinates all geometric relations at any regular $\bar{M}\in\partial\overline{\mathcal{M}}^{(k)}$ are now analogous to those given at $M\in\partial\mathcal{M}^{(k)}$, and are expressed by quantities marked by a dash: $\bar{\mathbf{a}}_\alpha,\bar{\mathbf{a}}^\beta,\bar{a}_{\alpha\beta},\bar{a}^{\alpha\beta},\bar{\mathbf{a}},\bar{a},\bar{b}_{\alpha\beta},\bar{\varepsilon}_{\alpha\beta},\bar{\boldsymbol{\nu}},\bar{\boldsymbol{\tau}}$, etc. The dashed quantities can be expressed through analogous quantities defined on $\mathcal{M}^{(k)}$ and the displacement field **u** with the help of formulae given in [14,21].

Components of the Green type surface deformation measures are defined by

$$\gamma_{\alpha\beta}(\mathbf{u})=\frac{1}{2}\left(\bar{a}_{\alpha\beta}-a_{\alpha\beta}\right),\quad \kappa_{\alpha\beta}(\mathbf{u})=-\left(\bar{b}_{\alpha\beta}-b_{\alpha\beta}\right), \tag{3}$$

where $\gamma_{\alpha\beta}(\mathbf{u})$ are quadratic polynomials of $\mathbf{u},\mathbf{u},_\alpha$, and $\kappa_{\alpha\beta}(\mathbf{u})$ are non-rational functions of $\mathbf{u},\mathbf{u},_\alpha,\mathbf{u},_{\alpha\beta}$.

In the neighbourhood of the regular surface elements $\mathcal{M}^{(k)}$ and $\overline{\mathcal{M}}^{(k)}$ the space $\mathcal{E}$ can be parameterised by the normal system of convected co-ordinates $(\theta^\alpha,\varsigma)$, where $\varsigma$ is the distance from $\mathcal{M}^{(k)}$ and $\overline{\mathcal{M}}^{(k)}$ along $\mathbf{n}$ and $\bar{\mathbf{n}}$, respectively. Extending the domain of $\chi$ to the neighbourhood of $\mathcal{M}^{(k)}$, the spatial deformation gradient $\mathbf{F}:E\to E$ taken at the surface element $\mathcal{M}^{(k)}$ has the form

$$\begin{aligned}&\mathbf{F}=\nabla\chi(\mathbf{r}+\varsigma\mathbf{n})\big|_{\varsigma=0}=\bar{\mathbf{a}}_\alpha\otimes\mathbf{a}^\alpha+\bar{\mathbf{n}}\otimes\mathbf{n},\quad \det\mathbf{F}=\sqrt{\frac{\bar{a}}{a}}>0,\\&\frac{\bar{a}}{a}=1+2\gamma^\alpha_\alpha+2\left(\gamma^\alpha_\alpha\gamma^\beta_\beta-\gamma^\alpha_\beta\gamma^\beta_\alpha\right),\end{aligned} \tag{4}$$

where $\otimes$ is the tensor product.

The left polar decomposition of **F** gives

$$\mathbf{F} = \mathbf{VR}, \quad \mathbf{r}_\alpha = \mathbf{Ra}_\alpha = \mathbf{V}^{-1}\bar{\mathbf{a}}_\alpha. \tag{5}$$

Here $\mathbf{R} \in SO(3)$ is the rotation tensor, **V** is the left spatial stretch tensor at $\bar{\mathcal{M}}^{(k)}$, and $\mathbf{r}_\alpha$ are the rotated surface non-holonomic base vectors. These fields satisfy the relations

$$\begin{aligned} &\mathbf{R} = \mathbf{r}_\alpha \otimes \mathbf{a}^\alpha + \bar{\mathbf{n}} \otimes \mathbf{n}, \quad \mathbf{R}^T = \mathbf{R}^{-1}, \quad \det\mathbf{R} = +1, \\ &\mathbf{V} = \bar{\mathbf{a}}_\alpha \otimes \mathbf{r}^\alpha + \bar{\mathbf{n}} \otimes \bar{\mathbf{n}}, \quad \mathbf{V}^T = \mathbf{V}, \quad \det\mathbf{V} = \sqrt{\frac{\bar{a}}{a}} > 1. \end{aligned} \tag{6}$$

The modified surface deformation measures associated with $\mathbf{r}_\alpha$ are introduced through the following formulae, [14]:

$$\begin{aligned} &\boldsymbol{\eta} = \mathbf{V} - \mathbf{1} = \boldsymbol{\eta}_\beta \otimes \mathbf{r}^\beta, \quad \boldsymbol{\eta}_\beta = \eta_{\alpha\beta}\mathbf{r}^\alpha, \\ &\boldsymbol{\mu} = \left(\bar{\mathbf{n}},_\beta - \mathbf{Rn},_\beta\right) \otimes \mathbf{r}^\beta = \boldsymbol{\mu}_\beta \otimes \mathbf{r}^\beta, \quad \boldsymbol{\mu}_\beta = \mu_{\alpha\beta}\mathbf{r}^\alpha, \\ &\eta_{\alpha\beta} = \eta_{\beta\alpha}, \quad \mu_{\alpha\beta} \neq \mu_{\beta\alpha}. \end{aligned} \tag{7}$$

Here $\mathbf{1} = \mathbf{a}_\alpha \otimes \mathbf{a}^\alpha + \mathbf{n} \otimes \mathbf{n} = \mathbf{r}_\alpha \otimes \mathbf{r}^\alpha + \bar{\mathbf{n}} \otimes \bar{\mathbf{n}}$ is the spatial identity tensor. The surface measures $\eta_{\alpha\beta}$ and $\mu_{\alpha\beta}$ satisfy useful kinematic relations given in [14].

Along the boundary $\partial\bar{\mathcal{M}}^{(k)}$ we have

$$\begin{aligned} &\bar{\mathbf{a}}_\tau \equiv \bar{\mathbf{r}}' = a_\tau \bar{\boldsymbol{\tau}}, \quad \bar{\mathbf{a}}_\nu = \bar{\mathbf{a}}_\tau \times \bar{\mathbf{n}} = a_\tau \bar{\boldsymbol{\nu}}, \quad \bar{\mathbf{n}} = \sqrt{\frac{a}{\bar{a}}}\bar{\mathbf{r}},_\nu \times \bar{\mathbf{r}}', \\ &a_\tau = |\bar{\mathbf{r}}'| = \sqrt{1 + 2\gamma_{\tau\tau}}, \quad \gamma_{\tau\tau} = \gamma_{\alpha\beta}\tau^\alpha\tau^\beta. \end{aligned} \tag{8}$$

The transformation of $(\boldsymbol{\nu}, \boldsymbol{\tau}, \mathbf{n})$ during deformation into $(\bar{\mathbf{a}}_\nu, \bar{\mathbf{a}}_\tau, \bar{\mathbf{n}})$ is performed in two steps: the rotation of $(\boldsymbol{\nu}, \boldsymbol{\tau}, \mathbf{n})$ into $(\bar{\boldsymbol{\nu}}, \bar{\boldsymbol{\tau}}, \bar{\mathbf{n}})$ by the total rotation tensor $\mathbf{R}_\tau$ with the subsequent extension of $\bar{\boldsymbol{\nu}}, \bar{\boldsymbol{\tau}}$ into $\bar{\mathbf{a}}_\nu, \bar{\mathbf{a}}_\tau$ by the factor $a_\tau$:

$$\begin{aligned} &\bar{\mathbf{a}}_\nu = a_\tau \mathbf{R}_\tau \boldsymbol{\nu}, \quad \bar{\mathbf{a}}_\tau = a_\tau \mathbf{R}_\tau \boldsymbol{\tau}, \quad \bar{\mathbf{n}} = \mathbf{R}_\tau \mathbf{n}, \\ &\mathbf{R}_\tau = \frac{1}{a_\tau}\left(\bar{\mathbf{a}}_\nu \otimes \boldsymbol{\nu} + \bar{\mathbf{a}}_\tau \otimes \boldsymbol{\tau} + \bar{\mathbf{n}} \otimes \mathbf{n}\right). \end{aligned} \tag{9}$$

Kinematic relations involving the tensors **R** and $\mathbf{R}_\tau$ are given in [14].

### 3. Principle of virtual work for thin irregular shells

A consistent formulation of the mechanical boundary value problem for thin irregular shell-like structures was developed in Makowski *et al.* [1,2], where the displacements were taken as the only independent field variables. The mechanical modelling of such structures was based on two postulates:

- *The deformation of the entire shell-like structure is determined by deformation of a distinguished surface-like continuum, called the shell reference network.*
- *The equilibrium conditions of the entire structure are determined by a suitable form of the principle of virtual work involving only the fields associated with the stretching and bending of the reference network.*

The undeformed reference network $\mathcal{M} \subset E$ introduced in [1] consists of a finite number of regular surface elements $\mathcal{M}^{(k)}, k = 1,2,\ldots,K,$ with the following properties:

a) No two distinct elements $\mathcal{M}^{(k)}$ have common interior points.

b) Two or more distinct elements may have a smooth spatial curve $\Gamma^{(a)}$ as a common part of the boundaries, which is defined by

$$\Gamma^{(a)} = \partial\mathcal{M}^{(k_1)} \cap \partial\mathcal{M}^{(k_2)} \cap \ldots \cap \partial\mathcal{M}^{(k_m)} \quad \text{if } k_1 \neq k_2 \neq \ldots \neq k_m. \tag{10}$$

c) Two or more distinct curves $\Gamma^{(a)}$ may have in common only single isolated points.

Each $\mathcal{M}^{(k)}$ represents a reference surface of a regular shell part. Each $\Gamma^{(a)}$ can be a surface curve across which some fields fail to be smooth. Examples of geometric irregularities along $\Gamma^{(a)}$ are surface folds or intersections of two or more regular surfaces. Shell parts can be made of different materials, or there may be stepwise thickness changes at $\Gamma^{(a)}$. However, $\Gamma^{(a)}$ can also represent a reference axis of a rod-like element, a technological junction, a plastic hinge developing during deformation process, etc.

The network $\mathcal{M}$ is then regarded as the union of all the closed elements $\mathcal{M}^{(k)} \cup \partial\mathcal{M}^{(k)}$, and the singular curve $\Gamma \in \mathcal{M}$ is regarded as the union of all the curves $\Gamma^{(a)}$:

$$\mathcal{M} = \bigcup_{k=1}^{K}\left(\mathcal{M}^{(k)} \cup \partial\mathcal{M}^{(k)}\right), \quad \Gamma = \bigcup_{a=1}^{A}\Gamma^{(a)}. \tag{11}$$

From (11) it is apparent that $\Gamma \subset \mathcal{M}$. The boundary $\partial\mathcal{M}$ of the entire network $\mathcal{M}$ defined by

$$\partial\mathcal{M} = \left(\bigcup_{k=1}^{K}\partial\mathcal{M}^{(k)}\right) \setminus \Gamma \tag{12}$$

consists of a finite number of spatial curves. Several examples of such networks are given in [2].

Deformation of $\mathcal{M}$ can be described by two deformation functions: $\chi: \mathcal{M}\backslash\Gamma \rightarrow \mathcal{E}$ and $\chi_\Gamma: \Gamma \rightarrow \mathcal{E}$, since the singular curve may be admitted to follow its own deformation, in general. In many cases the deformation $\chi$ may be defined on the entire $\mathcal{M}$, and then $\chi_\Gamma$ is a restriction of $\chi$ at $\Gamma$: $\chi_\Gamma = \chi|_\Gamma$. However, we do not assume such a restricted shell deformation at the moment.

The principle of virtual work compatible with the two postulates given above can be taken in the form

$$G \equiv G_{\text{int}} - G_{\text{ext}} - G_\Gamma = 0, \tag{13}$$

where $G_{\text{int}} = G_{\text{int}}(\mathbf{u};\delta\mathbf{u})$ represents the internal virtual work, $G_{\text{ext}} = G_{\text{ext}}(\mathbf{u};\delta\mathbf{u})$ is the external virtual work, and $G_\Gamma = G_\Gamma(\mathbf{u}_\Gamma;\delta\mathbf{u}_\Gamma)$ is the additional virtual work of the generalised forces acting along $\Gamma$. Here we have explicitly indicated that all the virtual works are functionals of the displacements as the only independent field variables. The individual parts of (13) are defined by

$$\begin{aligned}
G_{\text{int}} &= \sum_{k=1}^{K} \iint_{\mathcal{M}^{(k)}} \left(N^{\alpha\beta}\delta\gamma_{\alpha\beta} + M^{\alpha\beta}\delta\kappa_{\alpha\beta}\right)dA,\\
G_{\text{ext}} &= \sum_{k=1}^{K} \iint_{\mathcal{M}^{(k)}} \left(\mathbf{p}\cdot\delta\mathbf{u} + \mathbf{h}\cdot\delta\bar{\mathbf{n}}\right)dA + \int_{\partial\mathcal{M}_f} \left(\mathbf{N}^*\cdot\delta\mathbf{u} + \mathbf{H}^*\cdot\delta\bar{\mathbf{n}}\right)ds, \qquad (14)\\
G_\Gamma &= \int_\Gamma \sigma_\Gamma ds + \sum_{P_i\in\Gamma} \sigma_i .
\end{aligned}$$

Here $N^{\alpha\beta}$ and $M^{\alpha\beta}$ are components of the symmetric stress resultant and stress couple tensors of the Piola-Kirchhoff type, $\delta$ is the symbol of variation, $\delta\gamma_{\alpha\beta}$ and $\delta\kappa_{\alpha\beta}$ are virtual changes of the surface deformation measures (3), $\mathbf{p}$ and $\mathbf{h}$ are the external surface force and moment resultant vectors, $\mathbf{N}^*$ and $\mathbf{H}^*$ are the external boundary force and moment resultant vectors, whereas $\sigma_\Gamma$ and $\sigma_i$ are the external virtual work densities along regular parts of $\Gamma$ and at any singular point $P_i \in \Gamma$, respectively. The explicit forms of $\sigma_\Gamma$ and $\sigma_i$ depend on the type of irregularity assumed along $\Gamma$, [2]. Note that since $\bar{\mathbf{n}}\cdot\delta\bar{\mathbf{n}} = 0$, only the surface components of $\mathbf{h}$ and $\mathbf{H}^*$ can explicitly be taken into account in the non-linear theory of thin irregular shells discussed here. Transforming (13) and (14) with the help of Stokes' theorem, the local equilibrium equations, boundary conditions and jump conditions at singular curves were derived, [1,2].

If the rotation tensor $\mathbf{R}$ is supposed to be an independent field variable of the boundary value problem, some constraint conditions have to be introduced into the relations (14) and the virtual densities should be expressed in terms of modified surface deformation and stress measures.

Let us remind that the components $\gamma_{\alpha\beta}, \kappa_{\alpha\beta}$ and $\eta_{\alpha\beta}, \mu_{\alpha\beta}$ of the surface deformation measures are related by (see [14], formula (5.11))

$$\begin{aligned} \gamma_{\alpha\beta} &= \eta_{\alpha\beta} + \frac{1}{2}\eta^{\lambda}_{\alpha}\eta_{\lambda\beta}, \\ \kappa_{\alpha\beta} &= \frac{1}{2}\left[\left(\delta^{\lambda}_{\alpha} + \eta^{\lambda}_{\alpha}\right)\mu_{\lambda\beta} + \left(\delta^{\lambda}_{\beta} + \eta^{\lambda}_{\beta}\right)\mu_{\lambda\alpha}\right] - \frac{1}{2}\left(b^{\lambda}_{\alpha}\eta_{\lambda\beta} + b^{\lambda}_{\beta}\eta_{\lambda\alpha}\right). \end{aligned} \tag{15}$$

As a result, the internal surface virtual work density appearing in $(14)_1$ can be presented in an alternative form

$$\begin{aligned} & N^{\alpha\beta}\delta\gamma_{\alpha\beta} + M^{\alpha\beta}\delta\kappa_{\alpha\beta} = S^{\alpha\beta}\delta\eta_{\alpha\beta} + H^{\alpha\beta}\delta\mu_{\alpha\beta}, \\ & S^{\alpha\beta} = N^{\alpha\beta} + \frac{1}{2}\left(\eta^{\alpha}_{\lambda}N^{\lambda\beta} + \eta^{\beta}_{\lambda}N^{\alpha\lambda}\right) - \frac{1}{2}\left[\left(b^{\alpha}_{\lambda} - \mu^{\alpha}_{.\lambda}\right)M^{\lambda\beta} + \left(b^{\beta}_{\lambda} - \mu^{\beta}_{.\lambda}\right)M^{\alpha\lambda}\right], \\ & H^{\alpha\beta} = \left(\delta^{\alpha}_{\lambda} + \eta^{\alpha}_{\lambda}\right)M^{\lambda\beta}, \end{aligned} \tag{16}$$

where now $S^{\alpha\beta} = S^{\beta\alpha}$, but $H^{\alpha\beta} \neq H^{\beta\alpha}$, in general.

In the non-linear theory of thin shells the rotation tensor $\mathbf{R}$ is a non-rational function of $\mathbf{u}, \mathbf{u}_{,\alpha}$ (explicit formulae are given in [11,13]). This dependence of $\mathbf{R}$ upon $\mathbf{u}$ can also be expressed implicitly through three constraint conditions [22,14]

$$\varepsilon^{\alpha\beta}\mathbf{r}_{\alpha}\bullet\eta_{\lambda\beta}\mathbf{r}^{\lambda} = 0, \quad \bar{\mathbf{n}}\bullet\eta_{\lambda\beta}\mathbf{r}^{\lambda} = 0. \tag{17}$$

These constraints express the known property of the relative surface strain tensor $\boldsymbol{\eta}$, which in thin shell theory is symmetric and does not have out-of-surface components. The property was also confirmed by Libai and Simmonds [3] who used the constitutive Kirchhoff hypothesis to define the classical theory of thin shells as a special case of the general shell theory.

For a virtual deformation the relations (17) put the following constraints on the virtual changes $\delta\eta_{\alpha\beta}$ of $\eta_{\alpha\beta}$:

$$\varepsilon^{\alpha\beta}\mathbf{r}_{\alpha}\bullet\delta\eta_{\lambda\beta}\mathbf{r}^{\lambda} = 0, \quad \bar{\mathbf{n}}\bullet\delta\eta_{\lambda\beta}\mathbf{r}^{\lambda} = 0. \tag{18}$$

Inside of each $\mathcal{M}^{(k)}$ the constraints (18) can be introduced into the surface integral of $(14)_1$ with the help of the respective Lagrange multipliers $S$ and $Q^{\beta}$. It was shown in [14] that in order to express also the boundary terms at each $\partial\mathcal{M}^{(k)}$ explicitly through independent rotations it is necessary to introduce a line integral over $\partial\mathcal{M}^{(k)}$ into $(14)_1$ with the constraints $(18)_2$ multiplied by $B\tau^{\beta}$. Additionally, in $(14)_2$ the external virtual work done by the moments $\mathbf{h}$ and $\mathbf{H}^*$ should be expressed directly in terms of now independent virtual rotations. As a result, $(14)_{1,2}$ can be modified to the form

$$G_{\rm int}=\sum_{k=1}^{K}\left\{\iint_{\mathcal{M}^{(k)}}\left(\mathbf{N}^{\beta}\cdot\delta\eta_{\lambda\beta}\mathbf{r}^{\lambda}+H^{\alpha\beta}\mathbf{r}_{\alpha}\cdot\delta\mu_{\lambda\beta}\mathbf{r}^{\lambda}\right)dA+\int_{\partial\mathcal{M}^{(k)}}B\tau^{\beta}\overline{\mathbf{n}}\cdot\delta\eta_{\lambda\beta}\mathbf{r}^{\lambda}ds\right\},$$
$$G_{\rm ext}=\sum_{k=1}^{K}\iint_{\mathcal{M}^{(k)}}(\mathbf{p}\cdot\delta\mathbf{u}+\mathbf{m}\cdot\boldsymbol{\omega})dA+\int_{\partial\mathcal{M}_f}(\mathbf{N}^{*}\cdot\delta\mathbf{u}+\mathbf{M}^{*}\cdot\boldsymbol{\omega}_{\tau})ds, \tag{19}$$

where now $G_{\rm int}=G_{\rm int}(\mathbf{u},\mathbf{R};\delta\mathbf{u},\boldsymbol{\omega})$ and $G_{\rm ext}=G_{\rm ext}(\mathbf{u},\mathbf{R};\delta\mathbf{u},\boldsymbol{\omega})$, while other fields are defined by

$$\mathbf{N}^{\beta}=\left(S^{\alpha\beta}+\varepsilon^{\alpha\beta}S\right)\mathbf{r}_{\alpha}+Q^{\beta}\overline{\mathbf{n}},\quad \mathbf{M}^{\beta}=\overline{\mathbf{n}}\times H^{\alpha\beta}\mathbf{r}_{\alpha},$$
$$\mathbf{M}^{*}=\overline{\mathbf{n}}\times\mathbf{H}^{*},\quad \mathbf{m}=\overline{\mathbf{n}}\times\mathbf{h}, \tag{20}$$

$$\boldsymbol{\omega}=\frac{1}{2}(\mathbf{1}\times\mathbf{1})\cdot\left(\delta\mathbf{R}\mathbf{R}^{T}\right)=\frac{1}{2}\left(\mathbf{r}^{\alpha}\times\delta\mathbf{r}_{\alpha}+\overline{\mathbf{n}}\times\delta\overline{\mathbf{n}}\right),$$
$$\boldsymbol{\omega}_{\tau}=\frac{1}{2}(\mathbf{1}\times\mathbf{1})\cdot\left(\delta\mathbf{R}_{\tau}\mathbf{R}_{\tau}^{T}\right)=\frac{1}{2}\left(\overline{\mathbf{v}}\times\delta\overline{\mathbf{v}}+\overline{\boldsymbol{\tau}}\times\delta\overline{\boldsymbol{\tau}}+\overline{\mathbf{n}}\times\delta\overline{\mathbf{n}}\right). \tag{21}$$

Here $\boldsymbol{\omega}$ and $\boldsymbol{\omega}_{\tau}$ are the virtual rotation vectors in the interior of each $\mathcal{M}^{(k)}$ and along each $\partial\mathcal{M}^{(k)}$, respectively. Please note that all the couple vectors $\mathbf{M}^{\beta}$, $\mathbf{M}^{*}$ and $\mathbf{m}$ in (20) do not have normal components, that is $\mathbf{M}^{\beta}\cdot\overline{\mathbf{n}}=\mathbf{M}^{*}\cdot\overline{\mathbf{n}}=\mathbf{m}\cdot\overline{\mathbf{n}}\equiv 0$. This is the fundamental property of the theory of thin shells resulting from the two basic postulates given above.

## 4. Local field equations

Let us transform the virtual work principle (13) with (19) keeping in mind that both $\mathbf{R}$ and $\mathbf{u}$ are now the independent field variables subject to variation.

According to [14], the virtual deformation measures $\delta\eta_{\alpha\beta}$ and $\delta\mu_{\alpha\beta}$ are expressible through $\delta\mathbf{u}$ and $\boldsymbol{\omega}$ by the relations

$$\delta\eta_{\lambda\beta}\mathbf{r}^{\lambda}=\delta\mathbf{u},_{\beta}+\overline{\mathbf{a}}_{\beta}\times\boldsymbol{\omega},\quad \delta\mu_{\lambda\beta}\mathbf{r}^{\lambda}=\boldsymbol{\omega},_{\beta}\times\overline{\mathbf{n}}. \tag{22}$$

Introducing (22) into $(19)_1$, the virtual work principle (13) takes the form

$$\sum_{k=1}^{K}\left\{\iint_{\mathcal{M}^{(k)}}\left[\mathbf{N}^{\beta}\cdot\left(\delta\mathbf{u},_{\beta}+\overline{\mathbf{a}}_{\beta}\times\boldsymbol{\omega}\right)+\mathbf{M}^{\beta}\cdot\boldsymbol{\omega},_{\beta}\right]dA+\int_{\partial\mathcal{M}^{(k)}}B\overline{\mathbf{n}}\cdot\left(\delta\mathbf{u}'+\overline{\mathbf{a}}_{\tau}\times\boldsymbol{\omega}\right)ds\right.$$
$$\left.-\iint_{\mathcal{M}^{(k)}}(\mathbf{p}\cdot\delta\mathbf{u}+\mathbf{m}\cdot\boldsymbol{\omega})dA\right\}-\int_{\partial\mathcal{M}_f}(\mathbf{N}^{*}\cdot\delta\mathbf{u}+\mathbf{M}^{*}\cdot\boldsymbol{\omega}_{\tau})ds-\int_{\Gamma}\sigma_{\Gamma}ds-\sum_{P_i\in\Gamma}\sigma_i=0. \tag{23}$$

The fields $\mathbf{N}^{\beta}$ and $\mathbf{M}^{\beta}$ are assumed to be of class $C^1$ in the interior of each regular surface element $\mathcal{M}^{(k)}$ and to have extensions of the same class to the boundary with finite limits at any $M \in \partial \mathcal{M}^{(k)}$. Then the Stokes theorem allows one to transform the first two surface integrals of (23) for each $\mathcal{M}^{(k)}$ into

$$-\iint_{\mathcal{M}^{(k)}} \left\{ \mathbf{N}^{\beta}|_{\beta} \cdot \delta\mathbf{u} + \left( \mathbf{M}^{\beta}|_{\beta} + \bar{\mathbf{a}}_{\beta} \times \mathbf{N}^{\beta} \right) \cdot \boldsymbol{\omega} \right\} dA$$
$$+ \int_{\partial \mathcal{M}^{(k)}} \left\{ \mathbf{T}_{\nu} \cdot \delta\mathbf{u} + \mathbf{K}_{\nu} \cdot \boldsymbol{\omega} + (B\bar{\mathbf{n}} \cdot \delta\mathbf{u})' \right\} ds, \tag{24}$$

where

$$\mathbf{T}_{\nu} = \mathbf{N}^{\beta}\nu_{\beta} - (B\bar{\mathbf{n}})', \quad \mathbf{K}_{\nu} = \mathbf{M}^{\beta}\nu_{\beta} - B\bar{\mathbf{a}}_{\nu}. \tag{25}$$

Along each $\partial \mathcal{M}^{(k)}$ there may be singular points $P_c, c = 1,2,\ldots,C,$ described by $s = s_c$, at which the field $B\bar{\mathbf{n}} \cdot \delta\mathbf{u}$ is not differentiable. Such singular points are, for example, corners of the closed curves composing $\partial \mathcal{M}^{(k)}$ or points of singularities of $B$, $\bar{\mathbf{n}}$ and $\delta\mathbf{u}$. At such singular points we assume the existence of finite limits of $B\bar{\mathbf{n}} \cdot \delta\mathbf{u}$ defined by

$$B_c^{\pm}\bar{\mathbf{n}}_c^{\pm} \cdot \delta\mathbf{u}_c^{\pm} = \lim_{h \to 0} \left\{ B(s_c \pm h)\bar{\mathbf{n}}(s_c \pm h) \cdot \delta\mathbf{u}(s_c \pm h) \right\} \tag{26}$$

Then, the last term in the boundary line integral of (24) can be transformed further to give

$$\int_{\partial \mathcal{M}^{(k)}} (B\bar{\mathbf{n}} \cdot \delta\mathbf{u})' ds = - \sum_{P_c \in \partial \mathcal{M}^{(k)}} \left( B_c^{+}\bar{\mathbf{n}}_c^{+} \cdot \delta\mathbf{u}_c^{+} - B_c^{-}\bar{\mathbf{n}}_c^{-} \cdot \delta\mathbf{u}_c^{-} \right). \tag{27}$$

The second term of the boundary line integral in (24) contains the virtual rotation $\boldsymbol{\omega}$, which should still be expressed through the virtual rotation $\boldsymbol{\omega}_{\tau}$ of the boundary. Let us remind that along each $\partial \mathcal{M}^{(k)}$ the total rotation tensor $\mathbf{R}_{\tau}$ is defined as the superposition of two finite rotations, [14]:

$$\mathbf{R}_{\tau} = \mathbf{Q}_{V}\mathbf{R}, \quad \mathbf{Q}_{V} = \bar{\boldsymbol{\nu}} \otimes \mathbf{r}_{\nu} + \bar{\boldsymbol{\tau}} \otimes \mathbf{r}_{\tau} + \bar{\mathbf{n}} \otimes \bar{\mathbf{n}}, \tag{28}$$

where

$$\mathbf{r}_{\nu} = \mathbf{r}_{\alpha}\nu^{\alpha} = \mathbf{R}\boldsymbol{\nu} = \frac{1}{a_{\tau}} \left\{ \left(1 + \eta_{\tau\tau}\right)\bar{\boldsymbol{\nu}} + \eta_{\nu\tau}\bar{\boldsymbol{\tau}} \right\},$$
$$\mathbf{r}_{\tau} = \mathbf{r}_{\alpha}\tau^{\alpha} = \mathbf{R}\boldsymbol{\tau} = \frac{1}{a_{\tau}} \left\{ -\eta_{\nu\tau}\bar{\boldsymbol{\nu}} + \left(1 + \eta_{\tau\tau}\right)\bar{\boldsymbol{\tau}} \right\}, \tag{29}$$
$$\bar{\boldsymbol{\nu}} = \mathbf{Q}_{V}\mathbf{r}_{\nu}, \quad \bar{\boldsymbol{\tau}} = \mathbf{Q}_{V}\mathbf{r}_{\tau}, \quad \eta_{\nu\tau} = \eta_{\alpha\beta}\nu^{\alpha}\tau^{\beta}, \quad \eta_{\tau\tau} = \eta_{\alpha\beta}\tau^{\alpha}\tau^{\beta}.$$

Therefore, taking variations of $\mathbf{R}_{\tau}$ defined either by $(9)_2$ or by $(28)_1$ we obtain

$$\begin{aligned}\delta\mathbf{R}_\tau\mathbf{R}_\tau^T &= \boldsymbol{\omega}_\tau\times\mathbf{1}\\ &= \left(\delta\mathbf{Q}_V\mathbf{Q}_V^T\right)\mathbf{Q}_V\mathbf{R}\mathbf{R}^T\mathbf{Q}_V^T+\mathbf{Q}_V\left(\delta\mathbf{R}\mathbf{R}^T\right)\mathbf{Q}_V^T\\ &= \left(\mathbf{q}+\mathbf{Q}_V\boldsymbol{\omega}\right)\times\mathbf{1},\end{aligned}\tag{30}$$

where

$$\mathbf{q}=\frac{1}{2}(\mathbf{1}\times\mathbf{1})\cdot\left(\delta\mathbf{Q}_V\mathbf{Q}_V^T\right).\tag{31}$$

From $(30)_{1,2}$ it follows that

$$\boldsymbol{\omega}_\tau=\mathbf{q}+\mathbf{Q}_V\boldsymbol{\omega},\quad \boldsymbol{\omega}=\mathbf{Q}_V^T\left(\boldsymbol{\omega}_\tau-\mathbf{q}\right).\tag{32}$$

Let us evaluate more explicitly the formula $(32)_2$ for $\boldsymbol{\omega}$ at the boundary $\partial\overline{\mathcal{M}}^{(k)}$. Keeping in mind that $\mathbf{1}=\overline{\boldsymbol{\nu}}\otimes\overline{\boldsymbol{\nu}}+\overline{\boldsymbol{\tau}}\otimes\overline{\boldsymbol{\tau}}+\overline{\mathbf{n}}\otimes\overline{\mathbf{n}}$ along $\partial\overline{\mathcal{M}}^{(k)}$, and taking variation of $(28)_2$ we establish the relations

$$\begin{aligned}\mathbf{1}\times\mathbf{1} &= -\overline{\boldsymbol{\nu}}\otimes\left(\overline{\boldsymbol{\tau}}\otimes\overline{\mathbf{n}}-\overline{\mathbf{n}}\otimes\overline{\boldsymbol{\tau}}\right)-\overline{\boldsymbol{\tau}}\otimes\left(\overline{\mathbf{n}}\otimes\overline{\boldsymbol{\nu}}-\overline{\boldsymbol{\nu}}\otimes\overline{\mathbf{n}}\right)-\overline{\mathbf{n}}\otimes\left(\overline{\boldsymbol{\nu}}\otimes\overline{\boldsymbol{\tau}}-\overline{\boldsymbol{\tau}}\otimes\overline{\boldsymbol{\nu}}\right),\\ \delta\mathbf{Q}_V\mathbf{Q}_V^T &= \delta\overline{\boldsymbol{\nu}}\otimes\overline{\boldsymbol{\nu}}+\delta\overline{\boldsymbol{\tau}}\otimes\overline{\boldsymbol{\tau}}+\delta\overline{\mathbf{n}}\otimes\overline{\mathbf{n}}+\overline{\boldsymbol{\nu}}\otimes\overline{\boldsymbol{\tau}}\left(\delta\mathbf{r}_\nu\cdot\mathbf{r}_\tau\right)+\overline{\boldsymbol{\nu}}\otimes\overline{\mathbf{n}}\left(\delta\mathbf{r}_\nu\cdot\overline{\mathbf{n}}\right)\\ &\quad+\overline{\boldsymbol{\tau}}\otimes\overline{\boldsymbol{\nu}}\left(\delta\mathbf{r}_\tau\cdot\mathbf{r}_\nu\right)+\overline{\boldsymbol{\tau}}\otimes\overline{\mathbf{n}}\left(\delta\mathbf{r}_\tau\cdot\overline{\mathbf{n}}\right)+\overline{\mathbf{n}}\otimes\overline{\boldsymbol{\nu}}\left(\delta\overline{\mathbf{n}}\cdot\mathbf{r}_\nu\right)+\overline{\mathbf{n}}\otimes\overline{\boldsymbol{\tau}}\left(\delta\overline{\mathbf{n}}\cdot\mathbf{r}_\tau\right).\end{aligned}\tag{33}$$

Introducing (33) into (31) and taking into account that

$$\boldsymbol{\omega}_\tau=\overline{\boldsymbol{\nu}}\left(\boldsymbol{\omega}_\tau\cdot\overline{\boldsymbol{\nu}}\right)+\overline{\boldsymbol{\tau}}\left(\boldsymbol{\omega}_\tau\cdot\overline{\boldsymbol{\tau}}\right)+\overline{\mathbf{n}}\left(\boldsymbol{\omega}_\tau\cdot\overline{\mathbf{n}}\right),\tag{34}$$

after some transformations from $(32)_2$ we obtain

$$\boldsymbol{\omega}=\mathbf{r}_\nu\left(\boldsymbol{\omega}_\tau\cdot\mathbf{r}_\nu\right)+\mathbf{r}_\tau\left(\boldsymbol{\omega}_\tau\cdot\mathbf{r}_\tau\right)-\overline{\mathbf{n}}\left(\delta\mathbf{r}_\tau\cdot\mathbf{r}_\nu\right).\tag{35}$$

The relation (35) means that the virtual rotations $\boldsymbol{\omega}$ and $\boldsymbol{\omega}_\tau$ differ only by their normal components. But $\mathbf{K}_\nu$ in (25) does not have a normal component at all. Thus, using $(8)_1$, $(20)_1$, $(25)_2$ and (35) we are able to show that at the boundary

$$\mathbf{K}_\nu\cdot\boldsymbol{\omega}=\mathbf{K}_\nu\cdot\boldsymbol{\omega}_\tau.\tag{36}$$

The simple relation (36) just confirms that the theory of thin shells discussed here is insensitive to the virtual works done on the normal, drilling components of $\boldsymbol{\omega}$ and $\boldsymbol{\omega}_\tau$, for the corresponding drilling components of the couples are indefinite in this shell model. The virtual works done by the drilling couples can be taken into account only in the general theory of shells, [3,6,7].

With the help of (24), (27) and (36) the internal virtual work for the entire reference network $\mathcal{M}$ can be put in the form

$$G_{\text{int}} = -\iint_{\mathcal{M}\setminus\Gamma} \left\{ \mathbf{N}^{\beta}\,|_{\beta} \boldsymbol{\cdot} \delta\mathbf{u} + \left( \mathbf{M}^{\beta}\,|_{\beta} + \bar{\mathbf{a}}_{\beta} \times \mathbf{N}^{\beta} \right) \boldsymbol{\cdot} \boldsymbol{\omega} \right\} dA + \int_{\partial\mathcal{M}} \left( \mathbf{T}_{\nu} \boldsymbol{\cdot} \delta\mathbf{u} + \mathbf{K}_{\nu} \boldsymbol{\cdot} \boldsymbol{\omega}_{\tau} \right) ds$$
$$+ \int_{\Gamma} \left( [\![\mathbf{T}_{\nu} \boldsymbol{\cdot} \delta\mathbf{u}]\!] + [\![\mathbf{K}_{\nu} \boldsymbol{\cdot} \boldsymbol{\omega}_{\tau}]\!] \right) ds + \sum_{P_i \in \Gamma} [B\bar{\mathbf{n}} \boldsymbol{\cdot} \delta\mathbf{u}]_i + \sum_{P_b \in \partial\mathcal{M}} [B\bar{\mathbf{n}} \boldsymbol{\cdot} \delta\mathbf{u}]_b . \tag{37}$$

In (37) the jumps at each regular point $P \in \Gamma^{(a)}$ of the common curve $\Gamma^{(a)} = \partial\mathcal{M}^{(1)} \cap \partial\mathcal{M}^{(2)} \cap \ldots \cap \partial\mathcal{M}^{(n)}$ for $n \geq 2$ adjacent surface elements are defined by

$$[\![\mathbf{T}_{\nu} \boldsymbol{\cdot} \delta\mathbf{u}]\!] = \pm \mathbf{T}_{\nu}^{(1)\pm} \boldsymbol{\cdot} \delta\mathbf{u}^{(1)\pm} \pm \mathbf{T}_{\nu}^{(2)\pm} \boldsymbol{\cdot} \delta\mathbf{u}^{(2)\pm} \pm \ldots \pm \mathbf{T}_{\nu}^{(n)\pm} \boldsymbol{\cdot} \delta\mathbf{u}^{(n)\pm},$$
$$[\![\mathbf{K}_{\nu} \boldsymbol{\cdot} \boldsymbol{\omega}_{\tau}]\!] = \pm \mathbf{K}_{\nu}^{(1)\pm} \boldsymbol{\cdot} \boldsymbol{\omega}_{\tau}^{(1)\pm} \pm \mathbf{K}_{\nu}^{(2)\pm} \boldsymbol{\cdot} \boldsymbol{\omega}_{\tau}^{(2)\pm} \pm \ldots \pm \mathbf{K}_{\nu}^{(n)\pm} \boldsymbol{\cdot} \boldsymbol{\omega}_{\tau}^{(n)\pm} . \tag{38}$$

The numerical superscripts $(n)$ introduced into the right hand sides of (38) indicate explicitly that those functions are defined only along the particular $\partial\mathcal{M}^{(n)}$.

The signs in the definitions (38) must be chosen consistently with a fixed orientation of the curve $\Gamma^{(a)}$. If the orientation of $\Gamma^{(a)}$ coincides with the orientation of $\partial\mathcal{M}^{(n)}$, that is when the unit tangent vector $\boldsymbol{\tau}_{\Gamma}$ specifying the orientation of $\Gamma^{(a)}$ is related to $\boldsymbol{\nu}^{(n)}$ of $\partial\mathcal{M}^{(n)}$ by $\boldsymbol{\nu}^{(n)} = +\boldsymbol{\tau}_{\Gamma} \times \mathbf{n}^{(n)}$, the minus sign must be chosen in front of the corresponding term in (38), and the plus sign otherwise.

The jumps at all singular points of $\mathcal{M}$ have been divided in (37) into the jumps $[B\bar{\mathbf{n}} \boldsymbol{\cdot} \delta\mathbf{u}]_i$ at the internal points $P_i \in \Gamma$ and the jumps $[B\bar{\mathbf{n}} \boldsymbol{\cdot} \delta\mathbf{u}]_b$ at the boundary points $P_b \in \partial\mathcal{M}$. At each internal point $P_i$ being the common point of $m \geq 2$ adjacent branches $\Gamma^{(m)}$, as well as at each boundary point $P_b$ being the common point of $t \geq 2$ adjacent parts $\partial\mathcal{M}^{(t)}$ and $q$ adjacent branches $\Gamma^{(q)}$ approaching $P_b$ from inside of $\mathcal{M}$, the jumps are defined by

$$[B\bar{\mathbf{n}} \boldsymbol{\cdot} \delta\mathbf{u}]_i = \pm B_i^{(1)\pm} \bar{\mathbf{n}}_i^{(1)\pm} \boldsymbol{\cdot} \delta\mathbf{u}_i^{(1)\pm} \pm B_i^{(2)\pm} \bar{\mathbf{n}}_i^{(2)\pm} \boldsymbol{\cdot} \delta\mathbf{u}_i^{(2)\pm} \pm \ldots \pm B_i^{(m)\pm} \bar{\mathbf{n}}_i^{(m)\pm} \boldsymbol{\cdot} \delta\mathbf{u}_i^{(m)\pm},$$
$$[B\bar{\mathbf{n}} \boldsymbol{\cdot} \delta\mathbf{u}]_b = \pm B_b^{(1)\pm} \bar{\mathbf{n}}_b^{(1)\pm} \boldsymbol{\cdot} \delta\mathbf{u}_b^{(1)\pm} \pm B_b^{(2)\pm} \bar{\mathbf{n}}_b^{(2)\pm} \boldsymbol{\cdot} \delta\mathbf{u}_b^{(2)\pm} \pm \ldots \pm B_b^{(t)\pm} \bar{\mathbf{n}}_b^{(t)\pm} \boldsymbol{\cdot} \delta\mathbf{u}_b^{(t)\pm}$$
$$\pm B_i^{(1)\pm} \bar{\mathbf{n}}_i^{(1)\pm} \boldsymbol{\cdot} \delta\mathbf{u}_b^{(1)\pm} \pm B_i^{(2)\pm} \bar{\mathbf{n}}_i^{(2)\pm} \boldsymbol{\cdot} \delta\mathbf{u}_b^{(2)\pm} \pm \ldots \pm B_i^{(q)\pm} \bar{\mathbf{n}}_i^{(q)\pm} \boldsymbol{\cdot} \delta\mathbf{u}_b^{(q)\pm} . \tag{39}$$

Here the numerical superscripts indicate that these functions are defined only either on a particular internal branches $\Gamma^{(m)}$ and $\Gamma^{(q)}$, or on a particular $\partial\mathcal{M}^{(t)}$ composing a part of boundary $\partial\mathcal{M}$.

Introducing (37) with (38) and (39) into (23) we obtain

$$
\begin{aligned}
&-\iint_{\mathcal{M}\setminus\Gamma}\left\{\left(\mathbf{N}^{\beta}\,|_{\beta}+\mathbf{p}\right)\boldsymbol{\cdot}\delta\mathbf{u}+\left(\mathbf{M}^{\beta}\,|_{\beta}+\bar{\mathbf{a}}_{\beta}\times\mathbf{N}^{\beta}+\mathbf{m}\right)\boldsymbol{\cdot}\boldsymbol{\omega}\right\}ds \\
&+\int_{\partial\mathcal{M}_f}\left\{(\mathbf{T}_{\nu}-\mathbf{N}^*)\boldsymbol{\cdot}\delta\mathbf{u}+(\mathbf{K}_{\nu}-\mathbf{M}^*)\boldsymbol{\cdot}\boldsymbol{\omega}_{\tau}\right\}ds+\sum_{P_b\in\partial\mathcal{M}_f}[B\bar{\mathbf{n}}\boldsymbol{\cdot}\delta\mathbf{u}]_b \\
&+\int_{\partial\mathcal{M}_d}(\mathbf{T}_{\nu}\boldsymbol{\cdot}\delta\mathbf{u}+\mathbf{K}_{\nu}\boldsymbol{\cdot}\boldsymbol{\omega}_{\tau})ds+\sum_{P_b\in\partial\mathcal{M}_d}[B\bar{\mathbf{n}}\boldsymbol{\cdot}\delta\mathbf{u}]_b \\
&+\int_{\Gamma}\left\{[\![\mathbf{T}_{\nu}\boldsymbol{\cdot}\delta\mathbf{u}]\!]+[\![\mathbf{K}_{\nu}\boldsymbol{\cdot}\boldsymbol{\omega}_{\tau}]\!]\right\}ds+\sum_{P_i\in\Gamma}[B\bar{\mathbf{n}}\boldsymbol{\cdot}\delta\mathbf{u}]_i-\int_{\Gamma}\sigma_{\Gamma}ds-\sum_{P_i\in\Gamma}\sigma_i=0.
\end{aligned}
\tag{40}
$$

For an arbitrary, but kinematically admissible, virtual deformation the fields $\delta\mathbf{u}$ and $\boldsymbol{\omega}_{\tau}$ vanish identically along $\partial\mathcal{M}_d$, and the third line of (40) vanishes as well. Then the virtual work principle (40) requires the following local relations to be satisfied:

the local equilibrium equations

$$\mathbf{N}^{\beta}\,|_{\beta}+\mathbf{p}=\mathbf{0},\quad \mathbf{M}^{\beta}\,|_{\beta}+\bar{\mathbf{a}}_{\beta}\times\mathbf{N}^{\beta}+\mathbf{m}=\mathbf{0}\quad \text{at each regular } M\in\mathcal{M}\setminus\Gamma\,; \tag{41}$$

the static boundary conditions

$$\mathbf{T}_{\nu}-\mathbf{N}^*=\mathbf{0},\quad \mathbf{K}_{\nu}-\mathbf{M}^*=\mathbf{0}\quad \text{along regular parts of } \partial\mathcal{M}_f\,; \tag{42}$$

the jump conditions

$$[B\bar{\mathbf{n}}\boldsymbol{\cdot}\delta\mathbf{u}]_b=0\quad \text{at each singular boundary point } P_b\in\partial\mathcal{M}_f. \tag{43}$$

The corresponding work-conjugate geometric boundary conditions are:

$$\mathbf{u}-\mathbf{u}^*=\mathbf{0},\quad \mathbf{R}_{\tau}\mathbf{n}-\mathbf{R}_{\tau}^*\mathbf{n}=\mathbf{0}\quad \text{along regular parts of } \partial\mathcal{M}_d\,. \tag{44}$$

As it has been expected, the local equilibrium equations (41) as well as the boundary conditions (42) and (44) for thin irregular shell-like structures coincide with those derived within the theory of thin, regular shells expressed in terms of displacements and rotations as the primary variables (see [14], Section 5.2). However, in the jump conditions (43) the virtual displacements still remain coupled with the generalised forces, for in case of the general irregularity of deformation we may not be able to define a common $\delta\mathbf{u}_b$ associated with a singular boundary point $P_b\in\partial\mathcal{M}_f$.

## 5. Jump conditions along singular curves

If the local relations (41)-(44) are satisfied, the principle of virtual work still requires the last line of (40) to be satisfied identically for any part of $\Gamma$. This leads to the following local forms of the jump conditions:

$$[\![\mathbf{T}_{\nu}\boldsymbol{\cdot}\delta\mathbf{u}]\!]+[\![\mathbf{K}_{\nu}\boldsymbol{\cdot}\boldsymbol{\omega}_{\tau}]\!]-\sigma_{\Gamma}=0\quad \text{at regular points of } \Gamma\,; \tag{45}$$

$$[B\bar{\mathbf{n}}\boldsymbol{\cdot}\delta\mathbf{u}]_i-\sigma_i=0\quad \text{at each internal singular point } P_i\in\Gamma. \tag{46}$$

The jump conditions (45) and (46) constitute the additional set of basic relations that should be satisfied at the singular curves representing the irregularities of shell geometry, deformation, material properties and loading. The conditions are valid for unrestricted displacements, rotations, strains and/or bendings of the reference network $\mathcal{M}$.

The singular curves $\Gamma^{(a)}$ embedded into the shell reference network $\mathcal{M}$ may be of either geometric or physical type, in general. At the geometric curve some fields in the relations (45) or (46) fail to be continuous or smooth of the required class. With the physical curve we can additionally associate some mechanical properties by prescribing appropriate functions $\sigma_\Gamma = \sigma_\Gamma(\mathbf{u}_\Gamma, \mathbf{R}_\Gamma; \delta\mathbf{u}_\Gamma, \boldsymbol{\omega}_\Gamma)$ and $\sigma_i = \sigma_i(\mathbf{u}_i; \delta\mathbf{u}_i)$ along $\Gamma^{(a)}$. For special types of irregularities the jump conditions (45) and (46) can be simplified or presented in a more explicit, uncoupled form along the lines suggested in [2] for the displacement formulation of the non-linear theory of thin irregular shells. Such particular forms of the functions $\sigma_\Gamma$ and $\sigma_i$ as well as special cases of the jump conditions will be discussed separately.

## References

1. Makowski, J., Pietraszkiewicz, W. and Stumpf, H.: On the general form of jump conditions for thin irregular shells, *Archives of Mechanics* **50**(1998), 483-495.

2. Makowski, J., Pietraszkiewicz, W. and Stumpf, H.: Jump conditions in the non-linear theory of thin irregular shells, *J. Elasticity* **54**(1999), 1-26.

3. Libai, A. and Simmonds, J.G.: *The Non-linear Theory of Elastic Shells,* 2nd ed., Cambridge University Press, Cambridge, 1998.

4. Reissner, E.: Linear and nonlinear theory of shells, in Y.C. Fung and E.E. Sechler (eds.), *Thin-Shell Structures: Theory, Experiment and Design*, Prentice-Hall Inc., New Jersey, 1974, pp. 29-44.

5. Simmonds, J.G.: The nonlinear thermodynamical theory of shells: descent from 3-dimensions without thickness expansion, in E.L. Axelrad and F.A. Emmerling (eds.), *Flexible Shells, Theory and Applications*, Springer-Verlag, Berlin, 1984, pp. 1-11.

6. Chróścielewski, J., Makowski, J. and Stumpf, H.: Genuinely resultant shell finite elements accounting for geometric and material non-linearity, *Int. J. Numerical Methods in Engineering* **35**(1992), 63-94.

7. Chróścielewski, J., Makowski, J. and Stumpf, H.: Finite element analysis of smooth, folded and multi-shell structures, *Computer Methods in Applied Mechanics and Engineering* **141**(1997), 1-46.

8. Reissner, E.: On axisymmetric deformation of thin shells of revolution, *Proc. Symposia in Applied Mathematics* **3**(1950), 27-52.

9. Simmonds, J.G. and Danielson, D.A.: Nonlinear shell theory with a finite rotation vector, *Proc. Koninkl. Nederl. Akademie van Wetenschappen - Amsterdam*, Series B **73**(1970), 460-478.

10. Simmonds, J.G. and Danielson, D.A.: Nonlinear shell theory with finite rotation and stress function vectors, *Trans. ASME, J. Applied Mechanics* **39**(1972), 1085-1090.

11. Pietraszkiewicz, W.: *Introduction to the Non-linear Theory of Shells*, Ruhr-Universität, Mitt. IfMech. No 10, Bochum, 1977.

12. Pietraszkiewicz, W.: *Finite Rotations and Lagrangean Description in the Non-linear Theory of Shells*, Polish Sci. Publ., Warszawa – Poznań, 1979.

13. Pietraszkiewicz, W.: Finite rotations in the non-linear theory of thin shells, in W. Olszak (ed.), *Thin Shell Theory, New Trends and Applications*, Springer-Verlag, Wien - New York, 1980, pp. 151-208.

14. Pietraszkiewicz, W.: Geometrically nonlinear theories of thin elastic shells, *Advances in Mechanics* **12**(1989), 51÷130.

15. Shkutin, L.I.: Exact formulation of equations for non-linear deformability of thin shells, *Applied Problems of Strength and Plasticity* (in Russian) **7**(1977), 3-9; **8**(1978), 38-43; **9**(1978), 19-25. Gorkii University Press.

16. Valid, R.: Finite rotations, variational principles and buckling in shell theory, in W. Pietraszkiewicz (ed.), *Finite Rotations in Structural Mechanics*, Springer-Verlag, Berlin, 1986, pp. 317-332.

17. Valid, R.: *The Nonlinear Theory of Shells through Variational Principles*, John Wiley and Sons, Chichester *et al.*, 1995.

18. Atluri, S.N.: Alternate stress and conjugate strain measures, and mixed variational formulations involving rigid rotations, for computational analyses of finitely deformed solids, with applications to plates and shells – I. Theory, *Computers and Structures* **18**(1984), 93-116.

19. Libai, A. and Simmonds, J.G.: Nonlinear elastic shell theory, in J.W. Hutchinson and T.Y. Wu (eds.), *Advances in Applied Mechanics* **23**, Academic Press, New York, 1983, pp. 271-371.

20. Schieck, B., Pietraszkiewicz, W. and Stumpf, H.: Theory and numerical analysis of shells undergoing large elastic strains, *Int. J. Solids and Structures* **29**(1992), 689-709.

21. Pietraszkiewicz, W.: Lagrangian description and incremental formulation in the non-linear theory of thin shells, *Int. J. Non-Linear Mechanics* **19**(1984), 115-140.

22. Badur, J. and Pietraszkiewicz, W.: On geometrically non-linear theory of elastic shells derived from pseudo-Cosserat continuum with constrained micro-rotations, in W. Pietraszkiewicz (ed.), *Finite Rotations in Structural Mechanics*, Springer-Verlag, Berlin, 1986, pp. 19-32.

# ON SHEAR DEFORMATION PLATE SOLUTIONS: RELATIONSHIP TO THE CLASSICAL SOLUTIONS

J. N. REDDY

*Department of Mechanical Engineering*
*Texas A&M University, College Station, TX 77843-3123*

C. M. WANG

*Department of Civil Engineering*
*The National University of Singapore, Singapore, 119260*

## 1. A Review of Plate Theories

### 1.1 INTRODUCTORY COMMENTS

The primary objective of this section is to review two-dimensional theories of elastic plates in which both bending and transverse shear effects are taken into account. The plate theories are derived from the 3-D elasticity theory by making suitable assumptions concerning the kinematics of deformation or the stress state through the thickness of the laminate, thereby reducing three-dimensional elasticity problem to a two dimensional one. There exist a number of plate theories, and they differ in two principal ways: (1) choice of the field to be expanded in terms of the thickness coordinate, and (2) the choice of terms (i.e., powers of the thickness coordinate) in the expansion. The choice of the field is often restricted to either displacements or stresses, although a mixed approach is possible. The choice of terms in the displacement or stress expansions is limited, at most, to cubic in thickness coordinate. Thus, a component of the stress or displacement field, $\varphi_i$ is expanded in the form

$$\varphi_i(x,y,z,t) = \sum_{j=0}^{n} (z)^j \varphi_i^j(x,y,t) \tag{1.1}$$

where $\varphi_i$ is the $i$th component of displacement or stress field, $(x,y)$ are the in-plane coordinates, $z$ the thickness coordinate, $t$ the time, $n$ the number of terms in the expansion ($n$-1st degree polynomial in thickness coordinate $z$), and $\varphi_i^j$ are functions to be determined. This expansion is used to derive plate theories via an



*D. Durban et al. (eds.), Advances in the Mechanics of Plates and Shells,* 259–276.

appropriate principle of virtual work. Thus, a plate theory can be developed for any combination of the field variable and number of terms in the expansion of the variable. The number of theories further multiply if different order expansions are used for different components of the field.

A brief review of various plate theories is presented in the remainder of this section. The authors wish to acknowledge that no review of plate theories will be complete, as there are thousands of papers dealing with one aspect or the other of the many combinations mentioned above. It should be remarked that it is not uncommon to find researchers referring to any plate theory that accounts for transverse shear deformation as a Reissner-Mindlin plate theory. This is technically incorrect, as there is no such a theory as the 'Reissner-Mindlin' plate theory. The theories that Reissner and Mindlin developed individually are quite different, as will be shown in the sequel, and they are based on different field expansions, which were proposed by others.

## 1.2 CLASSICAL PLATE THEORY

To enable comparison with the refined theories, a brief review of the classical (or Kirchhoff-Love) plate theory is given here. The governing equations of the classical plate theory can be derived using either the equilibrium of stress resultants on a plate element or the assumed displacement field

$$u = -z\frac{\partial w_0}{\partial x}, \quad v = -z\frac{\partial w_0}{\partial y}, \quad w = w_0(x,y) \tag{1.2}$$

and the principle of virtual dispalcement. Here $(u, v, w)$ denote the displacement components of a point $(x, y, z)$. The governing equations of the classical plate theory (C) are given by

$$\frac{\partial M_{xx}^C}{\partial x} + \frac{\partial M_{xy}^C}{\partial y} = Q_x^C \tag{1.3a}$$

$$\frac{\partial M_{xy}^C}{\partial x} + \frac{\partial M_{yy}^C}{\partial y} = Q_y^C \tag{1.3b}$$

$$\frac{\partial Q_x^C}{\partial x} + \frac{\partial Q_y^C}{\partial y} = -q \tag{1.3c}$$

where $(M_{xx}^C, M_{yy}^C, M_{xy}^C)$ are moments per unit length and $(Q_x^C, Q_y^C)$ are the transverse shear forces per unit length

$$\begin{Bmatrix} M_{xx}^C \\ M_{yy}^C \\ M_{xy}^C \end{Bmatrix} = \int_{-\frac{h}{2}}^{\frac{h}{2}} \begin{Bmatrix} \sigma_{xx} \\ \sigma_{yy} \\ \sigma_{xy} \end{Bmatrix} zdz\,, \quad \begin{Bmatrix} Q_x^C \\ Q_y^C \end{Bmatrix} = \int_{-\frac{h}{2}}^{\frac{h}{2}} \begin{Bmatrix} \sigma_{xz} \\ \sigma_{yz} \end{Bmatrix} dz \tag{1.4}$$

where $h$ is the total thickness of the plate, $q$ the distributed transverse load, and $\sigma_{xx}, \sigma_{xy}$, and so on are the components of stress in the plate coordinates. The three equations (1.3a-c) can be combined into a single equation by eliminating $(Q_x^C, Q_y^C)$ (which are zero if computed using the constitutive equations)

$$\frac{\partial^2 M_{xx}^C}{\partial x^2} + 2\frac{\partial^2 M_{xy}^C}{\partial x \partial y} + \frac{\partial^2 M_{yy}^C}{\partial y^2} = -q \tag{1.5}$$

Suppose that the material of the plate is isotropic and obeys Hooke's law. Then the stress-strain relations are given by

$$\sigma_{xx} = \frac{E}{1-\nu^2}\left(\varepsilon_{xx} + \nu\varepsilon_{yy}\right) \tag{1.6a}$$

$$\sigma_{yy} = \frac{E}{1-\nu^2}\left(\varepsilon_{yy} + \nu\varepsilon_{xx}\right) \tag{1.6b}$$

$$\sigma_{xy} = G\gamma_{xy} = \frac{E}{2(1+\nu)}\gamma_{xy} \tag{1.6c}$$

where $E$ denotes Young's modulus, $G$ shear modulus, and $\nu$ Poisson's ratio. Using Eqs. (1.2) and (1.6a-c) in Eq. (1.4) and carrying out the indicated integration over the plate thickness, we arrive at

$$M_{xx}^C = -D\left(\frac{\partial^2 w_0^C}{\partial x^2} + \nu\frac{\partial^2 w_0^C}{\partial y^2}\right) \tag{1.7a}$$

$$M_{yy}^C = -D\left(\nu\frac{\partial^2 w_0^C}{\partial x^2} + \frac{\partial^2 w_0^C}{\partial y^2}\right) \tag{1.7b}$$

$$M_{xy}^C = -(1-\nu)D\frac{\partial^2 w_0^C}{\partial x \partial y} \tag{1.7c}$$

where $D$ is the flexural rigidity $D = \frac{Eh^3}{12(1-\nu^2)}$. The shear forces $(Q_x^C, Q_y^C)$ are computed using Eqs. (1.3a,b).

## 1.3 STRESS-BASED THEORIES

The plate theories based on the expansion of the stress field are due to Reissner [1,2] and Kromm [3,4], and the book by Panc [5] contains chapters devoted to these theories and their extensions. In the Reissner plate theory, the distribution of the stress components through the plate thickness is assumed to be

$$\sigma_{xx} = z\frac{12M_{xx}}{h^3}, \quad \sigma_{yy} = z\frac{12M_{yy}}{h^3}, \quad \sigma_{xy} = z\frac{12M_{xy}}{h^3} \tag{1.9a}$$

$$\sigma_{xz} = \frac{3Q_x}{2h}\left[1 - \left(\frac{2z}{h}\right)^2\right], \quad \sigma_{yz} = \frac{3Q_y}{2h}\left[1 - \left(\frac{2z}{h}\right)^2\right] \tag{1.9b}$$

$$\sigma_{zz} = -\frac{q}{4}\left[2 - 3\left(\frac{2z}{h}\right) + \left(\frac{2z}{h}\right)^3\right] \tag{1.9c}$$

where the load $q$ acts at the surface $z = -h/2$. The stress field in Eq. (1.9a) is the same as that of the classical plate theory; the transverse shear stress field in Eqs. (1.9b,c) is the same as that obtained from 3-D stress-equilibrium equations after using the in-plane stress field from (1.9a) (see Reddy [6-8]). Thus, the in-plane stresses are linear, the transverse shear stresses are quadratic, and transverse normal stress is cubic in $z$. Obviously, these stress components satisfy both the stress-equilibrium of equations of 3D elasticity as well as those in Eqs. (1.3a-c).

The transverse displacement $w^R$ of Reissner's theory is a function of $(x, y, z)$, and this complicates its determination. To make the theory tractable, Reissner introduced the thickness-integrated transverse displacement (a 'mean deflection with respect to the plate thckness')

$$w^R(x, y) = \frac{3}{2h} \int_{-\frac{h}{2}}^{\frac{h}{2}} w \left[1 - \left(\frac{2z}{h}\right)^2\right] dz \tag{1.10}$$

With the introduction of $w^R$, the governing equations of Reissner's plate theory for isotropic plates can be written as (see Panc [5] for details)

$$D\nabla^2\nabla^2 w^R = q - \frac{h^2}{10}\frac{2-\nu}{1-\nu}\nabla^2 q \tag{1.11a}$$

$$\frac{h^2}{10}\nabla^2 Q_x^R - Q_x^R = D\frac{\partial}{\partial x}\left(\nabla^2 w^R\right) + \frac{h^2}{10}\frac{1}{1-\nu}\frac{\partial q}{\partial x} \tag{1.11b}$$

$$\frac{h^2}{10}\nabla^2 Q_y^R - Q_y^R = D\frac{\partial}{\partial y}\left(\nabla^2 w^R\right) + \frac{h^2}{10}\frac{1}{1-\nu}\frac{\partial q}{\partial y} \tag{1.11c}$$

where $(Q_x^R, Q_y^R)$ of Eqs. (1.11b,c) satisfy Eq. (1.3c).

The moment-deflection relationships of Reissner's plate theory are

$$M_{xx}^R = -D\left(\frac{\partial^2 w^R}{\partial x^2} + \nu\frac{\partial^2 w^R}{\partial y^2}\right) + \frac{h^2}{10}\left(2\frac{\partial Q_x^R}{\partial x} - \frac{\nu}{1-\nu}q\right) \tag{1.12a}$$

$$M_{yy}^R = -D\left(\frac{\partial^2 w^R}{\partial y^2} + \nu\frac{\partial^2 w^R}{\partial x^2}\right) + \frac{h^2}{10}\left(2\frac{\partial Q_y^R}{\partial y} - \frac{\nu}{1-\nu}q\right) \tag{1.12b}$$

$$M_{xy}^R = -(1-\nu)D\frac{\partial^2 w^R}{\partial x \partial y} + \frac{h^2}{10}\left(\frac{\partial Q_x^R}{\partial y} + \frac{\partial Q_y^R}{\partial x}\right) \tag{1.12c}$$

The refined theory of Kromm [3,4] is based on a more general stress distributions, especially for the transverse shear stress components, across the thickness of the plate. For a complete description of theory and its governing equations one may consult Panc [5].

The stress-based theories have not received as much attention as the displacement-based plate theories. This might be due to the fact that the stress-based theories are relatively more complicated and inconsistencies between the actual (i.e. consistent with the assumed stress distributions) and adopted displacements exist.

## 1.4 DISPLACEMENT-BASED THEORIES

The simplest displacement-based plate theory is that uses the displacement field

$$u(x,y,z) = z\phi_x^F(x,y), \quad v(x,y,z) = z\phi_y^F(x,y), \quad w(x,y,z) = w_0^F(x,y) \tag{1.13}$$

where $\phi_x^F$ and $-\phi_y^F$ denote rotations about the $y$ and $x$ axes, respectively:

$$\phi_x^F(x,y) = \left(\frac{\partial u}{\partial z}\right)_{z=0}, \quad \phi_y^F(x,y) = \left(\frac{\partial v}{\partial z}\right)_{z=0} \tag{1.14}$$

The theory is known in the literature as the Mindlin plate theory, although the use of the displacement field (1.13) and associated plate theory were developed much earlier by Bassett [9], Hildebrand et al [10], and Hencky [11]. Mindlin [12] extended the theory developed by Hencky [11] to the vibration of crystal plates. Hence, it would be incorrect to attribute the theory to Mindlin (or Reissner). The theory is now being referred to as the first-order shear deformation plate theory (FSDT, see Reddy [6-8]). Note that the classical plate theory (CPT) is also a first-order theory but it is not a shear deformation theory.

The governing equations of the FSDT are

$$\left(\frac{\partial M_{xx}^F}{\partial x} + \frac{\partial M_{xy}^F}{\partial y}\right) = Q_x^F \tag{1.15a}$$

$$\left(\frac{\partial M_{xy}^F}{\partial x} + \frac{\partial M_{yy}^F}{\partial y}\right) = Q_y^F \tag{1.15b}$$

$$\left(\frac{\partial Q_x^F}{\partial x} + \frac{\partial Q_y^F}{\partial y}\right) = -q \tag{1.15c}$$

Using the stress-strain relations (1.6), we can express the moments and shear fores in terms of $(w_0^F, \phi_x^F, \phi_y^F)$

$$M_{xx}^F = D\left(\frac{\partial \phi_x^F}{\partial x} + \nu\frac{\partial \phi_y^F}{\partial y}\right), \quad M_{yy}^F = D\left(\nu\frac{\partial \phi_x^F}{\partial x} + \frac{\partial \phi_y^F}{\partial y}\right) \tag{1.16a}$$

$$M_{xy}^F = \frac{D(1-\nu)}{2}\left(\frac{\partial \phi_x^F}{\partial y} + \frac{\partial \phi_y^F}{\partial x}\right) \tag{1.16b}$$

$$Q_x^F = \frac{K_s Eh}{2(1+\nu)}\left(\phi_x^F + \frac{\partial w_0^F}{\partial x}\right), \quad Q_y^F = \frac{K_s Eh}{2(1+\nu)}\left(\phi_y^F + \frac{\partial w_0^F}{\partial y}\right) \tag{1.17}$$

where $K_s$ denotes the shear correction factor.

The FSDT extends the kinematics of the CPT by including the transverse shear strains in its kinematic assumptions by removing the normality restriction of the classical plate theory. Since the transverse shear strains in the FSDT are constant through the plate thickness, the transverse shear stresses are also constant through the plate thickness, whereas the the stress equilibrium equations [see Eqs. (1.9b)] predict them to be quadratic. To make the shear forces $(Q_x, Q_y)$ computed in the FSDT to be equal to those obtained using the transverse shear stresses from the stress-equilibrium equations, shear correction factors were introduced. In both CPT and FSDT, the plane-stress state assumption is used and plane-stress reduced form of the constitutive law is used. In both theories the inextensibility (i.e. $\varepsilon_{zz} = 0$) and/or straightness of transverse normals can be removed. Such extensions lead to second- and higher-order theories of plates.

Second- and higher-order plate theories[1] use higher-order polynomials [i.e., $n > 1$ in Eq. (1.1)] in the expansion of the displacement components through the thickness of the plate (see [7] for a additional references). The higher-order theories introduce additional unknowns that are often difficult to interpret in physical terms.

A second-order plate theory with transverse inextensibility is based on the displacement field

$$\begin{aligned} u(x,y,z) &= z\phi_x(x,y) + z^2\psi_x(x,y) \\ v(x,y,z) &= z\phi_y(x,y) + z^2\psi_y(x,y) \\ w(x,y,z) &= w_0(x,y) \end{aligned} \tag{1.18}$$

where $\psi_x$ and $\psi_y$ are the curvatures at the midplane (i.e. $z = 0$) of the plate

$$\psi_x(x,y) = \left(\frac{\partial^2 u}{\partial z^2}\right)_{z=0}, \quad \psi_y(x,y) = \left(\frac{\partial^2 v}{\partial z^2}\right)_{z=0} \tag{1.19}$$

The second-order plate theories are not used much as it incorrectly estimates the in-plane displacements of points $(x, y, z)$ and $(x, y, -z)$ to be the same irrespective whether the plate bends convex or concave.

A third-order plate theory with transverse inextensibility is based on the displacement field (see Reddy [13,14])

$$\begin{aligned} u(x,y,z) &= z\phi_x(x,y) + z^3\left(-\frac{4}{3h^2}\right)\left(\phi_x + \frac{\partial w_0}{\partial x}\right) \\ v(x,y,z) &= z\phi_y(x,y) + z^3\left(-\frac{4}{3h^2}\right)\left(\phi_y + \frac{\partial w_0}{\partial y}\right) \\ w(x,y,z) &= w_0(x,y) \end{aligned} \tag{1.20}$$

---

[1] The order referred here is to the degree $n$ of the thickness coordinate in the displacement expansion and not to the order of the governing differential equations.

The displacement field accommodates quadratic variation of transverse shear strains (and hence stresses) and vanishing of transverse shear stresses on the top and bottom surfaces of a plate. Thus there is no need to use shear correction factors in a third-order plate theory. There are a number of people who have used a displacement field of the form (1.20), but they differ in actual form (see Table 1) and hence the resulting governing equations have different looks. However, it can be shown that all third-order theories are special cases of that derived by Reddy [13,14,20].

**TABLE 1.** Relationship of the displacements of other third-order theories to the one in Eq. (1.20).

| References | Displacement Field† | Correspondence with Eq. (1.20)‡ |
|---|---|---|
| Vlasov [15], Jemielita[16] | $u_\alpha = f(z)\psi_\alpha - c_1 z^3 u^0_{3,\alpha}$ | $\varphi_\alpha = \psi_\alpha + u^0_{3,\alpha}$ |
| Schmidt [17] | $u_\alpha = -zu^0_{3,\alpha} + \frac{3}{2} f(z)\varepsilon_\alpha$ | $\varphi_\alpha = \frac{3}{2}\varepsilon_\alpha$ |
| Krishna Murty [18], Levinson [19], Reddy [13] | $u_\alpha = -zu^0_{3,\alpha} - c_1 f(z)\theta_\alpha$ | $\varphi_\alpha = -c_1\theta_\alpha$ |
| Reddy [14] | $u_\alpha = -zu^0_{3,\alpha} + f(z)\phi_\alpha$ | $\varphi_\alpha = \phi_\alpha$ |
| Reddy [20] | $u_\alpha = f(z)\hat{\phi}_\alpha - zu^b_{3,\alpha}$ $-c_1 h z^3 u^0_{3,\alpha}$ | $\varphi_\alpha = \hat{\phi}_\alpha + u^s_{3,\alpha}$ $w_0 = u^s_3 + u^b_3$ |

† $c_1 = \frac{4}{3h^2}$, $f(z) = z(1 - c_1 z^2)$. ‡ $\varphi_\alpha \equiv \phi_\alpha + u^0_{3,\alpha}$, $u^0_3 = w_0$.

The equilibrium equations of the third-order plate theory (T), based on the displacement field (1.18) and the principle of virtual displacements, are

$$\left(\frac{\partial \hat{M}^T_{xx}}{\partial x} + \frac{\partial \hat{M}^T_{xy}}{\partial y}\right) = \hat{Q}^T_x \tag{1.21a}$$

$$\left(\frac{\partial \hat{M}^T_{xy}}{\partial x} + \frac{\partial \hat{M}^T_{yy}}{\partial y}\right) = \hat{Q}^T_y \tag{1.21b}$$

$$\left(\frac{\partial \hat{Q}^T_x}{\partial x} + \frac{\partial \hat{Q}^T_y}{\partial y}\right) + \frac{4}{3h^2}\left(\frac{\partial^2 P_{xx}}{\partial x^2} + 2\frac{\partial^2 P_{xy}}{\partial x \partial y} + \frac{\partial^2 P_{yy}}{\partial y^2}\right) = -q \tag{1.21c}$$

where ( $\xi, \eta = x, y$)

$$\hat{M}_{\xi\eta}^T = M_{\xi\eta}^T - \frac{4}{3h^2}P_{\xi\eta}, \quad \hat{Q}_\xi^T = Q_\xi^T - \frac{4}{h^2}R_\xi \tag{1.22}$$

The third-order plate theory of Levinson is based on the same displacement field as in Eqs. (1.20), but he used the equilibrium equations of the FSDT, i.e., did not use the principle of virtual displacements to derive the equilibrium equations. The Levinson theory results in much simpler equations and does not involve the higher-order stress resultants. The equilibrium equations of Levinson's theory may be obtained as a special case from Eqs. (1.21a-c) by setting $\alpha$ and $\beta$ to zero in Eqs. (1.21a-c) and (1.22).

In both Reddy's and Levinson's theories, the moments and shear forces are related to the displacements $(w_0^T, \phi_x^T, \phi_y^T)$ by

$$M_{xx}^T = \frac{4D}{5}\left(\frac{\partial \phi_x^T}{\partial x} + \nu\frac{\partial \phi_y^T}{\partial y}\right) - \frac{D}{5}\left(\frac{\partial^2 w_0^T}{\partial x^2} + \nu\frac{\partial^2 w_0^T}{\partial y^2}\right) \tag{1.23a}$$

$$P_{xx} = \frac{4h^2 D}{35}\left(\frac{\partial \phi_x^T}{\partial x} + \nu\frac{\partial \phi_y^T}{\partial y}\right) - \frac{h^2 D}{28}\left(\frac{\partial^2 w_0^T}{\partial x^2} + \nu\frac{\partial^2 w_0^T}{\partial y^2}\right) \tag{1.23b}$$

$$M_{yy}^T = \frac{4D}{5}\left(\nu\frac{\partial \phi_x^T}{\partial x} + \frac{\partial \phi_y^T}{\partial y}\right) - \frac{D}{5}\left(\nu\frac{\partial^2 w_0^T}{\partial x^2} + \frac{\partial^2 w_0^T}{\partial y^2}\right) \tag{1.23c}$$

$$P_{yy} = \frac{4h^2 D}{35}\left(\nu\frac{\partial \phi_x^T}{\partial x} + \frac{\partial \phi_y^T}{\partial y}\right) - \frac{h^2 D}{28}\left(\nu\frac{\partial^2 w_0^T}{\partial y^2} + \frac{\partial^2 w_0^T}{\partial y^2}\right) \tag{1.23d}$$

$$M_{xy} = \left(\frac{1-\nu}{2}\right)\left[\frac{4D}{5}\left(\frac{\partial \phi_x^T}{\partial y} + \frac{\partial \phi_y^T}{\partial x}\right) - \frac{D}{5}\left(2\frac{\partial^2 w_0^T}{\partial x \partial y}\right)\right] \tag{1.23e}$$

$$P_{xy} = \left(\frac{1-\nu}{2}\right)\left[\frac{4h^2 D}{35}\left(\frac{\partial \phi_x^T}{\partial y} + \frac{\partial \phi_y^T}{\partial x}\right) - \frac{h^2 D}{28}\left(2\frac{\partial^2 w_0^T}{\partial x \partial y}\right)\right] \tag{1.21f}$$

$$Q_x = \frac{2hG}{3}\left(\phi_x^T + \frac{\partial w_0^T}{\partial x}\right), \quad R_x = \frac{h^3 G}{30}\left(\phi_x^T + \frac{\partial w_0^T}{\partial x}\right) \tag{1.23g}$$

$$Q_y = \frac{2hG}{3}\left(\phi_y^T + \frac{\partial w_0^T}{\partial y}\right), \quad R_y = \frac{h^3 G}{30}\left(\phi_y^T + \frac{\partial w_0^T}{\partial y}\right) \tag{1.23h}$$

It should be noted that the higher-order stress resultants $(P_{xx}, P_{yy}, P_{xy})$ and $(Q_x, Q_y)$ do not arise in Levinson's theory [19].

The third-order theories provide a slight increase in accuracy relative to the FSDT solution, at the expense of a significant increase in computational effort. In principle, it is possible to expand the displacement field in terms of the thickness coordinate up to any desired degree. However, due to the algebraic complexity and computational effort involved with higher-order theories in return for marginal gain in accuracy, theories higher than third order have not been attempted. The reason for expanding the displacements up to the cubic term in the thickness coordinate is to have quadratic variation of the transverse shear strains and transverse shear stresses through each layer. This avoids the need for shear correction factors used in the first-order theory.

## 2. Relationships Between Theories

### 2.1 BACKGROUND

Equations governing shear deformation theories of plates are typically more complicated than those of the classical plate theory. Hence it is desirable to have exact relationships between solutions of the classical plate theory and shear deformation plate theories so that whenever classical theory solutions are available, the corresponding solutions of a shear deformation theory can be readily obtained. Such relationships not only furnish benchmark solutions of shear deformation theories but also provide insight into the significance of shear deformation on the response. The relationships for beams and plates have been developed by the authors and their colleagues over the last several years, and the developments till 1999 were included in the authors' book [21]. Recent developments, especially those related to the Reissner's and Levinson's plate theories, are included here along with a summary of the prior developments.

### 2.2 DEFLECTION RELATIONSHIPS

#### *2.2.1 First-Order Shear Shear Deformation Plate Theory*

Consider the elastic bending problem of an isotropic plate of uniform thickness $h$, Poisson's ratio $\nu$ , modulus of elasticity $E$, and shear modulus $G$. The governing equations of static equilibrium of plates according to the CPT and FSDT can be expressed in terms of the deflection $w_0$ and moment sum $\mathcal{M}$ as

$$\nabla^2 \mathcal{M}^C = -q, \quad \nabla^2 w_0^C = -\frac{\mathcal{M}^C}{D} \tag{2.1a, b}$$

$$\nabla^2 \mathcal{M}^F = -q, \quad \nabla^2 \left( w_0^F - \frac{\mathcal{M}^F}{K_s G h} \right) = -\frac{\mathcal{M}^F}{D} \tag{2.2a, b}$$

where the superscripts $C$ and $F$ refer to quantities of the CPT and FSDT plates, respectively, $K_s$ the shear correction factor, the moment sum $\mathcal{M}$ is defined as

$$\mathcal{M} = \frac{M_{xx} + M_{yy}}{1+\nu} \tag{2.3}$$

and $D = Eh^3/[12(1-\nu^2)]$ is the flexural rigidity of the plate. The moment sum is related to the generalized displacements by the relations

$$\mathcal{M}^C = \frac{M_{xx}^C + M_{yy}^C}{1+\nu} = -D\left(\frac{\partial^2 w_0^C}{\partial x^2} + \frac{\partial^2 w_0^C}{\partial y^2}\right) = -D\nabla^2 w_0^C \tag{2.4a}$$

$$\mathcal{M}^F = \frac{M_{xx}^F + M_{yy}^F}{1+\nu} = D\left(\frac{\partial \phi_x^F}{\partial x} + \frac{\partial \phi_y^F}{\partial y}\right) \tag{2.4b}$$

From Eqs. (2.1a) and (2.2a), in view of the load equivalence, it follows that

$$\mathcal{M}^F = \mathcal{M}^C + \psi \tag{2.5}$$

where $\psi$ is a function such that

$$\nabla^2 \psi = 0 \tag{2.6}$$

Using this result in Eqs. (2.1b) and (2.2b), we arrive at the relationship

$$\begin{aligned} w_0^F &= w_0^C + \frac{\mathcal{M}^C}{K_s G h} + \Psi \\ &= w_0^C + \frac{h^2}{6K_s(1-\nu)} \nabla^2 w_0^C + \Psi \end{aligned} \tag{2.7}$$

where $\Psi$ is a (biharmonic) function that satisfies the condition

$$\nabla^4 \Psi = 0 \tag{2.8}$$

Note that the relationship (2.7) is valid for all plates with arbitrary boundary conditions and transverse load. One must determine $\Psi$ from Eq. (2.8) subject to the geometry and boundary conditions of the plate.

In cases where $w_0^F = w_0^C$ on the boundaries and $\mathcal{M}^C$ is either zero or equal to a constant $\mathcal{M}^{*C}$ (which can be zero) over the boundaries, $\Psi$ simply takes on the value of $-\mathcal{M}^{*C}/(K_s G h)$. However, if $\mathcal{M}^C$ varies over the boundaries, the function $\Psi$ must be determined separately.

### *2.2.2 The Third-Order Plate Theory*

The governing equations of static equilibrium of plates according to the Kirchhoff or classical (C) and Levinson (L) plate theories can be expressed in terms of the moment sum defined in Eq. (2.3). We have

$$\mathcal{M}^L = \frac{D}{5}\left[4\left(\frac{\partial \phi_x^L}{\partial x} + \frac{\partial \phi_y^L}{\partial y}\right) - \nabla^2 w_0^L\right] \tag{2.9}$$

The equilibrium equation of the Levinson plate theory

$$\frac{\partial^2 M_{xx}^L}{\partial x^2} + 2\frac{\partial^2 M_{xy}^L}{\partial x \partial y} + \frac{\partial^2 M_{yy}^L}{\partial y^2} + q = 0 \tag{2.10}$$

can be expressed in terms of the moment sum as

$$\nabla^2 \mathcal{M}^L = -q, \qquad \frac{\partial \phi_x^L}{\partial x} + \frac{\partial \phi_y^L}{\partial y} = \frac{1}{4}\left(\nabla^2 w_0^L + \frac{5\mathcal{M}^L}{D}\right) \tag{2.11a,b}$$

From Eqs. (2.1a) and (2.11a), in view of the load equivalence, it follows that

$$\mathcal{M}^L = \mathcal{M}^C + D\nabla^2 \Phi \tag{2.12}$$

where $\Phi$ is a function such that it satisfies the biharmonic equation

$$\nabla^4 \Phi = 0 \tag{2.13}$$

After a series of algebraic manipulations (see Reddy et al [22]), we can establish the following relationships:

$$w_0^L(x,y) = w_0^C(x,y) + \frac{\mathcal{M}^C}{\frac{5}{6}Gh} - \Phi + \Psi \tag{2.14a}$$

$$\phi_x^L(x,y) = -\frac{\partial w_0^C}{\partial x} + \frac{3}{10Gh}\frac{\partial \mathcal{M}^C}{\partial x} + \frac{\partial \Theta}{\partial x} + \frac{h^2}{10}\frac{\partial \Omega}{\partial y} \tag{2.14b}$$

$$\phi_y^L(x,y) = -\frac{\partial w_0^C}{\partial y} + \frac{3}{10Gh}\frac{\partial \mathcal{M}^C}{\partial y} + \frac{\partial \Theta}{\partial y} - \frac{h^2}{10}\frac{\partial \Omega}{\partial x} \tag{2.14c}$$

where $\Psi(x,y)$ is a function such that

$$\nabla^2\Psi = 0 \tag{2.15}$$

where

$$\Theta = \frac{3D}{2Gh}\nabla^2\Phi + \Phi - \Psi, \quad \Lambda = \frac{D}{\frac{5}{6}Gh}\nabla^2\Phi + \Phi - \Psi \tag{2.16}$$

and $\Omega$ is defined by

$$\Omega(x,y) = \frac{\partial \psi_x^L}{\partial y} - \frac{\partial \psi_y^T}{\partial x} \tag{2.17}$$

and it is the solution of

$$-\nabla^2\Omega + \frac{10}{h^2}\Omega = 0 \tag{2.18}$$

A differential (not algebraic) deflection relationship between the CPT and the third-order shear deformation theory (TSDT) of Reddy for polygonal plates was developed by Reddy and Wang [23], and it is not included here due to the space limitation.

## 2.3 BUCKLING LOAD RELATIONSHIPS

For a simply supported polygonal plate under isotropic inplane load $N$, the buckling load $N^F$ of FSDT is related to its corresponding CPT buckling load $N^C$ by (see Wang [24])

$$N^F = N^C\left(1 + \frac{N^C}{K_sGh}\right)^{-1} \tag{2.19}$$

where $G$ is the shear modulus, $K_s$ the shear correction factor, and $h$ the plate thickness. Wang [25] showed that the above buckling relationship applies to radially loaded circular plates with any homogeneous edge restraint such as simply supported, clamped, or simply supported with elastic rotational restraint.

For simply supported polygonal plates, the third-order shear deformation theory (TSDT) the relationship between the buckling loads is given by (see Wang and Reddy [26])

$$N^T = \frac{N^C\left(1 + \frac{N^C}{70Gh}\right)}{1 + \frac{N^C}{\frac{14}{17}Gh}} \tag{2.20}$$

It should be remarked that the relationships developed in this section are valid only for simply supported polygonal plates under uniform inplane forces (i.e., the same uniform load applied on all sides). For example, the relationships in Eqs. (2.19) and (2.20) do not hold for a simply supported rectangular plate subjected to biaxial loads

$$N_{xx} = N, \quad N_{yy} = \gamma N \tag{2.21}$$

For this case, a relationship between the Kirchhoff and Mindlin plate can be derived using the solutions of the Kirchhoff plate theory and the Mindlin plate theory (see Reddy [8])

$$N^C(m) = \frac{D\pi^2}{s^2 b^2} \frac{(s^2 + m^2)^2}{(\gamma s^2 + m^2)} \tag{2.22}$$

$$N^F(m) = \frac{D\pi^2}{s^2 b^2} \frac{(s^2 + m^2)^2}{(\gamma s^2 + m^2)} \left[ \frac{1}{1 + \kappa\pi^2(1 + \frac{m^2}{s^2})} \right] \tag{2.23}$$

where $m$ is the number of half waves in the $x$–direction, and

$$s = \frac{a}{b}, \quad \kappa = \frac{h^2}{6b^2 K_s(1-\nu)} \tag{2.24}$$

Although $N^F$ can be expressed in terms of $N^C$ as

$$N^F(m) = N^C(m) \left[ \frac{1}{1 + \kappa\pi^2(1 + \frac{m^2}{s^2})} \right] \tag{2.25}$$

they do not necessarily correspond, in general, to the same number of half waves $m$. This is because $N^F(m)$ contains an additional factor involving $m$. In cases in which both theories yield the critical buckling load (i.e. the minimum buckling load) for the same half wave number $m$, it is possible to arrive at the following relationship

$$N^F = \frac{N^C}{1 + \frac{N^C}{2K_s Gh} \left[ 1 + \sqrt{1 - \frac{4\pi^2(1-\gamma)D}{N^C b^2}} \right]} \tag{2.26}$$

Note that Eq. (2.26) is independent of the aspect ratio $s$ and the half-wave number $m$. When $\gamma = 1$ (i.e. uniform compression), the relationship in Eq. (2.26) reduces to the one in Eq. (2.19).

For buckling of rectangular plates under uniform inplane shear load, Wang, Xiang, and Kitipornchai [27] developed an approximate relationship in the same form as in Eq. (2.19):

$$N^F = \frac{N^C}{(1 + f\frac{N^C}{K_s Gh})} \tag{2.27}$$

where $f$ is a correction factor (see Wang, Reddy, and Lee [21]).

The relationship between the CPT buckling load of a simply supported polygonal plate and the FSDT buckling load of a simply supported polygonal sandwich plate is (see Wang [28])

$$N^F = \hat{N}^C \left(1 + \frac{\hat{N}^C}{K_s(G_c h_c + 2G_f h_f)}\right)^{-1}, \quad \hat{N}^C = \left(\frac{D_c + D_f}{D}\right) N^C \qquad (2.28)$$

where $D_c$ and $D_f$ denote the flexural rigidities of the core and face sheets, respectively, and $h_c$ and $h_f$ denote the thicknesses of the core and face sheets, respectively.

Tables 2 and 3 contain numerical results for buckling loads of skew plates and circular plates. These results were obtained with the help of Eqs. (2.19) and (2/26). When the skew angle $\alpha = 0°$, the case corresponds to rectangular plates. Note that the results for $h/b = 0.001$ and $h/R = 0.001$ represent the thin (or Kirchhoff) plate solutions.

**TABLE 2.** Critical buckling loads $\lambda = N_{cr}b^2/(\pi^2 D)$ of simply supported, isotropic ($\nu = 0.3$), skew plates (skew angle $\alpha$ from the vertical axis), according to FSDT ($K_s = 5/6$).

| $\frac{a}{b}$ | $\frac{h}{b}$ | $\alpha = 0°$ | $\alpha = 15°$ | $\alpha = 30°$ | $\alpha = 45°$ |
|---|---|---|---|---|---|
| 1.0 | 0.001 | 2.0000 | 2.1147 | 2.5240 | 3.5253 |
| 1.0 | 0.1 | 1.8932 | 1.9957 | 2.3563 | 3.2066 |
| 1.0 | 0.2 | 1.6319 | 1.7074 | 1.9646 | 2.5224 |
| 2.0 | 0.001 | 1.2499 | 1.3280 | 1.6098 | 2.3177 |
| 2.0 | 0.1 | 1.2074 | 1.2801 | 1.5399 | 2.1755 |
| 2.0 | 0.2 | 1.0955 | 1.1550 | 1.3625 | 1.8374 |

**TABLE 3.** Critical buckling loads $\lambda = N_{cr}R^2/D$ of simply supported and clamped, isotropic ($\nu = 0.3$), circular plates of radius $R$, according to FSDT ($K_s = 5/6$).

| $\frac{h}{R}$ | $\lambda_s$ | $\lambda_c$ |
|---|---|---|
| 0.001 | 4.1978 | 14.6819 |
| 0.05 | 4.1853 | 14.5296 |
| 0.10 | 4.1480 | 14.0909 |
| 0.15 | 4.0875 | 13.4157 |
| 0.20 | 4.0056 | 12.5725 |

## 2.4 NATURAL FREQUENCY RELATIONSHIPS

The relationship between vibration frequencies of simply supported, polygonal plates was derived by Wang [29,30]

$$(\omega_F^2)_N = \frac{6K_sG}{\rho h^2}\left\{\left[1+\frac{h^2}{12}(\omega_C)_N\sqrt{\frac{\rho h}{D}}\left(1+\frac{2}{K_s(1-\nu)}\right)\right]\right.$$
$$\left. - \sqrt{\left[1+\frac{h^2}{12}(\omega_C)_N\sqrt{\frac{\rho h}{D}}\left(1+\frac{2}{K_s(1-\nu)}\right)\right]^2 - \frac{\rho h^2}{3K_sG}(\omega_C^2)_N}\right\} \tag{2.29}$$

where $N = 1, 2, \cdots$, corresponds to the mode sequence number, $\omega$ the natural frequency, $G$ the shear modulus, $\rho$ the density, $K_s$ the shear correction factor, and $h$ the plate thickness.

If the rotary inertia effect is neglected, it can be shown that the frequency relationship (2.29) simplifies to

$$(\hat{\omega}_F^2)_N = \frac{(\omega_C^2)_N}{1+\frac{(\omega_C^2)_N h^2}{6(1-\nu)K_s}\sqrt{\frac{\rho h}{D}}} \tag{2.30}$$

where $\hat{\omega}_F$ is the frequency of Mindlin (FSDT) plate without the rotary inertia effect. This frequency value $\hat{\omega}_F$ is greater than its corresponding $\omega_F$ but smaller than $\omega_C$. Table 4 contains, for example, numerical results for various regular polygonal shapes with side $a$. Wang [29] showed that the relationship may also be used to predict quite accurately the vibration of frequencies of FSDT plates with simply supported curved edges. Similar relationships were developed for TSDT by Wang, and his colleagues [31,32].

**TABLE 4.** Fundamental natural frequencies $\omega a^2\sqrt{\rho h/D}$ of simply supported polygonal plates ($K_s = 5/6$, $\nu = 0.3$).

| Shape | $\omega^C$ | $\omega^F$ | | |
|---|---|---|---|---|
| | | $\frac{h}{a} = 0.05$ | $\frac{h}{a} = 0.10$ | $\frac{h}{a} = 0.15$ |
| Triangle | 52.638 | 51.414 | 48.279 | 44.275 |
| Square | 19.739 | 19.562 | 19.065 | 18.328 |
| Pentagon | 10.863 | 10.809 | 10.653 | 10.410 |
| Hexagon | 7.129 | 7.106 | 7.037 | 6.929 |
| Octagon | 3.624 | 3.618 | 3.600 | 3.571 |

## 5. Closing Remarks

In this paper, an overview of various shear deformation plate theories and the exact relationships between the solutions (i.e., deflections, buckling loads, and natural frequencies) of the classical and shear deformation plate theories for isotropic and sandwich plates of various shapes and boundary conditions are presented. The first order shear deformation theory (FSDT) solutions may be readily obtained from known classical plate theory (CPT) solutions of isotropic plates. The exact FSDT solutions obtained via these relationships should serve as useful benchmark values for researchers to check the validity, convergence and accuracy of their numerical results. The exact relationships also show clearly the intrinsic features of the effect of transverse shear deformation on the classical solutions.

Deflection and rotation-slope relationships between the classical and the Levinson plate theory are also developed. Similar relationships between the classical and the Reddy plate theory have been developed but not included here due to the space limitation. These and many other results can be found in the monograph by the authors (see Wang, Reddy, and Lee [21]).

Relationships between the Reissner and Mindlin (i.e., first-order) plate theories were also derived recently (see Wang et al. [33]). The deflection relationship is given by

$$w_0^R = w_0^F + \left[\frac{3(2-\nu)}{5Gh} - \frac{1}{K_s Gh}\right]\mathcal{M}^F - \Theta + \Lambda \tag{2.31}$$

where $\Theta(x,y)$ is a biharmonic function that satisfies the condition

$$\nabla^4\Theta = 0 \tag{2.32}$$

and $\Lambda(x,y)$ is another function that satisfies the Laplace equation

$$\nabla^2\Lambda = 0 \tag{2.33}$$

For a simply supported rectangular plate, these functions can be shown to be zero, and the relationship (2.31) becomes

$$w_0^R = w_0^F + \left[\frac{3(2-\nu)}{5Gh} - \frac{1}{K_s Gh}\right]\mathcal{M}^F \tag{2.34}$$

which clearly shows that the deflections predicted by the two theories are different. For simply supported polygonal plates , we have $\mathcal{M}^F = \mathcal{M}^C$ and the deflection relationship in Eq. (2.7) becomes (because $\Psi = 0$)

$$w_0^F = w_0^C + \frac{\mathcal{M}^C}{K_s Gh} \tag{2.35}$$

Consequently, we have

$$w_0^R = w_0^C + \frac{3(2-\nu)}{5Gh}\mathcal{M}^C \tag{2.36}$$

Extension of the present approach to determine relationships for orthotropic and laminated composite plates, for shells, and for transient problems awaits attention. In addition, the use of the relationships to construct finite element models of the FSDT using those of the CPT is a challenging task but proves to be very useful (see Reddy and his colleagues [34,35]).

## References

1. Reissner, E., "On the theory of bending of elastic plates," *Journal of Mathematical Physics,* **23,** 184–191 (1944).
2. Reissner, E., "The effect of transverse shear deformation on the bending of elastic plates," *Journal of Applied Mechanics,* **12,** 69–77 (1945).
3. Kromm, A., "Verallgemeinerte Theorie der Plattenstatik," *Ing.-Arch.,* **21** (1953).
4. Kromm, A., "Über die Randquerkräfte bei gestützten Platten.," *Z. angew. Math. Mech.,* **35** (1955).
5. Panc, V., *Theories of Elastic Plates,* Noordhoff, Leyden, Netherlands (1975).
6. Reddy, J. N., *Energy and Variational Methods in Applied Mechanics,* John Wiley and Sons, New York, 1984.
7. Reddy, J. N., *Mechanics of Laminated Composite Plates: Theory and Analysis,* CRC Press, Boca Raton, Florida (1997).
8. Reddy, J. N., *Theory and Analysis of Elastic Plates,* Taylor & Francis, Philadelphia, PA (1999).
9. Basset, A. B., "On the extension and flexure of cylindrical and spherical thin elastic shells," *Phil. Trans. Royal Soc.,* (London) Ser. A, **181** (6), 433–480 (1890).
10. Hildebrand, F. B., Reissner, E., and Thomas, G. B., "Notes on the foundations of the theory of small displacements of orthotropic shells," NACA TN–1833, Washington, D.C. (1949).
11. Hencky, H., "Uber die Berucksichtigung der Schubverzerrung in ebenen Platten," *Ing. Arch.,* **16**, 72–76 (1947).
12. Mindlin, R. D., "Influence of rotatory inertia and shear on flexural motions of isotropic, elastic plates," *Journal of Applied Mechanics, Transactions of ASME,* **18**, 31–38 (1951).
13. Reddy, J. N., "A simple higher-order theory for laminated composite plates," *Journal of Applied Mechanics,* **51**, 745–752 (1984).

14. Reddy, J. N., "A general non-linear third-order theory of plates with moderate thickness," *International Journal of Non-Linear Mechanics*, **25**(6), 677–686 (1990).

15. Vlasov, B. F., "Ob uravneniyakh teovii isgiba plastinok (On the Equations of the Theory of Bending of Plates)," *Izv. Akd. Nauk SSR*, OTN, **4**, 102–109 (1958).

16. Jemielita, G., "Techniczna Teoria Plyt Sredniееj Grubbosci (Technical theory of plates with moderate thickness)", *Rozprawy Insynierskie (Engineering Transactions), Polska Akademia Nauk.* **23**(3), 483-499 (1975).

17. Schmidt, R., "A refined nonlinear theory for plates with transverse shear deformation," *Journal of the Industrial Mathematics Society,* **27**(1), 23–38 (1977).

18. Krishna Murty, A. V., "Higher order theory for vibration of thick plates," *AIAA Journal,* **15**(12), 1823–1824 (1977).

19. Reddy, J. N., "A small strain and moderate rotation theory of laminated anisotropic plates," *Journal of Applied Mechanics,* **54**, 623–626 (1987).

20. Levinson, M., "An accurate, simple theory of the statics and dynamics of elastic plates", *Mechanics Research Communications*, **7**(6), 343-350 (1980).

21. Wang, C. M., Reddy, J. N., and Lee, K. H., *Shear Deformable Beams and Plates. Relationships with Classical Solutions,* Elsevier, U.K., 2000.

22. Reddy, J. N., Wang, C. M., Lim, G. T., and Ng, K. H., " Bending solutions of the Levinson beams and plates in terms of the classical theories," *Int. J. Solids & Structures,* in review.

23. Reddy, J. N. and Wang, C. M., "Deflection relationships between classical and third-order plate theories", *Acta Mechanica,* **130**(3-4), 199-208 (1998).

24. Wang, C. M., "Allowance for prebuckling deformations in buckling load relationship between Mindlin and Kirchhoff simply supported plates of general polygonal shape", *Engineering Structures*, **17**(6), 413-418 (1995).

25. Wang, C. M., Discussion on "Postbuckling of moderately thick circular plates with edge elastic restraint", *J. Engng. Mech., ASCE*, **122**(2), 181-182 (1996).

26. Wang, C. M. and Reddy, J. N., "Buckling load relationship between Reddy and Kirchhoff plates of polygonal shape with simply supported edges", *Mechanics Research Communications*, **24**(1), 103-108 (1997).

27. Wang, C. M., Xiang, Y., and Kitipornchai, S., "Buckling solutions of rectangular Mindlin plates under uniform shear", *Journal of Engineering Mechanics*, **120**(11), 2462-2470 (1994).

28. Wang, C. M., "Buckling of polygonal and circular sandwich plates" *AIAA Journal*, **33**(5), 962-964 (1995).

29. Wang, C. M., "Natural frequencies formula for simply supported Mindlin plates", *Trans. ASME, Journal of Vibration and Acoustics*, **116**(4), 536-540 (1994).

30. Wang, C. M., "Vibration frequencies of simply supported polygonal sandwich plates via Kirchhoff solutions", *Journal of Sound and Vibration*, **190**(2), 255-260 (1996).

31. Wang, C. M., Kitipornchai, S. and Reddy, J. N., "Relationship between vibration frequencies of Reddy and Kirchhoff polygonal plates with simply supported edges", *ASME Journal of Vibration and Acoustics* **122**(1), 77-81 (2000).

32. Wang, C. M., Wang, C., and Ang, K. K., "Vibration of Initially Stressed Reddy Plates on a Winkler-Pasternak Foundation", *Journal of Sound and Vibration*, **204**(2), 203-212 (1997).

33. Wang, C. M., Lim, G. T., Reddy, J. N., and Lee, K. H., "Relationships between bending solutions of Reissner and Mindlin plate theories", *Engineering Structures*, to appear.

34. Reddy, J. N., "On locking-free shear deformable beam finite elements", *Computer Meth. Appl. Mech. Engng.*, **149**, 113-132 (1997).

35. Reddy, J. N., Wang, C. M. and Lam, K. Y., "Unified finite elements based on the classical and shear deformation theories of beams and axisymmetric circular plates", *Communications in Numerical Methods in Engineering*, **13**, 495-510 (1997).

# A SIMPLE DERIVATION OF COSSERAT THEORIES OF SHELLS, RODS AND POINTS

*This paper is dedicated to Professor A. Libai in honor of his 70th birthday.*

M.B. RUBIN
*Faculty of Mechanical Engineering*
*Technion - Israel Institute of Technology*
*32000 Haifa, Israel*

## 1. Introduction

Cosserat theories of shells, rods and points are continuum theories which model the response of three-dimensional structures that have special geometrical properties. Specifically, the Cosserat theory of shells (Naghdi, 1972) models the response of a shell-like structure that is "thin" in one of its dimensions called the thickness; the Cosserat theory of rods (Green et al., 1974a,b) models a rod-like structure that is "thin" in two of its dimensions characterizing its cross-section; and the theory of a Cosserat point (Rubin, 1985a,b) models a structure that is "thin" in all three of its dimensions. These Cosserat theories are simpler than the three-dimensional theory, which introduces partial differential equations that depend on three spatial coordinates and time because: for shell theory the equations depend on only two spatial coordinates and time; for rod theory the equations depend on only one spatial coordinate and time; and for the theory of a Cosserat point the equations are ordinary differential equations that depend only on time. The utility of theories of shells and rods is well known, and the theory of a Cosserat point has been shown to be a continuum model of a finite element that can be used to formulate the numerical solution of problems in continuum mechanics (Rubin, 1987, 1995).

From a theoretical point of view it is most clear to develop these Cosserat theories by the direct approach which proposes balance laws representing: conservation of mass, balance of linear momentum, balances of director momentum and balance of angular momentum. Moreover, within the context of the purely mechanical theory an expression for the rate of material dissipation can be introduced to place restrictions on constitutive equations. This direct approach to the development of these Cosserat theories is similar to the development of three-dimensional continuum mechanics in the sense that the balance laws are valid for arbitrary constitutive equations and the theories are properly invariant under superposed rigid body motions. Furthermore, for structures made from elastic materials, the constitutive equations are hyperelastic in the sense that the response functions are determined by derivatives of a strain energy function.

The objective of the present paper is to present a simple derivation of these Cosserat theories. This is most easily accomplished by deriving the local forms of the balance laws



*D. Durban et al. (eds.), Advances in the Mechanics of Plates and Shells, 277–294.*

of the Cosserat theories starting with the local form of the balance of linear momentum in the three-dimensional theory, introducing a kinematic approximation for the three-dimensional position vector and then integrating the resulting equations over appropriate spatial domains. This procedure is very similar to the Galerkin method that is used to obtain approximate solutions of the three-dimensional equations. However, the Galerkin method determines the response functions by integrating the three-dimensional constitutive equations which employ the kinematic approximation pointwise in the three-dimensional region. In contrast, the Cosserat approach determines the constitutive equations and the form of the strain energy function by comparison with special exact three-dimensional solutions and/or experimental data. This procedure of postulating a form for the strain energy function to determine constitutive equations is similar to that advocated by Libai and Simmonds (1998, p.3).

In order to emphasize parallels with the three-dimensional theory a new notation is introduced for variables in these Cosserat theories. Specifically, quantities related to the three-dimensional theory are denoted using a superposed star $(^*)$ and similar quantities related to the Cosserat theory are denoted by the same symbol without a superposed star. Thus, for example, the position vector associated with the three-dimensional theory is denoted by $\mathbf{x}^*$, and the vector $\mathbf{x}$ is used to denote the position vector locating: the reference surface of the Cosserat shell; the reference curve of the Cosserat rod; and the reference point of the theory of a Cosserat point.

An outline of the paper is as follows. Section 2 reviews a convenient form of the three-dimensional equations in terms of convected Lagrangian coordinates, section 3 presents an averaged form of the balance of linear momentum which is used to develop the balances of director momentum equations in the Cosserat theory. Section 4 compares the development from the three-dimensional theory with the direct approach. Sections 5, 6 and 7 present the equations of motion of the theories of Cosserat shells, rods and points, respectively. These sections have been written to be independent of each other so that the reader can pass from section 4 to either of the sections 5, 6 or 7 without breaking continuity of the development. Appendices A, B and C present definitions that are relevant to the Cosserat theories of shells, rods and points, respectively.

Throughout the text, bold faced symbols are used to denote vector and tensor quantities. Also, $\mathbf{I}$ denotes the unity tensor; $\mathrm{tr}(\mathbf{A})$ denotes the trace of the second order tensor $\mathbf{A}$; $\mathbf{A}^T$ denotes the transpose of $\mathbf{A}$; $\mathbf{A}^{-1}$ denotes the inverse of $\mathbf{A}$; and $\mathbf{A}^{-T}$ denotes the inverse of the transpose of $\mathbf{A}$. The scalar $\mathbf{a} \bullet \mathbf{b}$ denotes the dot product between two vectors $\mathbf{a},\mathbf{b}$; the scalar $\mathbf{A} \bullet \mathbf{B}=\mathrm{tr}(\mathbf{A}\mathbf{B}^T)$ denotes the dot product between two second order tensors $\mathbf{A},\mathbf{B}$; the vector $\mathbf{a} \times \mathbf{b}$ denotes the cross product between $\mathbf{a}$ and $\mathbf{b}$; and the second order tensor $\mathbf{a}\otimes\mathbf{b}$ denotes the tensor product between $\mathbf{a}$ and $\mathbf{b}$. Moreover, the usual summation convention over repeated lower cased indices is implied with the range of Latin indices being (1,2,3) and the range of Greek indices being (1,2).

## 2. Balance laws in the three-dimensional theory

The motion of a simple continuum is characterized by a nonsingular mapping from an arbitrary fixed reference configuration to the present configuration at time t. Specifically, the vector $\mathbf{X}^*$ denotes the location of a material point in the reference configuration, $\mathbf{x}^*$ denotes the location of the same material point in the present configuration and the mapping is given by

$$\mathbf{x}^* = \mathbf{x}^*(\mathbf{X}^*,t) \ . \tag{2.1}$$

Moreover, the absolute velocity $\mathbf{v}^*$ of the material point becomes

$$\mathbf{v}^* = \dot{\mathbf{x}}^* \ , \tag{2.2}$$

where a superposed dot denotes material time differentiation holding the material point $\mathbf{X}^*$ fixed. Also, the deformation gradient $\mathbf{F}^*$, the velocity gradient $\mathbf{L}^*$, the rate of deformation tensor $\mathbf{D}^*$ and the spin tensor $\mathbf{W}^*$ are characterized by the equations

$$\mathbf{F}^* = \partial\mathbf{x}^*/\partial\mathbf{X}^* \ , \quad \dot{\mathbf{F}}^* = \mathbf{L}^*\,\mathbf{F}^* \ , \quad \mathbf{L}^* = \partial\mathbf{v}^*/\partial\mathbf{x}^* = \mathbf{D}^* + \mathbf{W}^* \ ,$$
$$\mathbf{D}^* = \frac{1}{2}(\mathbf{L}^* + \mathbf{L}^{*T}) = \mathbf{D}^{*T} \ , \quad \mathbf{W}^* = \frac{1}{2}(\mathbf{L}^* - \mathbf{L}^{*T}) = -\,\mathbf{W}^{*T} \ . \tag{2.3}$$

It is well known that the global forms of the conservation of mass and the balance law of linear momentum of a simple continuum can be expressed as

$$\frac{d}{dt}\int_{P^*} \rho^*\, dv^* = 0 \ ,$$
$$\frac{d}{dt}\int_{P^*} \rho^*\, \mathbf{v}^*\, dv^* = \int_{P^*} \rho^*\, \mathbf{b}^*\, dv^* + \int_{\partial P^*} \mathbf{t}^*\, da^* \ , \tag{2.4}$$

and the balance of angular momentum is given by

$$\frac{d}{dt}\int_{P^*} \mathbf{x}^* \times \rho^*\, \mathbf{v}^*\, dv^* = \int_{P^*} \mathbf{x}^* \times \rho^*\, \mathbf{b}^*\, dv^* + \int_{\partial P^*} \mathbf{x}^* \times \mathbf{t}^*\, da^* \ . \tag{2.5}$$

In these equations: $P^*$ denotes an arbitrary smooth material region with boundary $\partial P^*$; $da^*$ and $dv^*$ denote, respectively, the element of area and element of volume in the present configuration; $\rho^*(\mathbf{x}^*,t)$ denotes the mass density (mass per unit present volume); $\mathbf{b}^*(\mathbf{x}^*,t)$ denotes the specific (per unit mass) external body force; and $\mathbf{t}^*(\mathbf{x}^*,t;\, \mathbf{n}^*)$ denotes the stress vector (force per unit present area) acting at the point $\mathbf{x}^*$ on the surface whose unit outward normal is $\mathbf{n}^*$. Moreover, it can be shown that $\mathbf{t}^*$ is a linear function of $\mathbf{n}^*$ given by

$$\mathbf{t}^*(\mathbf{x}^*,t\,;\, \mathbf{n}^*) = \mathbf{T}^*(\mathbf{x}^*,t\,)\, \mathbf{n}^* \ , \tag{2.6}$$

where the Cauchy stress tensor $\mathbf{T}^*$ is explicitly independent of the normal $\mathbf{n}^*$. Also, for the purely mechanical theory, it is possible to define the rate of dissipation $\mathcal{D}^*$ in terms of the strain energy function $\Sigma^*$ such that

$$\int_{P^*} \mathcal{D}^*\, dv^* = \mathcal{W}^* - \dot{\mathcal{K}}^* - \dot{\mathcal{U}}^* \geq 0 \ ,$$
$$\mathcal{W}^* = \int_{P^*} \rho^*\, \mathbf{b}^* \bullet \mathbf{v}^*\, dv^* + \int_{\partial P^*} \mathbf{t}^* \bullet \mathbf{v}^*\, da^* \ ,$$
$$\mathcal{K}^* = \int_{P^*} \frac{1}{2}\rho^*\, \mathbf{v}^* \bullet \mathbf{v}^*\, dv^* \ , \quad \mathcal{U}^* = \int_{P^*} \rho^*\, \Sigma^*\, dv^* \ , \tag{2.7}$$

where $\mathcal{W}^*$ represents the total rate of external work applied to the body, $\mathcal{K}^*$ represents the total kinetic energy and $\mathcal{U}^*$ represents the total strain energy.

Using the usual continuity assumptions it can be shown that the local forms of (2.4) become

$$\dot{\rho}^* + \rho^*\, \mathrm{div}^*\, \mathbf{v}^* = 0 \ , \quad \rho^*\, \dot{\mathbf{v}}^* = \rho^*\, \mathbf{b}^* + \mathrm{div}^*\, \mathbf{T}^* \ , \tag{2.8}$$

where $\text{div}^*$ denotes the divergence operator with respect to the present position $\mathbf{x}^*$. Also, with the help of (2.8), the reduced form of the balance of angular momentum (2.5) requires the Cauchy stress to be symmetric

$$\mathbf{T}^{*T} = \mathbf{T}^* \ , \tag{2.9}$$

and the local form of the dissipation inequality $(2.7)_1$ reduces to

$$\mathcal{D}^* = \mathbf{T}^* \bullet \mathbf{D}^* - \rho^* \dot{\Sigma}^* \geq 0 \ , \tag{2.10}$$

which must be valid for all motions.

In the following developments it is convenient to express tensor quantities in terms of a set of convected Lagrangian coordinates $\theta^i$ (i=1,2,3)

$$\mathbf{x}^* = \mathbf{x}^*(\theta^i,t) \ , \quad \dot{\theta}^i = 0 \ , \tag{2.11}$$

and to define the covariant and contravariant base vectors, respectively, by $\mathbf{g}_i$ and $\mathbf{g}^i$, such that

$$\mathbf{g}_i = \mathbf{x}^*{}_{,i} \ , \quad g^{1/2} = \mathbf{g}_1 \times \mathbf{g}_2 \bullet \mathbf{g}_3 \ , \quad \mathbf{g}_i \bullet \mathbf{g}^j = \delta_i^j \ ,$$
$$g^{1/2}\,\mathbf{g}^1 = \mathbf{g}_2 \times \mathbf{g}_3 \ , \quad g^{1/2}\,\mathbf{g}^2 = \mathbf{g}_3 \times \mathbf{g}_1 \ , \quad g^{1/2}\,\mathbf{g}^3 = \mathbf{g}_1 \times \mathbf{g}_2 \ . \tag{2.12}$$

In these equations, a comma denotes partial differentiation with respect to $\theta^i$, $\delta_i^j$ is the Kronecker delta symbol, and $g^{1/2}$ is related to the element of volume $dv^*$ by the expression

$$dv^* = g^{1/2}\, d\theta^1 d\theta^2 d\theta^3 \ . \tag{2.13}$$

It then follows that similar quantities related to the reference configuration are defined by

$$\mathbf{G}_i = \mathbf{X}^*{}_{,i} \ , \quad G^{1/2} = \mathbf{G}_1 \times \mathbf{G}_2 \bullet \mathbf{G}_3 \ , \quad \mathbf{G}_i \bullet \mathbf{G}^j = \delta_i^j \ ,$$
$$G^{1/2}\,\mathbf{G}^1 = \mathbf{G}_2 \times \mathbf{G}_3 \ , \quad G^{1/2}\,\mathbf{G}^2 = \mathbf{G}_3 \times \mathbf{G}_1 \ , \quad G^{1/2}\,\mathbf{G}^3 = \mathbf{G}_1 \times \mathbf{G}_2 \ . \tag{2.14}$$

Moreover, using these definitions it can be shown that the deformation gradient, the velocity gradient and the divergence operator can be expressed in the alternative forms

$$\mathbf{F}^* = \mathbf{g}_i \otimes \mathbf{G}^i \ , \quad \mathbf{L}^* = \mathbf{v}^*{}_{,j} \otimes \mathbf{g}^j \ , \quad \text{div}^*\, \mathbf{T}^* = \mathbf{T}^*{}_{,i} \bullet \mathbf{g}^i \ . \tag{2.15}$$

Also, with the help of the result

$$(g^{1/2}\,\mathbf{g}^j)_{,j} = 0 \ , \tag{2.16}$$

it follows that

$$g^{1/2}\, \text{div}^*\, \mathbf{T}^* = (g^{1/2}\,\mathbf{T}^*\,\mathbf{g}^j)_{,j} = \mathbf{t}^{*j}{}_{,j} \ , \tag{2.17}$$

where the stress tensor has been expressed in terms of the three vectors $\mathbf{t}^{*j}$, such that

$$g^{1/2}\,\mathbf{T}^* = \mathbf{t}^{*j} \otimes \mathbf{g}_j \ , \quad \mathbf{t}^{*j} = g^{1/2}\,\mathbf{T}^* \bullet \mathbf{g}^j \ . \tag{2.18}$$

Next, introducing the scaled mass density $m^*$ and the reference value $\rho_0^*$ of $\rho$ by the equation

$$m^* = \rho^*\, g^{1/2} = \rho_0^*\, G^{1/2} = m^*(\theta^i) \ , \tag{2.19}$$

the local forms of the conservation of mass and the balance of linear momentum can be written in the alternative forms (see Green and Adkins, 1960; Green and Zerna, 1968)

$$\dot{m}^* = 0 \ , \quad m^*\, \dot{\mathbf{v}}^* = m^*\, \mathbf{b}^* + \mathbf{t}^{*j}{}_{,j} \ . \tag{2.20}$$

It will be shown in the following sections that the conservation of mass and the balances of linear and angular momentum of the Cosserat theories of shells, rods and points can be written in forms that are very similar to equations (2.19), (2.20) and (2.9). Moreover, by using the definitions of the vectors $\mathbf{t}^{*i}$, the effect of the divergence operator has been simplified to mere partial differentiation. This has advantages in handling the complicated geometry of shells and space rods where it is convenient to use curvilinear coordinates instead of rectangular Cartesian coordinates.

For a purely elastic material it is well known that the stress $\mathbf{T}^*$ and the strain energy $\Sigma^*$ are functions of the deformation gradient $\mathbf{F}^*$ only and are explicitly independent of time. Also, due to invariance under superposed rigid body motions, $\Sigma^*$ depends on $\mathbf{F}^*$ only through the right Cauchy-Green deformation tensor $\mathbf{C}^*$, such that

$$\mathbf{T}^* = \hat{\mathbf{T}}^*(\mathbf{F}^*) \ , \quad \Sigma^* = \hat{\Sigma}^*(\mathbf{C}^*) \ , \quad \mathbf{C}^* = \mathbf{F}^{*T}\mathbf{F}^* \ . \tag{2.21}$$

Moreover, for a purely elastic material the rate of dissipation $\mathcal{D}^*$ vanishes and (2.10) can be used to prove that $\mathbf{T}^*$ is restricted so that

$$\mathbf{T}^* = \hat{\mathbf{T}}^* \ , \quad \hat{\mathbf{T}}^* = 2\rho^* \, \mathbf{F}^* \frac{\partial \hat{\Sigma}^*}{\partial \mathbf{C}^*} \mathbf{F}^{*T} \ . \tag{2.22}$$

Once a form for the strain energy function $\Sigma^*$ is specified, the constitutive equation for stress $\mathbf{T}^*$ is determined by (2.22). Then, the equation of linear momentum $(2.20)_2$ represents a vector equation for the unknown kinematic quantity $\mathbf{x}^*$, which is a function of three spatial coordinates $\theta^i$ and time. Furthermore, a complete formulation of a problem requires specification of appropriate initial conditions and boundary conditions.

## 3. An averaged form of the balance of linear momentum

In the purely mechanical three-dimensional theory, the balance of angular momentum places restrictions on the constitutive equations which require the Cauchy stress tensor to be symmetric (2.9). Also, the conservation of mass and the balance of linear momentum are used to determine the mass density and the position of each material point in the continuum. For the Cosserat theories that will be developed in the next sections, equations representing the conservation of mass and the balances of linear and angular momentum will be used in a similar manner to determine the mass density and the position of each material point. However, the Cosserat theories introduce additional kinematical quantities called director vectors at each material point which also need to be determined by additional balance laws. In order to motivate the forms for these balance laws it is convenient to consider an averaged form of the balance of linear momentum.

To this end, let $\phi(\theta^i)$ be a general weighting function that depends on the convected coordinates $\theta^i$ only and is independent of time so that

$$\phi = \phi(\theta^i) \ , \quad \dot{\phi} = 0 \ . \tag{3.1}$$

Now, multiplying the local form $(2.20)_2$ of linear momentum by $\phi$ it follows that

$$\phi m^* \dot{\mathbf{v}}^* = \phi m^* \, \mathbf{b}^* + (\phi \, \mathbf{t}^{*j})_{,j} - \mathbf{t}^{*j} \, \phi_{,j} \ . \tag{3.2}$$

Moreover, with the help of the expression $(2.15)_3$ for the divergence operator and the formula (2.6) for the stress vector, it can be shown that integration of (3.2) over the region $P^*$ yields the global averaged form the balance of linear momentum

$$\frac{d}{dt}\int_{P^*} \phi\, \rho^*\, \mathbf{v}^*\, dv^* = \int_{P^*} \phi\, \rho^*\, \mathbf{b}^*\, dv^* + \int_{\partial P^*} \phi\, \mathbf{t}^*\, da^* - \int_{P^*} g^{-1/2}\, \mathbf{t}^{*j}\, \phi_{,j}\, dv^* \ . \quad (3.3)$$

Specifically, it is noted that, relative to the balance law $(2.4)_2$, equation (3.3) contains an extra term on the right hand side of the equation which characterizes the integrated average of the effect of the stresses $\mathbf{t}^{*j}$. Also, it is noted that for $\phi=1$ this equation reduces to the balance of linear momentum $(2.4)_2$.

## 4. Development from the three-dimensional theory and the direct approach

The balance laws for Cosserat theories of shells, rods and points can be developed by integration using the three-dimensional theory. Specifically, a kinematic assumption for the position vector $\mathbf{x}^*$ is introduced of the form

$$\mathbf{x}^*(\theta^i,t) = \mathbf{x} + \phi^j(\theta^i)\, \mathbf{d}_j \ , \quad (4.1)$$

where $\mathbf{x}$ is the position vector, $\mathbf{d}_j$ are called director vectors, and $\phi^j$ are functions of $\theta^i$ only. Depending on the theory being developed, the vectors $\mathbf{x}$ and $\mathbf{d}_i$ depend on time and two spatial coordinates (shell theory), one spatial coordinate (rod theory) or no spatial coordinates (point theory).

Next, by substituting this kinematic assumption into $(2.4)_1$ and (3.3) it is possible to derive the global forms of: the conservation of mass; the balance of linear momentum; and the balances of director momentum associated with the Cosserat theory. Moreover, interpreting the constant 1 and $\phi^j$ in (4.1) as shape functions and taking the weighting functions $\phi$ in (3.3) to be the same as these shape functions, it is possible to identify this development as a standard Galerkin procedure for obtaining an approximate solution of the partial differential equations (2.8). In this regard, it is important to emphasize that within the context of the Galerkin procedure, the three-dimensional constitutive equations and the kinematic assumption are assumed to be valid pointwise in the region occupied by the body. Consequently, the stiffness matrix of the resulting system of ordinary differential equations is uniquely determined for a given set of shape functions and weighting functions. In other words, the constitutive equations of the resultant forces and moments of the approximate equations are uniquely determined by the Galerkin method [e.g. in the appendices see equations (A2) for shells, (B2) for rods, and (C2) for points].

When nonlinear inhomogeneous deformations are considered it is often impossible to analytically integrate the expressions characterizing these resultant forces and moments. Consequently, approximate methods of integration are usually employed. However, even when the three-dimensional material is hyperelastic, with the stress being obtained by the derivative of a strain energy function (2.22), there is no guarantee that the resultant forces and moments obtained using this Galerkin procedure will satisfy integrability conditions needed to ensure that a strain energy function exists for the approximate equations. Also, there is no guarantee that the global form of the balance of angular momentum will be satisfied exactly.

In contrast, the Cosserat approach uses the kinematic assumption (4.1) merely to motivate the theoretical structure of the balance laws. Specifically, the Cosserat approach abandons the notion that the three-dimensional constitutive equations and the kinematic assumption are valid pointwise in the region occupied by the body. Instead, constitutive equations for the resultant forces and director couples (moments) are obtained by demanding that they satisfy restrictions which ensure that the global forms of the balance of angular momentum (2.5) and the dissipation inequality $(2.7)_1$ are satisfied for all possible motions. This procedure is identical to that associated with the three-dimensional theory and it ensures that for elastic response the resultant forces and director couples are related to derivatives of a strain energy function. Consequently, the Cosserat theory automatically preserves the fundamental properties of the three-dimensional theory (global conservation of mass and balances of linear momentum and angular momentum; the existence of a strain energy function; and invariance under superposed rigid body motions). Furthermore, specific forms for the strain energy function are obtained by comparison with exact solutions of the three-dimensional equations and/or with experimental data.

Development of the balance laws of the Cosserat theory using the kinematic assumption (4.1) and integration of the three-dimensional balance of linear momentum has the advantage that the procedure is straight forward and the starting point is known. However, this procedure does not expose the full potential of the Cosserat approach. In this regard, the Cosserat theory can be developed by a direct approach (see: Naghdi, 1972 for shells; Green, et al, 1974b for rods; and Rubin, 1985a for points) which proposes a continuum theory that often is considered to be independent of the full three-dimensional theory. Within the context of the direct approach, the kinematics of the continuum are characterized by the position vector $\mathbf{x}$ and the directors $\mathbf{d}_i$. The balance laws include: the conservation of mass and the balances of linear momentum, director momentum and angular momentum. Also, the dissipation inequality is proposed to place additional restrictions on the constitutive equations for the kinetic quantities like resultant forces and director couples. In this regard, the direct approach parallels the three-dimensional theory in that the balance laws are postulated and are presumed to be valid for all materials (solids, fluids, etc.).

As continuum theories, the Cosserat theories of shells, rods and points, which are developed by the direct approach, can remain exact nonlinear theories if no further approximations are made. However, in comparison with the three-dimensional theory these Cosserat theories represent simpler approximate models for the response of shell-like, rod-like and point-like structures.

Although the direct approach to Cosserat theories has the advantages described above, the following sections will develop balance laws by integration of the three-dimensional equations. However, the constitutive equations will be developed using the dissipation inequality instead of integration of the three-dimensional constitutive equations as is done for the Galerkin approximation.

## 5. Cosserat shells

A Cosserat shell is a continuum model for the response of a three-dimensional structure that is shell-like in the sense that it is "thin" in one of its dimensions. This structure is essentially a curved surface with some small thickness. Material points in the shell's stress-free reference configuration are located by the position vector $\mathbf{X}^*$

$$\mathbf{X}^* = \mathbf{X}^*(\theta^\alpha,\theta^3) = \mathbf{X}(\theta^\alpha) + \theta^3\,\mathbf{D}_3(\theta^\alpha) \ , \tag{5.1}$$

where $\theta^\alpha$ ($\alpha$=1,2) are convected coordinates that characterize material points on the reference surface $\mathbf{X}$ of the shell and $\theta^3$ is a convected coordinate that characterizes material points in the thickness of the shell. Moreover, the director vector $\mathbf{D}_3$ is restricted so that it is linearly independent of the tangent vectors $\mathbf{D}_\alpha$ to this reference surface

$$\mathbf{D}_\alpha = \mathbf{X}_{,\alpha} \ , \quad D^{1/2} = \mathbf{D}_1 \times \mathbf{D}_2 \bullet \mathbf{D}_3 > 0 \ . \tag{5.2}$$

For simplicity, the reference surface $\mathbf{X}$ is taken to be the middle surface of the shell [$\mathbf{X}=\mathbf{X}^*(\theta^\alpha,0)$] and the lateral surface of the shell is characterized by

$$f(\theta^\alpha) = 0 \ , \quad |\,\theta^3\,| \le \frac{H(\theta^\alpha)}{2} \ , \tag{5.3}$$

where $f(\theta^\alpha)$ characterizes a smooth closed curve and $H(\theta^\alpha)$ is the variable thickness of the shell. This thickness is limited by the condition that the representation (5.1) provides a one-to-one invertible mapping between $\theta^i$ and $\mathbf{X}^*$, which requires

$$\mathbf{G}_\alpha = \mathbf{D}_\alpha + \theta^3\,\mathbf{D}_{3,\alpha} \ , \quad \mathbf{G}_3 = \mathbf{D}_3 \ , \quad G^{1/2} = \mathbf{G}_1 \times \mathbf{G}_2 \bullet \mathbf{G}_3 > 0 \ , \tag{5.4}$$

where use has been made of the definitions (2.14).

In the present deformed configuration, the material points of the shell are assumed to be located by a position vector $\mathbf{x}^*$ which has a representation similar to (5.1)

$$\mathbf{x}^* = \mathbf{x}^*(\theta^\alpha,\theta^3,t) = \mathbf{x}(\theta^\alpha,t) + \theta^3\,\mathbf{d}_3(\theta^\alpha,t) \ , \tag{5.5}$$

with the reference surface $\mathbf{X}$ and director $\mathbf{D}_3$ being mapped to their present values $\mathbf{x}$ and $\mathbf{d}_3$, respectively. Again, the director vector $\mathbf{d}_3$ is restricted so that it is linearly independent of the tangent vectors $\mathbf{d}_\alpha$ to the present location $\mathbf{x}$ of the reference surface

$$\mathbf{d}_\alpha = \mathbf{x}_{,\alpha} \ , \quad d^{1/2} = \mathbf{d}_1 \times \mathbf{d}_2 \bullet \mathbf{d}_3 > 0 \ . \tag{5.6}$$

Also, the representation (5.5) remains invertible provided that

$$\mathbf{g}_\alpha = \mathbf{d}_\alpha + \theta^3\,\mathbf{d}_{3,\alpha} \ , \quad \mathbf{g}_3 = \mathbf{d}_3 \ , \quad g^{1/2} = \mathbf{g}_1 \times \mathbf{g}_2 \bullet \mathbf{g}_3 > 0 \ , \tag{5.7}$$

where use has been made of the definitions (2.12).

For elastic shells, the director $\mathbf{d}_3$ can be identified with the material line element through the thickness which in the reference configuration was oriented in the direction $\mathbf{D}_3$. It then follows that the representation (5.5) is only approximate for general deformations of the shell-like structure since it tacitly assumes that this material fiber remains straight. However, this approximation is reasonably accurate for a wide range of practical shell-like structures.

Within the context of the direct approach to Cosserat shell theory, the shell is modeled as a surface which is characterized by the position vector $\mathbf{x}(\theta^\alpha,t)$. This surface is endowed with an additional director vector $\mathbf{d}_3(\theta^\alpha,t)$ which provides limited information about deformation through the thickness of the shell-like body. Specifically, the kinematics of the shell are specified by the vectors

$$\{\ \mathbf{x}(\theta^\alpha,t)\ ,\ \mathbf{d}_3(\theta^\alpha,t)\ \}\ ,\quad \{\ \mathbf{v} = \dot{\mathbf{x}}\ ,\ \mathbf{w}_i = \dot{\mathbf{d}}_i\ \}\ , \tag{5.8}$$

where a superposed dot denotes material time differentiation holding $\theta^\alpha$ fixed and the restriction $(5.6)_2$ is imposed on $\mathbf{d}_3$. Moreover, it is noted that $\mathbf{d}_3$ is a general vector which models the effects of transverse shear deformation and normal extension. Here, local forms of the balance laws are obtained by direct integration of the local equations (2.20) and (3.2). Specifically, with the help of the representation (5.5) and integration over the thickness of the shell: $(2.20)_1$ yields the conservation of mass; $(2.20)_2$ yields the balance of linear momentum; and (3.2) with $\phi-\theta^3$, yields the balance of director momentum which, respectively, are given by

$$\dot{m} = 0 \ , \quad m\,(\dot{\mathbf{v}} + y^3\,\dot{\mathbf{w}}_3) = m\,\mathbf{b} + \mathbf{t}^\alpha{}_{,\alpha} \ ,$$

$$m\,(y^3\,\dot{\mathbf{v}} + y^{33}\,\dot{\mathbf{w}}_3) = m\,\mathbf{b}^3 - \mathbf{t}^3 + \mathbf{m}^\alpha{}_{,\alpha} \ . \tag{5.9}$$

In these equations: the scaled mass density m is related to the mass density $\rho$ (mass per unit present area $d\sigma$ of the surface $\mathbf{x}$) and its reference value $\rho_0$ through the equations

$$m = \rho\, a^{1/2} = \rho_0\, A^{1/2} = m(\theta^\alpha) \ , \quad a^{1/2} = |\, \mathbf{d}_1 \times \mathbf{d}_2 \,| \ ,$$

$$A^{1/2} = |\, \mathbf{D}_1 \times \mathbf{D}_2 \,| \ , \quad d\sigma = a^{1/2}\, d\theta^1 d\theta^2 \ , \tag{5.10}$$

$y^3$ and $y^{33}$ are director inertia coefficients, which are independent of time

$$\dot{y}^3 = 0 \ , \quad \dot{y}^{33} = 0 \ , \tag{5.11}$$

$\mathbf{b}$ is the specific external force; $\mathbf{b}^3$ is the specific external director couple; $\mathbf{t}^\alpha$ are related to the resultant contact forces applied to the boundary of the shell; $\mathbf{m}^\alpha$ are related to the resultant contact couples applied to the boundary of the shell; and $\mathbf{t}^3$ is the intrinsic director couple. Moreover, the assigned fields $\{\mathbf{b}, \mathbf{b}^3\}$ include contributions $\{\mathbf{b}_b, \mathbf{b}_b^3\}$ due to the three-dimensional body force and contributions $\{\mathbf{b}_c, \mathbf{b}_c^3\}$ due to contact forces on the major surfaces ($\theta^3$=$\pm$H/2) of the shell

$$\mathbf{b} = \mathbf{b}_b + \mathbf{b}_c \ , \quad \mathbf{b}^3 = \mathbf{b}_b^3 + \mathbf{b}_c^3 \ . \tag{5.12}$$

Expressions for these quantities in terms of related three-dimensional quantities are recorded in appendix A. Also, use of (2.9) and (5.5), and integration of $(2.18)_1$ yields the local form of the balance of angular momentum

$$\mathbf{T}^T = \mathbf{T} \ , \quad a^{1/2}\, \mathbf{T} = \mathbf{t}^i \otimes \mathbf{d}_i + \mathbf{m}^\alpha \otimes \mathbf{d}_{3,\alpha} \ . \tag{5.13}$$

Next, the rate of dissipation $\mathcal{D}$ is defined in terms of the strain energy function $\Sigma$ by the expression

$$a^{1/2}\, \mathcal{D} = [\, m\, \mathbf{b} \cdot \mathbf{v} + (\mathbf{t}^\alpha \cdot \mathbf{v})_{,\alpha} + m\, \mathbf{b}^3 \cdot \mathbf{w}_3 + (\mathbf{m}^\alpha \cdot \mathbf{w}_3)_{,\alpha} \,]$$

$$- \frac{d}{dt} [\, \frac{1}{2}\, m \{\, \mathbf{v} \cdot \mathbf{v} + 2\, y^3\, \mathbf{v} \cdot \mathbf{w}_3 + y^{33}\, \mathbf{w}_3 \cdot \mathbf{w}_3 \,\}\,] - m\, \dot{\Sigma} \geq 0 \ , \tag{5.14}$$

where the first term in square brackets on the right-hand side represents the rate of work of all external forces and couples applied to the shell and the second term in square brackets on the right-hand side represents the kinetic energy of the shell. Now, with the help of the equations of motion (5.9) it can be shown that

$$a^{1/2}\, \mathcal{D} = \mathbf{t}^i \cdot \mathbf{w}_i + \mathbf{m}^\alpha \cdot \mathbf{w}_{3,\alpha} - m\, \dot{\Sigma} \geq 0 \ . \tag{5.15}$$

Moreover, it is convenient to introduce the deformation tensor $\mathbf{F}$, the inhomogeneous strain measures $\boldsymbol{\beta}_\alpha$, and the rate of deformation tensors $\mathbf{L}$, $\mathbf{D}$ and $\mathbf{W}$, by the equations

$$\mathbf{F} = \mathbf{d}_i \otimes \mathbf{D}^i \ , \quad \boldsymbol{\beta}_\alpha = \mathbf{F}^{-1} \mathbf{d}_{3,\alpha} - \mathbf{D}_{3,\alpha} \ , \quad \dot{\mathbf{F}} = \mathbf{L}\mathbf{F} \ ,$$

$$\mathbf{L} = \mathbf{w}_i \otimes \mathbf{d}^i = \mathbf{D} + \mathbf{W} \ , \quad \mathbf{D} = \frac{1}{2}(\mathbf{L} + \mathbf{L}^T) = \mathbf{D}^T \ , \quad \mathbf{W} = \frac{1}{2}(\mathbf{L} - \mathbf{L}^T) = -\mathbf{W}^T , \tag{5.16}$$

where the reciprocal vectors $\mathbf{D}^i$ and $\mathbf{d}^i$ are defined by equations similar to (2.12) and (2.14), such that

$$\mathbf{D}_i \bullet \mathbf{D}^j = \delta_i^j \ , \quad \mathbf{d}_i \bullet \mathbf{d}^j = \delta_i^j \ . \tag{5.17}$$

Then, with the help of (5.10), (5.13), (5.14) and (5.16) it can be shown that the rate of dissipation (5.15) can be written in alternative form

$$\mathcal{D} = \mathbf{T} \bullet \mathbf{D} + a^{-1/2} (\mathbf{F}^T \mathbf{m}^\alpha) \bullet \dot{\boldsymbol{\beta}}_\alpha - \rho \dot{\Sigma} \geq 0 \ . \tag{5.18}$$

For a purely elastic shell the tensors $\mathbf{T}$ and $\mathbf{m}^\alpha$ and the strain energy $\Sigma$ are functions of the deformation quantities $\mathbf{F}$ and $\boldsymbol{\beta}_\alpha$ only and are explicitly independent of time. Also, due to invariance under superposed rigid body motions, $\Sigma$ depends on $\mathbf{F}$ only through the deformation tensor $\mathbf{C}$, such that

$$\mathbf{T} = \hat{\mathbf{T}}(\mathbf{F}, \boldsymbol{\beta}_\alpha) \ , \quad \mathbf{m}^\alpha = \hat{\mathbf{m}}^\alpha(\mathbf{F}, \boldsymbol{\beta}_\alpha) \ , \quad \Sigma = \hat{\Sigma}(\mathbf{C}, \boldsymbol{\beta}_\alpha) \ , \quad \mathbf{C} = \mathbf{F}^T \mathbf{F} \ . \tag{5.19}$$

Moreover, for a purely elastic shell the rate of dissipation $\mathcal{D}$ vanishes and (5.18) can be used to prove that $\mathbf{T}$ and $\mathbf{m}^\alpha$ are restricted so that

$$\mathbf{T} = \hat{\mathbf{T}} \ , \quad \hat{\mathbf{T}} = 2\rho \, \mathbf{F} \frac{\partial \hat{\Sigma}}{\partial \mathbf{C}} \mathbf{F}^T \ , \quad \mathbf{m}^\alpha = \hat{\mathbf{m}}^\alpha \ , \quad \hat{\mathbf{m}}^\alpha = \mathbf{F}^{-T} \, m \frac{\partial \hat{\Sigma}}{\partial \boldsymbol{\beta}_\alpha} \ . \tag{5.20}$$

Additional restrictions on the functional form for $\Sigma$ can be imposed to ensure that the Cosserat theory produces solutions which are consistent with exact solutions of the three-dimensional theory of a homogeneous material for all three dimensionally homogeneous deformations for which $\mathbf{F}^*=\mathbf{F}(t)$ is independent of position and $\boldsymbol{\beta}_\alpha=0$ vanish (Naghdi and Rubin, 1995).

Once a form for the strain energy function $\Sigma$ is specified, the constitutive equations for $\mathbf{T}$ and $\mathbf{m}^\alpha$ are determined by (5.20), and the constitutive equations for $\mathbf{t}^i$ are determined using the definition (5.14)

$$\mathbf{t}^i = a^{1/2} \, \mathbf{T} \bullet \mathbf{d}^i - \mathbf{m}^\alpha \, (\mathbf{d}_{3,\alpha} \bullet \mathbf{d}^i) \ . \tag{5.21}$$

In addition, it is necessary to specify the inertia quantities $\{m, y^3, y^{33}\}$ and the assigned fields $\{\mathbf{b}, \mathbf{b}^3\}$. Then, the equations of linear momentum $(5.9)_2$ and director momentum $(5.9)_3$ represent two vector equations for the two unknown kinematic quantities $\{\mathbf{x}, \mathbf{d}_3\}$, which are functions of only two spatial coordinates $\theta^\alpha$ and time. Furthermore, a complete formulation of a problem requires specification of appropriate initial conditions and boundary conditions.

## 6. Cosserat rods

A Cosserat rod is a continuum model for the response of a three-dimensional structure that is rod-like in the sense that it is "thin" in two of its dimensions. This structure is essentially a space curve with some small thickness. Material points in the rod's stress-free reference configuration are located by the position vector $\mathbf{X}^*$

$$\mathbf{X}^* = \mathbf{X}^*(\theta^\alpha, \theta^3) = \mathbf{X}(\theta^3) + \theta^\alpha \, \mathbf{D}_\alpha(\theta^3) \ , \tag{6.1}$$

where $\theta^3$ is a convected coordinate that characterizes material points on the reference curve $\mathbf{X}$ of the rod and $\theta^\alpha$ are convected coordinates that characterize material points in the cross-section of the rod. Moreover, the director vectors $\mathbf{D}_\alpha$ are restricted so that they are linearly independent of the tangent vector $\mathbf{D}_3$ to this reference curve

$$\mathbf{D}_3 = \mathbf{X}_{,3} \ , \quad D^{1/2} = \mathbf{D}_1 \times \mathbf{D}_2 \bullet \mathbf{D}_3 > 0 \ . \tag{6.2}$$

For simplicity, the reference curve $\mathbf{X}$ is specified by $[\mathbf{X}=\mathbf{X}^*(0,\theta^3)]$ and the cross-section is taken to be rectangular such that the lateral surface of the rod is characterized by

$$|\theta^1| \leq \frac{H(\theta^3)}{2} \ , \quad |\theta^2| \leq \frac{W(\theta^3)}{2} \ , \quad |\theta^3| \leq \frac{L}{2} \ , \tag{6.3}$$

where $H(\theta^3)$ is the variable thickness, $W(\theta^3)$ is the variable width of the cross-section and L is the length of the reference curve. The thickness and width are limited by the condition that the representation (6.1) provides a one-to-one invertible mapping between $\theta^i$ and $\mathbf{X}^*$, which requires

$$\mathbf{G}_\alpha = \mathbf{D}_\alpha \ , \quad \mathbf{G}_3 = \mathbf{D}_3 + \theta^\alpha \mathbf{D}_{\alpha,3} \ , \quad G^{1/2} = \mathbf{G}_1 \times \mathbf{G}_2 \bullet \mathbf{G}_3 > 0 \ , \tag{6.4}$$

where use has been made of the definitions (2.14).

In the present deformed configuration, the material points of the rod are assumed to be located by a position vector $\mathbf{x}^*$ which has a representation similar to (6.1)

$$\mathbf{x}^* = \mathbf{x}^*(\theta^\alpha,\theta^3,t) = \mathbf{x}(\theta^3,t) + \theta^\alpha \, \mathbf{d}_\alpha(\theta^3,t) \ , \tag{6.5}$$

with the reference curve $\mathbf{X}$ and directors $\mathbf{D}_\alpha$ being mapped to their present values $\mathbf{x}$ and $\mathbf{d}_\alpha$, respectively. Again, the director vectors $\mathbf{d}_\alpha$ are restricted so that they are linearly independent of the tangent vector $\mathbf{d}_3$ to the present location $\mathbf{x}$ of the reference curve

$$\mathbf{d}_3 = \mathbf{x}_{,3} \ , \quad d^{1/2} = \mathbf{d}_1 \times \mathbf{d}_2 \bullet \mathbf{d}_3 > 0 \ . \tag{6.6}$$

Also, the representation (6.5) remains invertible provided that

$$\mathbf{g}_\alpha = \mathbf{d}_\alpha \ , \quad \mathbf{g}_3 = \mathbf{d}_3 + \theta^\alpha \, \mathbf{d}_{\alpha,3} \ , \quad g^{1/2} = \mathbf{g}_1 \times \mathbf{g}_2 \bullet \mathbf{g}_3 > 0 \ , \tag{6.7}$$

where use has been made of the definitions (2.12).

For elastic rods, the directors $\mathbf{d}_\alpha$ can be identified with material line elements in the cross-section which in the reference configuration were oriented in the directions $\mathbf{D}_\alpha$. It then follows that the representation (6.5) is only approximate for general deformations of the rod-like structure since it tacitly assumes that these material fibers remains straight. However, this approximation is reasonably accurate for a wide range of practical rod-like structures.

Within the context of the direct approach to Cosserat rod theory, the rod is modeled as a space curve which is characterized by the position vector $\mathbf{x}(\theta^3,t)$. This curve is endowed with additional director vectors $\mathbf{d}_\alpha(\theta^3,t)$ which provide limited information about deformation through the cross-section of the rod-like body. Specifically, the kinematics of the rod are specified by the vectors

$$\{ \ \mathbf{x}(\theta^3,t) \ , \ \mathbf{d}_\alpha(\theta^3,t) \ \} \ , \quad \{ \ \mathbf{v} = \dot{\mathbf{x}} \ , \ \mathbf{w}_i = \dot{\mathbf{d}}_i \ \} \ , \tag{6.8}$$

where a superposed dot denotes material time differentiation holding $\theta^3$ fixed and the restriction $(6.6)_2$ is imposed on $\mathbf{d}_\alpha$. Moreover, it is noted that $\mathbf{d}_\alpha$ are general vectors which model the effects of tangential shear deformation, normal cross-sectional extension and normal cross-sectional shear deformation (see Naghdi and Rubin, 1984). Here, local forms of the balance laws are obtained by direct integration of the local equations (2.20) and (3.2). Specifically, with the help of the representation (6.5) and integration over the cross-section of the rod: $(2.20)_1$ yields the conservation of mass; $(2.20)_2$ yields the balance of linear momentum; and (3.2) with $\phi=(\theta^1$ or $\theta^2)$, yields the balances of director momentum which, respectively, are given by

$$\dot{m} = 0 \ , \quad m\,(\dot{\mathbf{v}} + y^\alpha\, \dot{\mathbf{w}}_\alpha) = m\,\mathbf{b} + \mathbf{t}^3{}_{,3} \ ,$$

$$m\,(y^\alpha\, \dot{\mathbf{v}} + y^{\alpha\beta}\, \dot{\mathbf{w}}_\beta) = m\,\mathbf{b}^\alpha - \mathbf{t}^\alpha + \mathbf{m}^\alpha{}_{,3} \ . \tag{6.9}$$

In these equations: the scaled mass density m is related to the mass density $\rho$ (mass per unit present arclength ds of the curve $\mathbf{x}$) and its reference value $\rho_0$ through the equations

$$m = \rho\, d_{33}^{1/2} = \rho_0\, D_{33}^{1/2} = m(\theta^3) \ , \quad d_{33}^{1/2} = (\mathbf{d}_3 \bullet \mathbf{d}_3)^{1/2} \ ,$$

$$D_{33}^{1/2} = (\mathbf{D}_3 \bullet \mathbf{D}_3)^{1/2} \ , \quad ds = d_{33}^{1/2}\, d\theta^3 \ ; \tag{6.10}$$

$y^\alpha$ and $y^{\alpha\beta}=y^{\beta\alpha}$ are director inertia coefficients, which are independent of time

$$\dot{y}^\alpha = 0 \ , \quad \dot{y}^{\alpha\beta} = 0 \ ; \tag{6.11}$$

$\mathbf{b}$ is the specific external force; $\mathbf{b}^\alpha$ are the specific external director couples; $\mathbf{t}^3$ is related to the resultant contact force applied to the ends of the rod; $\mathbf{m}^\alpha$ are related to the resultant contact couples applied to the ends of the rod; and $\mathbf{t}^\alpha$ are the intrinsic director couples. Moreover, the assigned fields $\{\mathbf{b},\ \mathbf{b}^\alpha\}$ include contributions $\{\mathbf{b}_b,\ \mathbf{b}_b^\alpha\}$ due to the three-dimensional body force and contributions $\{\mathbf{b}_c,\ \mathbf{b}_c^\alpha\}$ due to contact forces on the lateral surface (6.3) of the rod

$$\mathbf{b} = \mathbf{b}_b + \mathbf{b}_c \ , \quad \mathbf{b}^\alpha = \mathbf{b}_b^\alpha + \mathbf{b}_c^\alpha \ . \tag{6.12}$$

Expressions for these quantities in terms of related three-dimensional quantities are recorded in appendix B. Also, use of (2.9) and (6.5), and integration of $(2.18)_1$ yields the local form of the balance of angular momentum

$$\mathbf{T}^T = \mathbf{T} \ , \quad d_{33}^{1/2}\, \mathbf{T} = \mathbf{t}^i \otimes \mathbf{d}_i + \mathbf{m}^\alpha \otimes \mathbf{d}_{\alpha,3} \ . \tag{6.13}$$

Next, the rate of dissipation $\mathcal{D}$ is defined in terms of the strain energy function $\Sigma$ by the expression

$$d_{33}^{1/2}\, \mathcal{D} = [\ m\,\mathbf{b} \bullet \mathbf{v} + (\mathbf{t}^3 \bullet \mathbf{v})_{,3} + m\,\mathbf{b}^\alpha \bullet \mathbf{w}_\alpha + (\mathbf{m}^\alpha \bullet \mathbf{w}_\alpha)_{,3}\ ]$$

$$- \frac{d}{dt}\Big[\frac{1}{2}\, m\{\ \mathbf{v} \bullet \mathbf{v} + 2\, y^\alpha\, \mathbf{v} \bullet \mathbf{w}_\alpha + y^{\alpha\beta}\, \mathbf{w}_\alpha \bullet \mathbf{w}_\beta\}\Big] - m\,\dot{\Sigma} \ge 0 \ , \tag{6.14}$$

where the first term in square brackets on the right-hand side represents the rate of work of all external forces and couples applied to the rod and the second term in square brackets on the right-hand side represents the kinetic energy of the rod. Now, with the help of the equations of motion (6.9) it can be shown that

$$d_{33}^{1/2}\, \mathcal{D} = \mathbf{t}^i \bullet \mathbf{w}_i + \mathbf{m}^\alpha \bullet \mathbf{w}_{\alpha,3} - m\,\dot{\Sigma} \ge 0 \ . \tag{6.15}$$

Moreover, it is convenient to introduce the deformation tensor $\mathbf{F}$, the inhomogeneous strain measures $\boldsymbol{\beta}_\alpha$, and the rate of deformation tensors $\mathbf{L}$, $\mathbf{D}$ and $\mathbf{W}$, by the equations

$$\mathbf{F} = \mathbf{d}_i \otimes \mathbf{D}^i \ , \quad \boldsymbol{\beta}_\alpha = \mathbf{F}^{-1}\mathbf{d}_{\alpha,3} - \mathbf{D}_{\alpha,3} \ , \quad \dot{\mathbf{F}} = \mathbf{L}\mathbf{F} \ ,$$

$$\mathbf{L} = \mathbf{w}_i \otimes \mathbf{d}^i = \mathbf{D} + \mathbf{W} \ , \ \mathbf{D} = \frac{1}{2}(\mathbf{L} + \mathbf{L}^T) = \mathbf{D}^T \ , \ \mathbf{W} = \frac{1}{2}(\mathbf{L} - \mathbf{L}^T) = -\,\mathbf{W}^T, \tag{6.16}$$

where the reciprocal vectors $\mathbf{D}^i$ and $\mathbf{d}^i$ are defined by equations similar to (2.12) and (2.14), such that

$$\mathbf{D}_i \bullet \mathbf{D}^j = \delta_i^j \ , \quad \mathbf{d}_i \bullet \mathbf{d}^j = \delta_i^j \ . \tag{6.17}$$

Then, with the help of (6.10), (6.13), (6.14) and (6.16) it can be shown that the rate of dissipation (6.15) can be written in alternative form

$$\mathcal{D} = \mathbf{T} \bullet \mathbf{D} + d_{33}^{1/2} \, (\mathbf{F}^T \mathbf{m}^\alpha) \bullet \dot{\boldsymbol{\beta}}_\alpha \quad \rho \, \dot{\Sigma} \geq 0 \ . \tag{6.18}$$

For a purely elastic rod the tensors $\mathbf{T}$ and $\mathbf{m}^\alpha$ and the strain energy $\Sigma$ are functions of the deformation quantities $\mathbf{F}$ and $\boldsymbol{\beta}_\alpha$ only and are explicitly independent of time. Also, due to invariance under superposed rigid body motions, $\Sigma$ depends on $\mathbf{F}$ only through the deformation tensor $\mathbf{C}$, such that

$$\mathbf{T} = \hat{\mathbf{T}}(\mathbf{F},\boldsymbol{\beta}_\alpha) \ , \quad \mathbf{m}^\alpha = \hat{\mathbf{m}}^\alpha(\mathbf{F},\boldsymbol{\beta}_\alpha) \ , \quad \Sigma = \hat{\Sigma}(\mathbf{C},\boldsymbol{\beta}_\alpha) \ , \quad \mathbf{C} = \mathbf{F}^T\mathbf{F} \ . \tag{6.19}$$

Moreover, for a purely elastic rod the rate of dissipation $\mathcal{D}$ vanishes and (6.18) can be used to prove that $\mathbf{T}$ and $\mathbf{m}^\alpha$ are restricted so that

$$\mathbf{T} = \hat{\mathbf{T}} \ , \quad \hat{\mathbf{T}} = 2\rho \, \mathbf{F} \frac{\partial \hat{\Sigma}}{\partial \mathbf{C}} \mathbf{F}^T \ , \quad \mathbf{m}^\alpha = \hat{\mathbf{m}}^\alpha \ , \quad \hat{\mathbf{m}}^\alpha = \mathbf{F}^{-T} \, m \frac{\partial \hat{\Sigma}}{\partial \boldsymbol{\beta}_\alpha} \ . \tag{6.20}$$

Additional restrictions on the functional form for $\Sigma$ can be imposed to ensure that the Cosserat theory produces solutions which are consistent with exact solutions of the three-dimensional theory of a homogeneous material for all three dimensionally homogeneous deformations for which $\mathbf{F}^*=\mathbf{F}(t)$ is independent of position and $\boldsymbol{\beta}_\alpha=0$ vanish (Rubin, 1996).

Once a form for the strain energy function $\Sigma$ is specified, the constitutive equations for $\mathbf{T}$ and $\mathbf{m}^\alpha$ are determined by (6.20), and the constitutive equations for $\mathbf{t}^i$ are determined using the definition (6.14)

$$\mathbf{t}^i = d_{33}^{1/2} \, \mathbf{T} \bullet \mathbf{d}^i - \mathbf{m}^\alpha \, (\mathbf{d}_{\alpha,3} \bullet \mathbf{d}^i) \ . \tag{6.21}$$

In addition, it is necessary to specify the inertia quantities $\{m, y^\alpha, y^{\alpha\beta}\}$ and the assigned fields $\{\mathbf{b}, \mathbf{b}^\alpha\}$. Then, the equations of linear momentum $(6.9)_2$ and director momentum $(6.9)_3$ represent three vector equations for the three unknown kinematic quantities $\{\mathbf{x}, \mathbf{d}_\alpha\}$, which are functions of only one spatial coordinate $\theta^3$ and time. Furthermore, a complete formulation of a problem requires specification of appropriate initial conditions and boundary conditions.

## 7. Cosserat points

A Cosserat point is a continuum model for the response of a three-dimensional structure that is point-like in the sense that it is "thin" in three of its dimensions. This structure is essentially a point with some small thickness so it is like a finite element. Material points in this structure's stress-free reference configuration are located by the position vector $\mathbf{X}^*$

$$\mathbf{X}^* = \mathbf{X}^*(\theta^i) = \mathbf{X} + \theta^i \, \mathbf{D}_i \ , \tag{7.1}$$

where $\mathbf{X}$ is a constant vector which locates the reference point of the Cosserat point and $\theta^i$ are convected coordinates that characterize material points in the three-dimensional region $P^*$ (bounded by $\partial P^*$) occupied by the Cosserat point. Moreover, the director vectors $\mathbf{D}_i$ are constant vectors that are restricted to be linearly independent

$$D^{1/2} = \mathbf{D}_1 \times \mathbf{D}_2 \bullet \mathbf{D}_3 > 0 \ . \tag{7.2}$$

Next, using the definitions (2.14) it follows that (7.1) provides a one-to-one invertible mapping between $\theta^i$ and $\mathbf{X}^*$ since

$$\mathbf{G}_i = \mathbf{D}_i \ , \quad G^{1/2} = \mathbf{G}_1 \times \mathbf{G}_2 \bullet \mathbf{G}_3 = D^{1/2} > 0 \ . \tag{7.3}$$

In the present deformed configuration, the material points of the Cosserat point are assumed to be located by a position vector $\mathbf{x}^*$ which has a representation similar to (7.1)

$$\mathbf{x}^* = \mathbf{x}^*(\theta^i,t) = \mathbf{x}(t) + \theta^i \, \mathbf{d}_i(t) \ , \tag{7.4}$$

with the reference point $\mathbf{X}$ and directors $\mathbf{D}_i$ being mapped to their present values $\mathbf{x}$ and $\mathbf{d}_i$, respectively. Again, the director vectors $\mathbf{d}_i$ are restricted so that they are linearly independent

$$d^{1/2} = \mathbf{d}_1 \times \mathbf{d}_2 \bullet \mathbf{d}_3 > 0 \ . \tag{7.5}$$

Also, the representation (7.4) automatically remains invertible since

$$\mathbf{g}_i = \mathbf{d}_i \ , \quad g^{1/2} = \mathbf{g}_1 \times \mathbf{g}_2 \bullet \mathbf{g}_3 = d^{1/2} > 0 \ . \tag{7.6}$$

For elastic Cosserat points, the directors $\mathbf{d}_i$ can be identified with material line elements which in the reference configuration were oriented in the directions $\mathbf{D}_i$. It then follows that the representation (7.4) is only approximate for general deformations of the point-like structure since it tacitly assumes that these material fibers remains straight. However, this approximation is reasonably accurate when the region occupied by the Cosserat point is small enough that the deformation can be approximated as being nearly homogeneous.

Within the context of the direct approach to the theory of a Cosserat point, the Cosserat point is modeled as a point in space which is characterized by the position vector $\mathbf{x}(t)$. This point is endowed with additional director vectors $\mathbf{d}_i(t)$ which provide limited information about deformation in the region occupied by of the point-like body. Specifically, the kinematics of the Cosserat point are specified by the vectors

$$\{ \ \mathbf{x}(t) \ , \mathbf{d}_i(t) \ \} \ , \quad \{ \ \mathbf{v} = \dot{\mathbf{x}} \ , \ \mathbf{w}_i = \dot{\mathbf{d}}_i \ \} \ , \tag{7.7}$$

where a superposed dot denotes material time differentiation (which is indistinguishable from ordinary time differentiation since all quantities in the theory are functions of time only) and the restriction (7.5) is imposed on $\mathbf{d}_i$. Moreover, it is noted that $\mathbf{d}_i$ are general vectors which model the effects of general homogeneous deformations of the Cosserat point. Here, local forms of the balance laws are obtained by direct integration of the local equations (2.20) and (3.2). Specifically, with the help of the representation (7.4) and integration over the region $P^*$: $(2.20)_1$ yields the conservation of mass; $(2.20)_2$ yields the balance of linear momentum; and (3.2) with $\phi=(\theta^1$ or $\theta^2$ or $\theta^3)$, yields the balances of director momentum which, respectively, are given by

$$\dot{m} = 0 \ , \quad m \, (\dot{\mathbf{v}} + y^i \, \dot{\mathbf{w}}_i) = m \, \mathbf{b} \ , \quad m \, (y^i \, \dot{\mathbf{v}} + y^{ij} \, \dot{\mathbf{w}}_j) = m \, \mathbf{b}^i - \mathbf{t}^i \ . \tag{7.8}$$

In these equations: the scaled mass density m is related to the mass density $\rho$ (mass per unit present volume $dv^*$) and its reference value $\rho_0$ through the equations

$$m = \rho \, d^{1/2} = \rho_0 \, D^{1/2} = \text{constant} \ , \quad dv^* = d^{1/2} \, d\theta^1 d\theta^2 d\theta^3 \ ; \tag{7.9}$$

$y^i$ and $y^{ij}=y^{ji}$ are director inertia coefficients, which are independent of time

$$\dot{y}^i = 0 \ , \quad \dot{y}^{ij} = 0 \ ; \tag{7.10}$$

$\mathbf{b}$ is the specific external force; $\mathbf{b}^i$ are the specific external director couples; and $\mathbf{t}^i$ are the intrinsic director couples. Moreover, the assigned fields $\{\mathbf{b}, \mathbf{b}^i\}$ include contributions $\{\mathbf{b}_b, \mathbf{b}_b^i\}$ due to the three-dimensional body force and contributions $\{\mathbf{b}_c, \mathbf{b}_c^i\}$ due to contact forces on the boundary $\partial P^*$ of the Cosserat point

$$\mathbf{b} = \mathbf{b}_b + \mathbf{b}_c \ , \quad \mathbf{b}^\alpha = \mathbf{b}_b^\alpha + \mathbf{b}_c^\alpha \ . \tag{7.11}$$

Expressions for these quantities in terms of related three-dimensional quantities are recorded in appendix C. Also, use of (2.9) and (7.4), and integration of $(2.18)_1$ yields the reduced form of the balance of angular momentum

$$\mathbf{T}^T = \mathbf{T} \ , \quad d^{1/2}\,\mathbf{T} = \mathbf{t}^i \otimes \mathbf{d}_i \ . \tag{7.12}$$

Next, the rate of dissipation $\mathcal{D}$ is defined in terms of the strain energy function $\Sigma$ by the expression

$$\begin{aligned} d^{1/2}\,\mathcal{D} = &\left[\, m\,\mathbf{b} \bullet \mathbf{v} + m\,\mathbf{b}^i \bullet \mathbf{w}_i \,\right] \\ &- \frac{d}{dt}\left[\frac{1}{2}\, m\{\, \mathbf{v} \bullet \mathbf{v} + 2\,y^i\,\mathbf{v} \bullet \mathbf{w}_i + y^{ij}\,\mathbf{w}_i \bullet \mathbf{w}_j \}\right] - m\,\dot{\Sigma} \geq 0 \ , \end{aligned} \tag{7.13}$$

where the first term in square brackets on the right-hand side represents the rate of work of all external forces and couples applied to the Cosserat point and the second term in square brackets on the right-hand side represents the kinetic energy of the Cosserat point. Now, with the help of the equations of motion (7.8) it can be shown that

$$d^{1/2}\,\mathcal{D} = \mathbf{t}^i \bullet \mathbf{w}_i - m\,\dot{\Sigma} \geq 0 \ . \tag{7.14}$$

Moreover, it is convenient to introduce the deformation tensor $\mathbf{F}$, and the rate of deformation tensors $\mathbf{L}$, $\mathbf{D}$ and $\mathbf{W}$, by the equations

$$\begin{aligned} &\mathbf{F} = \mathbf{d}_i \otimes \mathbf{D}^i \ , \quad \dot{\mathbf{F}} = \mathbf{L}\mathbf{F} \ , \quad \mathbf{L} = \mathbf{w}_i \otimes \mathbf{d}^i = \mathbf{D} + \mathbf{W} \ , \\ &\mathbf{D} = \frac{1}{2}(\mathbf{L} + \mathbf{L}^T) = \mathbf{D}^T \ , \quad \mathbf{W} = \frac{1}{2}(\mathbf{L} - \mathbf{L}^T) = -\,\mathbf{W}^T \ , \end{aligned} \tag{7.15}$$

where the reciprocal vectors $\mathbf{D}^i$ and $\mathbf{d}^i$ are defined by equations similar to (2.12) and (2.14), such that

$$\mathbf{D}_i \bullet \mathbf{D}^j = \delta_i^j \ , \quad \mathbf{d}_i \bullet \mathbf{d}^j = \delta_i^j \ . \tag{7.16}$$

Then, with the help of (7.9), (7.12), (7.13) and (7.15) it can be shown that the rate of dissipation (7.14) can be written in alternative form

$$\mathcal{D} = \mathbf{T} \bullet \mathbf{D} - \rho\,\dot{\Sigma} \geq 0 \ . \tag{7.17}$$

For a purely elastic Cosserat point the tensor $\mathbf{T}$ and the strain energy $\Sigma$ are functions of the deformation quantity $\mathbf{F}$ only and are explicitly independent of time. Also, due to invariance under superposed rigid body motions, $\Sigma$ depends on $\mathbf{F}$ only through the deformation tensor $\mathbf{C}$, such that

$$\mathbf{T} = \hat{\mathbf{T}}(\mathbf{F}) \ , \quad \Sigma = \hat{\Sigma}(\mathbf{C}) \ , \quad \mathbf{C} = \mathbf{F}^T\mathbf{F} \ . \tag{7.18}$$

Moreover, for a purely elastic Cosserat point the rate of dissipation $\mathcal{D}$ vanishes and (7.17) can be used to prove that $\mathbf{T}$ is restricted so that

$$\mathbf{T} = \hat{\mathbf{T}} \ , \quad \hat{\mathbf{T}} = 2\rho\,\mathbf{F}\,\frac{\partial \hat{\Sigma}}{\partial \mathbf{C}}\,\mathbf{F}^T \ . \tag{7.19}$$

When the functional form for the strain energy $\hat{\Sigma}(\mathbf{C})$ is the same as that for the three-dimensional material $\Sigma^*(\mathbf{C})$ then it can be shown that the Cosserat theory produces solutions which are consistent with exact solutions of the three-dimensional theory of a homogeneous material for all three dimensionally homogeneous deformations for which $\mathbf{F}^*=\mathbf{F}(t)$ is independent of position.

Once a form for the strain energy function Σ is specified, the constitutive equations for **T** is determined by (7.19), and the constitutive equations for $\mathbf{t}^i$ are determined using the definition (7.13)

$$\mathbf{t}^i = d^{1/2}\, \mathbf{T} \bullet \mathbf{d}^i \quad . \tag{7.20}$$

In addition, it is necessary to specify the inertia quantities $\{m, y^i, y^{ij}\}$ and the assigned fields $\{\mathbf{b}, \mathbf{b}^i\}$. Then, the equations of linear momentum $(7.8)_2$ and director momentum $(7.8)_3$ represent four vector equations for the four unknown kinematic quantities $\{\mathbf{x}, \mathbf{d}_i\}$. Furthermore, a complete formulation of a problem requires specification of appropriate initial conditions.

## Acknowledgment

This research was partially supported by the fund for promotion of research at the Technion.

## References

Green, A.E. and Adkins, J.E. (1960) *Large elastic deformations*. Oxford University Press, Oxford.

Green, A.E. and Zerna, W. (1968) *Theoretical elasticity*. (second edition) Oxford University Press, Oxford.

Green, A.E., Naghdi, P.M. and Wenner, M.L. (1974a) On the theory of rods, I. Derivation from the three-dimensional equations. *Proc. R. Soc. Lond.* **A337**, 451-483.

Green, A.E., Naghdi, P.M. and Wenner, M.L. (1974b) On the theory of rods, II. Developments by direct approach. *Proc. R. Soc. Lond.* **A337**, 485-507.

Libai, A. and Simmonds, J.G. (1998) *The nonlinear theory of elastic shells*. Cambridge University Press, Cambridge.

Naghdi, P.M. (1972) *The theory of shells and plates*. In S. Flugge's Handbuch der Physik, Vol. VIa/2 (ed. Truesdell, C.), Springer-Verlag, Berlin, 425-640.

Naghdi, P.M. (1982) Finite deformation of elastic rods and shells. In D.E. Carlson and R.T. Shield (eds.), *Proc. IUTAM Symp. on Finite Elasticity*. Martinus Nijhoff Publishers, The Hague.

Naghdi, P.M. and Rubin, M.B. (1984) Constrained theories of rods, *J. of Elasticity* **14**, 343-361.

Naghdi, P.M. and Rubin, M.B. (1995) Restrictions on nonlinear constitutive equations for elastic shells, *J. Elasticity* **39**, 133-163.

Rubin, M.B. (1985a) On the theory of a Cosserat point and its application to the numerical solution of continuum problems, *J. Appl. Mech.* **52**, 368-372.

Rubin, M.B. (1985b) On the numerical solution of one-dimensional continuum problems using the theory of a Cosserat point, *J. Appl. Mech.* **52**, 373-378.

Rubin, M.B. (1987) On the numerical solution of nonlinear string problems using the theory of a Cosserat point, *Int. J. of Solids Structures* **23**, 335-349.

Rubin, M.B. (1995) Numerical solution of two- and three-dimensional thermomechanical problems using the theory of a Cosserat point, *J. of Math. and Physics (ZAMP)* **46**, Special Issue, S308-S334. In *Theoretical, Experimental, And Numerical Contributions To The Mechanics Of Fluids And Solids*, Edited by J. Casey and M. J. Crochet, Brikhauser Verlag, Basel (1995).

Rubin, M.B. (1996) Restrictions on nonlinear constitutive equations for elastic rods, *J. Elasticity* **44**, 9-36.

## Appendix A: Definitions for Cosserat shells

Inertia quantities

$$m = \int_{-H/2}^{H/2} m^* \, d\theta^3 \ , \quad m y^3 = \int_{-H/2}^{H/2} m^* \, \theta^3 \, d\theta^3 \ , \quad m y^{33} = \int_{-H/2}^{H/2} m^* \, \theta^3 \theta^3 \, d\theta^3 \ . \quad \text{(A1)}$$

Resultant forces, couples and intrinsic director couples

$$\mathbf{t}^i = \int_{-H/2}^{H/2} \mathbf{t}^{*i} \, d\theta^3 \ , \quad \mathbf{m}^\alpha = \int_{-H/2}^{H/2} \mathbf{t}^{*\alpha} \, \theta^3 \, d\theta^3 \ . \quad \text{(A2)}$$

Assigned fields due to body force

$$m \, \mathbf{b}_b = \int_{-H/2}^{H/2} m^* \mathbf{b}^* \, d\theta^3 \ , \quad m \, \mathbf{b}_b^3 = \int_{-H/2}^{H/2} m^* \mathbf{b}^* \, \theta^3 \, d\theta^3 \ . \quad \text{(A3)}$$

Assigned fields due to contact forces

$$m \, \mathbf{b}_c = \left[ \mathbf{t}^{*3} - \frac{1}{2} H_{,\alpha} \mathbf{t}^{*\alpha} \right] \Big|_{\theta^3 = H/2} - \left[ \mathbf{t}^{*3} + \frac{1}{2} H_{,\alpha} \mathbf{t}^{*\alpha} \right] \Big|_{\theta^3 = -H/2} \ ,$$

$$m \, \mathbf{b}_c^3 = \frac{H}{2} \left[ \mathbf{t}^{*3} - \frac{1}{2} H_{,\alpha} \mathbf{t}^{*\alpha} \right] \Big|_{\theta^3 = H/2} - \frac{H}{2} \left[ \mathbf{t}^{*3} + \frac{1}{2} H_{,\alpha} \mathbf{t}^{*\alpha} \right] \Big|_{\theta^3 = -H/2} \ . \quad \text{(A4)}$$

## Appendix B: Definitions for Cosserat rods

Inertia quantities

$$m = \int_{-W/2}^{W/2} \int_{-H/2}^{H/2} m^* \, d\theta^1 d\theta^2 \ , \quad m y^\alpha = \int_{-W/2}^{W/2} \int_{-H/2}^{H/2} m^* \, \theta^\alpha \, d\theta^1 d\theta^2 \ ,$$

$$m \, y^{\alpha\beta} = m \, y^{\beta\alpha} = \int_{-W/2}^{W/2} \int_{-H/2}^{H/2} m^* \, \theta^\alpha \theta^\beta \, d\theta^1 d\theta^2 \ . \quad \text{(B1)}$$

Resultant forces, couples and intrinsic director couples

$$\mathbf{t}^i = \int_{-W/2}^{W/2} \int_{-H/2}^{H/2} \mathbf{t}^{*i} \, d\theta^1 d\theta^2 \ , \quad \mathbf{m}^\alpha = \int_{-W/2}^{W/2} \int_{-H/2}^{H/2} \mathbf{t}^{*3} \, \theta^\alpha \, d\theta^1 d\theta^2 \ . \quad \text{(B2)}$$

Assigned fields due to body force

$$m \, \mathbf{b}_b = \int_{-W/2}^{W/2} \int_{-H/2}^{H/2} m^* \, \mathbf{b}^* \, d\theta^1 d\theta^2 \ , \quad m \, \mathbf{b}_b^\alpha = \int_{-W/2}^{W/2} \int_{-H/2}^{H/2} m^* \, \mathbf{b}^* \, \theta^\alpha \, d\theta^1 d\theta^2 \ . \quad \text{(B3)}$$

Assigned fields due to contact forces

$$m \, \mathbf{b}_c = \int_{-W/2}^{W/2} \left\{ \mathbf{t}^{*1}(H/2, \theta^2, \theta^3) - \frac{1}{2} H_{,3} \, \mathbf{t}^{*3}(H/2, \theta^2, \theta^3) \right\} d\theta^2$$

$$+ \int_{-H/2}^{H/2} \left\{ \mathbf{t}^{*2}(\theta^1, W/2, \theta^3) - \frac{1}{2} W_{,3} \, \mathbf{t}^{*3}(\theta^1, W/2, \theta^3) \right\} d\theta^1$$

$$- \int_{-W/2}^{W/2} \{ \mathbf{t}^{*1}(-H/2,\theta^2,\theta^3) + \frac{1}{2} H_{,3}\, \mathbf{t}^{*3}(-H/2,\theta^2,\theta^3) \} d\theta^2 \,]$$

$$- \int_{-H/2}^{H/2} \{ \mathbf{t}^{*1}(\theta^1,-W/2,\theta^3) + \frac{1}{2} W_{,3}\, \mathbf{t}^{*3}(\theta^1,-W/2,\theta^3) \} d\theta^1 \;,$$

$$m\,\mathbf{b}_c^1 = \frac{H}{2} \int_{-W/2}^{W/2} \{ \mathbf{t}^{*1}(H/2,\theta^2,\theta^3) - \frac{1}{2} H_{,3}\, [\mathbf{t}^{*3}(H/2,\theta^2,\theta^3) \} d\theta^2$$

$$+ \int_{-H/2}^{H/2} \{ \mathbf{t}^{*2}(\theta^1,W/2,\theta^3) - \frac{1}{2} W_{,3}\, \mathbf{t}^{*3}(\theta^1,W/2,\theta^3) \} \theta^1 d\theta^1$$

$$+ \frac{H}{2} \int_{-W/2}^{W/2} \{ \mathbf{t}^{*1}(-H/2,\theta^2,\theta^3)] + \frac{1}{2} H_{,3}\, \mathbf{t}^{*3}(-H/2,\theta^2,\theta^3) \} d\theta^2$$

$$- \int_{-H/2}^{H/2} \{ \mathbf{t}^{*2}(\theta^1,-W/2,\theta^3)] + \frac{1}{2} W_{,3}\, \mathbf{t}^{*3}(\theta^1,-W/2,\theta^3) \} \theta^1 d\theta^1] \;,$$

$$m\,\mathbf{b}_c^2 = \int_{-W/2}^{W/2} \{ \mathbf{t}^{*1}(H/2,\theta^2,\theta^3) - \frac{1}{2} H_{,3}\, \mathbf{t}^{*3}(H/2,\theta^2,\theta^3) \} \theta^2 d\theta^2$$

$$+ \frac{W}{2} \int_{-H/2}^{H/2} \{ \mathbf{t}^{*2}(\theta^1,W/2,\theta^3) - \frac{1}{2} W_{,3}\, \mathbf{t}^{*3}(\theta^1,W/2,\theta^3) \} d\theta^1$$

$$- \int_{-W/2}^{W/2} \{ \mathbf{t}^{*1}(-H/2,\theta^2,\theta^3) + \frac{1}{2} H_{,3}\, \mathbf{t}^{*3}(-H/2,\theta^2,\theta^3) \} \theta^2 d\theta^2$$

$$+ \frac{W}{2} \int_{-H/2}^{H/2} \{ \mathbf{t}^{*2}(\theta^1,-W/2,\theta^3) + \frac{1}{2} W_{,3}\, \mathbf{t}^{*3}(\theta^1,-W/2,\theta^3) \} d\theta^1 \;. \quad \text{(B4)}$$

## Appendix C: Definitions for Cosserat points

Inertia quantities

$$m = \int_{P^*} \rho^* \, dv^* \;, \quad m\, y^i = \int_{P^*} \rho^* \, \theta^i \, dv^* \;,$$

$$m\, y^{ij} = m\, y^{ji} = \int_{P^*} \rho^* \, \theta^i \theta^j \, dv^* \;. \quad \text{(C1)}$$

Intrinsic director couples

$$\mathbf{t}^i = \int_{P^*} g^{-1/2}\, \mathbf{t}^{*i} \, dv^* \;, \quad \text{(C2)}$$

Assigned fields due to body force

$$m\, \mathbf{b}_b = \int_{P^*} \rho^* \, \mathbf{b}^* \, dv^* \;, \quad m\, \mathbf{b}_b^i = \int_{P^*} \rho^* \, \mathbf{b}^* \, \theta^i \, dv^* \;. \quad \text{(C3)}$$

Assigned fields due to contact forces

$$m\, \mathbf{b}_c = \int_{\partial P^*} \mathbf{t}^* \, da^* \;, \quad m\, \mathbf{b}_c^i = \int_{\partial P^*} \mathbf{t}^* \, \theta^i \, da^* \;. \quad \text{(C4)}$$

# STRUCTURAL SIMILITUDE AND SCALING LAWS FOR PLATES AND SHELLS: A REVIEW

G.J. SIMITSES
University of Cincinnati
Cincinnati, Ohio 45221-0070

J.H. STARNES, JR
NASA Langley Research Center
Mail Stop 190, Hampton, Virginia 23665

J. REZAEEPAZHAND
Ferdossi University of Mashhad
Mashhad, Iran

## 1. Abstract

This paper deals with the development and use of scaled-down models in order to predict the structural behavior of large prototypes. The concept is fully described and examples are presented which demonstrate its applicability to beam-plates, plates and cylindrical shells of laminated construction. The concept is based on the use of field equations, which govern the response behavior of both the small model as well as the large prototype. The conditions under which the experimental data of a small model can be used to predict the behavior of a large prototype are called scaling laws or similarity conditions and the term that best describes the process is structural similitude. Moreover, since the term scaling is used to describe the effect of size on strength characteristics of materials, a discussion is included which should clarify the difference between "scaling law" and "size effect". Finally, a historical review of all published work in the broad area of structural similitude is presented for completeness.

## 2. Introduction

Aircraft and spacecraft comprise the class of aerospace structures that require efficiency and wisdom in design, sophistication and accuracy in analysis and numerous and careful experimental evaluations of components and prototype, in order to achieve the necessary system reliability, performance and safety.

Preliminary and/or concept design entails the assemblage of system mission requirements, system expected performance and identification of components and their connections as well as of manufacturing and system assembly techniques. This is accomplished through experience based on previous similar designs, and through the possible use of models to simulate the entire system characteristics.



*D. Durban et al. (eds.), Advances in the Mechanics of Plates and Shells,* 295–310.

Detail design is heavily dependent on information and concepts derived from the previous step. This information identifies critical design areas which need sophisticated analyses, and design and redesign procedures to achieve the expected component performance. This step may require several independent analysis models, which, in many instances, require component testing.

The last step in the design process, before going to production, is the verification of the design. This step necessitates the production of large components and prototypes in order to test component and system analytical predictions and verify strength and performance requirements under the worst loading conditions that the system is expected to encounter in service.

Clearly then, full-scale testing is in many cases necessary and always very expensive. In the aircraft industry, in addition to full-scale tests, certification and safety necessitate large component static and dynamic testing. The C-141A ultimate static tests include eight wing tests, 17 fuselage tests and seven empennage tests (McDougal, 1987). Such tests are extremely difficult, time consuming and definitely absolutely necessary. Clearly, one should not expect that prototype testing will be totally eliminated in the aircraft industry. It is hoped, though, that we can reduce full-scale testing to a minimum.

Moreover, crashworthiness aircraft testing requires full-scale tests and several drop tests of large components. The variables and uncertainties in crash behavior are so many that the information extracted from each test, although extremely valuable, is nevertheless small by comparison to the expense. Moreover, each test provides enough new and unexpected phenomena, to require new tests, specially designed to explain the new observations.

Finally, full-scale large component testing is necessary in other industries as well. Ship building, building construction, automobile and railway car construction all rely heavily on testing.

Regardless of the application, a scaled-down (by a large factor) model (scale model) which closely represents the structural behavior of the full-scale system (prototype) can prove to be an extremely beneficial tool. This possible development must be based on the existence of certain structural parameters that control the behavior of a structural system when acted upon by static and/or dynamic loads. If such structural parameters exist, a scaled-down replica can be built, which will duplicate the response of the full-scale system. The two systems are then said to be structurally similar. The term, then, that best describes this similarity is *structural similitude*.

## 3. Historical Review

Similarity of systems requires that the relevant system parameters be identical and these systems be governed by a unique set of characteristic equations. Thus, if a relation of equation of variables is written for a system, it is valid for all systems which are similar to it (Kline, 1965). Each variable in a model is proportional to the corresponding variable of the prototype. This ratio, which plays an essential role in predicting the relationship between the model and its prototype, is called the *scale factor*. In establishing similarity conditions between the model and prototype, two procedures can be used, dimensional analysis and direct use of governing equations.

Models, as a design aid, have been used for many years, but the use of scientific models which are based on dimensional analysis was first discussed in a paper by Rayleigh (1915). Similarity conditions based on dimensional analysis have been used since Rayleigh's time (Macagno, 1971), but the applicability of the theory of similitude to structural systems was first discussed by Goodier and Thomson (1944) and later by Goodier (1950). They presented a systematic procedure for establishing similarity conditions based on dimensional analysis.

There exist several books that refer to all elements of structural similitude. Murphy (1950), Langhaar (1951), Charlton (1954), Pankhurst (1964) and Gukhman (1965) all dealt with similitude and modeling principles, and most of them dealt with dimensional analysis. Kline (1965) gives a perspective of the method based on both dimensional analysis and the direct use of the government equations. Szucs (1980) is particularly thorough on the topic of similitude theory. He explains the method with emphasis on the direct use of the governing equations of the system. A recent book by Singer, Arbocz and Weller (1997) devotes an entire chapter on modeling with emphasis on dimensional analysis concepts.

A few studies concerning the use of scaled-down shell models have been conducted in the past. Ezra (1962) presented a study based on dimensional analysis, for buckling behavior subjected to impulse loads. A similar investigation was presented by Morgen (1964) for an orthotropic cylindrical shell subjected to a variety of static loads. Soedel (1971) investigated similitude for vibrating thin shells.

Due to special characteristics of advanced reinforced composite materials, they have been used extensively in weight efficient aerospace structures. Since reinforced composite components require extensive experimental evaluation, there is a growing interest in small scale model testing. Morton (1988) discusses the application of scaling laws for impact-loaded carbon-fiber composite beams. His work is based on dimensional analysis. Qian et al. (1990) conducted experimental studies of impact loaded composite plates, where the similarity conditions were obtained by considering the governing equations of the system. These works and many other experimental investigations have been conducted to characterize the size effect in material behavior for inelastic analysis (size effects are discussed in a later section).

In recent years, due to large dimensions and unique structural design of the proposed space station, small scale model testing and similitude analysis have been considered as the only option in order to gain experimental data. Shih et al. (1987), Letchworth et al. (1988), Hsu et al. (1989) and McGowan et al. (1990) discussed the possibility of scale model testing of space station geometries, especially for vibration analysis. Most of these studies have used complete similarity (defined in a later section) between model and prototype.

The present authors have published several papers [Simitses and Rezaeepazhand (1993), Rezaeepazhand et al. (1996), Rezaeepazhand et al. (1995), Simitses and Rezaeepazhand (1995), Simitses et al. (1997), and Rezaeepazhand and Simitses (1997)] that deal with the design of scaled-down models and the use of test data of these models to predict the behavior of large prototypes. The behavior includes displacements, stresses, buckling loads, and natural frequencies of laminated beam-plates, plates and shells. In these studies, in the absence of model test data, the authors theoretically analyzed the models, and they used the similarity conditions, obtained by the use of the governing equations, to predict the behavior of the prototype. They then theoretically analyzed the prototype and they compared these results to the predictions. In most cases, the compared results were very close to each other and they concluded that the designed model can accurately predict the behavior of the prototype. Very recently, Ochoa and her collaborators (1999a, 1999b) applied similitude theory to a laminated cylindrical tube under tensile, torsion and bending loads and under external and internal pressure. They demonstrated the validity of developing a scale model, testing it and use the similarity conditions to predict the behavior of the prototype.

## 4. Scaling Effects in Composites

Considerable renewed interest has been exhibited in the broad field of scaling in the recent years, as evidenced by the multitude of research papers that have appeared in the technical literature. Before discussing any and all efforts, we must have a good understanding, for clear discussion of the meaning of the words that have been used. These words are scaling or scale effects, similarity conditions or scaling laws and size effects.

Scaling effects mean the effect of changing the geometric dimensions of a structure or structural component on the response to external causes. The external causes include all types of forces. Examples of the above is a beam made out of metallic material or man-made composite and subjected to bending. The main questions associated with predicting the response of the beam are: Are stiffness and strength affected by scaling? This means is the effective Young's modulus (both in tension and compression), which is usually obtained from small specimens affected by scale. In addition, is the strength affected by scale? Recognizing that beams are primarily designed for strength, the answer to the second question is important. On the other hand, since columns are primarily designed for stiffness (buckling), the answer to the first question is important.

In this context, the use of the term size effect is similar to the term scale effect. On the other hand, one may wish to find the conditions under which the behavioral response of a small size beam and a large size beam are similar. In this case, the interest is to find the similarity conditions or scaling laws in order to achieve similarity in response. In this context, the primary interest is to be able to test a small scale model, obtain response characteristics (displacements, buckling loads, vibration frequencies, etc.) and use the scaling laws to predict the behavior of the large prototype. In this second case, one can still use the term scaling effects, if he clearly does not refer to size effects on strength and stiffness.

### 4.1. SIZE EFFECTS

There exist two main sources of recent studies of size effects. First, Jackson (1994) contains an outline of papers presented at a Workshop on Scaling Effects in Composite Materials and Structures, and second, Bazant and Rajapakse (1999) is a compilation of papers dealing with, primarily, fracture scaling.

From the conclusions, of virtually all presenters at the workshop (see Jackson, 1994) who dealt with size effects, one can say that the size effect on stiffness is almost nonexistent (Jackson and Kellas, 1994; Camponeshi, 1994; and Johnson et al., 1994). Similarly, the size effect on strength has created some controversy. Jackson and Kellas (1994) conclude that there is considerable size effect on strength. Grimes (1994) contends that for solid laminates, the largest size effect on static strength is less than 4.5%. Furthermore, he states that the cause of scale effects is not size but other factors such as poor quality tooling, differences in environmental exposure, etc. A similar conclusion was reached by O'Brien (1994) who claims that the effect of scaling is not because of size, but because different damage sequence occurs in two different sizes. In a private communication by L.B. Greszczuk (1999) of McDonnell Douglas Space Systems Co., he stated, quote "If the small and big parts are made by the same process, there is not size effect neither on stiffness nor on strength." He further explained that the tests performed at his company on specimens with twelve to one ratio in thickness (laminates), reveal that the effect on stiffness is nonexistent, while the effect on strength is less than 4%. The specimens used were carefully manufactured by the same process and they had the same filament volume fraction and porosity.

The objective of most papers Bazant and Rajapakse (1999) is to study the size effect on fracture of ice, concrete and notched composite beams. In these papers, the conclusion is that size does affect fracture and crack propagation.

One particular paper in Bazant and Rajapakse (1999), that by Daniel and Hsiao (1999), dealt with the thickness effect on compressive strength of unnotched laminates. It is an experimental study that used various sizes and layups and it concluded that the size effect is extremely small. Further evidence that size has negligible effect on stiffness is provided by the tests performed by Jackson (1990) on graphite/epoxy beams at NASA Langley. The scale varied from one-sixth to full and she employed unidirectional and quasi-isotropic layups.

Clearly, then one can at this junction say with confidence that size effect on stiffness is negligibly small and that more work on strength needs to be done in order to explain the reasons for the conflicting conclusions (if there is an effect, what causes it).

In view of the above, the authors embarked into a research program on structural similitude based on the following premises: (a) both model and prototype are governed by the same field equations (equilibrium, kinematic relations and constitutive equations, subject to boundary conditions), (b) the only set of equations that may be affected by size are the constitutive relations. It has already been concluded though that stiffness is not affected by size and therefore one is safe to use the same constitutive relations for model and prototype up to but not in the vicinity of strength limits, (c) damage accumulation for both model and prototype is minimal. On this basis one can use similitude theory and obtain the similarity conditions.

## 5. Theory of Similitude

Similitude theory is concerned with establishing necessary and sufficient conditions of similarity between two phenomena. Establishing similarity between systems helps to predict the behavior of a system from the results of investigating other systems which have already been investigated or can be investigated more easily than the original system. Similitude among systems means similarity in behavior in some specific aspects. In other words, knowing how a given system responds to a specific input, the response of all similar systems to similar input can be predicted.

The behavior of a physical system depends on many parameters, i.e. geometry, material behavior, dynamic response and energy characteristics of the system. The nature of any system can be modeled mathematically in terms of its variables and parameters. A prototype and its scale model are two different systems with similar but not necessarily identical parameters. The necessary and sufficient conditions of similitude between prototype and its scale model require that the mathematical model of the scale model can be transformed to that of the prototype by a bi-unique mapping or vice versa (Szucs, 1980). It means, if vectors $\mathbf{X}_p$ and $\mathbf{X}_m$ are the characteristic vectors of the prototype and model, then we can find a transformation matrix $\Lambda$ such that:

$$\mathbf{X}_p = \Lambda \mathbf{X}_m \quad \text{or} \quad \mathbf{X}_m = \Lambda^{-1} \mathbf{X}_p \tag{1}$$

The elements of vector **X** are all the parameters and variables of the system. A diagonal form of the transformation matrix Λ is the simplest form of transformation. The diagonal elements of the matrix are the scale factors of the pertinent elements of the characteristic vector **X**.

$$\Lambda = \begin{bmatrix} \lambda x1 & 0 \cdots 0 \\ 0 & \lambda x2 \cdots 0 \\ \vdots & \vdots \ \ddots \ \vdots \\ 0 & 0 \cdots \lambda xn \end{bmatrix} , \qquad (2)$$

where $\lambda_{xi} = x_{ip}/x_{im}$ denotes the scale factor of $x_i$. In general the transformation matrix is not diagonal.

In establishing similarity conditions between the model and prototype two procedures can be used, dimensional analysis and direct use of governing equations. The similarity conditions can be established either directly from the field equations of the system or, if it is a new phenomenon and the mathematical model of the system is not available, through dimensional analysis. In the second case, all of the variables and parameters, which affect the behavior of the system, must be known. By using dimensional analysis, an incomplete form of the characteristic equation of the system can be formulated. This equation is in terms of nondimensional products of variables and parameters of the system. Then, similarity conditions can be established on the basis of this equation.

In our studies, we consider only direct use of the governing equations procedure. This method is more convenient than dimensional analysis, since the resulting similarity conditions are more specific. When governing equations of the system are used for establishing similarity conditions, the relationships among variables are forced by the governing equations of the system.

The field equations of a system with proper boundary and initial conditions characterize the behavior of the system in terms of its variables and parameters. If the field equations of the scale model and its prototype are invariant under transformation Λ and $\Lambda^{-1}$, then the two systems are completely similar. This transformation defines the scaling laws (similarity conditions) among all parameters, structural geometry and cause and response of the two systems. Examples of the direct use of governing equations is offered below.

## 5.1. BENDING OF LAMINATED BEAM PLATES

Consider a laminated beamplate of length a and width b and simply supported at both ends. We desire to find the maximum deflection of this beamplate. The beamplate is subjected to a transverse line load. By assuming that the displacement functions are independent of y, or u=u(x), v=0, w=w(x) (cylindrical bending), from Ashton and Whitney (1970), the governing differential equations and boundary conditions are reduced to:

$$\frac{d^4w}{dx^4}=\frac{qA_{11}}{A_{11}D_{11}-B_{11}^2} \quad , \tag{3}$$

$$\frac{d^3u}{dx^3}=\frac{B_{11}d^4w}{A_{11}dx^4} \quad , \tag{4}$$

and the B.C.s at $x=0$, $a$ are:

$$w=0 \tag{5}$$

$$N_{xx}=A_{11}\frac{du}{dx}-B_{11}\frac{d^2w}{dx^2}=0 \tag{6}$$

$$M_{xx}=B_{11}\frac{du}{dx}-D_{11}\frac{d^2w}{dx^2}=0 \tag{7}$$

Equation (1) can be written as:

$$(A_{11}D_{11}-B_{11}^2)\frac{d^4w}{dx^4}=qA_{11} \tag{8}$$

By applying similitude theory, the resulting similarity conditions are:

$$\lambda_{A_{11}}\lambda_{D_{11}}\lambda_w=\lambda_{B_{11}}^2\lambda_w=\lambda_{A_{11}}\lambda_x^4\lambda_q \quad , \tag{9}$$

or

$$\lambda_{A_{11}}\lambda_{D_{11}}\lambda_w=\lambda_{B_{11}}^2 \quad , \tag{10}$$

$$\lambda_w\ \lambda_{D_{11}}=\lambda_x^4\lambda_q \quad . \tag{11}$$

Similarly from Eqs. (5), (6) and (7) we have:

$$\lambda_{A_{11}}\lambda_u\lambda_x=\lambda_w\lambda_{B_{11}} \quad , \tag{12}$$

$$\lambda_{B_{11}}\lambda_u\lambda_x=\lambda_w\lambda_{D_{11}} \quad , \tag{13}$$

The condition depicted by Eq. (13) can be obtained by combining Eqs. (10) and (12). So, Eqs. (10) through (12) denote the necessary conditions for complete similarity between the scale model and its prototype, as far as deflectional response is concerned.

Note that the similarity conditions, Eqs. (10)-(12) are three, while the number of geometric and material parameters, cause parameter (load) and response parameters (u and w) is much larger than three. This means that there is freedom in designing models for a given prototype. In addition, if, in projecting the data of the model to predict the behavior of the prototype, all three scaling laws are used, then we have complete similarity. If only one (or two) scaling laws are used, then we have partial similarity.

For this particular application, experimental data was supplied by Professor Sierakowski (1994) for tests performed on beam plates. The total number of laminates used is ten. In Simitses (1999), some beam plates are considered as models and some as prototypes. Similitude theory is used and the results are compared to the test results of the prototypes (see Simitses, 1999 for details). Partial similarity is used in the comparison.

In addition to the above, similitude theory is employed in a case where experimental results do not exist. In this case, the theoretical results of the model are treated as test data, then a scaling law (partial similarity) is used to predict the behavior of the prototype and the predictions are compared to the theoretical results of the prototype. If these two compare well, success has been achieved in designing the model and in using similitude theory.

Consider a cross-ply laminated E-Glass/Epoxy plate composed of 96 orthtropic layers $(0/90/0/\ldots)_{96}$ as the prototype. We desire to find the maximum deflection of the prototype by extrapolating the pertinent values of a small scale model. The model has the same stacking sequence as the prototype but with a smaller number of layers. The prototype and its scale model have the following characteristics:

| | | |
|---|---|---|
| Prototype $(0(90/0/\ldots)_{96}$: | a = 90 in.<br>h = 0.858 in. | b = 100 in.<br>N = 96, |
| model $(0(90/0/\ldots)_{16}$: | a = 5.0 in.<br>h = 0.143 in. | b = 6.139 in.<br>N = 16, |
| scale factors: | $\lambda_a = 18$<br>$\lambda_h = 6$ | $\lambda_b = 16.29$<br>$\lambda_N = 6$. |

In designing the model, we assume that it is made of the same material as the prototype and that $\lambda_q = \lambda_b$. By employing only the similarity condition of Eq. (11) (partial similarity), the results are plotted on Fig. 1. For details, see Simitses and Rezaeepazhand (1993).

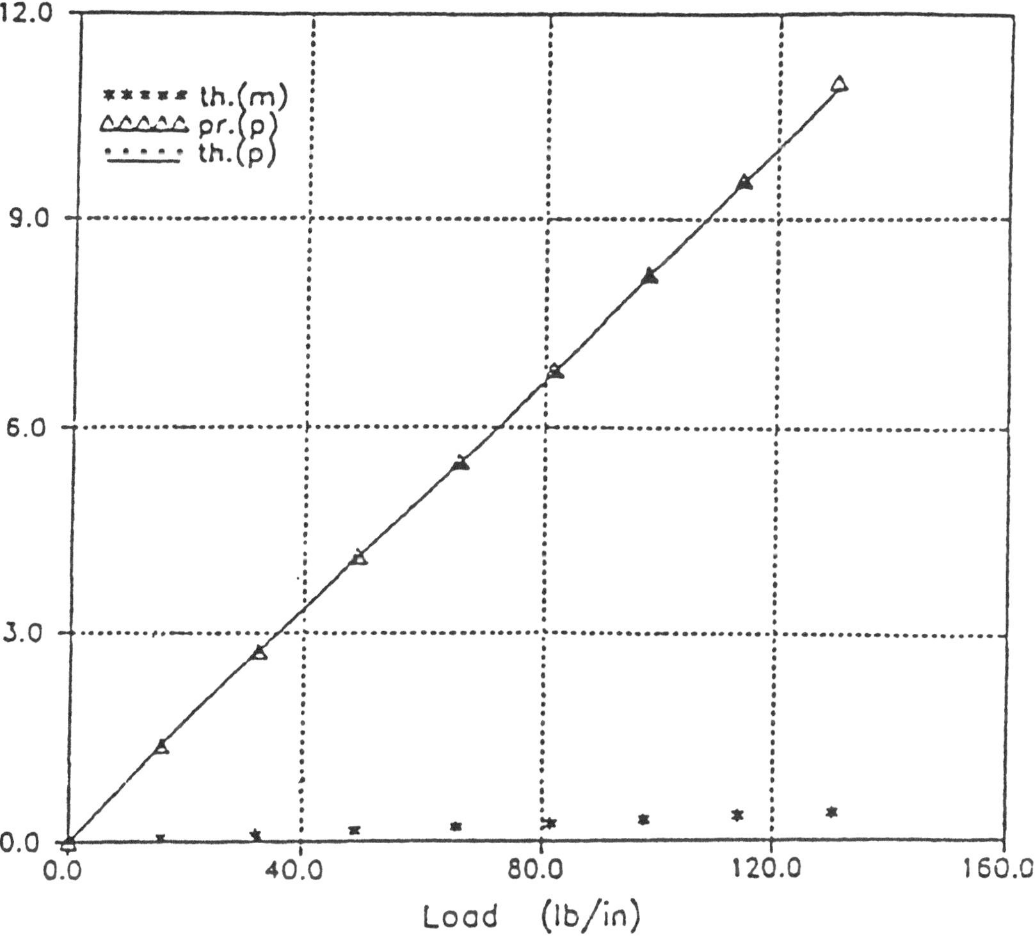


Fig. 1. Theoretical and predicted maximum deflections of prototype (0°/90°/0°...)$_{96}$ when the model is (0°/90°/0°...)$_{16}$ $[\lambda_{E_{11}} = \lambda_{E_{22}} = \lambda_{\nu_{12}} = 1;$ $\lambda_a = 18; \lambda_b = \lambda_q = 16.92; \lambda_h = \lambda_N = 6]$.

## 5.2. BUCKLING AND VIBRATIONS OF PLATES

Consider a simply supported, rectangular, symmetric, cross-ply laminated plate. The governing differential equation for buckling and vibration analyses is given by:

$$D_{11}w_{,xxxx} + 2\overline{D}_{12}w_{,xxyy} + D_{22}w_{,yyyy}\ \overline{N}_x w_{,xx} = \rho w_{,tt} \tag{14}$$

For buckling alone the characteristic equation is:

$$\overline{N}_x = \pi^2[D_{11}\left(\frac{m}{a}\right)^2 + 2\overline{D}_{12}\left(\frac{n}{b}\right)^2 + D_{22}\left(\frac{n}{b}\right)^4\left(\frac{a}{m}\right)^2] \tag{15}$$

For free vibrations the characteristic equation is

$$\rho\omega^2 = \pi^4\left[D_{11}\left(\frac{m}{a}\right)^4 + 2\overline{D}_{12}\left(\frac{mn}{ab}\right)^2 + D_{22}\left(\frac{n}{b}\right)^4\right] \tag{16}$$

By applying similitude theory to Eq. (14), we obtain:

$$\lambda_\rho\lambda_\omega^2 = \lambda_{D_{11}}\frac{\lambda_m^4}{\lambda_a^4} = \lambda_{D_{12}}\frac{\lambda_m^2\lambda_n^2}{\lambda_a^2\lambda_b^2} = \lambda_{D_{22}}\frac{\lambda_n^4}{\lambda_b^b} \tag{17}$$

which yield the following scaling laws:

$$\lambda_\Omega^2 = \frac{\lambda_{D_{11}}}{\lambda_{E_{22}}\lambda_h^3}\frac{\lambda_m^4}{\lambda_R^4}, \qquad \lambda_\Omega^2 = \frac{\lambda_{D_{12}}}{\lambda_{E_{22}}\lambda_h^3}\frac{\lambda_m^2\lambda_n^2}{\lambda_R^2},$$

$$\lambda_\Omega^2 = \frac{\lambda_{D_{22}}}{\lambda_{E_{22}}\lambda_h^3}\lambda_n^4 \tag{18}$$

$$\Omega^2 = \frac{b^4\omega^2}{\pi^4}\frac{\rho}{E_{22}h^3} \tag{19}$$

For details and results, see Rezaeepazhand et al. (1995a), Simitses and Rezaeepazhand (1995) and Rezaeepazhand et al. (1995b).

## 6. Application to Shell Configurations

Complete similarity and partial similarity were applied to laminated cylindrical shell configurations (Rezaeepazhand et al., 1996; Simitses et al., 1997; and Rezaeepazhand and Simitses, 1997). Details can be found in these references, but some basic equations and steps are presented, herein, for completeness. The buckling equation for a symmetric, laminated, cross-ply ($B_{ij} = D_{16} = D_{26} = A_{16} = A_{26} = 0$), cylindrical shell (see Jones and Morgan, 1975) is given by:

$$T_{33}+\frac{(2T_{12}T_{13}T_{23}-T_{11}T_{23}^2-T_{22}T_{13}^2)}{(T_{11}T_{22}-T_{12}^2)} \tag{20}$$
$$=-\overline{N}_{xx}\eta^2-N_{yy}\xi^2$$

where $\eta=\overline{m}\pi/L$, $\xi=\overline{n}/R$,

$$T_{11}=A_{11}\eta^2+A_{66}\xi^2$$

$$T_{12}=(A_{12}+A_{66})\xi\eta$$

$$T_{13}=\frac{A_{12}}{R}\eta$$

$$T_{22}=A_{22}\xi^2+A_{66}\eta^2$$

$$T_{23}=\frac{A_{22}}{R}\xi$$

$$T_{33}=D_{11}\eta^4+2\overline{D}_{12}\xi^2\eta^2+D_{22}\xi^4+\frac{A_{22}}{R^2}$$

and

$$\overline{D}_{12}=D_{12}+2D_{66}$$

The lowest eigenvalue corresponds to the buckling load, and minimization with respect to integer values of m and n yields the critical load.

Equation (20) represents the buckling response of both prototype and its models. Applying similitude theory to the preceding equation, Eq. (20) yields the following scaling laws for symmetric, cross-ply, laminated cylinders:

$$\left(\frac{\lambda_\eta}{\lambda_\xi}\right)^4=\frac{\lambda_{A_{22}}}{\lambda_{A_{11}}} \tag{22}$$

$$\left(\frac{\lambda_\eta}{\lambda_\xi}\right)^2=\frac{\lambda_{A_{22}}}{\lambda_{A_{66}}} \tag{23}$$

$$\left(\frac{\lambda_\eta}{\lambda_\xi}\right)^2=\frac{\lambda_{A_{66}}}{\lambda_{A_{11}}} \tag{24}$$

$$\left(\frac{\lambda_\eta}{\lambda_\xi}\right)^2 = \frac{\lambda_{A_{66}}}{\lambda_{A_{11}}\lambda_{A_{66}}} \tag{25}$$

$$\lambda_\psi = \lambda_\eta^4 \lambda_{D_{11}} \tag{26}$$

$$\lambda_\psi = \lambda_{\overline{D}_{12}} \lambda_\eta^2 \lambda_\xi^2 \tag{27}$$

$$\lambda_\psi = \lambda_{D_{22}} \lambda_\xi^4 \tag{28}$$

$$\lambda_\psi = \frac{\lambda_{A_{22}}}{\lambda_R^2} \tag{29}$$

$$\lambda_\psi = \frac{\lambda_{A_{12}}^2}{\lambda_{A_{11}}\lambda_R^2} \tag{30}$$

where $\psi = -\overline{N}_{xx}\eta^2, -\overline{N}_{yy}\xi^2, \text{or} -(pR/2)(\eta^2 + 2\xi^2)$,

and $\overline{A}_{12} = A_{12} + A_{66}$,

$\overline{D}_{12} = D_{12} + 2D_{66}$.

The nine scaling laws, Eqs. (22)-(30) are the necessary scaling laws for cross-ply laminated cylindrical shells for axial compression, lateral pressure, and hydrostatic pressure. The conditions that represent structural geometries and mode shapes, Eqs. (22)-(25) are the necessary scaling laws for symmetric cross-ply laminated cylinders regardless of the destabilizing load.

As is apparent, the scaling laws are arranged in the form of different scale factors for each load case ($\psi$). It should be pointed out that the presented form of arranging the scaling laws is not unique. However, previous experience of establishing scaling laws (Rezaeepazhand et al., 1995a) strongly recommends this type of representation.

## 6.1. SCALING LAWS FOR LATERAL PRESSURE LOAD

For the case of a cylinder subjected to lateral pressure p, $N_{yy} = pR$ and Eqs. (26)-(30) assume the following form:

$$\lambda_{K_{yy}} = \frac{\lambda_\eta^4}{\lambda_\xi^2}\lambda_L^2 \tag{31}$$

$$\lambda_{K_{yy}} = \frac{\lambda_{\overline{D}_{12}}}{\lambda_{D_{11}}}\lambda_\eta^2\lambda_L^2 \tag{32}$$

$$\lambda_{K_{yy}} = \frac{\lambda_{A_{22}}}{\lambda_{D_{11}}}\frac{\lambda_L^2}{\lambda_R^2\lambda_\xi^2} \tag{33}$$

$$\lambda_{K_{yy}} = \frac{\lambda^2_{A_{12}}}{\lambda_{A_{11}}\lambda_{D_{11}}}\frac{\lambda_L^2}{\lambda_R^2\lambda_\xi^2} \tag{34}$$

$$\lambda_{K_{yy}} = \frac{\lambda_{D_{22}}}{\lambda_{D_{11}}}\lambda_\xi^2\lambda_L^2 \tag{35}$$

where $K_{yy} = -\overline{N}_{yy}L^2/\pi^2 D_{11}$, $\overline{A}_{12}$, and $\overline{D}_{12}$, have already been defined, and $\lambda_{x_i} = x_{i_p}/x_{i_m}$ denotes the scale factor of parameter $x_i$.

*Parenthetical remarks:* For the case of lateral pressure $\psi = -\overline{N}_{yy}\xi^2$. Therefore, $\lambda_\psi = \lambda_{\overline{N}_{yy}}\lambda_\xi^2$. Similarly, from the definition of $K_{yy}$ ($K_{yy} = \overline{N}_{yy}L^2/\pi^2 D_{11}$), one can write $\lambda_{K_{yy}} = -\lambda_{\overline{N}_{yy}}\lambda_L^2/\lambda_{D_{11}}$. Use of these two expressions in Eq. (26) yields Eq. (31). In a similar manner one can derive Eqs. (32)-(35).

Equations (31)-(35) are the necessary scaling laws for symmetric, cross-ply, laminated cylinders subjected to uniform lateral pressure.

The interested reader is referred to Rezaeepazhand et al. (1996) and Simitses et al. (1997) for results with primarily partial similarity with distortion in number of plies, stacking sequence and cylinder length, radius and thickness. Distortion here means that prototype and model have different parameters (as mentioned above).

## 7. Discussion

It has been demonstrated through the studies reported herein, that structural similitude is a powerful tool in minimizing the need for full scale and large component testing of structural systems. Future work should include the study of systems that exhibit imperfection sensitivity, extension to sandwich configurations and validation of the process through an experimental program for laminated plates and shells as well as beam plates, plates and shells of sandwich construction.

Through this review, the authors have demonstrated a procedure that can be used in designing small scale and easily testable models to predict the behavior of large prototypes through the use of scaling laws. These laws are based on the premise that both model and prototype are governed by the same field equations and that the systems behave in a linearly elastic manner and they are free of damage (delaminations, fiber breaks, matrix microcracking, etc.). This last premise guarantees that no size effects are present.

## 8. Acknowledgement

Work was supported by NASA Langley Research Center, under Grant No. NASA NAG-1-2071. The financial support provided by NASA is gratefully acknowledged.

## 9. References

Ashton, J.E. and Whitney, J.M. (1970), *Theory of Laminated Plates,* Technomic Publ., Stamford, CT.

Bazant, Z.P. and Rajapakse, Y.D.S. (editors) (1999), *Fracture Scaling*, Kluwer Academic Publishers, Dordrecht.

Camponeschi, G. (1994), "The Effects of Specimen Scale on the Compression Strength of Composite Materials," NASA CP 3271, pp. 81-99.

Charlton, T.M. (1954), *Model Analysis of Structures,* John Wiley & Sons, New York.

Chouchaoui, C.S. and Ochoa, O.O. (1999a), "Similitude Study for a Laminated Cylindrical Tube Under Tensile, Torsion, Bending, Internal and External Pressure, Part I: Governing Equations," *Composite Structures*, Vol. 44, pp. 221-229.

Chouchaoui, C.S., Parks, P. and Ochoa, O.O. (1999b), "Similitude Study for a Laminated Cylindrical Tube Under Tensile, Torsion, Bending, Internal and External Pressure. Part II: Scale Models," *Composite Structures*, Vol. 44, pp. 231-236.

Daniel, I.M. and Hsiao, H.M. (1999), "Is There a Thickness Effect on Compressive Strength of Unnotched Composite Laminates?" in *Fracture Scaling*, Z.P. Bazant and Y.D.S. Rajapakse (editors), Klumer Academic Publishers, pp. 143-158.

Ezra, A.A. (1962), "Similitude Requirements for Scale Model Determination of Shell Buckling under Impulsive Pressure," NASA TN D-1510, pp. 661-670.

Goodier, J.N. and Thomson, W.T. (1944), "Applicability of Similarity Principles to Structural Models," NACA Tech. Note 993.

Goodier, J.N. (1950), "Dimensional Analysis," *Handbook of Experimental Stress Analysis* (Edited by M. Hetenyi), pp. 1035-1045, John Wiley & Sons, NY.

Greszczuk, L.B. (1999), "Private Communication," September 23, 1999.

Grimes, G.C. (1994), "Experimental Observations of Scale Effects on Bonded and Bolted Joints in Composite Structures," NASA CP 3271, pp. 57-80.

Gukhman, A.A. (1965), *Introduction to the Theory of Similarity,* Academic Press, New York.

Hsu, C.S., Griffin, J.H. and Bielak, J. (1989), "How Gravity and Joint Scaling Affect Dynamic Response," *AIAA Journal*, Vol. 27, No. 9, pp. 1280-87.

Jackson, K.E. (compiler) (1994), "Workshop on Scaling Effects in Composite Materials and Structures," NASA Conference Publication 3271, July 1994.

Jackson, K.E. and Kellas, S. (1994), "Sub-Ply Level Scaling Approach Investigated for Graphite-Epoxy Composite Beam-Columns," NASA CP 3271, pp. 19-36.

Jackson, K.E. (1990), "Scaling Effects in the Static and Dynamic Response of Graphite-Epoxy Beam-Columns," NASA TM 102697.

Johnson, D.P., Morton, J., Kellas, S. and Jackson, K.E. (1994), "Scaling Effects in the Tensile and Flexure Response of Laminated Composite Coupons," NASA CP 3271, pp. 265-282.

Jones, R.M. and Morgan, H.S. (1975), "Buckling and Vibration of Cross-Ply Laminated Circular Cylindrical Shells," *AIAA Journal*, Vol. 13, No. 5, pp. 664-671.

Kline, S.J. (1965), *Similitude and Approximation Theory*. McGraw-Hill, NY.

Langhaar, H.L. (1951), *Dimensionless Analysis and Theory of Models*, John Wiley & Sons, New York.

Letchworth, R., McGowan, P.E. and Gronet, J.J. (1988), "Space Station: A Focus for the Development of Structural Dynamics Scale Model for Large Flexible Space Structures," presented at AIAA/ASME/ASCE/AHS 29th SDM Conference (not included in Proceedings), Williamsburg, VA, April 18-20, 1988.

Macagno, E.O. (1971), "Historico-Critical Review of Dimensional Analysis," *J. Franklin Inst.*, Vol. 292, No. 6, pp. 391-402.

McDougal, R.L. (1987), Private Communications, Structural Division, the Lockheed-Georgia Company, Marietta, GA.

McGowan, P.E., Edighoffer, H.E. and Wallace, J.W. (1990), "Development of an Experimental Space Station Model for Structural Dynamics Researches," NASA Technical Memorandum 102601.

Morgen, G.W. (1964), Scaling Techniques for Orthotropic Cylindrical Aerospace Structures," *Proceedings* of AIAA 5th Structures and Materials Conference, Pal Springs, pp. 333-343.

Morton, J. (1988), "Scaling of Impact Loaded Carbon Fiber Composites," *AIAA Journal*, Vol. 26, No. 8, pp. 989-994.

Murphy, G. (1950), *Similitude in Engineering*, Ronald Press, New York.

O'Brien, T.K. (1994), "Damage and Strength of Composite Materials: Trends, Predictions and Challenges," NASA CP 3271, pp. 145-160.

Pankhust, R.C. (1964), *Dimensional Analysis and Scale Factors*, Chapman & Hall, London, Reinhold, New York.

Qian, Y., Swanson, S.R., Nuismer, R.J. and Bucinell, R.B. (1990), "An Experimental Study of Scaling Rules for Impact Damage in Fiber Composites," *J. Composite Materials*, Vol. 24, No. 5, pp. 559-570.

Rayleigh, Lord (1915), "The Principle of Similitude," *Nature*, 95, 66-68.

Rezaeepazhand, J. and Simitses, G.J. (1997), "Structural Similitude for Vibration Response of Laminated Cylindrical Shells with Double Curvature," *Composites, Part B*, Vol. 28B, pp. 195-200.

Rezaeepazhand, J., Simitses, G.J. and Starnes, J.H. Jr. (1996), "Scale Models for Laminated Cylindrical Shells Subjected to Axial Compression," *Composite Structures*, Vol. 34, No. 4, pp. 371-379.

Rezaeepazhand, J., Simitses, G.J. and Starnes, J.H., Jr. (1995a), "Design of Scaled Down Models for Stability of Laminated Plates," *AIAA Journal*, Vol. 33, No. 3, pp. 515-519.

Rezaeepazhand, J., Simitses, G.J. and Starnes, J.H., Jr. (1995b), "Use of Scaled Down Models for Predicting Vibration Response of Laminated Plates," *Composite Structures*, Vo. 30, pp. 419-426.

Shih, C., Chen, J.C. and Garba, J. (1987), "Verification of Large Beam-Type Space Structures," NASA Report No. 87-22712.

Sierakowski, R.L. (1994), "Private Communication," Ohio State University, Columbus, OH.

Simitses, G.J. (1999), "Structural Similitude for Flat Laminated Surfaces," Composite Structures, accepted for publication.

Simitses, G.J. and Rezaeepazhand, J. (1995), "Structural Similitude and Scaling Laws for Buckling of Cross-Ply Laminated Plates," *J. Thermoplastic Composite Materials*, Vol. 8, pp. 240-251.

Simitses, G.J. and Rezaeepazhand, J. (1993), "Structural Similitude for Laminated Structures," *J. Composites Engineering*, Vol. 3, Nos. 7-8, pp.; 751-765.

Simitses, G.J., Rezaeepezhand, J. and Sierakowski, R.L. (1997), "Scaled Models for Laminated Cylindrical Shells Subjected to External Pressure," *Mechanics of Comp. Materials & Structures*, Vol. 4, pp. 267-280.

Singer, J., Arbocz, I. And Weller, T. (1997), *Buckling Experiments*, John Wiley and Sons, Chickester, England.

Soedel, W. (1971), "Similitude Approximations for Vibrating Thin Shells," *J. Acoustical Society of America*, Vol. 49, No. 5, pp. 1535-41.

Szucs, E. (1980), *Similitude and Modelling*, Elsevier, NY.

# Reduction of the Linear Sanders-Koiter Equations for Fully Anisotropic Non-Circular Cylindrical Shells to Two Coupled Fourth-Order Equations

by

J. G. Simmonds

Department of Civil Engineering

University of Virginia, Charlottesville, VA 22903

## Abstract

With the aid of the static-geometric duality of Goldenveiser (1940) and Lure (1940), order of magnitude estimates, and ideas developed by Sanders (1967) for the reduction of the governing equilibrium and compatibility equations for elastically *isotropic* shells, it is shown that the governing Sanders-Koiter equations for fully *anisotropic* (21 elastic constants) general (non-circular) cylindrical shells can be reduced to two coupled fourth-order equations for a dimensionless resultant function and its static-geometric dual, a dimensionless rotation function. The equations contain pointwise formal errors of $O(\sqrt{\varepsilon})$, where $\varepsilon = h/R$ is the ratio of the constant shell thickness to the mean cross sectional radius. For a circular cylindrical shell the equations have, except for negligible terms, the same form as those derived recently by McDevitt & Simmonds (1999), but involve different unknowns.



*D. Durban et al. (eds.), Advances in the Mechanics of Plates and Shells*, 311–326.

## Introduction

Recently, McDevitt & Simmonds (1999) showed that the linear first-approximation shell theory developed independently and simultaneously by Sanders (1959) and Koiter (1959), when specialized to fully anisotropic *circular* cylindrical shells, can be reduced to two coupled equations for a dimensionless stress function $F$ and its static-geometric dual $G$, a dimensionless curvature function. Furthermore, it was shown that this reduction could be achieved *exactly* provided one added to the stress-strain relations certain small terms that, by arguments due to Simmonds (1971) and Koiter & Clément (1979), could be shown to lead to mean-square ($L_2$) errors of the order of $\varepsilon$. Such errors are of the same order as others unavoidably contained in any first-approximation shell theory, as emphasized by Koiter (1959). For non-circular cylindrical shells, the variable cross-sectional curvature prevents the introduction of stress and curvature functions. However, by suitably differentiating and combining the equilibrium and compatibility conditions and making reasonable (and standard) order of magnitude arguments, I shall show that these equations may be reduced to two fourth-order partial differential equations for $\chi$, a dimensionless *resultant function*, and $\phi$, its static-geometric dual, a dimensionless *rotation function*. Save for a few negligible terms, these equations have the same *form* as those of McDevitt & Simmonds (1999), although the unknowns are different ($\chi$ and $\phi$ instead of $F$ and $G$).

I note that, for elastically *isotropic* general cylindrical shells, Sanders (1967) has shown that the governing equations can be reduced to a single complex-valued fourth-order partial differential equation for an unknown whose real part is the trace of the stress resultant tensor and whose imaginary part is its static-geometric dual, the trace of the bending strain tensor. Unfortunately, this elegant approach no longer works for elastically anisotropic shells.

**Geometry**

Let $\{\mathbf{e}_r(\theta), \mathbf{e}_\theta(\theta), \mathbf{e}_z\}$ denote the standard orthonormal triad of base vectors associated with a set of circular cylindrical coordinates $\{r, \theta, z\}$. Then the midsurface of a general cylinder may be given the vector parametric form

$$C: \ \mathbf{x} = R\,[x\mathbf{e}_z + \rho(y)\tilde{\mathbf{e}}_r(y)] \ , \ 0 \leq x \leq l \ , \ 0 \leq y \leq 2\pi \, , \tag{1}$$

Here, $x$ and $y$ are, respectively, dimensionless axial and circumferential distances, derivatives with respect to which will be denoted by a prime ($'$) and dot ($\bullet$), respectively. Further, $\rho(y)$ is a dimensionless radius and

$$\tilde{\mathbf{e}}_r(y) = \mathbf{e}_r(\theta) \ \text{ where } \ \theta = \pm\int_0^y \frac{\sqrt{1 - \dot{\rho}^2(\bar{y})}}{\rho(\bar{y})}\, d\bar{y} \, , \tag{2}$$

the $\pm$-sign allowing for the possibility that the cylinder may fold back on itself. (If the cross-section is star-shaped with respect to a point, then the + sign can be taken. For a circular cylinder, $\theta = y$.) In addition,

$$R = (1/2\pi)\int_0^{2\pi} \mathbf{x}\bullet\tilde{\mathbf{e}}_r(y)dy \tag{3}$$

is the *mean* cross-sectional radius.

In Cartesian tensor notation (which we shall use intermittently for conciseness), let $R^2 a_{\alpha\beta}$ and $Rb_{\alpha\beta}$ denote, respectively, the metric and curvature tensors of $C$. It then follows from (1) and (2) that

$$a_{\alpha\beta} = \delta_{\alpha\beta} \ \text{ and } \ b_{\alpha\beta} = f(y)\delta_{2\alpha}\delta_{2\beta} \ , \ \text{ where } \ f = \pm\frac{1 - \tfrac{1}{2}(\rho^2)^{\bullet\bullet}}{\rho\sqrt{1 - \dot{\rho}^2}} \tag{4}$$

and Greek indices range from 1 to 2.

## The governing equations

Let

$$
\begin{aligned}
E_{\alpha\beta} &= \{E_x, E, E_y\} \ , \ h^{-1}K_{\alpha\beta} = h^{-1}\{K_x, K, K_y\} \\
\bar{E}hN_{\alpha\beta} &= \bar{E}h\{N_x, N, N_y\} \ , \ \bar{E}h^2 M_{\alpha\beta} = \bar{E}h^2\{M_x, M, M_y\}
\end{aligned}
\tag{5}
$$

denote, respectively, the symmetric extensional and bending strains and the (modified) symmetric stress resultants and couples of the Sanders-Koiter first-approximation shell theory for a general cylindrical shell. In (5), $\bar{E}$ denotes some nominal Young's modulus. Further, let $\bar{E}hR^{-1}\{p_x, p_y, p\}$ denote any external loads in the axial, circumferential, and *outward* normal directions, respectively, and let

$$\varepsilon = h/R \tag{6}$$

denote the fundamental small parameter of shell theory. Then the equilibrium and compatibility conditions of the Sanders-Koiter theory may be written as

$$-\dot{N} = N_x' - (1/2)\varepsilon(fM)^{\bullet} + p_x \tag{7}$$

$$-N' = \dot{N}_y + \varepsilon f\,[(3/2)M' + \dot{M}_y] + p_y \tag{8}$$

$$-\varepsilon(M_x'' + 2\dot{M}' + \ddot{M}_y) + fN_y - p = 0 \tag{9}$$

$$\dot{K} = K_y' - (1/2)\varepsilon(fE)^{\bullet} \tag{7*}$$

$$K' = \dot{K}_x + \varepsilon f\,[(3/2)E' - \dot{E}_x] \tag{8*}$$

$$\varepsilon(E_y'' - 2\dot{E}' + \ddot{E}_x) + fK_x = 0\,. \tag{9*}$$

These equations display the static-geometric duality of Goldenveiser (1940) and Lure (1940).

That is, the first set of equations, (7)-(9), with $p_x = p_y = p$, goes over into the second set, (7*)-(9*), if the sets of variables below on the left are replaced by those on the right:

$$\hat{N}_{\alpha\beta} = \{N_y, -N, N_x\} : K_{\alpha\beta} = \{K_x, K, K_y\}$$
$$M_{\alpha\beta} = \{M_x, M, M_y\} : -\hat{E}_{\alpha\beta} = \{-E_y, E, -E_x\}\,. \tag{10}$$

**Resultant and Rotation Functions**

The first of the three equilibrium equations, (7), may be satisfied identically by the introduction of a resultant function $\chi(x, y)$ such that

$$N_x = \dot{\chi} - \int p_x dx \;\;,\;\; -N = \chi' - (1/2)\varepsilon f M\,. \tag{11}$$

Then (8) reduces to

$$\chi'' = \dot{N}_y + \varepsilon f\,(2M' + \dot{M}_y) + p_y\,. \tag{12}$$

Likewise, through the introduction of the dual rotation function $\phi(x, y)$ such that

$$K_y = \dot{\phi}\;\;,\;\; K = \phi' - (1/2)\varepsilon f E\,, \tag{11*}$$

(7*) is satisfied identically and (8*) reduces to

$$\phi'' = \dot{K}_x + \varepsilon f\,(2E' - \dot{E}_x)\,. \tag{12*}$$

Finally, I use (9) and (9*) to eliminate, respectively, $N_y$ from (12) and $K_x$ from (12*), so obtaining

$$\chi'' = \varepsilon\{[f^{-1}(M_x'' + 2\dot{M}' + \ddot{M}_y)]^{\bullet} + f\,(2M' + \dot{M}_y)\} + (p/f)^{\bullet} + p_y \tag{13}$$

$$\phi'' = \varepsilon\{[f^{-1}(-E_y'' + 2\dot{E}' - \ddot{E}_x)]^{\bullet} + f\,(2E' - \dot{E}_x)\}\,. \tag{13*}$$

With the introduction of stress-strain relations and the neglect of certain terms, these equations will reduce to the two coupled equations for $\chi$ and $\phi$ that are the goal of this paper.

**Stress-strain relations**

To exploit fully the static-geometric duality, I follow McDevitt & Simmonds (1999) and take the stress-strain relations in the partially inverted forms

$$M_{\alpha\beta} = A^*_{\alpha\beta\lambda\mu} K_{\lambda\mu} + C^*_{\alpha\beta\lambda\mu} \hat{N}_{\lambda\mu} \, , \tag{14}$$

and

$$-\hat{E}_{\alpha\beta} = -A_{\alpha\beta\lambda\mu} \hat{N}_{\lambda\mu} + C_{\alpha\beta\lambda\mu} K_{\lambda\mu} \tag{14*}$$

where $C^*_{\alpha\beta\lambda\mu} = C_{\lambda\mu\alpha\beta}$ and $A^*_{\alpha\beta\lambda\mu}$ is the static-geometric dual of $-A_{\alpha\beta\lambda\mu}$. The 21 independent elastic constants in (14) and (14*) display the symmetries

$$A_{\alpha\beta\lambda\mu} = A_{\lambda\mu\alpha\beta} = A_{\beta\alpha\lambda\mu} \; , \quad C_{\alpha\beta\lambda\mu} = C_{\beta\alpha\lambda\mu} = C_{\alpha\beta\mu\lambda} \, . \tag{15}$$

A derivation of (14), (14*), and (15) from the conventional stress-strain relations (stress resultants and couples in terms of extensional and bending strains) is given by McDevitt & Simmonds (1999).

**Special cases**

I now consider three special cases: (a) inextensional bending and its static-geometric dual, membrane theory; (b) semi membrane-inextensional bending theory; and (c) quasi-shallow shell theory. For simplicity, I set henceforth $p_x = p_y = p = 0$.[1] I then present a form of (13) and (13*) involving $\chi$ and $\phi$ only that embraces these three cases and I conjecture that this equation is, in fact, a good approximation for *any* type of deformation, with a formal pointwise relative error of $O(\sqrt{\varepsilon})$.

---

[1]At the expense of some additional algebra, such loading terms may be included without difficulty.

*(a) inextensional bending and membrane theory*

By definition, inextensional bending means that in (14*)

$$A_{\alpha\beta\lambda\mu} - C_{\alpha\beta\lambda\mu} = 0 \tag{16}$$

so that one obtains immediately from (13*),

$$\phi_{inext} \equiv \Phi = xP(y) + Q(y)\,, \tag{17}$$

where $P$ and $Q$ are sufficiently smooth but otherwise arbitrary functions of $y$. The corresponding values of $K_x$, $K_y$, and $K$ follow from (9*) and (11*) as

$$K_x = 0 \;\;,\;\; K = \Phi' \;\;,\;\; K_y = \dot{\Phi}\,. \tag{18}$$

The stress resultants are statically determinant and follow from (11) and (12), where $\chi_{inext} \equiv X$ comes from (13). Thus,

$$X'' = \varepsilon\{[f^{-1}(4\overset{*}{A}_{1222}\Phi' + \overset{*}{A}_{2222}\dot{\Phi})^{\bullet\bullet}]^{\bullet} + f(4\overset{*}{A}_{1222}\Phi' + \overset{*}{A}_{2222}\dot{\Phi})^{\bullet}\}\,. \tag{19}$$

This equation may be integrated easily.

The dual of (19), which applies to membrane theory, that is, which applies when

$$\overset{*}{A}_{\alpha\beta\lambda\mu} = C_{\lambda\mu\alpha\beta} = 0\,, \tag{20}$$

follows immediately as

$$\Phi'' = -\varepsilon\{[f^{-1}(4A_{1222}X' + A_{2222}\dot{X})^{\bullet\bullet}]^{\bullet} + f(4A_{1222}X' + A_{2222}\dot{X})^{\bullet}\}\,. \tag{19*}$$

*(b) Semi membrane-inextensional bending theory*

This type of behavior, first identified by Vlasov—see Novozhilov (1970, § 49)—, represents very slow axial decay in very long shells. The governing equations may be obtained by scaling the axial distance and certain of the stress resultants and bending strains as follows:

$$\sqrt{\varepsilon}x = \zeta \ , \ N = \sqrt{\varepsilon}n \ , \ N_y = \varepsilon n_y \ , \ K = \sqrt{\varepsilon}k \ , \ K_x = \varepsilon k_x \, . \tag{21}$$

It then follows from (11), (11*), (13), (14), and (21), with $\partial(\,)/\partial\zeta = (\,),_\zeta$, that

$$\begin{aligned} M_x'' &= \varepsilon(\varepsilon \overset{*}{A}_{1111} k_x + 2\sqrt{\varepsilon}\overset{*}{A}_{1112} k + \overset{*}{A}_{1122} K_y + \varepsilon C_{1111} n_y - 2\sqrt{\varepsilon}C_{1211} n + C_{2211} N_x),_{\zeta\zeta} \\ &= \varepsilon(\overset{*}{A}_{1122}\phi + C_{2211}\chi),\dot{}_{\zeta\zeta} + O(\varepsilon^{3/2}) \, . \end{aligned} \tag{22}$$

Likewise,

$$\dot{M}' = \sqrt{\varepsilon}(\overset{*}{A}_{1222}\phi + C_{2212}\chi),\ddot{}_{\zeta} + O(\varepsilon) \ , \ \ddot{M}_y = (\overset{*}{A}_{2222}\phi + C_{2222}\chi)^{\dddot{}} + O(\sqrt{\varepsilon}) \tag{23}$$

and

$$M' = \sqrt{\varepsilon}(\overset{*}{A}_{1222}\phi + C_{1222}\chi),\dot{}_{\zeta} + O(\varepsilon) \ , \ \dot{M}_y = (\overset{*}{A}_{2222}\phi + C_{2222}\chi)^{\ddot{}} + O(\sqrt{\varepsilon}) \, . \tag{24}$$

When (21)-(24) are inserted into (13) there follows

$$\chi,_{\zeta\zeta} = [f^{-1}(\overset{*}{A}_{2222}\phi + C_{2222}\chi)^{\dddot{}}]^{\dot{}} + f(\overset{*}{A}_{2222}\phi + C_{2222}\chi)^{\ddot{}} + O(\sqrt{\varepsilon}) \, . \tag{25}$$

By the static-geometric duality I obtain immediately its companion,

$$\phi,_{\zeta\zeta} = [f^{-1}(-A_{2222}\chi + C_{2222}\phi)^{\dddot{}}]^{\dot{}} + f(-A_{2222}\chi + C_{2222}\phi)^{\ddot{}} + O(\sqrt{\varepsilon}) \, . \tag{25*}$$

Note that if (25*) is multiplied by $i\sqrt{\overset{*}{A}_{2222}/A_{2222}}$ and added to (25), one obtains the complex-valued equation

$$\psi,_{\zeta\zeta} = (C_{2222} - i\sqrt{A_{2222}\overset{*}{A}_{2222}})[(f^{-1}\dddot{\psi})^{\dot{}} + f\ddot{\psi}] + O(\sqrt{\varepsilon}) \, , \tag{26}$$

where $\psi = \chi + i\sqrt{\overset{*}{A}_{2222}/A_{2222}}\,\phi$. Note also that if $\zeta$ is replaced by $x$ and $y$ is replaced by $\varepsilon^{1/4}\alpha$, then (26) also applies to what Goldenveiser (1961, pp. 428 ff.) calls a "degenerate" edge effect which, on a general cylindrical shell, exists near an edge $y = \text{constant}$ and is of dimensionless width $O(\varepsilon^{1/4})$.

*(c) quasi shallow shell theory*

The well-known approximations of Donnell-Mushtari-Vlasov apply to a *geometrically* shallow shell, which a general cylindrical shell is not. However, Libai (1962)—and later, independently, Koiter (1966)—introduced the fruitful idea of a "quasi-shallow shell." Their purpose was to characterize shell solutions which vary rapidly relative to some characteristic geometrical dimension of the shell, for example, the mean radius $R$ for general cylindrical shells.

To see what form (13) and (13*) take for quasi-shallow shell theory, let

$$x = \sqrt{\varepsilon}\,\xi \ , \ y = \sqrt{\varepsilon}\,\eta \ \text{ and } \ \partial(\,)/\partial\xi = (\,)_{,\xi} \ , \ \partial(\,)/\partial\eta = (\,)_{,\eta} \ , \tag{27}$$

and note that if $[f^{-1}(y)]' = O(1)$, then

$$(f^{-1})_{,\eta} = O(\sqrt{\varepsilon})\,. \tag{28}$$

Thus,

$$\chi_{,\xi\xi} = \sqrt{\varepsilon}\,[f^{-1}(M_{x,\xi\xi} + 2M_{,\xi\eta} + M_{y,\eta\eta})_{,\eta} + O(\sqrt{\varepsilon})] \tag{29}$$

and

$$\phi_{,\xi\xi} = \sqrt{\varepsilon}\,[f^{-1}(-E_{y,\xi\xi} + 2E_{,\xi\eta} - E_{x,\eta\eta})_{,\eta} + O(\sqrt{\varepsilon})]\,. \tag{29*}$$

From (11)-(12*), (14), and (14*) follows

$$\chi_{,\xi\xi} = f^{-1}(\sqrt{\varepsilon}\,\eta)(\mathcal{L}_A^{*}\phi + \mathcal{L}_C\chi) + O(\sqrt{\varepsilon}) \tag{30}$$

and

$$\phi_{,\xi\xi} = f^{-1}(\sqrt{\varepsilon}\,\eta)(\mathcal{L}_A\chi + \mathcal{L}_C\phi) + O(\sqrt{\varepsilon})\,. \tag{30*}$$

Here, with $\bar{A}^{*}_{1212} = 2(A^{*}_{1122} + 2A^{*}_{1212})$ and $\bar{C}_{1212} = C_{1122} + C_{2211} + 4C_{1212}$,

$$\begin{aligned}\mathcal{L}_A^{*}(\,) = A^{*}_{1111}(\,)_{,\xi\xi\xi\xi} + 4A^{*}_{1112}(\,)_{,\xi\xi\xi\eta} + \bar{A}^{*}_{1212}(\,)_{,\xi\xi\eta\eta} \\ + 4A^{*}_{1222}(\,)_{,\xi\eta\eta\eta} + A^{*}_{2222}(\,)_{,\eta\eta\eta\eta}\end{aligned} \tag{31}$$

$$\mathcal{L}_C(\,) = C_{1111}(\,)_{,\xi\xi\xi\xi} + 2(C_{1211} + C_{1112})(\,)_{,\xi\xi\xi\eta} + \bar{C}_{1212}(\,)_{,\xi\xi\eta\eta}$$
$$+ 2(C_{2212} + C_{1222})(\,)_{,\xi\eta\eta\eta} + C_{2222}(\,)_{,\eta\eta\eta\eta}\,, \tag{32}$$

and $\mathcal{L}_A$ is the static-geometric dual of $\mathcal{L}_A^*$, obtained by replacing $A^*_{\alpha\beta\lambda\mu}$ everywhere in (31) by its dual, $-A_{\alpha\beta\lambda\mu}$.

## Conjecture

Examining the special forms taken by the basic differential equations (13) and (13*) for the three special cases of (a) inextensional bending and membrane theory, (b) semi membrane-inextensional bending theory, and (c) quasi shallow-shell theory, I note that all are special cases of the following two coupled equations:

$$\chi'' = \varepsilon\{f^{-1}(\mathcal{M}_A^*\phi + \mathcal{M}_C\chi) + [f^{-1}(\mathcal{N}_A^*\phi + \mathcal{N}_C\chi)^{\bullet}]^{\bullet} + f(\mathcal{N}_A^*\phi + \mathcal{N}_C\chi) + O(\sqrt{\varepsilon})\} \tag{33}$$

and

$$\phi'' = \varepsilon\{f^{-1}(\mathcal{M}_A\chi + \mathcal{M}_C\phi) + [f^{-1}(\mathcal{N}_A\chi + \mathcal{N}_C\phi)^{\bullet}]^{\bullet} + f(\mathcal{N}_A\chi + \mathcal{N}_C\phi) + O(\sqrt{\varepsilon})\}\,, \tag{33*}$$

where

$$\mathcal{M}_A^*(\,) = [A^*_{1111}(\,)'' + 4A^*_{1112}(\overset{\bullet}{\,})' + \bar{A}^*_{1212}(\overset{\bullet\bullet}{\,})]'' \tag{34}$$

$$\mathcal{M}_C(\,) = [C_{1111}(\,)'' + 2(C_{1211} + C_{1112})(\overset{\bullet}{\,})' + \bar{C}_{1212}(\overset{\bullet\bullet}{\,})]'' \tag{35}$$

$$\mathcal{N}_A^*(\,) = [4A^*_{1222}(\,)' + A^*_{2222}(\overset{\bullet}{\,})]^{\bullet} \tag{36}$$

$$\mathcal{N}_C(\,) = [2(C_{1222} + C_{2212})(\,)' + C_{2222}(\overset{\bullet}{\,})]^{\bullet}\,, \tag{37}$$

and $\mathcal{M}_A$ and $\mathcal{N}_A$ are the static-geometric duals, respectively, of $\mathcal{M}_A^*$ and $\mathcal{N}_A^*$.

Equations (33) and (33*) may be replaced by others of equal accuracy but, perhaps, more symmetry, by adopting an observation by Simmonds (1966) for elastically isotropic circular cylindrical shells, namely, that once stress-strain relations have been introduced into the right sides (13) and (13*), then there are inherent errors of relative order $\varepsilon$. Let me exploit this flexibility by forming the combination $(1-\varepsilon a)\times$Eq.(13) $-\ \varepsilon b\times$Eq.(13*), where $a$ and $b$ are arbitrary O(1) constants. Setting $p_y = p = 0$, I thus obtain an equation of the form

$$\chi'' = \varepsilon\{[f^{-1}(\underline{M}_x'' + 2\dot{\underline{M}}' + \ddot{\underline{M}}_y)]^{\bullet} + f(2\underline{M}' + \dot{\underline{M}}_y) + (a\chi + b\phi)'' + O(\sqrt{\varepsilon})\}\,, \tag{38}$$

where

$$\underline{M}_x = M_x + \varepsilon b E_y = M_x + O(\varepsilon) \tag{39}$$

$$\underline{M} = M - \varepsilon b E = M + O(\varepsilon) \tag{40}$$

$$\underline{M}_y = M_y + \varepsilon b E_x = M_y + O(\varepsilon) \tag{41}$$

and where it is understood that the dimensionless stress couples in (38)-(41) have been replaced by the stress-strain relations (14). The conclusion is that the accuracy of (33) is unchanged if the term $(a\chi + b\phi)''$ is added to the right side. By the static-geometric duality, the same is true if the term $(a^*\phi + b^*\chi)''$ is added to the right of (33*). In particular, if we take

$$a = C_{1122} + 2C_{1212} \quad \text{and} \quad b = \overset{*}{A}_{1122} + 2\overset{*}{A}_{1212}\,, \tag{42}$$

then, for a circular cylindrical shell ($f = 1$), (33), less the error term, reduces to an equation of the same form as equation (39) of McDevitt & Simmonds (1999). (As noted in the introduction, their unknowns are a dimensionless stress function $F$ and a dimensionless curvature function $G$.) Likewise, (33*), less the error term plus the term $(a^*\phi + b^*\chi)''$, reduces to the same form as their equation (39*), with

$$a^* = a\ ,\quad b^* = -(A_{1122} + 2A_{1212})\,. \tag{42*}$$

**Displacements**

In the Sanders-Koiter theory, the tangential displacements $R(u, v) = Ru_\alpha$ and the *outward* normal displacement $Rw$ are given by the displacement-strain-stress relations

$$\hat{E}_{\alpha\beta} = \delta_{\alpha\beta}(u_{\gamma,\gamma} + b_{\gamma\gamma}w) - \tfrac{1}{2}(u_{\alpha,\beta} + u_{\beta,\alpha}) - b_{\alpha\beta}w = A_{\alpha\beta\lambda\mu}\hat{N}_{\lambda\mu} - C_{\alpha\beta\lambda\mu}K_{\lambda\mu}\,. \tag{43}$$

As Koiter (1959) emphasized, the errors made in the stress-strain relations (43) by replacing the dimensionless stress resultants $N_{\alpha\beta}$ by terms of the form $N_{\alpha\beta} + \varepsilon D_{\alpha\beta\lambda\mu}M_{\lambda\mu}$ and the dimensionless bending strains by terms of the form $K_{\alpha\beta} + \varepsilon D^*_{\alpha\beta\lambda\mu}E_{\lambda\mu}$ are of the same order of magnitude as the inherent errors in these equations. If there are no surface loads, this means that (11)-(12*) may be used in (43) with only the $\chi$ and $\phi$-terms retained. Thus, with $\alpha = \beta = 2$,

$$\begin{aligned}\hat{E}_{22} = E_{11} = u' = A_{2211}\int\chi''\,dy + 2A_{2212}\chi' + A_{2222}\dot{\chi}\\ - (C_{2211}\int\phi''\,dy + 2C_{2212}\phi' + C_{2222}\dot{\phi})\,.\end{aligned} \tag{44}$$

An integration with respect to $x$ yields

$$u = A_{2211}\int\chi'\,dy + 2A_{2212}\chi + A_{2222}\int\dot{\chi}\,dx - (C_{2211}\int\phi'\,dy + 2C_{2212}\phi + C_{2222}\int\dot{\phi}\,dx)\,. \tag{45}$$

It is understood that the indefinite integrals contain (by virtue of their definition) arbitrary functions of integration.

Next, setting $\alpha = 1$, $\beta = 2$ in (43), I obtain

$$\begin{aligned}\hat{E}_{12} = -E_{12} = -\tfrac{1}{2}(\dot{u} + v') = A_{1211}\int\chi''\,dy + 2A_{1212}\chi' + A_{1222}\dot{\chi}\\ - (C_{1211}\int\phi''\,dy + 2C_{1212}\phi' + C_{1222}\dot{\phi})\,.\end{aligned} \tag{46}$$

Solving for $v'$, using (45), and integrating with respect to $x$, I find that

$$v = -[(A_{2211} + 4A_{1212})\chi + 2A_{1211}\int\chi' dy + 4A_{1222}\int\dot{\chi}dx + A_{2222}\int\int\ddot{\chi}dxdx]$$
$$+ (C_{2211} + 4C_{1212})\phi + 2C_{1211}\int\phi' dy + 2(C_{1222} + C_{2212})\int\dot{\phi}dx + C_{2222}\int\int\ddot{\phi}dxdx\,. \tag{47}$$

Finally, setting $\alpha = \beta = 1$ in (43) and using $(3)_2$, I get

$$\hat{E}_{11} = E_{22} = \dot{v} + fw = A_{1111}\int\chi'' dy + 2A_{1112}\chi' + A_{1122}\dot{\chi}$$
$$- (C_{1111}\int\phi'' dy + 2C_{1112}\phi' + C_{1122}\dot{\phi})\,. \tag{48}$$

Inserting (47) and solving for $w$, I get

$$fw = 4A_{1112}\chi' + 2(A_{1122} + 2A_{1212})\dot{\chi} + A_{1111}\int\chi'' dy + 4A_{1222}\int\ddot{\chi}dx + A_{2222}\int\int\dddot{\chi}dxdx$$
$$- [2(C_{1211} + C_{1112})\phi' + (C_{2211} + C_{1122} + 4C_{1212})\dot{\phi} + C_{1111}\int\phi'' dy \tag{49}$$
$$+ 2(C_{1222} + C_{2212})\int\ddot{\phi}dx + C_{2222}\int\dddot{\phi}dxdx]\,.$$

**Boundary conditions**

Canonical (classical) boundary conditions can be inferred from an expression for the virtual work of the edge forces and consist of prescribing either displacements and a rotation or the conjugate reduced forces and edge moment they multiply. For a general edge, these can be deduced from equations (14)-(17) and (23) of Budiansky & Sanders (1963). Here, I consider the specific, non-tensorial form of these boundary conditions for two types of edges: $x = \text{constant}$ and $y = \text{constant}$.

*On x = constant prescribe either*

$$
\begin{aligned}
u \;&\text{ or }\; N_x = \dot{\chi} \\
v \;&\text{ or }\; N + (3/2)\varepsilon f M = -\chi' + 2\varepsilon f M \\
w \;&\text{ or }\; M_x' + 2\dot{M} \\
w' \;&\text{ or }\; M_x
\end{aligned}
\tag{50}
$$

*On y = constant prescribe either*

$$
\begin{aligned}
u \;&\text{ or }\; N - (1/2)\varepsilon f M = -\chi' \\
v \;&\text{ or }\; N_y = \int [\chi'' - \varepsilon f\,(2M' + \dot{M}_y)]dy \\
w \;&\text{ or }\; 2M' + \dot{M}_y \\
\dot{w} - f v \;&\text{ or }\; M_y
\end{aligned}
\tag{51}
$$

Note that in the boundary conditions, one cannot, in general, neglect terms of $O(\varepsilon\sqrt{M_{\alpha\beta}M_{\alpha\beta}})$ compared to $\sqrt{N_{\alpha\beta}N_{\alpha\beta}}$. Thus, for example, in the second of the four boundary conditions listed in (50), one must retain the term $2\varepsilon f M$ and compute it, as with the other stress couples that appear in (50) and (51), by using the stress-strain relation (14) together with the expressions (11)-(12*) for the stress resultants and bending strains in which, however, only the $\phi$- and $\chi$-terms need be retained.

## Conclusions

The *analytical* advantages (as opposed to *numerical* ones) of working with fewer differential equations of higher order are well attested in the literature, especially for the application of perturbation methods. In particular, this is reflected in the many papers reviewed by Simmonds (1966) and Sanders (1983) devoted to reducing the linear field equations for elastically

*elastically isotropic, circular* cylindrical shells to two coupled fourth-order equations. I hope that the present paper has shown how these efforts may be extended to *elastically anisotropic, general* cylindrical shells.

However, my derivation of the two, reduced fourth-order partial differential equations (33) and (33*) for a dimensionless resultant function $\chi$ and a dimensionless rotation function $\phi$ leaves open the important question: Can these equations be shown *rigorously* to contain pointwise errors of $O(\sqrt{\varepsilon})$ or smaller? I commend this challenge to my colleagues who work in shell theory.

**Acknowledgement**

This work was supported by the National Aeronautics and Space Administration under Grant NAG-1-1854.

## References

B. Budiansky and J. L. Sanders, Jr. (1963), "On the 'best' first-order linear shell theory," *Progress in Applied Mechanics* (Prager Anniversary Volume), pp. 129-140, Macmillan, New York.

A. L. Goldenveiser (1940), "Equations of the theory of thin shells" (in Russian), *Prikl. Mat. Mekh.* **4,** 35-42.

A. L. Goldenveiser (1961), *Theory of Elastic Thin Shells*, Pergamon, New York.

W. T. Koiter (1959), "A consistent first approximation in the general theory of thin elastic shells," *The Theory of Thin Elastic Shells,* Proc. IUTAM Sympos. Delft (ed. W. T. Koiter), North-Holland, 1960, pp. 12-33.

W. T. Koiter (1966), "On the nonlinear theory of thin elastic shells," *Proc. Kon. Ned. Ak. Wet.* **B69,** 1-54.

W. T. Koiter and Ph. Clément (1979), "On the so-called comparison theorem in the linear theory of elasticity," *ZAMP* **30,** 534-536.

A. Libai (1962), "On the nonlinear elastokinetics of shells and beams," *J. Aerospace Sci.* **29,** 1190-1195.

A. I. Lure (1940), "General theory of thin elastic shells" (in Russian), *Prikl. Mat. Mekh.* **4,** 7-34.

T. J. McDevitt and J. G. Simmonds (1999), "Reduction of the Sanders-Koiter equations for fully anisotropic circular cylindrical shells to two coupled equations for a stress and a curvature function," (submitted for publication).

V. V. Novozhilov (1970), *Thin Shell Theory*, 2nd ed (transl. P. G. Lowe, ed J. R. M. Radok), Wolters-Noordhoff, Groningen.

J. L. Sanders, Jr. (1959), "An improved first-approximation theory for thin shells," NASA Rept. No. 24.

J. L. Sanders, Jr. (1967), "On the shell equations in complex form," Proc. IUTAM Sympos. Copenhagen (ed F. I. Niordson), Springer-Verlag, Berlin, 1969.

J. L. Sanders, Jr. (1983), "Analysis of circular cylindrical shells," *J. Appl. Mech.* **50,** 1165-1170.

J. G. Simmonds (1966), "A set of simple, accurate equations for circular cylindrical elastic shells," *Int. J. Solids Struct.* **2,** 524-541.

J. G. Simmonds (1971), "Extension of Koiter's $L_2$-error estimate to approximate shell solutions with no strain energy functional," *ZAMP* **22,** 339-345.

# LARGE DEFORMATION OF A PRESSURIZED TUBE

CHARLES R. STEELE
*Division of Mechanics and Computation*
*Department of Mechanical Engineering*
*Stanford University*
*Stanford, CA 94305-4035*

## 1. Introduction

The motivation for this work is the need for efficient means of computing the deployment of large, inflatable, space structures. Recent work has been on the load-deformation properties of folded and rolled pressurized tubes. The approach is to use basic properties of thin shells to determine appropriate approximations for the deformed shape. This is then used in an energy formulation, in which the work of the pressure and the external loads dominates. For some problems, the stretching of the tube wall is also significant. In the present work the behavior of a tube with symmetric rigid blade loading is considered. Previous results are inadequate for the moderate values of the blade displacement. A satisfactory solution is obtained with a modified shape function for the tube cross section and with elastic stretching of the wall.

## 2. Background

The large deformation of thin-walled structures is of importance in many areas. Generally, the analysis requires rather complex equations and intensive computation. However, there are key problems that have been handled with minimal computation by appealing to basic thin shell behavior. One example is the analysis of post–buckling behavior by Porgorelov (1960). A second is the notion of an inverted dimple in a spherical shell initiated by Ashwell (1960) and extended to the fluid–filled shell by Taber (1982). Possibly the best example of a difficult problem reduced to a few line



*D. Durban et al. (eds.), Advances in the Mechanics of Plates and Shells,* 327–342.

analysis is the instability of the measure band by Rimrott (1970). The present work is in this spirit.

The concern is the large deformation of tubes with internal pressure. For thin walls of high modulus material, the deformation can be considered as primarily inextensional with negligible bending resistance of the wall. The behavior of thin membranes has been the subject of many investigations, as discussed by Jenkins (1991) and by Libai and Simmonds (1998). Our motivation has been the deployment of large, inflatable, space structures. General structural considerations are surveyed by Szyszkowski and Glockner (1990), and specific current plans are outlined by Lou and Feria (1998). Because high degrees of accuracy and reliability are desired, and ground based simulations are often not possible, numerical simulation of deployment is important. Simulations of the dynamic deployment of large antennas with z-folded tubes are in Tsoi (1997), in which a crude approximation for the joint moments is made. However, results are highly dependent on the joint moments, so a better determination is necessary. Direct numerical computation of the long-time transient response of a structure with several hundred joints in unfolding tubes, including detailed calculation of the joints, is prohibitive. Consequently, we were motivated to find an approach requiring minimum computation.

## 3. Recent Work

A basic problem is the symmetric blade loading of a pressurized tube, as shown in Fig. 1. This problem has the difficulties associated with large displacement and is a good beginning because of the relatively simple geometry. One practical application of this specific configuration may the direct determination of the internal pressure in a biological cell, such as the outer hair cell in the cochlea (Tolomeo, et al., 1996). The bending of a tube, discussed in Fay and Steele (2000), is of importance, since this is the fundamental mechanism in the deployment of z-folded tubes. Another basic problem is the long tube in a rolled configuration, discussed in Fay and Steele (1999) and Steele and Fay (1998). As pointed out by J.F.V. Vincent (pers. comm.) these systems are also used in the insect world. Pressure rather than muscle is used for rapid deployment of wings (Glaser and Vincent,1979) and links, as well as coils (Bänziger,1971).

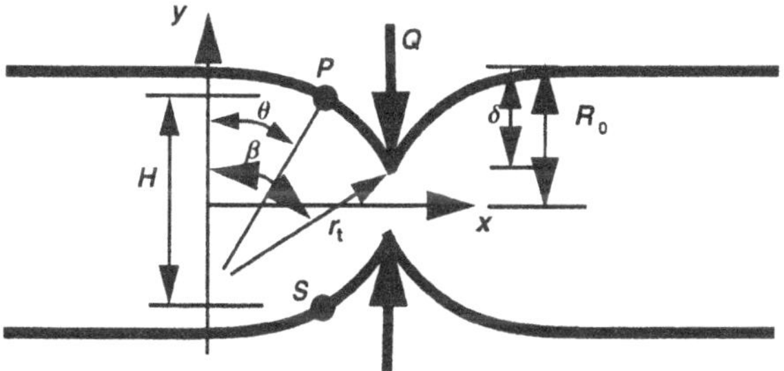


*Figure 1.* Geometry for symmetric pinch load acting on cylindrical tube of radius with tube of radius $R_0$. The region near the blade load $Q$ has the constant radius of curvature $r_t$ and subtends the angle $\beta$. The blade displacement is $\delta$. The points $P$ and $S$ are equidistant from the tube end and the distance $H$ apart.

Generally, large displacement of thin structures is associated with wrinkling. Substantial progress has been made on the direct analysis of the wrinkling, as by Epstein and Forcinito (1999). The interaction of bending and wrinkling is discussed by Cao and Boyse (1997) and Qui, et al. (1994). However, for the high internal pressure in Fig. 1, an alternate approach is used in Fay and Steele (1999).

### 3.1. APPROXIMATE THEORY

The main assumption is that all the work done by the applied loads goes into changing the volume and not into stretching and bending the wall of the tube. Such an approximation was used by Lukasiewicz and Glockner (1984) and undoubtedly has been used by many others. Consequently the total potential energy consists of only the work of the internal pressure and the external loads:

$$\Pi = -pV - Q\delta - M\phi \tag{1}$$

in which $V$ is the volume of the tube, $Q$ and $M$ are external loads, $\delta$ and $\phi$ are the corresponding displacements. The difficulty is in determining the volume when kinks and contacts between different portions of the wall, or between the wall and other surfaces, occur that are not known *a priori*. However, guidance can be found from the conditions used in Fay and Steele (1999).

### 3.1.1. *Basic principles for geometry.*

The principles for the determination of the geometry of the deformed tube are:

(1) An inextensional surface of zero Gaussian curvature must remain a surface of zero Gaussian curvature in a region of biaxial tension. Thus an initially flat surface can have a curvature in one direction or the other, but not both, in a region of biaxial tension.

(2) In a region of wrinkling, an "averaged" surface may be defined, for which the Gaussian curvature can be positive or negative. This occurs in a region with a nonpositive component of principal stress.

(3) For local equilibrium, constant pressure loading will be carried by constant curvature.

(4) Discontinuity in slope occurs only when an external line load is present, or in a direction of wrinkling.

### 3.1.2. *Deformed tube configuration.*

The foregoing principles can be used for a construction of the deformed shape. The blade loading (Fig. 1) causes local deformation. The top and bottom surfaces are in biaxial tension and by principle (1) must remain surfaces of zero Gaussian curvature. Consequently, there is a transition from a nonzero component of curvature in the circumferential direction in the main portion of the tube to a nonzero component in the plane of the figure in the deformed region near the load. By principle (3), the radius of curvature is constant in this region with the value $r_t$. By principle (4), the slope of the top and bottom generators of the cylindrical surface must be continuous except at the concentrated external load. The coordinates $x$, $y$ are at the beginning of the deformed region, while the angle $\beta$ is the total angle subtended by the deformed region, and $\theta$ is the angle from the edge of the deformed region to the general point $P$ on the top generator. The point $S$ is the point on the bottom generator that has the same arc length along the generator to the tube end as point $P$. The top and bottom generators are symmetric.

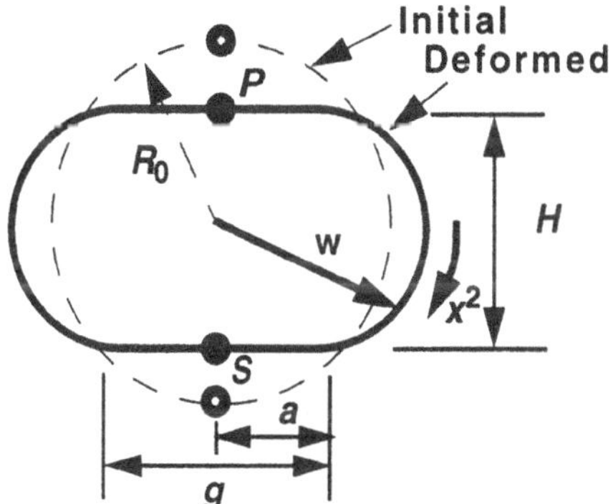


*Figure 2.* Cross section of tube (Shape 1) on the plane through points $P$ and $S$ which are the distance $H$ apart. The condition that the arc length remains unchanged provides the width of the reverse curvature region $g$ and the area in terms of $H$ and the original radius $R_0$. The vector **W** is from the center to a point on the perimeter, around which the arc length is $x^2$.

The cross section, which will be referred to as Shape 1, is shown in Fig. 2. The initial inflated tube with the circular cross section with radius $R_0$ becomes flattened with the height $H$, the distance between points $P$ and $S$. The distance $g$ is the width of the region of reversed curvature. The membrane must carry the pressure load by curvature in one direction or the other. Therefore by principle (3), the sides must have a constant radius of curvature. By principle (4), there cannot be a slope discontinuity, so the radius of the sides must be $H/2$. The width of the flat is:

$$g = \pi\left(R_0 - \frac{H}{2}\right) \tag{2}$$

and the cross sectional area is:

$$A_{section} = \pi\left(R_0^2 - \left(R_0 - \frac{H}{2}\right)^2\right) \tag{3}$$

Therefore, from the side view in Fig. 1, the coordinates of points $P$ and $S$ can be obtained. Then the distance $H$ between $P$ and $S$ provides the cross-sectional area.

### 3.1.3. *Volume integral.*

For a tube with a symmetrical cross section and planar center line, the volume integral can be reduced to (Fay and Steele, 2000):

$$V = \int \mathbf{R}_{c,1} \cdot \mathbf{n} A_{\text{section}} dx^1 \tag{4}$$

in which $\mathbf{R}_t$ is the position vector to the center line, $A_{\text{section}}$ is the area of the cross section Eq. 3, $\mathbf{n}$ is the unit normal to the cross section, and $x^1$ is the parameter along the center line. The coordinates of points $P$ and $S$ provide all that is needed for the numerical calculation of the volume from Eq. 4.

### 3.2. Symmetric blade loading

For the symmetric blade loading of the tube Fig. 1, the computation simplifies greatly. Because of the symmetry, the points at the top and bottom of the tube remain at the same axial distance. Thus the cross-sectional area in the deformed region at the angle $\theta$ is:

$$A_{\text{section}} = \pi[R_0^2 - r_t^2(1 - \cos\theta)^2] \tag{5}$$

so the change in volume from Eq. 4 is:

$$\Delta V = -\pi r_t^3 \int_0^\beta (1 - \cos\theta)^2 \cos\theta d\theta - \pi R_0^2 r_t(\beta - \sin\beta) \tag{6}$$

and the displacement under the load and the axial displacement are:

$$\delta = r_t(1 - \cos\beta) \tag{7}$$

$$\Delta = r_t(\beta - \sin\beta) \tag{8}$$

Thus the potential energy depends on $\beta$ and the displacement $\delta$, after $r_t$ is eliminated with Eq. 7. The form is best seen from the expansion for small angles $\beta$, for which the potential Eq. 1 becomes:

$$\Pi = p\pi\left(\frac{2}{5}\frac{\delta^3}{\beta^2} + R_0^2\frac{\delta\beta}{3}\right) - Q\delta \tag{9}$$

The derivative with respect to $\beta$ must be zero which gives:

$$\beta \approx \left(\frac{6}{5}\right)^{\frac{1}{2}}\delta^* \tag{10}$$

where $\delta^* = \delta/R_0$, and the derivative with respect to $\delta$ must be zero which gives the result:

$$Q^* \approx \frac{4}{3}\left(\frac{6}{5}\right)^{\frac{1}{2}}\delta^* = 1.46\delta^* \tag{11}$$

Fay and Steele (2000) give the results from experiments on a cylinder made from a sheet of urethane covered fabric with the thickness $t = 0.25$mm. The inflated radius was $R_0 = 30$mm and the length $L = 1.380$m. The small $\beta$ approximation Eq. 11 is surprising close to the experimental values for $0.3 < \delta^* < 0.9$. So the general behavior is rather tame and captured by the approximation (Shape 1) for the volume change. However, the experimental results for small displacement are substantially less than indicated from Eq. 11.

## 4. Need for Improved Theory

The development of the volume approximation from the basic principles of geometry and the cross section Fig. 2 works quite well for the problems considered, and can be enhanced by elastic effects for some problems. Since so much is gained for so little computational effort, it should be possible to improve the results with just a little more effort. We report some success with this for the symmetric blade loading problem. For small values of the indentation $\delta^* < 0.3$, the preceding calculations, based on the deformation of the cross section in Fig. 2, provide a load much too high. The shape of the surface is shown in Fig. 3. For the inextensional surface attached to rigid rings on the ends, an indentation of the top and bottom surfaces means that the sides must bulge out and form a surface of negative gaussian curvature. The axial strain in this region is negative, corresponding to wrinkling and a loss of axial load carrying capability. Thus all the axial load of the end pressure is carried at the top and bottom regions. For small indentation, these regions are narrow. So the indentation of these narrow regions forces the main portion of the shell into compression. This is an implausible situation if the elasticity of the wall is considered.

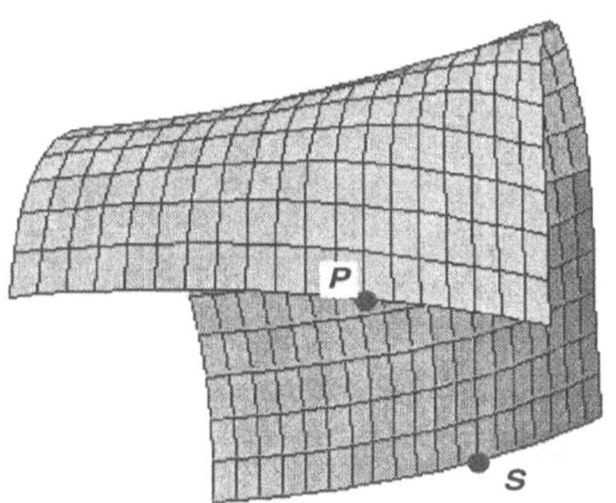


*Figure 3.* Shape of approximate surface with symmetric blade loading using Shape 1. The section is made on the plane passing through the points $P$ and $S$ in Fig. 6. The negative Gaussian curvature of the side can be seen in this view.

## 5. Hertzean contact

In Fay and Steele (2000), the very local indentation was considered as a region of a membrane with known pretension, with the equation:

$$N_x w_{,xx} + \frac{N_\varphi}{R^2} w_{,\varphi\varphi} = 0 \tag{12}$$

in which $\varphi$ is the circumferential angle, $R$ is the radius of the tube, $N_x$ and $N_\varphi$ are the axial and circumferential force resultants, and $w$ is the radial displacement. The blade contact is analogous to the two-dimensional contact problem in elasticity with an indenter of radius $R$. Following the analysis in Barber (1992), the relation of the load $Q$ to the half width of the contact region $a$ is found to be:

$$Q = \pi R \sqrt{N_x N_\varphi} \left(\frac{a}{R}\right)^2 \tag{13}$$

If Fig. 2 is used to relate the half width to the displacement:

$$\delta = \frac{2a}{\pi} \tag{14}$$

then the load-displacement relation for the pressured tube becomes:

$$Q^* = \frac{1}{\sqrt{2}} \left(\frac{\pi}{2}\right)^2 \delta^{*2} = 1.744\ \delta^{*2} \tag{15}$$

This appears to have the correct behavior for $\delta^* < 0.1$, for which the approximations have some justification. However, the transition for $0.2 < \delta^* < 0.3$ is missed.

An improved version of Eq. 12 can be derived from moderate rotation theory of shells:

$$N_x w_{,xx} + \frac{N_\varphi}{R^2}(w_{,\varphi\varphi} + w) = 0 \tag{16}$$

With this a Fourier series in the circumferential direction can be used to obtain a solution for the prescribed contact which has the correct behavior for $\delta^* < 0.3$. However, the axisymmetric solution of Eq. 16 is not valid. Using the exact axisymmetric solution of the bending equations produces a local band of circumferential compression that exceeds the buckling load. Omitting the axisymmetric solution leaves a discontinuity of slope on the sides, as in Fig. 3, which we are trying to avoid. The picture is further muddied when the validity of Eq. 16 is examined by considering the edge stiffness for a circumferential harmonic of line loading. The linear, elastic shell equations with prestress have the significant prestress parameter which is the ratio of the axial resultant to the magnitude of the classical axial buckling value in compression, which translates to pressure divided by the value:

$$p_{\text{cr Axial}} = \frac{4Et^2}{R^2\sqrt{12(1-\nu^2)}} \tag{17}$$

The results from the linear equations with bending stiffness indicate that the approximation Eq. 16 is valid only for pressures high in comparison with Eq. 17. However, for the experiments:

$$\frac{p}{p_{\text{cr Axial}}} = \begin{cases} 0.00039 & \text{for } p = 2\text{ kPa} \\ 0.023 & \text{for } p = 12\text{ kPa} \end{cases} \tag{18}$$

A calculation with the full linearized equations for the rigid blade loading indicates a high compressive stress in the region of the loading, much higher than that due to the internal pressure and enough to cause local buckling. So it seems that for displacements larger than the wall thickness, the local buckling will reduce the effectiveness of the bending stiffness and make Eq. 16 a reasonable approximation. Thus the behavior in for $0.01 < \delta^* < 0.3$ is in the post-buckling range, not easily attained in small steps from the full linearized equation.

It appears that the relevant pressure parameter for the blade loading is not Eq. 17, but rather is the magnitude for circumferential buckling of the infinite cylinder (ring buckling):

$$p_{\text{cr Circum}} = \frac{3Et^3}{R^3 12(1-\nu^2)} \tag{19}$$

For the experiments:

$$\frac{p}{p_{\text{cr Circum}}} = \begin{cases} 2.1 & \text{for } p = 2\text{ kPa} \\ 12.6 & \text{for } p = 12\text{ kPa} \end{cases} \tag{20}$$

When the internal pressure reaches the value of Eq. 19, the tube feels "stiff" to the touch.

We conclude that the prestressed membrane equation Eq. 12, or better Eq. 16, has some relation to the experimental behavior for $\delta^* < 0.3$. However, there is a yet a contradiction in the use of the equations. If the wall material is inextensional, the radial displacement of the wall at the blade causes an axial displacement of the axial fibers. Thus a different axial displacement of each axial fiber must be permitted. This would correspond to a diaphragm supported end, with a constant value of $N_x$ at each point of the circumference. For a rigid ring attached to the end, as in the experiments, the effect of the elasticity of the wall is significant, and a substantial redistribution of $N_x$ occurs. The conclusion is that for $\delta^* < 0.3$, neither the approximate shape Fig. 2, nor the approximate equation Eq. 16 with constant $N_x$, nor a perturbation solution from the full linear shell equations will provide a valid solution.

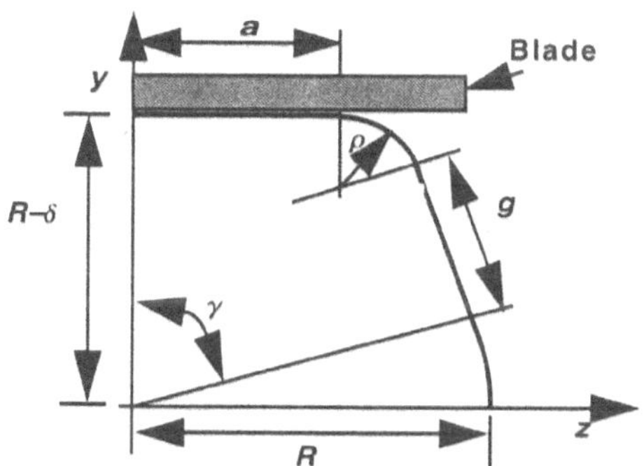


*Figure 4.* Assumed shape (Shape 2) for one quadrant of cross section. The length $a$ is the half width of the region of reversed curvature. At the section with the blade loading as shown, $a$ is the half width of the contact with the blade. The region with the radius $\rho$ provides the smooth transition to the straight segment of length $g$. For the displacement $\delta$ prescribed, and the arc length unchanged, the angle $\gamma$ is the only free parameter.

## 6. A New Shape Approximation

A better approach is to consider a different shape for the cross section, as shown in Fig. 4. This consists of the flat region on the top of width $a$, that is the region of axial curvature as before, a transition region with the radius $\rho$, a straight segment of length $g$, and finally a region of the original radius $R$. For a fixed value of the angle $\gamma$, the straight region and the region of radius $R$ are the same for all cross sections. Thus the outward curvature on the side seen in Fig. 3 does not occur. The displacement of the top at the general distance along the axis is denoted by $\delta_x$. For small displacement $\delta_x$, the deformation can remain more local to the top and most of the side can carry the axial load without wrinkling. The conditions for continuity of geometry and for no change in the circumferential arc length give the equations:

$$\begin{bmatrix} 1 & 1 & \gamma \\ 1 & \cos\gamma & \sin\gamma \\ 0 & \sin\gamma & 1-\cos\gamma \end{bmatrix} \begin{bmatrix} a(x) \\ g(x) \\ R-\rho(x) \end{bmatrix} = \begin{bmatrix} 0 \\ 0 \\ -\delta_x \end{bmatrix} \tag{21}$$

and the solution:

$$\begin{bmatrix} a \\ g \\ R-\rho \end{bmatrix} = \frac{\delta_x}{2\tan\frac{\gamma}{2}-\gamma} \begin{bmatrix} 1-\gamma\tan\gamma \\ \frac{\gamma}{\sin\gamma}-1 \\ \tan\frac{\gamma}{2} \end{bmatrix} \tag{22}$$

The area of the cross section is:

$$\text{Area}(x) = \pi R^2 - 2\frac{1-\gamma\cot\gamma}{2\tan\frac{\gamma}{2}-\gamma}\delta_x^{\ 2} \tag{23}$$

As mentioned before, for the top and bottom to deform and the sides not to wrinkle, there must be some elastic stretching of the top and bottom regions. For an approximation of this, we assume: (1) that the $x$-displacement of each point of the cross section is the same (no warping) and (2) that the axial strain consists of that due to the distortion of the cross section plus a constant component. As before, the axial coordinate Fig. 1 is:

$$x = r_t \sin\theta \tag{24}$$

and the $y$-coordinate at the top is:

$$y = R - \delta_x = R - \delta(1-\cos\theta) \tag{25}$$

The $y$– and $z$– coordinates at each point of the cross section Fig. 4 can be easily determined as a function of the arc length around the circumference. The integral for the arclength:

$$s(s_\varphi) = \int_0^\beta \sqrt{\left(\frac{\partial}{\partial\theta}x\right)^2 + \left(\frac{\partial}{\partial\theta}y\right)^2 + \left(\frac{\partial}{\partial\theta}z\right)^2}\, d\theta \tag{26}$$

can be computed numerically for each axial fiber. The arc length around the circumference to a given fiber is $s_\phi$. Therefore the elongation of an axial fiber due to the distortion of the cross sections with zero axial displacement is:

$$u(s_\varphi) = s(s_\varphi) - r_t \sin\beta \tag{27}$$

The potential energy expression that includes the wall elasticity is:

$$\Pi = \frac{Et}{2(1-\nu^2)}\int_0^L\int_0^{2\pi}(\varepsilon_x^2 + 2\nu\varepsilon_x\varepsilon_\varphi + \varepsilon_\varphi^2)Rd\varphi dx - p[Vol_{CS} + (2\bar{\varepsilon}_\varphi + \bar{\varepsilon}_x)\pi R^2 L] - 2Q\delta \tag{28}$$

in which $E$ is Young's modulus and $\nu$ is Poisson's ratio. The volume due to the distortion of the cross section is denoted by $Vol_{CS}$, and the overlines denote the values averaged around the circumference. The axial strain is split into the average and distortional part:

$$\varepsilon_x = 2\frac{u(s_\varphi) - \bar{u}}{L} + \bar{\varepsilon}_x \tag{29}$$

This axial strain is with the assumption of negligible coupling between the axial fibers, which is a reasonable approximation for a thin cylinder that is not too long. Taking the variation of Eq. 28 with respect to the average strain components yields the usual membrane resultants. What remains is the reduced potential energy:

$$\Pi_R = \frac{Et}{2(1-\nu^2)}RL\int_0^{2\pi}(\varepsilon_x - \bar{\varepsilon}_x)^2 d\varphi - pVol_{CS} - 2Q\delta \tag{30}$$

which can be written in the dimensionless form:

$$\frac{\Pi_R(\delta^*,\gamma,r_t)}{2p\pi R^3} = \frac{1}{2p_f(1-V^2)}\overline{\left(\frac{u-\bar{u}}{R}\right)^2} - \frac{Vol_{CS}}{2\pi R^3} - Q^*\delta^* \tag{31}$$

in which the elastic factor is:

$$p_f = \frac{pL}{2Et} \tag{32}$$

For fixed $\gamma$, the volume $Vol_{CS}$ is easily calculated from the cross sectional area Eq. 23 and the displacement Eq. 25, while the displacement $u$ is calculated from the numerical integration of Eq. 26. For a fixed value of the blade displacement $\delta^*$, the minimum of Eq. 31 is found with respect to changes in $r_t$ and $\gamma$. Somewhat surprising is that the minimum is always found at $\gamma = \pi/2$. The results are shown in Fig. 5 for the high and low values of pressure used in the experiment. It appears that the correct general behavior is captured by the cross–sectional shape in Fig. 4.

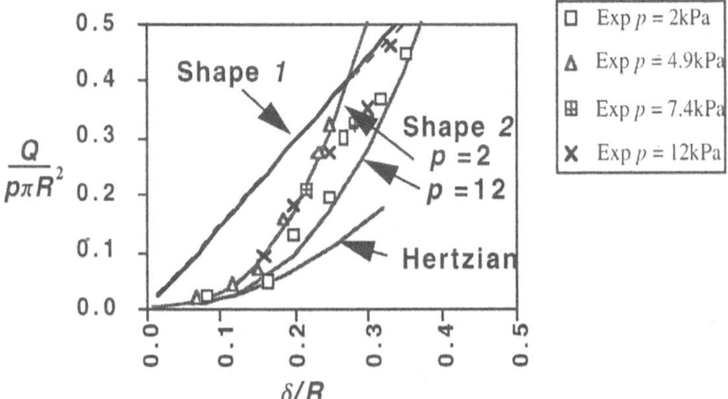


*Figure 5.* Blade loading of tube. The experimental results for four levels of pressure are shown, with the results from cross section approximation Shape 1 and Shape 2. The later calculation includes elastic energy. For small displacements, Shape 1 is far too stiff, while the local "Hertzian" solution is too soft. Shape 2 provides the correct transition to the Shape 1 behavior for $\delta/R > 0.3$.

The theoretical values show a small dependence on pressure, meaning that the elastic factor in Eq. 31 is significant. However, the experimental values do not show a clear dependence on pressure. For $\delta^* > 0.3$, the results from Shape 1 become the lower energy solution.

## 7. Conclusions

The large deformation of a tube with internal pressure is a challenging problem. However, in the recent work (Fay and Steele, 1999, 2000), it was found that deformation can be characterized in a simple and straight–forward manner by considering basic principles of the behavior of thin shells. Subsequently, little numerical work is needed for theoretical results that are in reasonable agreement with the experiments. In the present study, we have focussed on the problem of symmetric blade loading for relatively small displacements, for which the inextensional solution in Fay and Steele (2000) is too stiff. The primary reason is that the proper behavior consists of some elastic stretching of the wall, which we find can be included with a little extra work and a modified shape function. So an important conclusion is that in the class of problems of large deformation of inflated tubes with a high modulus wall, the deformation is primarily inextensional, but there can be situations in which some elasticity of the wall must be included. The present problem may serve as an interesting benchmark problem for finite element computations.

The present approach for a constant value of internal pressure can be easily extended to the case of a tube filled completely with either a compressible or incompressible fluid. Further generalizations in the loading and constraint seem possible. An interesting case is the tube on a rigid surface, loaded from one side by a curved surface. This could be a useful addition to the techniques of manipulation of cochlear hair cells for the determination of mechanical properties in Tolomeo, et al. (1996).

The general conclusion is that the volume calculation, based on the observations and basic principles of shell behavior, does a good job in reducing the large displacement problems under consideration to a trivial computation.

## 8. References


Ashwell, D.G. (1960). On the large deflection of a spherical shell with an inward point load, in *Theory of Elastic Shells*, W. Koiter, ed., (North Holland, Amsterdam) 44-63.

Barber, J.R. (1992). *Elasticity*, (Kluwer Academic Publishers, Dordrecht).

Bänziger, H. (1971). Extension and coiling of the lepidopterous proboscis, *Mitt. Schweiz. Ent. Ges.*, **43** (3,4): 226-239.

Cao, J. and Boyse, M.C. (1997). Wrinkling behavior of rectangular plates under lateral constraint, *Inter. J. Solids Structures*, **34** (2): 153-176.

Epstein, M., and Forcinito, M.A. (1999). Numerical analysis of wrinkling anisotropic elastic membranes, *Research Report,* University of Calgary.

Fay, J. P., and Steele, C. R. (1999). Forces for rolling and asymmetric pinching of pressurized cylindrical tubes, *J. Spacecraft and Rockets*, **36** (4): 531-537.

Fay, J.P., and Steele, C.R. (2000). Bending and symmetric pinching of pressurized tubes, to appear in *Inter. J. Solids Structures* (Koiter Special Issue).

Glaser, A.E., and Vincent, J.F.V. (1979). The autonomous inflation of insect wings, *J. Insect Physiol.* **25**:315-318.

Greschik, G., and Mikulas, M.M. (1998). Scale model testing of nonlinear phenomena with emphasis on thin film depopulates, to appear in Proceedings, Symposium on Deployable Structures: Theory and Applications, Cambridge, (Kluwer).

Haseganu, E.M., and Steigmann, D.J. (1994). Theoretical flexural response of a pressurized cylindrical membrane, *Int. J. Solids Struct.* **31**: 27-50.

Jenkins, C.H. (1991). Nonlinear dynamic response of membranes: State of the art-Update, *Appl. Mech. Rev.* **44**(7), S41-48.

Libai, A., and Simmonds, J.G. (1998). *The nonlinear theory of elastic shells*, Cambridge University Press.

Lou, M.C., and Feria, V.A. (1998). Development of space inflatable/rigidizable structures technology, to appear in Proceedings, Symposium on Deployable Structures: Theory and Applications, Cambridge, (Kluwer)

Lukasiewicz, S.A., and Glockner, P.G. (1984). Stability of lofty air-supported cylindrical membranes, *J. Struct. Mech.*, **12** (4): 543-555.

Porgorelov, A.V. (1960). *Geometrical Theory of Stability of Shells* (Nauka, Moscow) (in Russian).

Qui, Y., Kim, S., and Pence, T. (1994). Plane strain buckling and wrinkling of neo-Hookean laminates, *Inter. J. Solids Structures*, **31** (8): 1149-1178.

Rimrott, F.P.J. (1970). Querschnittsverformung bei Torsion offener Profile, *ZAMM* **50**, 775-778.

Steele, C.R., and Fay, J.P. (1998). Inflation of rolled tubes, to appear in Proceedings, Symposium on Deployable Structures: Theory and Applications, Cambridge, (Kluwer).

Stein, M., and Hedgepeth, J.M. (1961). Analysis of partly wrinkled membranes, NASA TN D-813.

Szyszkowski, W., and Glockner, P.G. (1990). The use of membrane structures in space, *Inter. J. Space Structures* **5** (2): 106-129.

Taber, L.A. (1982). Large deflection of a fluid-filled spherical shell under a point load, *J. App. Mech.* **49**, 121-128

Tolomeo, J.A., Steele, C.R., and Holley, M.C. (1996). "Mechanical properties of the lateral cortex of mammalian auditory outer hair cells", *Biophysical Journal*, 71, 421-429.

Tsoi, S.H.H, (1997). Modeling and simulation of inflatable space structures, Engineer's Thesis, Stanford University.

# ON LATERAL BUCKLING OF END-LOADED CANTILEVERS WITH TRANSVERSE SHEAR DEFORMATIONS

FREDERIC Y.M. WAN
*Department of Mathematics*
*University of California*
*Irvine, CA 92697*

## 1. Introduction

The main features of the problem of lateral buckling of transversely loaded beams were successfully treated by A.G. M. Michell [1] and L. Prandtl [2] using appropriate ad hoc considerations. Subsequently, H. Reissner [3] derived the equations governing this problem by appropriate specialization of Kirchhoff's general theory of space-curved beams. In separate ways, Reissner [3] and his assistant M.K. Grober [4] reduced the lateral buckling problem to the solution of a boundary value problem for a third order linear differential equation. Specific application of this theory to the case of a narrow rectangular cross-section beam was carried out by K. Federhofer [5]. More recent development of the lateral buckling problem of beams can be found in [6-9] and references therein. In [9], one-dimensional theories of beams were derived from a three-dimensional theory of elasticity by way of the principle of minimum potential energy. The various cases analyzed in that publication to study the effect of finite deformation with or without warping stiffness all assumed the cantilever is rigid with respect to transverse shear deformation. The present paper complements [9] by studying the impact of non-vanishing transverse shear strains (including nonlinear terms) on the critical load for lateral buckling of cantilevers. For the special case of the vanishing warping deformation, it is found that the nonlinear terms in transverse shear strains and pre-buckling deformation effects should be retained as long as transverse shear deformation is important.

## 2. Variationally Derived Equations for Cantilevers

Similar to the approach of [9], a set of one-dimensional differential equations governing finite deformations of prismatic elastic bodies is derived from three-dimensional theory of elasticity through the use of the variational equation

$$\delta\left\{\frac{1}{2}\iiint(E\epsilon_z^2+G\gamma_x^2+G\gamma_y^2)dxdydz\right\}=0 \qquad (2.1)$$

where the usual Young's modulus $E$ and shear modulus $G$ are known functions of the cross-sectional coordinates $x$ and $y$. In adopting the particular variational functional (2.1) as the appropriate strain energy of the prismatic body, it is tacitly assumed that the other three strain components of the elastic body are either negligibly small or identically zero for the class of problems of interest.



*D. Durban et al. (eds.), Advances in the Mechanics of Plates and Shells*, 343–356.

The three non-vanishing strain components $\epsilon_z, \gamma_x$, and $\gamma_y$ are defined in terms of the axial displacement component $\tilde{w}$ in the direction of the axial coordinate $z$ along the length of the prismatic body and the two cross-sectional displacement components $\tilde{u}$ and $\tilde{v}$ in the direction of $x$ and $y$, respectively. For lateral deformation and buckling of cantilevers, we take the following to be the approximate strain-displacement relations

$$\epsilon_z = \tilde{w}_{,z} + \frac{1}{2}(\tilde{u}_{,z}^2 + \tilde{v}_{,z}^2) \tag{2.2}$$

$$\gamma_x = \tilde{u}_{,z} + \tilde{w}_{,x} + \eta \tilde{v}_{,x}\tilde{v}_{,z}, \qquad \gamma_y = \tilde{v}_{,z} + \tilde{w}_{,y} + \eta \tilde{u}_{,y}\tilde{u}_{,z} \tag{2.3}$$

where the displacement components are approximated by

$$\tilde{w} = w(z) + \alpha(z)x + \beta(z)y + \lambda(z)\psi(x,y) \tag{2.4}$$

$$\tilde{u} = u(z) - \theta(z)y - \frac{1}{2}\theta^2(z)x, \qquad \tilde{v} = v(z) + \theta(z)x - \frac{1}{2}\theta^2(z)y \tag{2.5}$$

with $w$, $u$, $v$, $\alpha$, $\beta$, $\theta$ and $\lambda$ being functions of $z$ only and the Saint-Venant warping function $\psi$ being a function of $x$ and $y$ only. Analogous to [9], we use the constant parameter $\eta$ to allow for the inclusion or exclusion of the relevant nonlinear terms in the approximate displacement expressions. In [9], it was assumed that the prismatic body under consideration is rigid with respect to transverse shear with the conditions $\gamma_x = \gamma_y = 0$ incorporated in the variational equations as constraints through the Lagrange multipliers $Q_x$ and $Q_y$ (with the quantity $[G(\gamma_x)^2 + G(\gamma_y)^2]$ replaced by $[Q_x(\tilde{u}_{,z} + \tilde{w}_{,x} + \eta\tilde{v}_{,x}\tilde{v}_{,z}) + Q_y(\tilde{v}_{,z} + \tilde{w}_{,y} + \eta\tilde{u}_{,y}\tilde{u}_{,z})]$). The multipliers $Q_x$ and $Q_y$ have the interpretation of transverse force over the cross section of the prismatic body.

In this paper, we allow for transverse shear deformations so that there are no constraints in the extremization of the energy functional. With the effect of warping already studied extensively in [9], we will be concerned here mainly with a study of the effect of transverse shear deformability, with the nonlinear terms in $\gamma_x$ and $\gamma_y$ and with the pre-buckling deformations since their effects on the buckling load of prismatic bodies were not previously treated anywhere.

With the only non-vanishing stress-strain relations

$$\sigma_z = E\epsilon_z, \qquad \tau_x = G\gamma_x, \qquad \tau_y = G\gamma_y \tag{2.6}$$

giving the three non-vanishing stress components, the Euler differential equations of (2.1) are the following seven one-dimensional differential equations of equilibrium:

$$F' = 0, \quad M_x' - Q_x = 0, \quad M_y' - Q_y = 0, \quad R' - S = 0. \tag{2.7}$$

$$(Q_x - \eta\theta Q_y + u'F - \theta' M_y - \theta\theta' M_x)' = 0, \tag{2.8}$$

$$(Q_y + \eta\theta Q_x + v'F + \theta' M_x - \theta\theta' M_y)' = 0, \tag{2.9}$$

$$(1+\eta\theta^2)T' + (\theta' N - u'M_y + v'M_x)' + \eta(u'Q_y - v'Q_x) - \theta[v'M_y + u'M_x - \theta\theta' N + (1-\eta)Z]' = 0 \quad (2.10)$$

where $(\ \ )' = d(\ \ )/dz$ and where the resultant force and moment quantities in these equations are defined by the relations

$$(F, M_x, M_y, R, N) = \iint (1, x, y, \psi, x^2+y^2)\sigma_z dx\, dy \quad (2.11)$$

$$(Q_x, Q_y, T, S, Z) = \iint (\tau_x, \tau_y, x\tau_y - y\tau_x, \psi_{,x}\tau_x + \psi_{,y}\tau_y, x\tau_x + y\tau_y) dx\, dy \quad (2.12)$$

When the $x$-axis and $y$-axis are axes of geometrical an material symmetry of the cross section (so that $E(-x,y)\psi(-x,y) = -E(x,y)\psi(x,y)$, etc.), we have the following one-dimensional constitutive relations for these resultant quantities (see [6]):

$$F = A_E\epsilon_F + I_{pE}\epsilon_N, \quad N = I_{pE}\epsilon_F + I_{pp}\epsilon_N \quad (2.13)$$

$$M_x = I_x\kappa_x, \quad M_y = I_y\kappa_y, \quad R = A_\psi\kappa_R \quad (2.14)$$

$$Q_x = A_G\Gamma_x, \quad Q_y = A_G\Gamma_y \quad (2.15)$$

$$T = I_{pG}\kappa_T - J\kappa_S, \quad S = J(-\kappa_T + \kappa_S), \quad Z = I_{pG}\kappa_Z \quad (2.16)$$

where

$$\epsilon_F = w' + \frac{1}{2}[(u')^2 + (v')^2], \quad \epsilon_N = \frac{1}{2}(\theta')^2(1+\theta^2) \quad (2.17)$$

$$\kappa_x = \alpha' + (v'\theta' - u'\theta'\theta), \quad \kappa_y = \beta' - (u'\theta' + v'\theta'\theta) \quad (2.18)$$

$$\Gamma_x = u' + \alpha + \eta\theta v', \quad \Gamma_y = v' + \beta - \eta\theta u', \quad \kappa_R = \lambda' \quad (2.19)$$

$$\kappa_T = \theta'(1+\eta\theta^2), \quad \kappa_S = \lambda, \quad \kappa_Z = -(1-\eta)\theta\theta' \quad (2.20)$$

and

$$\{A_E, I_x, I_y, I_{pE}, I_{pp}, A_\psi\} = \iint \{1, x^2, y^2, x^2+y^2, (x^2+y^2)^2, \psi^2\} E dx\, dy \quad (2.21a)$$

$$\{A_G, I_{pG}, J\} = \iint \{1, x^2+y^2, \psi_{,x}^2 + \psi_{,y}^2\} G dx\, dy \quad (2.21b)$$

Note that the relations (2.7)-(2.10) and (2.13)-(2.20) are exact consequences of (2.1) - (2.5) while some higher order nonlinear terms in the displacement variables have been neglected in the corresponding relations in [9]. As to be seen later in section 6, at least one term in (2.10), $(-\theta v' M_y)$, which appears to be small of higher order, will actually contribute significantly to the buckling load not anticipated in [7,9].

At the clamped end of the cantilever, $z = L$, we have the no displacement conditions of

$$w(L) = \alpha(L) = \beta(L) = \theta(L) = \lambda(L) = u(L) = v(L) = 0 \quad (2.22)$$

At the loaded end, z = 0 , the Euler boundary conditions require

$$F(0) = M_x(0) = M_y(0) = R(0) = 0$$

$$[Q_x - \eta\theta Q_y]_{z=0} = F_x, \quad [Q_y + \eta\theta Q_x]_{z=0} = F_y \tag{2.23}$$

$$[(1+\eta\theta^2)T + (1+\theta^2)\theta' N + (\eta - 1)\theta Z]_{z=0} = 0$$

## 3. Equations for Buckling

For buckling of the prismatic body due to a lateral end force in the $y$ direction ($F_z = F_x = 0, F_y = P$), the non-vanishing stress and displacement measures of the corresponding pre-buckled state are given by

$$Q_y = P, \qquad M_y = Pz \tag{3.1}$$

and

$$Q_y = A_G(v_p' + \beta_p), \qquad M_y = I_y \beta_p' \tag{3.2}$$

From (3.2) and the end conditions $v_p(L) = \beta_p(L) = 0$, we obtain

$$\beta_p = \frac{P}{2I_y}(z^2 - L^2), \quad v_p = \frac{P}{6I_y}\left\{(L^3 - z^3) - 3L^2(L-z) - \frac{6I_y}{A_G}(L-z)\right\} \tag{3.3}$$

The corresponding (linearized) buckling equations are

$$[Q_x - P(\eta\theta + z\theta'])]' = 0, \quad M_x' - Q_x = 0, \quad R' - S = 0.$$

$$[T + P\{\eta u - (zu')\} + v_p' M_x]' - (v_p' Pz)'\theta - \eta v_p' Q_x = 0 \tag{3.4}$$

where

$$Q_x = A_G(u' + \alpha + \eta v_p'\theta), \quad M_x = I_x(\alpha' + v_p'\theta')$$

$$T = I_{pG}\theta' - J\lambda, \quad S = J(\lambda - \theta'), \quad R = A_\psi \lambda'. \tag{3.5}$$

Note that terms involving pre-buckling deformations are retained in (3.4) while they were neglected in [9]. Though the influence of pre-buckling displacement was analyzed in [7], it was for the case of no transverse shear deformation and the simple linear relations $u' = -\alpha$ and $v' = -\beta$. The first equation of (3.4) may be integrated once immediately to obtain

$$Q_x = P(\eta\theta + z\theta') \tag{3.6}$$

where the relevant boundary condition in (2.33) (with $F_x = 0$) has been used to determine the constant of integration. With (3.6), the second equation of (3.4) may be written as

$$M_x' - P(\eta\theta + z\theta') = 0 \tag{3.7}$$

Upon substituting (3.5) into (3.4), we obtain an eighth order system of differential equations for the four displacement measures $u, \theta, \alpha$, and $\lambda$ :

$$\{A_G(u'+\alpha+\eta v_p'\theta)-P(\eta\theta+z\theta')\}'=0, \qquad \{I_x(\alpha'+v_p'\theta')\}'-P(\eta\theta+z\theta')\}=0$$

$$\{I_{pG}\theta'-J\lambda+P(\eta u-zu')+v_p'I_x(\alpha'+v_p'\theta')\}-(Pzv_p')'\theta-\eta v_p'P(\eta\theta+z\theta')=0 \quad (3.8)$$

$$\{A_\psi\lambda'\}'+J(\theta'-\lambda)=0$$

where we have used (3.7) instead of the second equation of (3.4). The eighth order system (3.8) is supplemented by the eight buckling boundary conditions:

$$u(L)=\theta(L)=\alpha(L)=\lambda(L)=0 \qquad (3.9a)$$

$$[Q_x-\eta P\theta]_{z=0}=M_x(0)=T(0)=R(0)=0. \qquad (3.9b)$$

The homogeneous boundary value problem defined by (3.8) and (3.9) constitutes an eigenvalue problem with the buckling (end force) load $P$ as the eigenvalue parameter.

In addition to omitting terms involving pre-buckling deformations, the analyses in [9] were limited to the special case $\Gamma_x=\Gamma_y=0$, i.e., the cross section of the prismatic body is rigid and does not allow transverse shear deformations. In that case, we may eliminate $u$ from the first three equations of (3.5) to obtain a sixth order system for $\theta,\alpha$ and $\lambda$. In this paper, we complement the study in [9] by allowing for transverse shear deformations but not cross-sectional warping so that $\psi=0$. The latter restriction implies $A_\psi=J=0$ and the eighth order system (3.4) reduces to a sixth order system:

$$\{A_G(u'+\alpha+\eta v_p'\theta)-P(\eta\theta+z\theta')\}'=0$$

$$\{I_x(\alpha'+v_p'\theta')\}'-P(\eta\theta+z\theta')=0 \qquad (3.10)$$

$$\{I_{pG}\theta'+P(\eta u-zu')\}'-P(zv_p')'\theta+(1-\eta)v_p'P(\eta\theta+z\theta')+v_p''I_x(\alpha'+v_p'\theta)=0$$

while the last differential equation in (3.8) is trivially satisfied. Note that terms involving the pre-buckling displacement $v_p$ in (3.8) were omitted in the buckling analysis of [9]. We keep them in (3.8) to allow for an analysis of the effect of these terms later.

## 4. Linear Transverse Shear Deformability

We wish to investigate first the effect of transverse shear deformability on the buckling of cantilevers. For this purpose, we will omit terms involving the pre-buckling displacement $v_p$. In that case, the system (3.10) further reduces to

$$A_G(u'+\alpha)-P(\eta\theta+z\theta')=0 \qquad (4.1a)$$

$$\{I_x\alpha'\}'-P(\eta\theta+z\theta')=0, \qquad \{I_{pG}\theta'+P(\eta u-zu')\}'=0 \qquad (4.1b,c)$$

The fifth order system (4.1) is supplemented by the boundary conditions

$$u(L)=\theta(L)=\alpha(L)=M_x(0)=T(0)=0 \qquad (4.2)$$

For $\eta = 0$ so that $\Gamma_x$ and $\Gamma_y$ are linearly related to the displacement measures, (4.1) may be written as

$$A_G(u' + \alpha) - Pz\theta' = 0, \quad I_{pG}\theta' - Pzu' = 0, \quad (I_x\alpha')' - Pz\theta' = 0. \tag{4.3}$$

where $T(0) = 0$ has been used to determine the unknown constant in the second equation in (4.3) after integration. The first equation is then used to eliminate $u'$ from the second equation to get

$$\left(I_{pG} - \frac{P^2z^2}{A_G}\right)\theta' + Pz\alpha = 0. \tag{4.4}$$

The coupled system of (4.4) and the third equation in (4.3) may be further reduced to a single equation for

$$\left(I_{pG} - \frac{P^2z^2}{A_G}\right)(I_x\alpha')' + P^2z^2\alpha = 0. \tag{4.5}$$

To cast (4.5) in dimensionless form, we set

$$\zeta = \frac{z}{L}, \quad \sigma = \frac{PL^2}{\sqrt{I_xI_{pG}}}, \quad \epsilon_G = \frac{I_x}{A_GL^2}. \tag{4.6}$$

In terms of these dimensionless quantities, (4.5) becomes

$$(1 - \epsilon_G\sigma^2\zeta^2)\alpha^{\bullet\bullet} + \sigma^2\zeta^2\alpha = 0 \tag{4.7}$$

with $(\ )^{\bullet} = d(\ )/d\zeta$ . The second order differential equation (4.7) is supplemented by the two boundary conditions

$$\alpha^{\bullet}(0) = \alpha(1) = 0 \tag{4.8}$$

as previously noted in (4.2) with $\alpha^{\bullet}(0) = 0$ corresponding to $M_x(0) = 0$ (see (3.5)). The parameter $\epsilon_G$ is of the order of $Ea^2/GL^2$, where $a$ is a typical lineal dimension of the cross-section of the cantilever. For a long prismatic body, we have typically $a^2/L^2 << 1$ and, with $E/G$ not large compared to unity, we have $I_x/A_GL^2 << 1$. A perturbation solution of the eigenvalue problem (4.5) and (4.6) is therefore appropriate.

Let

$$\{\alpha(\zeta;\epsilon_G), \sigma(\epsilon_G)\} = \sum_{n=0}^{\infty}\{\alpha_n(\zeta), \sigma_n\}\epsilon_G^n. \tag{4.9}$$

We have from (4.7) and (4.8)

$$[\alpha_0^{\bullet\bullet} + \sigma_0^2\zeta^2\alpha_0] + \epsilon_G[\alpha_1^{\bullet\bullet} + \sigma_0^2\zeta^2\alpha_1 + 2\sigma_0\sigma_1\zeta^2\alpha_0 - \sigma_0^2\zeta^2\alpha_0^{\bullet\bullet}] + O(\epsilon_G^2) = 0 \tag{4.10}$$

and

$$[\alpha_0^{\bullet}(0) + \epsilon_G\alpha_1^{\bullet}(0) + \cdots] = [\alpha_0(1) + \cdots] = 0. \tag{4.11}$$

Since (4.10) and (4.11) must be satisfied for all $\epsilon_G(<< 1)$, we must have

$$\alpha_0^{\bullet\bullet} + \sigma_0^2\zeta^2\alpha_0 = 0 \tag{4.12}$$

$$\alpha_0^{\bullet}(0) = \alpha_0(1) = 0 \tag{4.13}$$

and

$$\alpha_1^{\bullet\bullet} + \sigma_0^2\zeta^2\alpha_1 + [2\sigma_0\sigma_1\zeta^2 + \sigma_0^4\zeta^4]\alpha_0 = 0 \tag{4.14}$$

$$\alpha_1^{\bullet}(0) = \alpha_1(1) = 0 \tag{4.15}$$

and so on.

The solution to the $O(1)$ problem (4.12)-(4.13) is known to be

$$\alpha_0 = \alpha_0^{(n)} \equiv \sqrt{\zeta}\, J_{-1/4}(\sigma_0^{(n)}\zeta^2/2) \qquad (n = 1, 2, 3, \cdots) \tag{4.16}$$

where $\sigma_o^{(n)}/2$ is the nth zero of the relevant Bessel function:

$$J_{-1/4}(t) = 0 \tag{4.17}$$

with $\sigma_o^{(1)} \cong 0.4013$. Since, for any particular eigenvalue $\sigma_0^{(n)}$, the solution of the $O(1)$ problem given by (4.16) is also a solution of the homogeneous differential equation corresponding to (4.14), we must satisfy the solvability condition

$$\int_0^1 \{2\sigma_0^{(n)}\sigma_1\zeta^2 + [\sigma_0^{(n)}]^4\zeta^4\}\{\alpha_0^{(n)}\}^2 d\zeta = 0 \tag{4.18}$$

in order for the $O(\epsilon_G)$ problem (4.14)-(4.15) to have a solution. The condition (4.17) determines $\sigma_1$ to be

$$\frac{\sigma_1}{\sigma_0^{(n)}} = \frac{\sigma_1^{(n)}}{\sigma_0^{(n)}} \equiv -\frac{1}{2}\left[\sigma_0^{(n)}\right]^2 \frac{\int_0^1 [\alpha_0^{(n)}]^2\zeta^4 d\zeta}{\int_0^1 [\alpha_0^{(n)}]^2\zeta^2 d\zeta} \tag{4.19}$$

From the results for the weighted integrals of $I_{-1/4}(t)$ (with weight $t^n$) obtained in [7], we have

$$\frac{\sigma_1^{(1)}}{\sigma_0^{(1)}} = -\sigma_0^{(1)} \frac{\int_0^{\sigma_0^{(1)}/2} [J_{-1/4}(t)]^2 t^2 dt}{\int_0^{\sigma_0^{(1)}/2} [J_{-1/4}(t)]^2 dt} \cong -\sigma_0^{(1)} \frac{0.483}{0.650} \cong -0.743\sigma_0^{(1)} \cong -2.98 \tag{4.20}$$

Thus, the buckling load (corresponding to the lowest eigenvalue $\sigma_0^{(1)}$) of the linear transverse shear strain model is given by

$$\sigma = \sigma^{(1)} \cong \sigma_0^{(1)}[1 - 2.98\epsilon_G + O(\epsilon_G^2)]. \tag{4.21}$$

The $O(\epsilon_G)$ correction term is not insignificant for a cantilever aspect ratio of 1/10.

The results of this section are identical to those obtained in [7], as they should, since the latter study worked with the same set of govering differential equations as (4.3). We reproduce the analysis and results here for subsequent comparisons with those for $\eta = 1$ and those for a model which includes the influence of pre-buckling deformations. The analysis of this section also allows us to omit the details of subsequent calculations for the new studies with $\eta = 1$.

## 5. Non-Linear Transverse Shear Deformability

For $\eta = 1$ so that $\Gamma_x$ and $\Gamma_y$ are nonlinear in the displacement measures, (4.1) may be written as

$$A_G(u' + \alpha) - P(z\theta)' = 0, \quad I_x\alpha' - Pz\theta = 0, \quad (I_{pG}\theta')' - Pzu'' = 0 \tag{5.1}$$

where $M_x(0) = 0$ has been used to determine the unknown constant in the second equation in (5.1) after integration. The first equation is then used to eliminate $u$ from the third equation to get

$$(I_{pG}\theta')' + Pz\left[\alpha - \frac{P}{A_G}(z\theta)'\right]' = 0. \tag{5.2}$$

The coupled system of (5.2) and the second equation in (5.1) may be further reduced to a single equation for

$$\left[\left(I_{pG} - \frac{P^2}{A_G}z^2\right)\theta'\right]' + \frac{P^2}{I_x}z^2\theta = 0. \tag{5.3}$$

In terms of the dimensionless quantities in (4.6), equation (5.3) becomes

$$[(1 - \epsilon_G\sigma^2\zeta^2)\theta^\bullet]^\bullet + \sigma^2\zeta^2\theta = 0 \tag{5.4}$$

with $(\ )^\bullet = d(\ )/d\zeta$. The second order differential equation (5.4) is supplemented by the two boundary conditions

$$\theta^\bullet(0) = \theta(1) = 0 \tag{5.6}$$

as previously noted in (4.2) with $\theta^\bullet(0) = 0$ corresponding to $T(0) = 0$. Whenever $E/G$ is not large compared to unity, we have for a long prismatic body $\epsilon_G = I_x/A_GL^2 << 1$. A perturbation solution of the eigenvalue problem (5.4) and (5.5), denoted by $\{\bar{\theta}(\zeta, \epsilon_G), \bar{\sigma}(\epsilon_G)\}$, is again appropriate.

With the help of the parametric expansions

$$\{\bar{\theta}(\zeta; \epsilon_G), \bar{\sigma}(\epsilon_G)\} = \sum_{n=0}^{\infty}\{\theta_n(\zeta), \bar{\sigma}_n\}\epsilon_G^n \tag{5.7}$$

we have the following sequence of simpler eigenvalue problems for $\{\theta_n(\zeta), \sigma_n\}$:

<u>The $O(1)$ Problem:</u>

$$\theta_0^{\bullet\bullet} + \bar{\sigma}_0^2 \zeta^2 \theta_0 = 0 \tag{5.8}$$

$$\theta_0^{\bullet}(0) = \theta_0(1) = 0$$

<u>The $O(\epsilon_G)$ Problem:</u>

$$\theta_1^{\bullet\bullet} + \bar{\sigma}_0^2 \zeta^2 \theta_1 + \{2\bar{\sigma}_1 \bar{\sigma}_0 \zeta^2 \theta_0 - \bar{\sigma}_0^2 (\zeta^2 \theta_0^{\bullet})^{\bullet}\} = 0 \tag{5.9}$$

$$\theta_1^{\bullet}(0) = \theta_1(1) = 0$$

and so on.

The solution of the $O(1)$ problem is again

$$\theta_0(\zeta) = \theta_0^{(n)}(\zeta) \equiv \sqrt{\zeta}\; J_{-1/4}(\bar{\sigma}_0^{(n)} \zeta^2/2) \qquad (n = 1, 2, \cdots) \tag{5.10}$$

where $\sigma_0^{(n)}/2$ is the $n^{th}$ zero of $J_{-1/4}(t) = 0$. For buckling, we are interested in the lowest eigenvalue so that $\bar{\sigma}_0 = \sigma_0^{(1)} \approx 4.013$. Similar to the linear transverse shear strain model, $\theta_0^{(1)}$ is also the solution for the homogeneous ODE corresponding to (5.9) for $\bar{\sigma}_0 = \sigma_0^{(1)}$. For the inhomogeneous ODE (5.9) to have a solution, $\bar{\sigma}_1$ must be chosen to satisfy the conventional solvability condition:

$$\int_0^1 \{2\bar{\sigma}_1 \sigma_0^{(1)} \zeta^2 [\theta_0^{(1)}]^2 - (\sigma_0^{(1)})^2 \theta_0^{(1)} [\zeta^2 (\theta_0^{(1)})^{\bullet}]^{\bullet}\} d\zeta = 0 \tag{5.11}$$

or

$$\frac{\bar{\sigma}_1}{\sigma_0^{(1)}} = \frac{1}{2} \frac{\int_0^1 \theta_0^{(1)} [\zeta^2 (\theta_0^{(1)})^{\bullet}]^{\bullet} d\zeta}{\int_0^1 [\theta_0^{(1)}]^2 \zeta^2 d\zeta} \tag{5.12}$$

With

$$\theta_0^{(1)} [\zeta^2 (\theta_0^{(1)})^{\bullet}]^{\bullet} = [\zeta (\theta_0^{(1)})^2]^{\bullet} - \{1 + [\sigma_0^{(1)}]^2 \zeta^4\} [\theta_0^{(1)}]^2 \tag{5.13}$$

we may re-write (5.12) as

$$\frac{\bar{\sigma}_1}{\sigma_0^{(1)}} = -\frac{1}{2} \frac{\int_0^1 \zeta [1 + (\sigma_0^{(1)})^2 \zeta^4] \left[J_{-1/4}\left(\frac{1}{2}\sigma_0^{(1)} \zeta^2\right)\right]^2 d\zeta}{\int_0^1 \zeta^3 \left[J_{-1/4}\left(\frac{1}{2}\sigma_0^{(1)} \zeta^2\right)\right]^2 d\zeta} \tag{5.14}$$

$$= -\frac{\sigma_0^{(1)}}{4} \frac{\int_0^{\sigma_0^{(1)}/2} \{1 + 4t^2\} [J_{-1/4}(t)]^2 dt}{\int_0^{\sigma_0^{(1)}/2} [J_{-1/4}(t)]^2 t dt}$$

It follows from the results of [7] that

$$\bar{\sigma}_1 \simeq -\frac{1}{4} [\sigma_0^{(1)}]^2 \left\{\frac{1.806 + 4(0.483)}{0.650}\right\} \simeq -1.438 [\sigma_0^{(1)}]^2 \tag{5.15}$$

We have then the following perturbation solution for the buckling load of the nonlinear transverse shear strain model:

$$\bar{\sigma} \simeq \sigma_0^{(1)}\{1 - 1.438\sigma_0^{(1)}\epsilon_G + O(\epsilon_G^2)\} = \sigma_0^{(1)}\{1 - 5.771\epsilon_G + O(\epsilon_G^2)\}. \quad (5.16)$$

Thus, when nonlinear terms are included in the strain-displacement relations for the transverse shearing strains, the $O(\epsilon_G)$ correction to the Michell-Prandtl solution for the buckling load is twice the value when the nonlinear terms are not included. For relatively long, homogeneous, isotropic, elastic prismatic bodies with $a/L = 1/10$, the correction term is now about 15% of the Michell-Prandtl solution.

## 6. The Effect of Pre-Buckling Deformations

In this section, we analyze the effect of pre-buckling deformations in the nonlinear transverse shear strain model for prismatic bodies without warping. For this case, the system (3.10) with $\eta = 1$ applies. Upon integrating the first two equations of (3.10) and oberving the relevant boundary conditions in (3.9), we obtain

$$u' = -(\alpha + v_p'\theta) + \frac{P}{A_G}(z\theta)', \qquad \alpha' + v_p'\theta' = \frac{Pz}{I_x}\theta \quad (6.1a, b)$$

$$(I_{pG}\theta' - Pzu')' - P(zv_p')\theta + v_p''Pz\theta + Pu' = 0. \quad (6.1c)$$

We now use the first and second equation of (6.1) to eliminate $u'$ and $\alpha'$ from the third leaving us with a single second order differential equation for $\theta$:

$$\left(I_{pG}\theta' - \frac{P^2z^2}{A_G}\,\theta'\right)' + \frac{P^2z^2}{I_x}\,\theta - P\theta(v_p' - zv_p'') = 0 \quad (6.2)$$

where $v_p'$ is as given in (3.3). This second order differential equation is supplemented by the boundary conditions $\theta'(0) = \theta(L) = 0$. Equation (6.2) can be written in dimensionless form with the help of the dimensionless quantities in (4.6) and

$$\epsilon_v = I_x/I_y. \quad (6.3)$$

The primary unknown $\theta$ is therefore determined by the dimensionless differential equation

$$[(1 - \epsilon_G\sigma^2\zeta^2)\theta^\bullet]^\bullet + \sigma^2\left[\left(1 - \frac{1}{2}\epsilon_v\right)\zeta^2 - \left(\frac{1}{2}\epsilon_v + \epsilon_G\right)\right]\theta = 0 \quad (6.4)$$

subject to $\theta^\bullet(0) = \theta(1) = 0$.

By setting $\epsilon_G = 0$, we recover the governing differential equation for cantilever buckling obtained in [7] when the prismatic bodies are known to be not transverse shear deformable. By allowing for transverse shear deformability while concurrently retaining terms involving pre-buckling displacement field $v_p$ and its derivatives, we are able to make the following observations for the first time:

1. While the effect of terms involving pre-buckling displacement $v_p$ terms may be $O(1)$ or small of higher order depending on the aspect ratio of the cantilevers cross section, $\epsilon_v$, (whether it is $O(1)$ or small by an order of magnitude), the effect of the transverse shear defomrability is always small of higher order for prismatic bodies as long as $E$ and $G$ are of the same order of magnitude (including homogeneous and isotropic elastic cantilevers) since we always have $a/L << 1$ for prismatic bodies.

2. When the effect of transverse shear deformability is significant and should be included in the model, then terms involving pre-buckling deformation must also be retained for consistency since these contribute a term in the final governing differential equation involving the effect of transverse shear strains, namely, the last term in the coefficient of $\theta$ proportional to $\epsilon_G$.

3. When we consider the effect of transverse shear deformability alone, there is a significant difference between the correction to the Michell-Prandtl buckling load given by the linear expression for the transverse shear strains and that by the nonlinear relations. However, the additional contribution from transverse shear deformability indirectly through terms involving pre-buckling displacements has the effect of offsetting the contribution of the nonlinear terms in the expressions for the transverse shear strains.

The validity of the first observation can be seen from the governing differential equation (6.4) for the problem. More specifically, we have from (6.3) the order of magnitude relation $\epsilon_v = I_x/I_y = O(a^2/b^2)$ for homogeneous, isotropic, elastic prismatic bodies where $a$ and $b$ are the lineal dimension in the $x$-direction and $y$-direction, respectively. The contribution of terms involving the pre-buckling displacement $v_p$ to (6.4) and the buckling load is small of higher order only if the cantilever cross section has a small aspect ration so that $a^2/b^2 << 1$.

To demonstrate the necessity to retain terms involving pre-buckling deformation when the effect of transverse shear deformability is considered significant, it suffices to limit our analysis to the case $\epsilon_v << 1$. The leading term solution for a perturbation solution in $\epsilon_v$ , denoted by $\{\tilde{\theta}, \tilde{\sigma}\}$, is determined by the eigenvalue problem

$$[(1 - \epsilon_G \tilde{\sigma}^2 \zeta^2)\tilde{\theta}^\bullet]^\bullet + \tilde{\sigma}^2(\zeta^2 - \epsilon_G)\tilde{\theta} = 0 \tag{6.5}$$

$$\tilde{\theta}^\bullet(0) = \tilde{\theta}(1) = 0 \tag{6.6}$$

Note that this problem differs from the one when terms involving the pre-buckling displacement $v_p$ are neglected.

We now seek a regular perturbation solution of this problem in the second small parameter $\epsilon_G$. The leading term solution is again determined by (5.8) and given by (5.10). Instead of (5.9), the $O(\epsilon_G)$ correction term is now determined by

$$\tilde{\theta}_1^{\bullet\bullet} + (\tilde{\sigma}_0^{(1)})^2\zeta^2\tilde{\theta}_1 + [2\sigma_0^{(1)}\tilde{\sigma}_1\zeta^2 - (\sigma_0^{(1)})^2]\theta_0^{(1)} - (\sigma_0^{(1)})^2[\zeta^2(\theta_0^{(1)})^\bullet]^\bullet = 0 \quad (6.7a)$$

$$\tilde{\theta}_0^\bullet(0) = \tilde{\theta}_1(1) = 0. \quad (6.7b)$$

The differential equation in (6.7) differs from that of (5.9) by a term associated with the pre-buckling displacement $v_p$ even if the aspect ratio $a/b$ is small so that $\epsilon_v = I_x/I_y$ is negligible. This confirms the second observation above indicating the necessity of retaining terms involving the pre-buckling displacement $v_p$ whenever the effect of transverse shear strains on the buckling load is significant.

Regarding the third observation above, we note that the solvability condition for (6.7) now leads to the following expression for $\tilde{\sigma}_1$:

$$\frac{\tilde{\sigma}_1}{\sigma_0^{(1)}} = \frac{\int_0^1[\theta_0^{(1)}]^2 d\zeta + \int_0^1[\zeta^2(\theta_0^{(1)})^\bullet]^\bullet\theta_0^{(1)} d\zeta}{2\int_0^1 \zeta^2[\theta_0^{(1)}]^2 d\zeta}$$

$$= -\sigma_0^{(1)} \frac{\int_0^{\sigma_0^{(1)}/2}[J_{-1/4}(t)]^2 t^2 dt}{\int_0^{\sigma_0^{(1)}/2}[J_{-1/4}(t)]^2 t dt} \cong -0.743\sigma_0^{(1)} \cong -2.98. \quad (6.8)$$

The condition (6.8) is identical to the corresponding expression for the $O(\epsilon_G)$ correction term for $\sigma_1$ in (4.19) with $n = 1$ and in (4.20). Hence, for the first two terms of a perturbation solution in $\epsilon_G$ without pre-buckling deformations, the linear transverse shear strain model gives a accurate approximation for the correction of the Michell-Prandtl buckling load than the nonlinear model, at least in the case where the aspect ratio $a/b$ is small so that $\epsilon_v = I_x/I_y$ is negligible.

## 7. On the Moment Equilibrium Equation (2.10)

We noted in section 2 that the seven equilibrium equations (2.7)-(2.10) are the exact consequence of the assumed strain-displacement relations (2.2)-(2.5) and the variational equation (2.1). In previous treatments of this lateral buckling problem for cantilevers, third and higher order nonlinear terms in these equilibrium equations are neglected. In particular, the terms $-\theta[v'M_y + u'M_x - \theta\theta' N]'$ are not included in the analysis of [7], [8] and [9]. While these terms appear to be of higher order in the unknowns, $-\theta[v'M_y]'$, for example, leads to a term $-\theta[v_p'Pz]'$ in the last of the equilibrium equation for the buckled state in (3.4). We see from (6.4) that this term contributes in a qualitatively significant way to the coefficient of $\theta$ of the governing differential equation for the determination of the buckling load. In fact, in the absence of the $-\theta[v_p'Pz]'$term, we would have

$$[(1 - \epsilon_G\tilde{\sigma}^2\zeta^2)\tilde{\theta}^\bullet]^\bullet + \tilde{\sigma}^2[\zeta^2(1 - 2\epsilon_v)]\tilde{\theta} = 0 \quad (7.1)$$

instead of (6.4). Thus, that portion of the effect of transverse shear deformation on the buckling load through pre-buckling deformation would be lost. We saw in section 6 that this effect is comparable to that induced by transverse shear deformability on the buckled state directly. For cases where $\epsilon_v = I_x/I_y = O(a^2/b^2)$ is not small by an order of magnitude, omitting the term $-\theta[v_p'Pz]'$ would change the buckling load significantly, not just a small perturbation since the difference between the differential equation in (6.4) and (7.1) is $-\sigma^2\theta[\epsilon_G - \epsilon_v(3\zeta^2 - 1)/2]$, which is $O(\sigma^2\theta)$ whenever $\epsilon_v$ is $O(1)$.

When the term $-\theta[v_p'Pz]'$ is included, the method of reduction of the buckling equations (3.10) and the relevant boundary conditions (3.9) for the linear transverse shear strain model (corresponding to $\eta = 0$) to an eigenvalue problem for a second order differential equation used effectively in [7] no longer applies. It appears that we would have to work with an eigenvalue problem for a fourth order system of two differential equations to determine the buckling load. In contrast, the reduction to a second order differential equation is still possible as shown in section 6 in the nonlinear transverse shear strain model corresponding to the case $\eta = 1$. Since both the nonlinear transverse shear strain terms and pre-buckling deformations contribute significantly to the buckling load when transverse shear strain effects are important, there is no reason to pursue a linear transverse shear strain model that includes pre-buckling deformations.

## 8. Concluding Remarks

The present study was intended to analyze the relevance or significance of (i) the nonlinear terms in the strain-displacement relations for the transverse shear strains, and (ii) the prebuckling deformation, on the buckling load of cantilevers subject to a transverse end force when the effect of the transverse shear deformations is sufficiently significant to necessitate a correction of the Michell-Prandtl solution. It was found (in sections 4 and 5) that the nonlinear terms in $\gamma_x$ and $\gamma_y$ contribute significantly to the buckling load in that they effectively double the magnitude of the correction to the Michell-Prandtl solution whenever warping and pre-buckling deformation are neglected. However, it was shown in section 6 that the the inclusion of terms involving the pre-buckling displacement $v_p$ in the buckling analysis further modifies this correction term, even if the aspect ratio of the cross section of the prismatic body is small so that terms involving $\epsilon_v = I_x/I_y$ may be neglected in the governing differential equation for the buckling problem. To order $\epsilon_G$, the additional modification resulting from the retention of $v_p$ terms is sufficiently substantial that it effectively offsets the effects (on the buckling load) of the nonlinear terms in the expressions for the transverse shear strains, at least for cases when the terms involving $\epsilon_v = I_x/I_y$ are negligible.

## REFERENCES

1. Michell, A.G.M. (1899) Elastic stability of long beams under transverse forces, *Phil. Mag.* (5th Series) 48, 298-309.

2. Prandtl, L. (1900) *Kipperscheinungen*, Dissertation der Universitat Munchen.

3. Reissner, H. (1904) Uber die Stabilitat der Biegung, *Sitz.-Ber. der Berliner Math. Gesellschaft* **3**, 53-56.

4. Grober, M.K. (1914) Ein Beispiel fur die Kirchhoff'schen Stabgleichungen, *Phys. Z.* **15**, 889-892.

5. Federhofer, K. (1931) Berechnung der Kipplasten gerader Stabe, *Sitz.-Ber. Akad. Wiss. Wien* **140**, 237-270.

6. Reissner, E. (1983) On a simple variational analysis of small finite deformations of prismatical beams, ZAMP **34**, 642-647.

7. Reissner, E. (1983) On some problems of buckling of prismatical beams under the influence of axial and transverse loads, ZAMP **34**, 649-667.

8. Reissner, E. (1984) On a variational analysis of finite deformations of prismatical beams and on the effect of warping stiffness on buckling loads, ZAMP **35**, 247-251.

9. Reissner, E., Reissner, J.E. and Wan, F.Y.M. (1987) On lateral buckling of end-loaded cantilevers, including the effect of warping stiffness, *Comp. Mech.* **2**, 137-147

# Mechanics

## *FLUID* MECHANICS AND ITS APPLICATIONS

*Series Editor*: R. Moreau

*Aims and Scope of the Series*

The purpose of this series is to focus on subjects in which fluid mechanics plays a fundamental role. As well as the more traditional applications of aeronautics, hydraulics, heat and mass transfer etc., books will be published dealing with topics which are currently in a state of rapid development, such as turbulence, suspensions and multiphase fluids, super and hypersonic flows and numerical modelling techniques. It is a widely held view that it is the interdisciplinary subjects that will receive intense scientific attention, bringing them to the forefront of technological advancement. Fluids have the ability to transport matter and its properties as well as transmit force, therefore fluid mechanics is a subject that is particularly open to cross fertilisation with other sciences and disciplines of engineering. The subject of fluid mechanics will be highly relevant in domains such as chemical, metallurgical, biological and ecological engineering. This series is particularly open to such new multidisciplinary domains.

1. M. Lesieur: *Turbulence in Fluids.* 2nd rev. ed., 1990 ISBN 0-7923-0645-7
2. O. Métais and M. Lesieur (eds.): *Turbulence and Coherent Structures.* 1991 ISBN 0-7923-0646-5
3. R. Moreau: *Magnetohydrodynamics.* 1990 ISBN 0-7923-0937-5
4. E. Coustols (ed.): *Turbulence Control by Passive Means.* 1990 ISBN 0-7923-1020-9
5. A.A. Borissov (ed.): *Dynamic Structure of Detonation in Gaseous and Dispersed Media.* 1991 ISBN 0-7923-1340-2
6. K.-S. Choi (ed.): *Recent Developments in Turbulence Management.* 1991 ISBN 0-7923-1477-8
7. E.P. Evans and B. Coulbeck (eds.): *Pipeline Systems.* 1992 ISBN 0-7923-1668-1
8. B. Nau (ed.): *Fluid Sealing.* 1992 ISBN 0-7923-1669-X
9. T.K.S. Murthy (ed.): *Computational Methods in Hypersonic Aerodynamics.* 1992 ISBN 0-7923-1673-8
10. R. King (ed.): *Fluid Mechanics of Mixing.* Modelling, Operations and Experimental Techniques. 1992 ISBN 0-7923-1720-3
11. Z. Han and X. Yin: *Shock Dynamics.* 1993 ISBN 0-7923-1746-7
12. L. Svarovsky and M.T. Thew (eds.): *Hydroclones.* Analysis and Applications. 1992 ISBN 0-7923-1876-5
13. A. Lichtarowicz (ed.): *Jet Cutting Technology.* 1992 ISBN 0-7923-1979-6
14. F.T.M. Nieuwstadt (ed.): *Flow Visualization and Image Analysis.* 1993 ISBN 0-7923-1994-X
15. A.J. Saul (ed.): *Floods and Flood Management.* 1992 ISBN 0-7923-2078-6
16. D.E. Ashpis, T.B. Gatski and R. Hirsh (eds.): *Instabilities and Turbulence in Engineering Flows.* 1993 ISBN 0-7923-2161-8
17. R.S. Azad: *The Atmospheric Boundary Layer for Engineers.* 1993 ISBN 0-7923-2187-1
18. F.T.M. Nieuwstadt (ed.): *Advances in Turbulence IV.* 1993 ISBN 0-7923-2282-7
19. K.K. Prasad (ed.): *Further Developments in Turbulence Management.* 1993 ISBN 0-7923-2291-6
20. Y.A. Tatarchenko: *Shaped Crystal Growth.* 1993 ISBN 0-7923-2419-6
21. J.P. Bonnet and M.N. Glauser (eds.): *Eddy Structure Identification in Free Turbulent Shear Flows.* 1993 ISBN 0-7923-2449-8
22. R.S. Srivastava: *Interaction of Shock Waves.* 1994 ISBN 0-7923-2920-1
23. J.R. Blake, J.M. Boulton-Stone and N.H. Thomas (eds.): *Bubble Dynamics and Interface Phenomena.* 1994 ISBN 0-7923-3008-0

# Mechanics

## *FLUID* MECHANICS AND ITS APPLICATIONS

*Series Editor*: R. Moreau

24. R. Benzi (ed.): *Advances in Turbulence V.* 1995 ISBN 0-7923-3032-3
25. B.I. Rabinovich, V.G. Lebedev and A.I. Mytarev: *Vortex Processes and Solid Body Dynamics.* The Dynamic Problems of Spacecrafts and Magnetic Levitation Systems. 1994 ISBN 0-7923-3092-7
26. P.R. Voke, L. Kleiser and J.-P. Chollet (eds.): *Direct and Large-Eddy Simulation I.* Selected papers from the First ERCOFTAC Workshop on Direct and Large-Eddy Simulation. 1994 ISBN 0-7923-3106-0
27. J.A. Sparenberg: *Hydrodynamic Propulsion and its Optimization.* Analytic Theory. 1995 ISBN 0-7923-3201-6
28. J.F. Dijksman and G.D.C. Kuiken (eds.): *IUTAM Symposium on Numerical Simulation of Non-Isothermal Flow of Viscoelastic Liquids.* Proceedings of an IUTAM Symposium held in Kerkrade, The Netherlands. 1995 ISBN 0-7923-3262-8
29. B.M. Boubnov and G.S. Golitsyn: *Convection in Rotating Fluids.* 1995 ISBN 0-7923-3371-3
30. S.I. Green (ed.): *Fluid Vortices.* 1995 ISBN 0-7923-3376-4
31. S. Morioka and L. van Wijngaarden (eds.): *IUTAM Symposium on Waves in Liquid/Gas and Liquid/Vapour Two-Phase Systems.* 1995 ISBN 0-7923-3424-8
32. A. Gyr and H.-W. Bewersdorff: *Drag Reduction of Turbulent Flows by Additives.* 1995 ISBN 0-7923-3485-X
33. Y.P. Golovachov: *Numerical Simulation of Viscous Shock Layer Flows.* 1995 ISBN 0-7923-3626-7
34. J. Grue, B. Gjevik and J.E. Weber (eds.): *Waves and Nonlinear Processes in Hydrodynamics.* 1996 ISBN 0-7923-4031-0
35. P.W. Duck and P. Hall (eds.): *IUTAM Symposium on Nonlinear Instability and Transition in Three-Dimensional Boundary Layers.* 1996 ISBN 0-7923-4079-5
36. S. Gavrilakis, L. Machiels and P.A. Monkewitz (eds.): *Advances in Turbulence VI.* Proceedings of the 6th European Turbulence Conference. 1996 ISBN 0-7923-4132-5
37. K. Gersten (ed.): *IUTAM Symposium on Asymptotic Methods for Turbulent Shear Flows at High Reynolds Numbers.* Proceedings of the IUTAM Symposium held in Bochum, Germany. 1996 ISBN 0-7923-4138-4
38. J. Verhás: *Thermodynamics and Rheology.* 1997 ISBN 0-7923-4251-8
39. M. Champion and B. Deshaies (eds.): *IUTAM Symposium on Combustion in Supersonic Flows.* Proceedings of the IUTAM Symposium held in Poitiers, France. 1997 ISBN 0-7923-4313-1
40. M. Lesieur: *Turbulence in Fluids.* Third Revised and Enlarged Edition. 1997 ISBN 0-7923-4415-4; Pb: 0-7923-4416-2
41. L. Fulachier, J.L. Lumley and F. Anselmet (eds.): *IUTAM Symposium on Variable Density Low-Speed Turbulent Flows.* Proceedings of the IUTAM Symposium held in Marseille, France. 1997 ISBN 0-7923-4602-5
42. B.K. Shivamoggi: *Nonlinear Dynamics and Chaotic Phenomena.* An Introduction. 1997 ISBN 0-7923-4772-2
43. H. Ramkissoon, *IUTAM Symposium on Lubricated Transport of Viscous Materials.* Proceedings of the IUTAM Symposium held in Tobago, West Indies. 1998 ISBN 0-7923-4897-4
44. E. Krause and K. Gersten, *IUTAM Symposium on Dynamics of Slender Vortices.* Proceedings of the IUTAM Symposium held in Aachen, Germany. 1998 ISBN 0-7923-5041-3
45. A. Biesheuvel and G.J.F. van Heyst (eds.): *In Fascination of Fluid Dynamics.* A Symposium in honour of Leen van Wijngaarden. 1998 ISBN 0-7923-5078-2

# Mechanics

## *FLUID* MECHANICS AND ITS APPLICATIONS

*Series Editor*: R. Moreau

46. U. Frisch (ed.): *Advances in Turbulence VII*. Proceedings of the Seventh European Turbulence Conference, held in Saint-Jean Cap Ferrat, 30 June–3 July 1998. 1998 ISBN 0-7923-5115-0
47. E.F. Toro and J.F. Clarke: *Numerical Methods for Wave Propagation*. Selected Contributions from the Workshop held in Manchester, UK. 1998 ISBN 0-7923-5125-8
48. A. Yoshizawa: *Hydrodynamic and Magnetohydrodynamic Turbulent Flows*. Modelling and Statistical Theory. 1998 ISBN 0-7923-5225-4
49. T.L. Geers (ed.): *IUTAM Symposium on Computational Methods for Unbounded Domains*. 1998 ISBN 0-7923-5266-1
50. Z. Zapryanov and S. Tabakova: *Dynamics of Bubbles, Drops and Rigid Particles*. 1999 ISBN 0-7923-5347-1
51. A. Alemany, Ph. Marty and J.P. Thibault (eds.): *Transfer Phenomena in Magnetohydrodynamic and Electroconducting Flows*. 1999 ISBN 0-7923-5532-6
52. J.N. Sørensen, E.J. Hopfinger and N. Aubry (eds.): *IUTAM Symposium on Simulation and Identification of Organized Structures in Flows*. 1999 ISBN 0-7923-5603-9
53. G.E.A. Meier and P.R. Viswanath (eds.): *IUTAM Symposium on Mechanics of Passive and Active Flow Control*. 1999 ISBN 0-7923-5928-3
54. D. Knight and L. Sakell (eds.): *Recent Advances in DNS and LES*. 1999 ISBN 0-7923-6004-4
55. P. Orlandi: *Fluid Flow Phenomena*. A Numerical Toolkit. 2000 ISBN 0-7923-6095-8
56. M. Stanislas, J. Kompenhans and J. Westerveel (eds.): *Particle Image Velocimetry*. Progress towards Industrial Application. 2000 ISBN 0-7923-6160-1
57. H.-C. Chang (ed.): *IUTAM Symposium on Nonlinear Waves in Multi-Phase Flow*. 2000 ISBN 0-7923-6454-6
58. R.M. Kerr and Y. Kimura (eds.): *IUTAM Symposium on Developments in Geophysical Turbulence* held at the National Center for Atmospheric Research, (Boulder, CO, June 16–19, 1998) 2000 ISBN 0-7923-6673-5
59. T. Kambe, T. Nakano and T. Miyauchi (eds.): *IUTAM Symposium on Geometry and Statistics of Turbulence* held at the Shonan International Village Center, Hayama (Kanagawa-ken, Japan November 2–5, 1999). 2001 ISBN 0-7923-6711-1
60. V.V. Aristov: *Direct Methods for Solving the Boltzmann Equation and Study of Nonequilibrium Flows*. 2001 ISBN 0-7923-6831-2

Kluwer Academic Publishers – Dordrecht / Boston / London

## ICASE/LaRC Interdisciplinary Series in Science and Engineering

1. J. Buckmaster, T.L. Jackson and A. Kumar (eds.): *Combustion in High-Speed Flows.* 1994 ISBN 0-7923-2086-X
2. M.Y. Hussaini, T.B. Gatski and T.L. Jackson (eds.): *Transition, Turbulence and Combustion.* Volume I: Transition. 1994 ISBN 0-7923-3084-6; set 0-7923-3086-2
3. M.Y. Hussaini, T.B. Gatski and T.L. Jackson (eds.): *Transition, Turbulence and Combustion.* Volume II: Turbulence and Combustion. 1994 ISBN 0-7923-3085-4; set 0-7923-3086-2
4. D.E. Keyes, A. Sameh and V. Venkatakrishnan (eds): *Parallel Numerical Algorithms.* 1997 ISBN 0-7923-4282-8
5. T.G. Campbell, R.A. Nicolaides and M.D. Salas (eds.): *Computational Electromagnetics and Its Applications.* 1997 ISBN 0-7923-4733-1
6. V. Venkatakrishnan, M.D. Salas and S.R. Chakravarthy (eds.): *Barriers and Challenges in Computational Fluid Dynamics.* 1998 ISBN 0-7923-4855-9
7. M.D. Salas, J.N. Hefner and L. Sakell (eds.): *Modeling Complex Turbulent Flows.* 1999 ISBN 0-7923-5590-3

KLUWER ACADEMIC PUBLISHERS – DORDRECHT / BOSTON / LONDON

# *ERCOFTAC* SERIES

1. A. Gyr and F.-S. Rys (eds.): *Diffusion and Transport of Pollutants in Atmospheric Mesoscale Flow Fields.* 1995 ISBN 0-7923-3260-1
2. M. Hallbäck, D.S. Henningson, A.V. Johansson and P.H. Alfredsson (eds.): *Turbulence and Transition Modelling.* Lecture Notes from the ERCOFTAC/IUTAM Summerschool held in Stockholm. 1996 ISBN 0-7923-4060-4
3. P. Wesseling (ed.): *High Performance Computing in Fluid Dynamics.* Proceedings of the Summerschool held in Delft, The Netherlands. 1996 ISBN 0-7923-4063-9
4. Th. Dracos (ed.): *Three-Dimensional Velocity and Vorticity Measuring and Image Analysis Techniques.* Lecture Notes from the Short Course held in Zürich, Switzerland. 1996 ISBN 0-7923-4256-9
5. J.-P. Chollet, P.R. Voke and L. Kleiser (eds.): *Direct and Large-Eddy Simulation II.* Proceedings of the ERCOFTAC Workshop held in Grenoble, France. 1997 ISBN 0-7923-4687-4

KLUWER ACADEMIC PUBLISHERS – DORDRECHT / BOSTON / LONDON

Zeitfracht Medien GmbH
Ferdinand-Jühlke-Straße 7
99095 Erfurt, Deutschland
produktsicherheit@kolibri360.de